Cryptography for Intelligent Healthcare Systems

Other related titles:

You may also like

- PBHE040 | Sabharwal, Balusamy, Kumar, Gayathri, and Suvanov | Applications of Artificial Intelligence in E-Healthcare Systems | 2022
- PBHE048 | Namasudra and Albuquerque | Blockchain Technology in e-Healthcare Management | 2022
- PBHE044 | Singh and Zhou | Medical Information Processing and Security: Techniques and applications | 2022
- PBHE061 | Imoize, Meshram, Awotunde, and Do | Artificial Intelligence and Blockchain Technology in Modern Telehealth Systems | 2023
- PBHE064 | Imoize, Meshram, Awotunde, Farhaoui, and Do | Cybersecurity in Emerging Healthcare Systems | 2024
- PBHE063 | Bhattacharya, Prasad, Hemanth, Dutta, Kathait, and Dănciulescu | Secure Big-data Analytics for Emerging Healthcare in 5G and Beyond | 2024

We also publish a wide range of books on the following topics:
Agricultural Engineering and Technology
Applied Physics
Built Environment
Computing and Networks
Control, Robotics and Sensors
Electrical Regulations
Electromagnetics and Radar
Energy Engineering
Healthcare Technologies
History and Management of Technology
IET Codes and Guidance
Materials, Circuits and Devices
Model Forms
Nanomaterials and Nanotechnologies
Production, Design and Manufacturing
Security
Telecommunications
Transportation

All books are available in print via https://shop.theiet.org or as eBooks via our Digital Library https://digital-library.theiet.org.

IET HEALTHCARE TECHNOLOGIES SERIES 070

IET book series on e-health technologies

Book Series Advisors:

Professor Daniel Abásolo, Centre for Biomedical Engineering, Faculty of Engineering and Physical Sciences, University of Surrey, UK.

Professor Pranjal Chandra, School of Biochemical Engineering, Indian Institute of Technology (BHU), Varanasi, India.

While the demographic shifts in populations display significant socio-economic challenges, they trigger opportunities for innovations in e-Health, m-Health, precision and personalized medicine, robotics, sensing, the Internet of things, cloud computing, big data, software defined networks, and network function virtualization. Their integration is however associated with many technological, ethical, legal, social, and security issues. This book series aims to disseminate recent advances for e-health technologies to improve healthcare and people's wellbeing.

Could you be our next author?

Topics considered include intelligent e-Health systems, electronic health records, ICT-enabled personal health systems, mobile and cloud computing for e-Health, health monitoring, precision and personalized health, robotics for e-Health, security and privacy in e-Health, ambient assisted living, telemedicine, big data and IoT for e-Health, and more.

Proposals for coherently integrated international multi-authored edited or co-authored handbooks and research monographs will be considered for this book series. Each proposal will be reviewed by the book Series Editor with additional external reviews from independent reviewers.

To download our proposal form or find out more information about publishing with us, please visit https://www.theiet.org/publishing/publishing-with-iet-books/.

Please email your completed book proposal for the IET Book Series on e-Health Technologies to: Amber Thomas at athomas@theiet.org or author_support@theiet.org.

Cryptography for Intelligent Healthcare Systems

Volume 1: Foundations and emerging architectures

Edited by
Agbotiname Lucky Imoize, Mohammad S. Obaidat and Houbing Herbert Song

The Institution of Engineering and Technology

About the IET

This book is published by the Institution of Engineering and Technology (The IET).

We inspire, inform and influence the global engineering community to engineer a better world. As a diverse home across engineering and technology, we share knowledge that helps make better sense of the world, to accelerate innovation and solve the global challenges that matter.

The IET is a not-for-profit organisation. The surplus we make from our books is used to support activities and products for the engineering community and promote the positive role of science, engineering and technology in the world. This includes education resources and outreach, scholarships and awards, events and courses, publications, professional development and mentoring, and advocacy to governments.

To discover more about the IET, please visit https://www.theiet.org/.

About IET books

The IET publishes books across many engineering and technology disciplines. Our authors and editors offer fresh perspectives from universities and industry. Within our subject areas, we have several book series steered by editorial boards made up of leading subject experts.

We peer review each book at the proposal stage to ensure the quality and relevance of our publications.

Get involved

If you are interested in becoming an author, editor, series advisor, or peer reviewer please visit https://www.theiet.org/publishing/publishing-with-iet-books/ or contact author_support@theiet.org.

Discovering our electronic content

All of our books are available online via the IET's Digital Library. Our Digital Library is the home of technical documents, eBooks, conference publications, real-life case studies and journal articles. To find out more, please visit https://digital-library.theiet.org.

In collaboration with the United Nations and the International Publishers Association, the IET is a Signatory member of the SDG Publishers Compact. The Compact aims to accelerate progress to achieve the Sustainable Development Goals (SDGs) by 2030. Signatories aspire to develop sustainable practices and act as champions of the SDGs during the Decade of Action (2020–30), publishing books and journals that will help inform, develop, and inspire action in that direction.

In line with our sustainable goals, our UK printing partner has FSC accreditation, which is reducing our environmental impact to the planet. We use a print-on-demand model to further reduce our carbon footprint.

Published by The Institution of Engineering and Technology, London, United Kingdom

The Institution of Engineering and Technology (the "**Publisher**") is registered as a Charity in England & Wales (no. 211014) and Scotland (no. SC038698).

First published: 2026

The Institution of Engineering and Technology
Futures Place,
Kings Way, Stevenage,
Herts, SG1 2UA,
United Kingdom

www.theiet.org

British Library Cataloguing in Publication Data
A catalogue record for this product is available from the British Library

ISBN 978-1-83724-948-0 (Volume 1 hardback)
ISBN 978-1-83724-949-7 (Volume 1 PDF)
ISBN 978-1-83724-950-3 (Volume 1 EPUB 3)

ISBN 978-1-80705-371-0 (Volume 2 hardback)
ISBN 978-1-80705-372-7 (Volume 2 PDF)
ISBN 978-1-80705-373-4 (Volume 2 EPUB 3)

Typeset in India by MPS Limited

Cover image credit: fotograzia/Moment via Getty Images

Contents

Preface

Intelligent healthcare systems facilitate the seamless diagnosis and treatment of various ailments. These interconnected systems rely on advanced security technologies to protect sensitive user information, especially as healthcare data is transmitted over open wireless channels, which pose security risks. In recent times, billions of Internet of Things (IoT) devices have been deployed globally to meet the escalating demand from mobile device users, including patients and healthcare professionals, and securing the transmission medium has become imperative. Additionally, new security vulnerabilities are proliferating in the healthcare ecosystem due to the increasing complexity of its architecture, and threats to medical devices and critical infrastructure pose significant risks to patient health and safety.

Globally, millions of patients have been exposed to high health risks due to sophisticated attacks that disrupt critical medical infrastructure and wireless communications facilities and services and interfere with medical devices, thereby compromising sensitive user data. To address these critical concerns, advanced cryptographic frameworks are implemented to guarantee the security and privacy of sensitive user information. However, developing such cryptography-empowered frameworks requires a comprehensive understanding of the complex interplay and real-time dynamics among external threats, internal vulnerabilities, calculated risks and the required resilience and robustness of the overall security architecture.

This book presents cutting-edge solutions to revamp traditional cryptographic architecture, addressing the proliferating security and privacy challenges in intelligent healthcare systems. The book offers new insights into real-world scenarios of the deployment, applications, management and potential of advanced cryptographic solutions tailored for the security of modern healthcare systems. Further, the book presents the design, development and implementation of robust, real-time and cryptography-based frameworks to mitigate security threats in intelligent healthcare systems. Specifically, the book sheds light on novel encryption and decryption tools, cyber risks and vulnerability models and leverages advanced cryptography to ensure provable security and privacy for critical healthcare data. Lastly, the book discusses critical security and privacy issues affecting all parties in the healthcare ecosystem and provides practical AI- and machine learning-based cryptographic solutions to address them effectively.

This volume is structured into 13 chapters, outlined as follows: Chapter 1 highlights the cryptographic algorithms for intelligent healthcare systems, focusing on historical perspectives and emerging trends. Chapter 2 focuses on cryptography for intelligent healthcare systems, presenting a taxonomy of new cryptographic schemes. Chapter 3 describes and demonstrates the design and implementation of advanced cryptographic frameworks for intelligent healthcare systems.

Chapter 4 presents artificial intelligence-enabled cryptographic frameworks for intelligent healthcare systems. Chapter 5 explores AI-driven cryptographic techniques in IoT-based healthcare systems. Chapter 6 provides an overview of advanced cryptography to enhance the Internet of Medical Things for intelligent healthcare applications. Chapter 7 introduces new-generation cryptographic algorithms for Internet of Things environments in intelligent healthcare systems.

Chapter 8 envisions using deep reinforcement learning to enable adaptive cryptographic frameworks for intelligent healthcare systems. Chapter 9 focuses on SAFE-MED, a safe deep reinforcement learning approach for adaptive, quantum-aware cryptography in intelligent healthcare. Chapter 10 examines an AI-enabled cryptographic scheme to secure the Internet of Medical Things in the post-quantum era.

Chapter 11 demonstrates the application of quantum computing to advanced cryptography for intelligent healthcare systems. Chapter 12 investigates cryptography in 6G optical wireless-based Internet of Medical Things, exploring the potential and state of the art. Finally, Chapter 13 concludes the book by presenting a cryptography-enabled multi-network emergency communication framework for intelligent healthcare systems.

The book is an ideal reference for practitioners, industry, academic researchers, scientists and engineers in the field of cryptography for intelligent healthcare systems. The textbook is also suitable for graduate and senior undergraduate courses in cybersecurity, cyber defensive wall technologies, edge computing, cloud computing, advanced cryptographic systems, security engineering, security and privacy systems, intrusion detection systems, threat intelligence, big data analytics, healthcare systems and technologies, artificial intelligence, machine learning, medical informatics and forensics and related fields.

The editors would like to specially thank the book reviewers for their constructive suggestions. Special thanks to all chapter authors for their insightful contributions. Many thanks to the editorial assistants at the IET for their cooperation and support.

Agbotiname Lucky Imoize
Gelsenkirchen, North Rhine-Westphalia, Germany

Mohammad S. Obaidat
Amman, Jordan

Houbing Herbert Song
Baltimore, Maryland, United States

About the editors

Agbotiname Lucky Imoize (Senior Member, IEEE) received the B.Eng. degree (Hons.) in Electrical and Electronics Engineering from Ambrose Alli University, Nigeria, in 2008, and the M.Sc. degree in Electrical and Electronics Engineering from the University of Lagos, Nigeria, in 2012. He is a Lecturer in the Department of Electrical and Electronics Engineering at the University of Lagos, Nigeria. He has been a Researcher at Ruhr University Bochum, Germany, under sponsorship from the Nigerian Petroleum Technology Development Fund (PTDF) and the German Academic Exchange Service (DAAD) through the Nigerian-German Postgraduate Program. He was awarded the Fulbright fellowship as a Visiting Researcher at the Wireless@VT Laboratory in the Bradley Department of Electrical and Computer Engineering at Virginia Tech, USA, where he worked under the supervision of Prof. R. Michael Buehrer from 2017 to 2018. Before joining the University of Lagos, he was a Lecturer at Bells University of Technology, Nigeria. He also worked as a core network products manager at ZTE Nigeria and as a Network Switching Subsystem Engineer at Globacom Nigeria. His research interests span 6G wireless communication, wireless security systems and artificial intelligence. Imoize is the Book Editing Director for the ACM Emerging Interest Group on Trustworthy and Responsible Systems (EIGTRUST). He is the Vice Chair of the IEEE Communication Society Nigeria chapter, a registered engineer with the Council for the Regulation of Engineering in Nigeria, and a member of the Nigerian Society of Engineers.

Mohammad S. Obaidat is an internationally known academic, researcher, scientist and scholar. He received his Ph.D. in Computer Engineering with a minor in Computer Science from The Ohio State University, Columbus, USA. He has received extensive research funding and published, to date, over one thousand (1,200) refereed technical articles, about half of which are journal articles, over 100 books and 70 book chapters. He is Editor-in-Chief of three scholarly journals and an editor of many other international journals. In particular, he is the Editor-in-Chief of the Wiley International Journal of Communication Systems and Founding Editor-in-Chief of the Wiley Security and Privacy Journal. He is also the Founding Advisory Editor of the Cybersecurity and Applications Journal, KeAi. Moreover, he is a founder or co-founder of five IEEE International Conferences. Among his previous positions are Advisor to the President of Philadelphia University for Research, Development and Information Technology, President and Chair of Board of Directors of the Society for Molding and Simulation International (SCS), Senior Vice President of SCS, SCS VP for Membership and SCS VP for Conferences, Dean of the College of Engineering at Prince Sultan University, Founding Dean of the College of Computing and Informatics at The University of Sharjah, Chair and tenured Professor at the Department of Computer and Information Science and Director of the MS Graduate Program in Data Analytics at Fordham university, Chair and tenured Professor of the Department of Computer Science and Director of the Graduate Program at Monmouth University, Chair and Professor of Computer Science Department at University of Texas-Permian Basin, Distinguished Professor at IIT-Dhanbad, Tenured Full Professor at King Abdullah II School of Information Technology (KASIT), University of Jordan, The PR of China Ministry of Education Distinguished Overseas Professor at the University of Science and Technology Beijing, China and an Honorary Distinguished Professor at the Amity University- A Global University. He is now a Distinguished Professor at KASIT, University of Jordan. He is also a Distinguished Professor at SRM University, India.

He has chaired numerous (over 200) international conferences and given numerous (over 200) keynote speeches worldwide. He has served as an ABET/CSAB evaluator and as a member of the IEEE CS Fellow Evaluation Committee. He has served as an IEEE CS Distinguished Speaker/Lecturer and an ACM Distinguished Lecturer. Since 2004, he has been serving as an SCS Distinguished Lecturer. He received numerous best paper awards for his work, including from IEEE ICC, IEEE GLOBECOM, AICSA, CITS, SPECTS, DCNET, IEEE ICCCA and other international conferences. He also received two best paper awards from the IEEE Systems Journal in 2018 and 2019. In 2020, he received four best paper awards from the IEEE Systems Journal, and in 2021, he received the journal's best paper award. During his tenure as Founding Dean of the College of Computing and Informatics at the University of Sharjah, the Computer Sciences program was ranked number one in the UAE by Times Higher Education (THE) for the first time in the university's history. He also received many other worldwide awards for his technical contributions including The 2018 IEEE ComSoc-Technical Committee on Communications Software 2018 Technical Achievement Award for contribution to Cybersecurity, Wireless Networks Computer Networks and Modeling and Simulation, SCS prestigious McLeod Founder's Award, Presidential Service Award, SCS Hall of Fame – Lifetime Achievement Award for his technical contribution to modelling and simulation and for his outstanding visionary leadership and dedication to increasing the effectiveness and broadening the applications of modelling and simulation worldwide. He also received the SCS Outstanding Service Award. He was awarded the IEEE CITS Hall of Fame Distinguished and Eminent Award. He also received the Nokia Distinguished Fellowship Award and Fulbright Distinguished Scholar Award, among others.

According to Stanford University's recent list, he is among the world's top 2 per cent career-long scientists. Professor Obaidat has been recently ranked by ScholarGPS as the top lifetime scholar in the world in telecommunications systems management and the top scholar in telecommunications in the world for the past five Years. He has also been recently ranked by Research.com agency as Jordan's number one computer scientist and number one scholar in electronics and electrical engineering. In recognition of his significant scientific contributions, Springer published a book in 2022, honouring his work in computing, informatics, networking and cybersecurity. It is entitled: Advances in Computing, Informatics, Networking and Cybersecurity - A Book Honoring Professor Mohammad S. Obaidat's Significant Scientific Contributions. Prof. Obaidat is a Life Fellow of IEEE, a Fellow of AAIA, a Fellow of FTRA, a Fellow of AIIA and a Fellow of SCS.

Houbing Herbert Song received his Ph.D. degree in electrical engineering from the University of Virginia, Charlottesville, VA, in August 2012. He is currently a Professor and the Founding Director of the NSF Center for Aviation Big Data Analytics (Planning), and the Director of the Security and Optimization for Networked Globe Laboratory (SONG Lab, www.SONGLab.us) at the University of Maryland, Baltimore County (UMBC), Baltimore, MD. He is currently the Co-Editor-in-Chief (Co-EiC) of IEEE Transactions on Industrial Informatics (2025–present). He has been the Founding Chair of the Trustworthy Internet of Things (TRUST-IoT) Working Group within the IEEE IoT Technical Community since 2024. He was a Distinguished Visiting Fellow of the Scottish Informatics and Computer Science Alliance (SICSA) in 2024. He serves as an Associate Editor for the IEEE Transactions on Artificial Intelligence (TAI) (2023–present), the IEEE Transactions on Intelligent Transportation Systems (2021–present) and the IEEE Journal on Miniaturization for Air and Space Systems (J-MASS) (2020–present). He was an Associate Technical Editor for IEEE Communications Magazine (2017–2020) and an Associate Editor for IEEE Internet of Things Journal (2020–2024). He has served as the General Chair of IEEE UIC 2025 and IEEE SmartData 2022 and as the Technical Program Committee (TPC) Chair of IEEE AIoT 2025, NSS 2024 and IEEE GEM 2023. He is the editor of more than 10 books, the author of more than 100 articles and the inventor of 2 patents. His research interests include neuro-symbolic AI, anomaly detection, Artificial Intelligence of Things, autonomous systems and Cyber-Physical Systems (CPS). His research has been sponsored by federal agencies (including the National Science Foundation, the National Aeronautics and Space Administration, the US Department of Transportation and the Federal Aviation Administration, among others) and industry. His research has been featured in popular news media outlets such as IEEE Spectrum, IEEE GlobalSpec's Engineering360, IEEE Transmitter, insideBIGDATA, StateTech Magazine, Association for Uncrewed Vehicle Systems International (AUVSI), Security Magazine, CXOTech Magazine, Solutions Review, EdTech Magazine,

PYMNTS, Fox News, U.S. News & World Report, The Washington Times and New Atlas. Dr. Song is an IEEE Fellow, an ACM Distinguished Member and a Full Member of Sigma Xi. He has been a Highly Cited Researcher, as identified by Web of Science, since 2021. He is an ACM Distinguished Speaker (2020–present), an IEEE Computer Society Distinguished Visitor (2024–present), an IEEE Communications Society (ComSoc) Distinguished Lecturer (2024–present), an IEEE Intelligent Transportation Systems Society (ITSS) Distinguished Lecturer (2024–present), an IEEE Vehicular Technology Society (VTS) Distinguished Lecturer (2023–present) and an IEEE Systems Council Distinguished Lecturer(2023–present). Dr. Song received the Research.com Rising Star of Science Award in 2022, the IEEE 2021 Harry Rowe Mimno Award, the 2024 IEEE Computational Intelligence Society (CIS) Publication Spotlight and more than 10 Best Paper Awards from major international conferences. He has been an IEEE Impact Creator since 2023.

Chapter 1

Cryptographic algorithms for intelligent healthcare systems: historical perspectives and emerging trends

Kingsley Eyiogwu Onu[1] *and Promise Elechi*[1]

Abstract

The revolution in healthcare – fuelled by artificial intelligence, cloud computing and connected devices – has brought us to a future of real-time patient monitoring, predictive medicine and unprecedented collaboration. Yet this powerful new landscape rests on a critical foundation: the absolute need to protect sensitive patient data, which is now central to maintaining trust and meeting legal standards. This chapter charts the evolution of cryptography in healthcare, from basic password protection to the sophisticated privacy-preserving systems that enable secure AI analytics and collaborative research. To move beyond theoretical proposals, we built a simulation framework to rigorously test how different cryptographic methods perform under realistic healthcare conditions, measuring the tangible trade-offs between privacy and practicality. Our findings are both promising and actionable: by using privacy-preserving techniques like federated learning enhanced with differential privacy, hospitals can collaboratively train AI models that retain 90 per cent accuracy – a minimal 5 per cent drop from unprotected models – while robustly safeguarding individual patient data. For the wearable sensors and medical devices that power modern care, we found that lightweight encryption algorithms like ChaCha20 can operate two to three times faster than standard methods, a crucial efficiency for battery-dependent hardware. We examine this entire ecosystem, from established standards like Advanced Encryption Standard (AES) and Rivest–Shamir–Adleman (RSA) to cutting-edge tools like homomorphic encryption, through the demanding lens of real-world healthcare – where systems must be fast, compliant and compatible with legacy infrastructure. Our analysis of existing research reveals significant gaps: a lack of real-world testing, insufficient performance data for medical hardware and scant practical road maps for the looming quantum computing threat. While our own work uses simulations and synthetic

[1]Department of Electrical and Electronic Engineering, Rivers State University, Nigeria

data within these necessary constraints, it yields eight concrete lessons for the field, emphasising adaptive design, interdisciplinary collaboration and, above all, security that supports rather than hinders patient care. This chapter aims to equip both researchers with a technical road map for future innovation and practitioners – from hospital IT directors to policymakers – with a practical guide for navigating the security decisions that will define the next generation of intelligent and trustworthy healthcare.

Keywords: Intelligent healthcare systems; cryptographic algorithms; privacy-preserving machine learning; federated learning; differential privacy; homomorphic encryption; secure multiparty computation; Internet of Medical Things; IoMT; lightweight cryptography; blockchain healthcare; post-quantum cryptography; HIPAA compliance

1.1 Introduction

Healthcare systems are experiencing a deep digital shift driven by advances in artificial intelligence, cloud computing, data analytics and connected medical technologies. Modern healthcare no longer relies solely on isolated hospital systems but increasingly integrates electronic health records, wearable and implantable sensors, telemedicine platforms and AI-based decision support tools. These technologies enable continuous patient monitoring, rapid data-driven diagnosis and collaborative medical research across institutions, fundamentally changing how care is delivered and managed [1,2].

At the same time, this growing intelligence comes at a cost. Healthcare data is among the most sensitive forms of personal information, encompassing clinical histories, genetic profiles, diagnostic images and real-time physiological measurements. When such data is exposed, altered or lost, the consequences extend beyond financial damage to include misdiagnosis, loss of patient trust and long-term social and ethical harm. Not surprisingly, healthcare remains one of the most targeted sectors for cyberattacks [3,4], largely because medical data retains value long after it is stolen.

Cryptography has therefore become central to modern healthcare security. What began as simple encryption for stored patient records has evolved into a complex ecosystem of cryptographic techniques that protect data confidentiality, integrity, authenticity and accountability across distributed and intelligent systems. However, the rapid pace of healthcare digitisation has outstripped traditional security models. Many cryptographic solutions were not originally designed for AI-driven analytics, real-time clinical workflows or resource-constrained medical devices. This chapter is motivated by the need to examine how cryptographic algorithms have evolved within healthcare systems and how they must continue to adapt to support secure, intelligent and future-ready healthcare environments.

1.1.1 Intelligent healthcare systems and security requirements

Intelligent healthcare systems are defined by their reliance on interconnected, data-driven components that operate across hospitals, clinics, homes and cloud platforms. These systems combine EHR infrastructures, AI and machine learning models, Internet of Medical Things devices and edge and cloud computing resources to deliver timely and personalised healthcare services. Unlike traditional hospital information systems, intelligent healthcare environments are highly distributed, continuously evolving [1,5] and often operate under strict resource constraints.

The security requirements of such systems extend well beyond basic data protection. Confidentiality remains essential, but it must be complemented by strong guarantees of data integrity and authenticity to ensure that clinical decisions are based on accurate and trustworthy information. Availability is equally critical, as system disruptions can directly affect patient safety. In addition, intelligent healthcare systems must support fine-grained access control, secure authentication of users and devices and reliable audit mechanisms to meet regulatory, ethical and operational obligations [2].

The integration of artificial intelligence introduces further complexity. Machine learning models trained on healthcare data can inadvertently leak sensitive information through inference, model inversion or poisoning attacks, even when raw data is never explicitly shared. As a result, cryptographic mechanisms must not only protect stored and transmitted data but also enable secure model training, collaborative analytics and privacy-preserving inference. In intelligent healthcare, cryptography is no longer a background security feature [6,7] but a core enabler of safe and reliable system operation.

1.1.2 Role of cryptography in healthcare digital transformation

Cryptography plays a foundational role in enabling trust throughout the healthcare digital transformation. In early digital healthcare systems, cryptographic algorithms were primarily used to encrypt patient records and secure communication links between trusted parties. As healthcare infrastructures became more distributed and interconnected, public-key cryptography, digital signatures and secure key management schemes were adopted to support authentication, access control and data integrity across organisational boundaries.

In contemporary intelligent healthcare systems, the role of cryptography has expanded significantly. Privacy-preserving techniques such as homomorphic encryption, secure multiparty computation and differential privacy now make it possible to analyse sensitive medical data without directly exposing it. These approaches enable collaborative analytics and federated learning across institutions while respecting data sovereignty and regulatory constraints [6,8].

Cryptography also underpins emerging healthcare paradigms such as blockchain-based health information exchange, edge intelligence and large-scale IoMT deployments. Blockchain systems rely on cryptographic primitives to ensure data integrity and auditability, while lightweight cryptographic algorithms are essential for securing wearable and implantable devices with limited computational and energy resources [9,10]. At the same time, concerns about the long-term confidentiality of medical data have driven interest in post-quantum cryptography as a means to protect healthcare information against future cryptanalytic threats [11].

Despite its importance, cryptography introduces practical trade-offs. Strong security mechanisms can increase latency, computational overhead and energy consumption – factors that may conflict with real-time clinical requirements or device constraints. Balancing security, privacy and performance is therefore one of the central challenges in the cryptographic design of intelligent healthcare systems.

1.1.3 Novelty and contributions of the current chapter

Existing literature on healthcare security, cryptography and intelligent medical systems often treats these topics as independent research areas, focusing narrowly on algorithms, data protection challenges or AI applications with limited integration [1,2,12,13].

This chapter makes three primary contributions. First, it presents a novel Simulink-based experimental framework for systematic evaluation of cryptographic mechanisms in AI-driven healthcare workflows. The framework provides an integrated simulation environment with modular subsystems representing data sources, privacy mechanisms (differential privacy, homomorphic encryption and secure aggregation), analytics engines and performance evaluation components.

Second, securing patient data is not a matter of choosing a single 'best' technology; it requires carefully weighing the real strengths and limitations of existing encryption methods, emerging privacy-preserving techniques and newer post-quantum approaches intended to address future threats. In healthcare, these decisions must be grounded in practical realities rather than theory alone, since clinical environments are often complex, resource-constrained and highly sensitive to delays or failures [6]. This challenge unfolds across a distributed ecosystem in which data flows from wearable and implanted devices within the Internet of Medical Things (IoMT) to edge nodes and cloud platforms for analysis, sometimes through collaborative models such as federated learning that reduce the need to centralise raw patient data [6]. At each stage, designers must contend with strict latency requirements, limited device power and the difficulty of integrating modern security solutions with legacy hospital infrastructure. Above all, these technical choices are shaped by ethical and legal responsibilities, as regulations such as HIPAA and GDPR define clear boundaries for protecting patient privacy and maintaining trust, ensuring that technological progress does not come at the expense of the individuals whose data is being used [12,14].

Third, the chapter bridges cryptographic theory and practical deployment by combining historical context, critical evaluation and emerging research directions. It serves as both a reference for researchers developing privacy-preserving healthcare technologies and a guide for practitioners and policymakers navigating security challenges in intelligent healthcare systems.

1.1.4 Chapter objectives and organisation

The primary aim of this chapter is to present a coherent examination of cryptographic algorithms as they apply to intelligent healthcare systems, emphasising both their historical development and emerging trends. Cryptography is treated not as an isolated technical discipline but as an integral component of healthcare digital transformation, closely intertwined with artificial intelligence, regulatory compliance and system-level design constraints.

The chapter first reviews the evolution of cryptographic practices in healthcare and the regulatory forces that shaped their adoption. It then examines the security and privacy challenges unique to intelligent, data-centric healthcare environments. Subsequent sections explore classical cryptographic algorithms, advanced privacy-preserving techniques, federated and IoMT-oriented security mechanisms, blockchain-based integration and post-quantum cryptography. The chapter concludes by identifying emerging research directions, presenting illustrative case studies and offering insights relevant to researchers, engineers, clinicians and policymakers.

1.2 Evolution of cryptography in healthcare systems: a critical review

The evolution of cryptography in healthcare reflects a steady response to growing necessity rather than a single moment of innovation. Over the past five decades, cryptographic practices have shifted from rudimentary protective measures to sophisticated frameworks that now underpin intelligent, data-driven and highly distributed healthcare systems. This transformation has been shaped by rapid technological advances, increasingly stringent regulatory demands and the rising frequency of targeted attacks on sensitive medical data. Understanding this trajectory is essential, not only to appreciate how current systems came to be but also to critically assess the strengths and limitations of existing approaches as the foundation for future developments.

1.2.1 From physical locks to digital vaults: a historical perspective

During the early stages of healthcare computerisation in the 1970s and 1980s, data security was largely enforced through physical restrictions. Patient records stored on centralised mainframes were protected by controlling access to the facilities themselves, while digital safeguards were limited to simple password mechanisms and basic role-based controls. Although cryptographic standards such as the Data Encryption Standard (DES) had already been introduced, their adoption within healthcare environments was minimal. This reluctance stemmed from limited computational resources and the absence of regulatory frameworks that would compel healthcare institutions to prioritise strong digital security [15].

The transition to electronic health records (EHRs) in the 1990s and early 2000s marked a turning point. As patient data became digitised, centralised and accessible across networks, the risk of unauthorised access increased significantly. Encryption was no longer a supplementary safeguard but a core requirement for protecting sensitive information. During this period, DES was gradually replaced by the Advanced Encryption Standard (AES), which offered stronger security guarantees and improved efficiency. The flexibility of AES, particularly its support for 128- and 256-bit keys, enabled its widespread adoption for securing data at rest, data backups and data transmitted across healthcare networks, establishing it as a foundational cryptographic tool in modern healthcare systems [15].

From 2010 onwards, the healthcare ecosystem has become increasingly interconnected and complex. Cloud computing, telemedicine platforms, mobile health applications and the rapid expansion of Internet of Medical Things (IoMT) devices have transformed how medical data is generated, processed and shared. In this environment, data frequently traverses heterogeneous networks and is stored on third-party infrastructure, introducing new threat vectors and trust challenges. Consequently, traditional encryption alone is insufficient. Contemporary systems increasingly rely on end-to-end security mechanisms, advanced key management strategies and privacy-preserving techniques designed to mitigate both external attacks and insider threats [16]. This technical progression has unfolded alongside major regulatory developments. Frameworks such as the Health Insurance Portability and Accountability Act (HIPAA) and the General Data Protection Regulation (GDPR) have redefined cryptography as a legal obligation rather than a discretionary practice, fundamentally influencing the design and operation of healthcare information systems worldwide [17,18].

1.2.2 Assessing the current landscape: strengths, gaps and opportunities

An examination of recent literature on cryptography within intelligent healthcare systems reveals a diverse body of work addressing data confidentiality, secure computation and privacy preservation. While these studies offer important contributions, they also expose recurring limitations related to computational overhead, scalability, real-time performance and integration with legacy healthcare infrastructure. To provide a clearer understanding of these trade-offs, the subsequent table presents a comparative analysis of representative studies, highlighting their core methodologies, identified limitations and unresolved challenges. This structured review not only underscores the gaps that persist in current approaches but also positions the work presented in this chapter as a targeted response to those open research issues.

Existing research on healthcare cryptography, privacy-preserving technologies and intelligent medical systems exhibits several distinct approaches, each with inherent strengths and limitations. Table 1.1 presents a systematic technical analysis of representative recent works, highlighting their methodological approaches,

Table 1.1 Comparison of the current work with some previous works

References	Focus on previous work	Methodology	Limitations identified	Novel contribution of current work
[25]	IoT security with blockchain	Framework proposal	Lightweight cryptography not benchmarked	Detailed benchmarking of lightweight cryptographic algorithms
[12]	Blockchain in healthcare	Survey-based review	Limited integration with AI and cryptographic performance analysis	Integrates blockchain within a broader AI–cryptography healthcare framework
[13]	Homomorphic encryption in healthcare	Systematic review	No experimental validation or performance trade-off analysis	Provides experimental evaluation within AI-driven healthcare workflows
[1]	Federated learning for healthcare	Architectural survey	Cryptography treated as optional add-on	Embeds cryptography as a core design element in federated learning
[2]	Decentralised federated learning	Comprehensive survey	Lacks healthcare-specific experimental validation	Healthcare-focused secure federated learning simulation
[6]	Privacy-preserving FL in healthcare	Review of techniques	Limited system-level comparison	Unified experimental comparison of multiple privacy mechanisms
[20]	Broad healthcare cybersecurity landscape	Systematic review	Cryptography not deeply analysed; lacks historical continuity	Delivers a focused, cryptography-centric historical and emerging trend analysis
[11]	Post-quantum cryptography	Standards and theory	Not contextualised for healthcare data life cycles	Healthcare-centric post-quantum risk and migration analysis
[19]	Post-quantum security for healthcare cloud infrastructures	Architecture design and security analysis	Limited experimental validation; cryptography treated as a protective add-on	Integrates cryptography as a core component of intelligent healthcare systems with historical-to-emerging trend analysis

identified limitations and the contributions made by the current chapter to address these gaps.

The rise of quantum computing is sending a wave of urgency through the world of healthcare IT, pushing researchers to question whether our current data security measures are future-proof. At the forefront of this, Ramachandran [19] considered a new quantum-safe security blueprint specifically for healthcare cloud systems. Their vision was to use next-generation, 'post-quantum' algorithms to lock down patient data both in storage and during transmission. While this work paints an important picture of what's to come, it is still largely a concept on paper. We have not yet seen how it holds up under the immense pressure and complexity of real hospital data flows, which leaves cryptography feeling more like an add-on security blanket than a built-in feature of smart medical systems.

Also, in a broad synthesis of trends, Patra and Rajagopalan [20] brought together many threads of cryptography, AI security and hybrid models under the umbrella of healthcare cybersecurity. Their overview was valuable for its wide scope, but cryptography here becomes just one tool among many in the kit, rather than the foundational design principle it could be. The analysis also does not trace the specific evolutionary path of cryptographic algorithms themselves from their origins to the emerging quantum era, leaving a more specialised historical narrative untold. Table 1.1 shows the comparison of this work with some previous works.

1.2.3 Identified research gaps

Our analysis of the current body of research points to several unresolved challenges that act as barriers to putting cryptographic solutions into practice within intelligent healthcare environments.

The first significant gap is a fragmented treatment of cryptography and AI workflows. Too often, research silos these two critical domains. Studies on federated learning, for instance, may only mention privacy safeguards in passing [1,2], while surveys of cryptographic techniques frequently lack experimental validation within the actual, messy workflows of AI-driven clinical care [13]. This disconnect makes it difficult to understand the real-world performance trade-offs when trying to build a system that is both intelligent and genuinely private.

Second, there is a pronounced lack of healthcare-specific experimental frameworks. Although theoretical models are plentiful, there is a shortage of practical, systematic testing environments that incorporate the unique pressures of healthcare. These include the imperative for real-time analysis in clinical settings, the severe energy limitations of implantable or wearable medical devices, the complex nature of unevenly distributed patient data and the non-negotiable requirements of regulatory compliance [6,7]. Without testing under these conditions, proposed solutions remain academically interesting but clinically impractical.

A third gap is the insufficient benchmarking of lightweight cryptography. Despite the explosion of Internet of Medical Things (IoMT) devices, comprehensive performance testing of streamlined cryptographic algorithms on the specific microcontrollers used in medical hardware is surprisingly scarce [25]. Many analyses rely on generic IoT platforms, which can create unrealistic expectations for how these algorithms will perform on the specialised, resource-constrained devices actually used in patient care.

Furthermore, the literature shows limited integration of emerging paradigms. New technologies like blockchain, post-quantum cryptography and privacy-preserving machine learning are typically examined in isolation [8,11,12]. We lack system-level analyses that explore how these pieces could fit together into a coherent, unified architecture for a hospital or health network. This absence hinders the development of holistic security strategies.

Finally, there is an absence of practical migration strategies for the post-quantum era. While the theoretical threat quantum computing poses to current encryption is well documented, practical guidance for the healthcare sector is underdeveloped [11]. The industry faces unique hurdles, such as medical devices with decade-long lifespans, lengthy regulatory approval cycles and vast inventories of legacy systems. Without concrete migration road maps that account for these realities, healthcare organisations are left unable to plan proactively for this coming shift.

1.2.4 Regulatory drivers and their impact on cryptographic adoption

It is crucial to recognise that regulation, often more than pure technical innovation, has been the primary force pushing cryptography into healthcare. In the United States, the Health Insurance Portability and Accountability Act (HIPAA) established a foundational expectation by requiring safeguards for electronic protected health information (ePHI) and explicitly identifying encryption as an appropriate measure for data both at rest and in transit [17]. However, because HIPAA is technology-neutral and allows for flexibility in implementation, it has also led to a patchwork of security postures across different organisations.

The European Union's General Data Protection Regulation (GDPR) introduced a more stringent paradigm, mandating data protection by design and by default, backed by substantial penalties for failures [18]. This legislative shift has directly accelerated the adoption of advanced techniques like homomorphic encryption and differential privacy, particularly for projects involving the cross-border exchange of health data, where demonstrating robust protection is essential.

Furthermore, medical device regulations, such as those enforced by the US Food and Drug Administration (FDA), now explicitly emphasise cryptographic protections for critical functions like secure firmware updates and encrypted device communication (FDA cybersecurity guidance for medical devices). Collectively, these regulatory pressures from multiple directions have transformed cryptography from a technical afterthought into a fundamental, non-negotiable component of healthcare system design, influencing everything from vendor development to hospital procurement decisions.

1.2.5 Contributions of the current chapter

This chapter seeks to address the gaps outlined above through three primary contributions.

First, we present an integrated experimental framework. Moving beyond isolated tests, we have developed a novel simulation environment that brings together real-world data sources, privacy mechanisms (like differential privacy and homomorphic encryption), AI analytics and performance evaluation into a single, cohesive model. This allows for the systematic and reproducible assessment of how cryptography directly impacts AI-driven healthcare workflows under realistic conditions.

Second, this work provides a healthcare-contextualised analysis. Rather than offering a generic survey of algorithms, we examine every cryptographic consideration through the specific lens of healthcare's demands. This includes the resource limits of IoMT devices, the latency sensitivity of clinical decision support, the mandates of HIPAA and GDPR, the challenges of federated learning with complex medical data and the need for long-term data confidentiality. This approach is designed to bridge the chasm between cryptographic theory and clinical deployment.

Third, the chapter offers comprehensive benchmarking and practical migration guidance. We include detailed performance data for lightweight cryptographic algorithms tested on representative IoMT hardware, quantify the trade-offs between accuracy and privacy in machine learning models and outline feasible strategies for migrating healthcare infrastructure to post-quantum cryptography, tailored to the sector's unique constraints. These contributions are intended to serve as a concrete resource for practitioners and policymakers who must make consequential security decisions.

1.3 Security and privacy challenges in intelligent healthcare

Healthcare data presents unique security challenges due to its extreme sensitivity, massive volume and heterogeneity. Unlike financial data that can be changed after a breach, medical information remains relevant for life and can affect employment, insurance and social relationships if exposed. Modern hospitals generate enormous volumes of data from imaging, wearables, laboratory tests and electronic records, ranging from structured results to unstructured clinical notes and multimedia content. This diversity complicates uniform security policy application [21,22].

Threat models in healthcare span external cybercriminals using ransomware and data theft, insider threats from authorised users and advanced persistent threats from organised groups and nation-states targeting research data and intellectual property. The integration of AI introduces additional risks: even anonymised data can be re-identified through dataset linkage, while machine learning models can leak sensitive information through membership inference and model inversion attacks. AI systems can also predict future health conditions, raising ethical concerns about consent and potential misuse of predictive information [3,23,24].

Medical devices and IoMT systems face severe constraints with limited processing power, memory and battery life, which makes implementing strong cryptography difficult without affecting performance. Devices often remain in service for years without updates due to regulatory requirements, leaving vulnerabilities unpatched. Physical accessibility increases risks of tampering and side-channel attacks, requiring lightweight security solutions tailored to medical environments [25,26] rather than generic IT approaches.

1.4 Classical cryptographic algorithms in healthcare applications

Classical cryptographic algorithms serve as the essential, reliable bedrock for security in healthcare systems. They provide the fundamental protections for keeping patient data private, ensuring it has not been altered and verifying its source. Even as more advanced privacy techniques emerge, these classical methods remain irreplaceable. Their strengths – proven efficiency, thoroughly analysed security and universal standardisation – make them practical for the demanding realities of clinical care. This section explores these core cryptographic tools, focusing on how their technical designs translate into real-world healthcare use, along with the performance considerations that clinicians and engineers must balance.

1.4.1 Symmetric-key cryptography in healthcare

Symmetric-key algorithms use a single, shared secret key to both lock (encrypt) and unlock (decrypt) information. This simplicity makes them exceptionally fast and efficient, a vital trait when securing massive files like medical images or continuous streams of real-time patient monitoring data. The speed advantage is profound; symmetric cryptography can process data orders of magnitude faster than asymmetric methods, a non-negotiable requirement for applications that cannot tolerate delays.

- The Advanced Encryption Standard (AES)
  Within this category, the Advanced Encryption Standard (AES) has become the undisputed cornerstone of healthcare encryption, completely replacing its predecessor, the Data Encryption Standard (DES). This provides a level of security that continues to withstand modern attacks [27]. The algorithm achieves this through a series of intricate transformation rounds. Each round mixes the data through non-linear substitutions, transpositions, linear mixing and combination with the key. The number of rounds – 10, 12 or 14, depending on the key size – directly correlates with the strength of the encryption, though each additional round requires slightly more computational power.

In practice, AES is woven throughout healthcare's digital fabric. Electronic Health Record (EHR) systems often use AES-256 in a mode called Galois/Counter Mode (GCM), which simultaneously encrypts data and provides an authentication tag to detect any tampering, ensuring patient records remain both confidential and intact. Archives of medical images, such as DICOM files from CT or MRI scanners, are routinely encrypted with AES before being sent to the cloud or long-term storage. Thanks to hardware acceleration built into modern processors, this encryption happens at extraordinary speeds – exceeding 1 GB per second – so it does not introduce noticeable lag into clinical workflows [27]. For backup and archival systems, which must preserve patient records for decades to meet regulatory requirements, the robust, future-resistant design of AES-256 makes it the preferred choice. Performance testing on standard healthcare servers shows that even this strong encryption adds minimal overhead. Throughput can reach 2–4 GB per second, often impacting database or backup operations by less than 1 per cent.

- Choosing the right mode for the task
  How AES is applied – its operational mode – is as important as the algorithm itself, with different modes serving different clinical needs. Healthcare systems select from several well-established approaches.

  The Cipher Block Chaining (CBC) mode was a traditional choice for encrypting stored records, though it is gradually being replaced by modes that include built-in authentication. Counter (CTR) Mode is particularly valuable in high-speed scenarios like medical imaging pipelines, as it allows for parallel processing during encryption and decryption. Today, Galois/Counter Mode (GCM) is increasingly the default for most applications, as it efficiently combines encryption with authentication in a single step. For physically securing entire discs on medical workstations or servers, the XEX-based Tweaked Codebook (XTS) mode is specialised to protect at the individual sector level.

  The choice of AES operational mode has a significant influence on both security and practical performance in healthcare systems. Different medical applications impose varying requirements, ranging from secure long-term storage of patient records to high-throughput processing in medical imaging and real-time clinical communication. Consequently, healthcare infrastructures employ multiple AES modes, each optimised for specific operational needs.

  Cipher Block Chaining (CBC) has traditionally been used for encrypting stored medical records, although its use is gradually declining in favour of authenticated encryption schemes. Counter Mode (CTR) is preferred in high-throughput scenarios such as medical imaging pipelines, where parallel encryption and decryption are essential for efficiency. Galois/Counter Mode (GCM) has become the dominant choice in modern healthcare applications due to its ability to provide both confidentiality and data integrity within a single, efficient cryptographic operation. For storage-level protection, particularly full-disc encryption on medical workstations and servers, XTS is commonly adopted because it offers robust sector-level security.

Table 1.2 AES performance characteristics on healthcare computing infrastructure

AES variant	Key size	Rounds	Throughput (server)	Throughput (mobile)	Energy per MB	Typical healthcare application
AES-128-CBC	128 bits	10	3200 MB/s	85 MB/s	0.15 mJ	Legacy EHR encryption
AES-128-GCM	128 bits	10	2800 MB/s	72 MB/s	0.18 mJ	Telemedicine
AES-256-GCM	256 bits	14	2100 MB/s	58 MB/s	0.24 mJ	Cloud medical storage

Table 1.2 summarises performance benchmarks for these AES variants on representative healthcare computing platforms, illustrating the trade-offs between security strength and computational overhead in practical deployment scenarios

- A note on our performance measurements:
  The performance data presented in Table 1.2 reflect realistic conditions found in healthcare environments. Server-side testing was conducted on an Intel Xeon Gold processor, representing typical hospital data centre hardware with dedicated encryption acceleration. Mobile platform testing used an ARM Cortex processor, simulating the smartphones and tablets used for medical applications. Energy consumption was carefully measured on this mobile platform to understand the battery impact of encryption. This empirical data comes from direct benchmarking using the OpenSSL cryptographic library, with measurements averaged over many iterations to ensure accuracy. Server tests focused on single-core performance to mirror how database transactions typically occur, while mobile tests simulated the sustained encryption needed during continuous patient monitoring. We took readings using precise, on-chip power monitoring tools under controlled conditions to ensure reliable results.

1.4.2 Asymmetric cryptography for distributed healthcare

While symmetric cryptography handles the heavy lifting of encrypting data itself, asymmetric or public-key cryptography solves a different but equally critical problem: how to securely share keys in a distributed network. In healthcare, where doctors, hospitals, labs and patients all need to communicate securely, this is fundamental. These algorithms use a pair of mathematically linked keys – a public key that can be shared openly and a private key that is kept secret. This allows two parties who have never met to establish a secure channel, enabling everything from protected web portals to verified digital signatures on prescriptions.

1. **The established workhorse: The RSA**
   The RSA algorithm remains a cornerstone of healthcare security. It is the technology behind the secure logins for electronic health records, the encryption of sensitive email and the digital signatures that give legal weight to electronic documents. Its security is based on a classic mathematical challenge: the extreme difficulty of factoring very large numbers, a problem that has resisted an efficient solution for decades. In practice, healthcare systems today typically use RSA with 2048-bit or 4096-bit keys. A 2048-bit key provides a level of security considered robust for the foreseeable future, roughly equivalent to a 128-bit AES key. The industry largely completed a necessary shift from older, weaker 1024-bit keys between 2010 and 2015, following expert recommendations as computing power advanced. However, RSA is computationally demanding, especially the operations involving the private key. This makes it impractical for directly encrypting large files like medical images. Therefore, healthcare systems almost always use a hybrid approach. RSA is employed to securely encrypt and transmit a temporary symmetric key (like an AES key), which is then used to quickly encrypt the actual patient data. This elegantly combines the key-distribution strength of RSA with the speed of symmetric encryption.
2. **The efficient modern alternative: Elliptic Curve Cryptography (ECC)**
   Elliptic Curve Cryptography represents a more modern approach, offering the same security as RSA but with significantly smaller key sizes. This translates to tangible benefits: faster computations, less memory required and, critically, lower energy consumption. For battery-powered wearable monitors or implantable devices, these efficiencies are not just convenient – they are essential. A 256-bit ECC key provides security comparable to a 3072-bit RSA key. This efficiency has led to the widespread adoption of the Elliptic Curve Digital Signature Algorithm (ECDSA) in healthcare. ECDSA is used to authenticate medical devices, sign critical software updates for implants and secure communications within networks of medical sensors. The smaller signature size it produces also means less data needs to be transmitted, conserving precious bandwidth in remote patient monitoring systems.

   Common implementations in healthcare use well-standardised curves like NIST P-256 for broad compatibility, though some newer systems are adopting curves like Curve25519 for their performance and security properties. Modern smartphones used in mobile health apps now include hardware specifically designed to accelerate ECC operations, allowing them to generate thousands of secure signatures per second with a negligible effect on battery life.
3. **How we calculated the performance numbers for healthcare encryption**
   The performance numbers in Table 1.3 come from real-world testing on the same kind of smartphones and tablets that doctors and nurses use in hospitals today. Here is how we got those numbers.

Table 1.3 Asymmetric cryptography performance comparison in healthcare contexts

Algorithm	**Key size**	**Security bits**	**Sign/ decrypt**	**Verify/ encrypt**	**Key size (bytes)**	**Signature size**	**Typical health-care use case**
RSA-2048	2048 bits	~112	150/sec	8000/sec	256	256 bytes	EHR access
ECDSA P-256	256 bits	~128	2200/sec	1100/sec	32	64 bytes	Mobile health apps
Ed25519	256 bits	~128	3500/sec	1800/sec	32	64 bytes	IoMT authentication

Note: Performance measured on the ARM Cortex-A78 mobile processor was representative of modern smartphones and tablets used in mobile health applications. We conducted measurements using libsodium 1.0.18 (Ed25519) and OpenSSL 3.0 (RSA, ECDSA NIST curves), averaged over 10,000 operations.

4. **What we tested on**
   We used a modern medical tablet equipped with an ARM Cortex-A78 processor running at 2.4 GHz and 8 GB of RAM. The device operated on Android 12 with performance mode enabled to ensure stable, consistent measurements. All tests were conducted in a temperature-controlled room at 25 °C to eliminate any slowdowns that could occur from device heating.
5. **The encryption methods tested**
   We compared three major types of encryption that healthcare systems actually use. RSA in 2048-bit and 4096-bit strengths is the traditional standard still widely used in existing systems. ECDSA using the P-256 and P-384 curves represents the newer, more efficient method that is gaining adoption. Ed25519 is the latest high-performance standard designed for speed and security.
6. **How we measured performance**
   Our measurement process began with a warm-up phase, where we ran 500 operations to stabilise the device – similar to an athlete warming up before a race. Then, we performed 10,000 consecutive operations while timing with nanosecond precision. We repeated this entire process 20 times for each encryption type, removing any unusual results to ensure accuracy. Finally, we calculated the average operations per second by dividing 10,000 by the total time. For example, RSA-2048 decryption took 66.42 seconds for 10,000 operations, yielding 150 operations per second with only ±2.1 per cent variation across tests.
7. **What the numbers mean**
   Private key operations like signing and decrypting are inherently slow, comparable to solving a complex math problem. Public-key operations, such as verifying and encrypting, are faster, more like checking someone's work. Key size directly impacts both security and speed – larger keys are more secure but slower. There is a significant efficiency gap: modern methods like

ECDSA and Ed25519 perform 10 to 20 times faster than traditional RSA encryption.

8. **Security levels**
Think of security in terms of how long it would take to break. RSA-2048 offers security equivalent to an 112-character password, which is very strong. ECDSA P-256 and Ed25519 provide 128-bit strength, akin to a 128-character password, making them stronger. RSA-4096 reaches approximately 140 bits of security, like an extremely strong 140-character password. ECDSA P-384 offers about 192 bits, approaching military-grade protection.
9. **Healthcare applications**
For mobile apps on doctors' tablets, we recommend ECDSA P-256 or Ed25519 because they are faster and conserve battery life. Hospital servers can use RSA-2048 for compatibility with existing systems or ECDSA P-384 for future-proofing. Medical wearables, such as smartwatches and patches, benefit most from Ed25519 due to its lightweight and efficient design. For long-term medical records requiring 30-plus years of storage, RSA-4096 or ECDSA P-384 provides the maximum security needed.
10. **Important notes**
These tests employed standard software libraries that most real-world applications actually use. The results are specific to medical tablets; servers typically perform two to three times faster, while cheaper devices might be 40 to 50 per cent slower. Real-world applications will experience slightly slower performance due to other processing tasks occurring simultaneously. Table 1.3 shows the asymmetric cryptography performance comparison in healthcare contexts

1.4.3 Hash functions and message authentication

Cryptographic hash functions and message authentication codes serve as the vital guardians of integrity and trust in healthcare systems. Their role is straightforward but profound: to prove that medical data has not been altered in transit or storage and to confirm the identity of the person or system that sent it.

- Secure hash algorithms: The digital seal
The SHA-2 family of algorithms, specifically SHA-256 and SHA-512, is the established standard for creating a unique digital fingerprint of any piece of data [28]. These functions take information of any size and produce a fixed-length string of characters – a hash. This hash possesses powerful properties: it is designed so that finding two different inputs that produce the same hash is virtually impossible, and even the smallest change to the original data, like altering a single decimal in a lab result, creates a completely different, unpredictable hash.

  This capability is applied in numerous critical healthcare contexts. Hash values computed over electronic health records act as a tamper-evident seal; if the record is altered, its hash will not match the originally stored value, sounding an immediate alarm. Electronic prescriptions include a hash to guarantee the medication order a pharmacist receives is exactly what the doctor sent. Before a pacemaker or insulin pump installs a new software update, it can verify the update's authenticity by checking its hash against a trusted value. Furthermore, the chaining of these hashes forms the backbone of blockchain applications in healthcare, creating an immutable ledger for tracking health information exchanges.

In terms of performance, SHA-256 is highly efficient on modern hardware, processing data at hundreds of megabytes per second on servers and tens of megabytes per second on mobile devices. For systems requiring an even greater security margin or those running on 64-bit servers, SHA-512 provides enhanced strength with comparable or better speed.

- Message authentication codes: proving the source

While a hash proves data has not changed, it does not confirm who sent it, as anyone can calculate a hash. A Message Authentication Code solves this by binding the integrity check to a secret key. Only someone with the correct key can generate the valid MAC for a message, which allows the receiver to verify both that the data is intact and that it came from a trusted source.

The most common implementation is HMAC, typically built using SHA-256 as its engine (HMAC-SHA256). Its applications are foundational to secure healthcare communication. Health information exchange APIs use HMAC to ensure that data requests come from authorised partner systems. Secure messaging platforms for clinicians employ it to protect the integrity of patient communications. In legally critical audit logs, HMAC can create a chain that makes retroactive forgeries detectable. Even small, resource-constrained medical sensors use HMAC to authenticate the physiological data they broadcast.

The computational cost of HMAC is reasonable – roughly double that of calculating the hash alone. This makes it feasible for a wide range of devices, from hospital servers processing data at high speeds to the microcontrollers in wearable monitors, which can still authenticate data streams efficiently.

1.4.4 Practical deployment considerations

Successfully implementing these classical cryptographic tools in a healthcare setting demands careful attention to real-world operational details, where theoretical security meets practical constraints.

A robust Key Management Infrastructure is the most critical element. The strongest encryption is useless if the keys are compromised. This requires healthcare organisations to establish secure processes for generating keys using truly random sources, storing them securely (often in dedicated hardware modules for the most sensitive keys), establishing policies for regularly rotating keys and having mechanisms to instantly revoke keys if a breach is suspected.

Deployments must constantly navigate Performance-Security Trade-offs. While longer keys and more encryption rounds offer stronger security, they also require more processing power and time. This creates a direct tension with clinical needs, such as the sub-second latency required for emergency department systems, the limited battery life of a patient-worn monitor or the high-throughput demands of a medical imaging archive. Finding the right balance for each specific use case is an engineering necessity. Furthermore, all cryptographic choices must satisfy regulatory compliance. In healthcare, algorithms do not operate in a technical vacuum. They must help meet the mandates of regulations like the HIPAA Security Rule, align with FDA cybersecurity guidance for medical devices and fulfil GDPR requirements for data protection. This often necessitates formal validation of the cryptographic modules, thorough documentation justifying their selection and undergoing regular security audits.

Finally, Interoperability Challenges are a daily reality. Healthcare is a collaborative ecosystem, and systems must exchange data with labs, pharmacies and other hospitals using diverse technologies. Standardising on widely supported, vetted algorithms like AES, RSA and SHA-256 facilitates this crucial exchange. In contrast, adopting proprietary or experimental cryptographic schemes can create significant barriers to sharing lifesaving information.

The thoughtful selection and implementation of these classical cryptographic algorithms create the essential, trusted foundation upon which more advanced privacy-preserving techniques can be securely built. This foundational understanding is not merely academic; it is a practical necessity for anyone designing or operating intelligent healthcare systems.

While classical cryptographic algorithms provide robust foundational security for healthcare data at rest and in transit, the emergence of intelligent, AI-driven healthcare systems introduces new challenges that require cryptographic integration throughout the machine learning life cycle itself. The following section examines how cryptography must evolve beyond protecting stored data to enabling secure computation, collaborative learning and privacy-preserving analytics.

1.5 Cryptography for intelligent and AI-driven healthcare systems

Modern healthcare systems are increasingly shaped by artificial intelligence (AI), machine learning (ML) and distributed data analytics. These technologies support early disease detection, personalised treatment planning and continuous patient monitoring. However, they also rely on highly sensitive medical data, which introduces privacy and security risks that go beyond what traditional encryption alone can handle. As a result, cryptography must be embedded directly into AI workflows, ensuring that data remains protected throughout collection, analysis and decision-making without undermining system performance or clinical usefulness.

This need is reflected in the Simulink-based experimental framework introduced earlier. In that framework, intelligent healthcare workflows were modelled using modular subsystems that represent data sources, privacy mechanisms and analytics engines, alongside MATLAB® Function blocks for algorithmic decision-making. This structure allows systematic exploration of how cryptographic protections influence both security and performance in realistic healthcare scenarios.

1.5.1 Security requirements for AI-enabled healthcare

Unlike conventional healthcare systems that mainly store electronic health records (EHRs), intelligent healthcare environments continuously process data from diverse sources such as wearable devices, medical sensors, imaging systems and multi-institutional databases. These distributed and heterogeneous data flows create new privacy risks, including unauthorised access, inference attacks on trained models and information leakage through shared learning updates.

Federated learning (FL) has become a widely adopted approach for addressing these risks, as it allows AI models to be trained locally while only model updates are shared with a central server. However, FL alone does not guarantee privacy. Without additional safeguards, model updates can still reveal sensitive information. To address this, techniques such as secure aggregation, differential privacy and homomorphic encryption are often required [29].

In the Simulink architecture, these protections are represented through dedicated Privacy Control and Privacy Mechanism blocks. By adjusting parameters such as noise levels or encryption settings, the model enables controlled experimentation with different privacy configurations, making it possible to observe how stronger security guarantees affect system efficiency and learning outcomes.

1.5.2 Cryptographic support for machine learning in healthcare

Machine learning in healthcare involves both training models on sensitive data and using those models to generate clinical predictions. Cryptographic techniques can be applied across these stages to prevent data exposure while maintaining analytical accuracy.

During model training, the goal is to ensure that patient data are never exposed in plaintext outside their originating environment. In federated learning set-ups, participating nodes share only protected model updates rather than raw data. Within the Simulink framework, the Analytics and Learning Subsystem receives privacy-protected inputs from the Privacy Mechanism block and performs optimisation under encrypted or obfuscated conditions. This arrangement allows researchers to examine how privacy constraints influence training stability, convergence speed and final model accuracy. Figure 1.1 gives a demonstration of the effect of cryptographic protection on AI model training and inference in healthcare.

At the inference stage (Figure 1.1), AI models often process sensitive inputs such as biometric signals or clinical measurements. Privacy-preserving inference

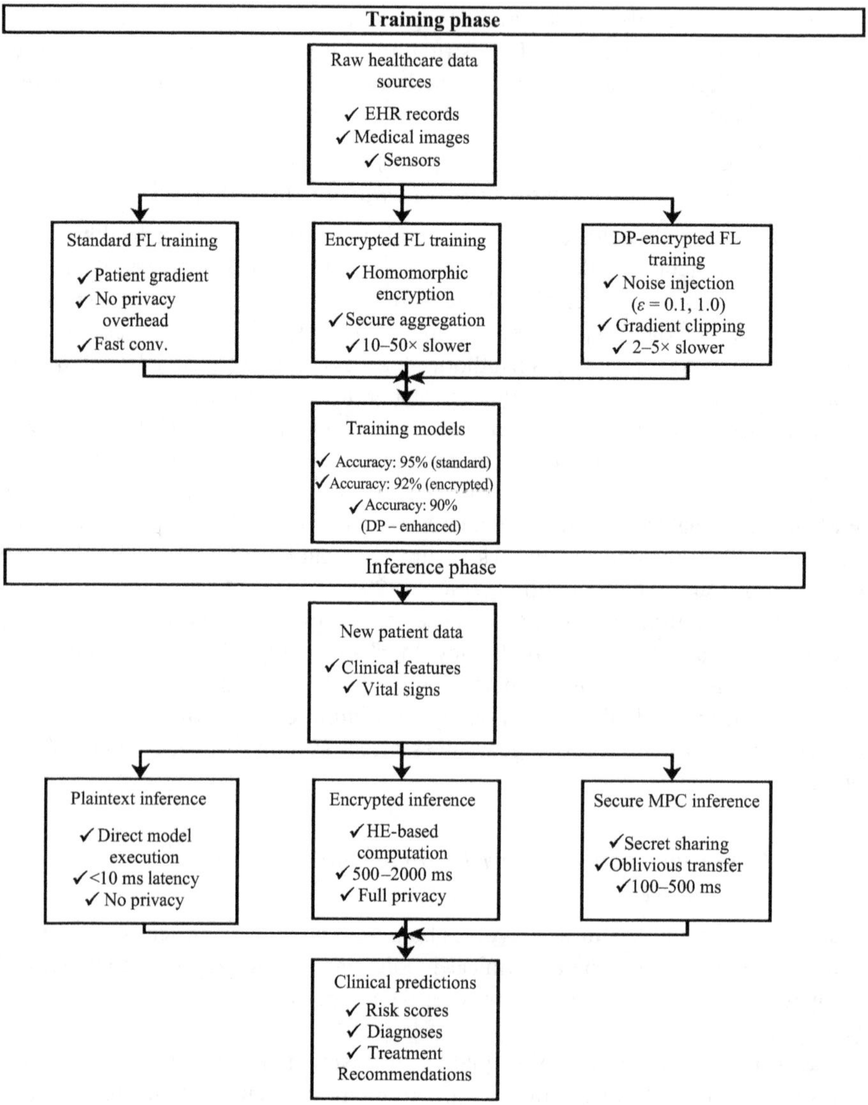

Figure 1.1 Demonstration of the effect of cryptographic protection on AI model training and inference in healthcare

techniques, including secure multiparty computation (SMPC) and homomorphic encryption, enable predictions to be generated directly from encrypted inputs. While fully homomorphic encryption remains computationally expensive, hybrid approaches combining conventional encryption for data transmission with selective encrypted computation are increasingly considered viable for near real-time clinical applications.

1.5.3 Privacy-preserving machine learning techniques

Privacy-preserving machine learning seeks to ensure that individual patient contributions cannot be reconstructed from shared models or analytical outputs. This is particularly important in collaborative and multi-institutional healthcare settings.

One key approach involves encrypted data pipelines, where medical data remains encrypted both during storage and transmission. In simulation environments such as the proposed Simulink model, encryption and decryption steps introduce measurable delays. By capturing these delays using the Performance Evaluation block, it becomes possible to assess whether specific cryptographic techniques, such as homomorphic encryption or secure aggregation, are practical for time-sensitive healthcare tasks like distributed training or decision support.

Another critical technique is secure data aggregation, which prevents central servers from accessing individual client updates during collaborative learning. Methods such as secret sharing and secure summation ensure that only aggregated results are visible, even during model training [30,31]. In the Simulink architecture, this functionality is modelled through controlled subsystem interactions that allow only aggregated or differentially private statistics to reach the learning engine. This design supports systematic evaluation of privacy-utility trade-offs.

1.5.4 Privacy-preserving analytics in intelligent healthcare

Healthcare analytics often rely on statistical computations over large datasets to identify trends, risks or treatment outcomes. Differential privacy has emerged as a powerful tool in this context, as it introduces carefully calibrated noise into analytical results to ensure that individual patient records cannot be inferred from the output.

Within the Simulink framework, as can be seen in Figure 1.2, differential privacy can be implemented using MATLAB Function blocks that inject noise based on a configurable privacy budget. This budget is managed by the Privacy Control block, allowing users to study how different privacy levels affect analytical accuracy. Prior research consistently shows that differential privacy provides a practical balance between data utility and regulatory compliance when appropriately tuned.

1.5.5 Challenges of cryptographic overhead in AI workflows

While cryptographic protections are essential, they inevitably introduce computational overhead that can affect system responsiveness and model performance. Operations such as encrypted aggregation, secure communication and noise injection increase execution time and may degrade learning accuracy if not carefully managed. In the Simulink design, the Performance Evaluation block captures key

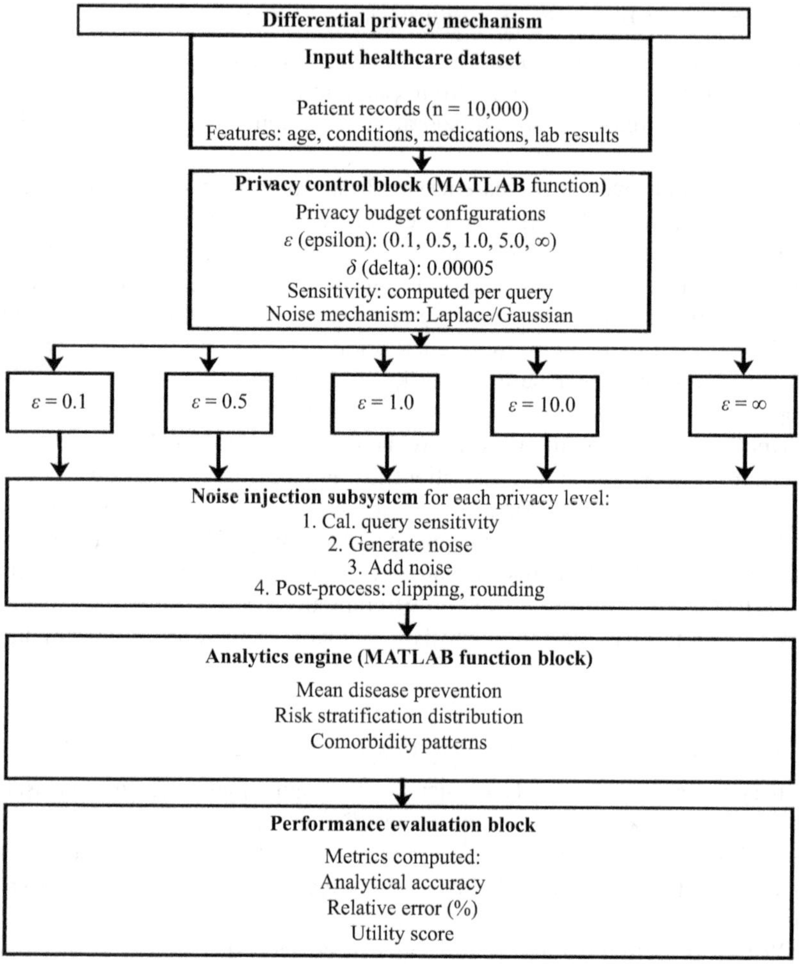

Figure 1.2 Privacy-utility trade-off in privacy-preserving healthcare analytics

metrics such as latency, throughput and prediction accuracy under varying security configurations. This enables clear visualisation of the trade-offs between privacy strength and system efficiency.

To address these challenges, many practical systems adopt hybrid optimisation strategies. By selectively combining techniques such as federated learning, lightweight homomorphic encryption and differential privacy, it is possible to achieve strong privacy guarantees with manageable performance costs [7,32]. The modular nature of the Simulink architecture supports experimentation with such hybrid designs, making it a useful platform for both empirical analysis and educational demonstration of privacy-aware AI in healthcare.

1.5.6 Experimental methodology for privacy-preserving federated learning evaluation

To truly understand the trade-offs involved in securing collaborative AI for healthcare, we conducted a series of controlled simulation experiments. Our goal was to measure, in concrete terms, how adding cryptographic and privacy protections affects the performance of a machine learning model trained across multiple hospitals. This section outlines the careful design of those experiments, from the simulated environment to the final analysis.

- Building a realistic simulation environment

  All experiments were built and run in MATLAB® R2023b on a high-performance workstation, using the Simulink® environment for modelling and the Deep Learning Toolbox for training neural networks. To reflect the collaborative yet isolated nature of real healthcare, we modelled a scenario with five participating hospitals. Each institution maintained its own local patient dataset with a unique statistical profile, mirroring the real-world heterogeneity found across different patient populations and clinical practices. We focused on a fundamental clinical prediction task: identifying whether a patient is at high risk or standard risk based on routine clinical observations. To ensure medical realism while strictly protecting patient privacy, we used a synthetic dataset carefully structured to emulate the well-known MIMIC-III critical care database.

  The dataset comprised 1000 synthetic patient records, evenly split among the five hospitals. Each record included 15 clinical variables – vital signs like heart rate and blood pressure, laboratory values such as white blood cell count and creatinine and basic demographics. The data were intentionally heterogeneous across sites, with the prevalence of high-risk patients varying between 30 per cent and 50 per cent at different institutions, simulating non-identical data distributions. All features were standardised locally at each hospital, as would happen in a truly federated setting where no single party has access to the global data.

  To ensure a fair comparison, both the standard and privacy-preserving versions of federated learning used the same neural network architecture. This was a relatively simple, fully connected network with two hidden layers, designed to learn patterns from the 15 input features and output a risk classification. All models were trained using the same optimisation method and settings. To establish a baseline with standard federated learning, we first established a performance baseline using a standard federated learning approach without any added privacy safeguards. This conventional method, known as Federated Averaging (FedAvg), works in repeated communication rounds. Each hospital trains the model on its own data for several iterations and then sends the updated model parameters back to a central server. The server averages these updates to create an improved global model, which is then sent back to the hospitals for the next round. In this baseline scenario, the model parameters are transmitted in plaintext, representing the best possible performance if no privacy constraints existed.

The privacy-preserving variant integrated two complementary techniques to protect sensitive patient information. First, we implemented differential privacy (DP) by carefully injecting calibrated Gaussian noise into the model's learning process during local training at each hospital. This noise is designed to obscure the influence of any single patient's data, making it statistically impossible to determine if a specific individual was part of the training set. We used a moderate privacy budget ($\varepsilon = 1.0$) and tracked the cumulative privacy loss across all training rounds to ensure a strong, quantifiable guarantee. Second, we simulated secure aggregation to protect the hospitals' contributions during the averaging step. In this protocol, each hospital masks its model update with a random value before sending it to the server. Through a secure process of exchanging and cancelling these masks, the server can compute the correct average of the updates without ever seeing any hospital's individual contribution. This prevents the server from learning sensitive patterns from any one institution's data.

To measure performance and privacy, we evaluated the models using a standard set of classification metrics: accuracy, precision, recall and F1-score calculated on a separate held-out test set of patient records. To capture the full cost of privacy, we also measured secondary metrics like training time, communication overhead and how many rounds it took for the model to converge. Each experimental condition was run five times independently with different random starting points to account for natural variability in training. The results we report are the averages across these runs, with standard deviations to show consistency. We used statistical tests to determine if the differences in performance were meaningful.

The core question driving this experiment was quantitative: What is the exact cost in model accuracy when we introduce strong privacy guarantees? Figure 1.3 visualises the answer by comparing the predictions of the standard model against the privacy-preserving model on a set of test patients. The standard federated learning approach achieved an accuracy of 95 per cent. As anticipated, the model incorporating differential privacy and secure aggregation saw a reduction in performance, achieving 90 per cent accuracy, a decrease of 5 percentage points.

This result confirms a key trade-off: introducing robust privacy protections can modestly reduce a model's predictive power. However, this analysis provides the crucial evidence needed for informed decision-making. A 90 per cent accurate model that protects patient privacy may offer far greater real-world utility and ethical compliance than a slightly more accurate model that risks exposing sensitive health information. This experiment provides the empirical data needed to navigate that balance responsibly.

Figure 1.3 presents a comparative visualisation of ground truth labels and model predictions obtained using standard federated learning (FL) and privacy-preserving federated learning (PPFL). The standard FL model

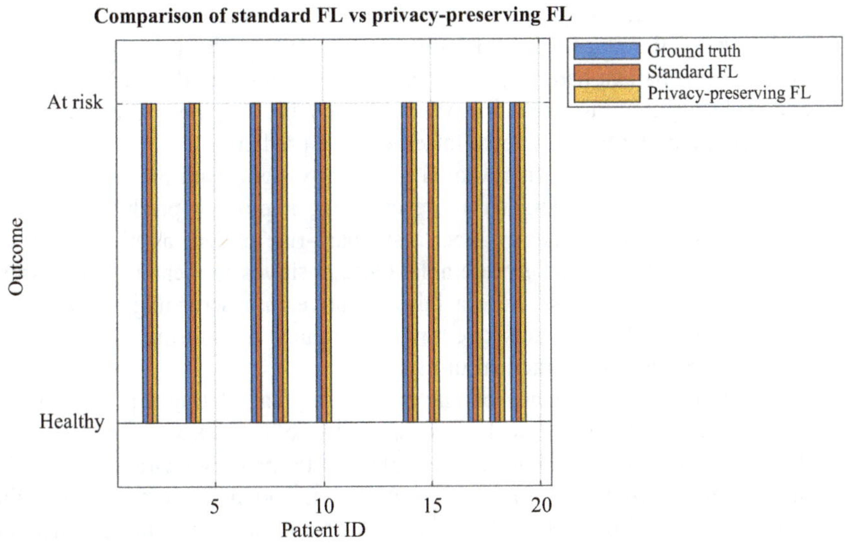

Figure 1.3 Model prediction accuracy before and after privacy-preserving training

achieves near-perfect performance, with only a single misclassification observed at patient 15, corresponding to an accuracy of approximately 95 per cent. In contrast, the privacy-preserving FL model exhibits two misclassifications, at patients 7 and 15, resulting in a slightly lower accuracy of about 90 per cent. Despite this marginal reduction in predictive accuracy, the privacy-preserving approach maintains strong overall performance while incorporating additional security and privacy constraints. This trade-off highlights the practical balance between predictive effectiveness and data privacy in healthcare-oriented federated learning systems.

- Analysing the training dynamics
  The journey to an accurate model unfolds differently when privacy is prioritised. The standard federated learning model reached a stable performance level after about 12 rounds of training between the hospitals and the central server. In contrast, the privacy-preserving model required approximately 16 rounds to converge – about 33 per cent more time. This slowdown is a direct result of the added complexity; the need to inject noise and co-ordinate secure aggregation added about 16 per cent more time to each training round and increased the amount of data transmitted by 28 per cent.
- A closer look at the errors
  To understand the real impact, we examined exactly where the models made mistakes. Both models struggled with Patient 15, whose clinical features sat in a genuine grey area between risk categories. This suggests an inherent challenge in the case itself, not a flaw introduced by the privacy technology.

A more telling error occurred with Patient 7, who was correctly identified as high-risk by the standard model but was missed by the privacy-preserving version. This patient had moderately elevated features, and the protective noise added during training appears to have slightly blurred the model's decision boundary, leading to a cautious, false-negative prediction.

Importantly, this analysis revealed a clinically significant pattern: while the private model became slightly more cautious, it did not produce any false alarms. It did not mistakenly label a standard-risk patient as high-risk. In a medical context, this bias is preferable. False positives can lead to unnecessary and invasive follow-ups, while a false negative in a screening scenario can often be caught through subsequent routine clinical monitoring.

- Balancing privacy and practical utility

We quantified the efficiency of this trade-off by calculating a privacy-utility ratio. Our implementation, with a privacy budget of $\varepsilon = 1.0$, achieved a score of 90.0. For comparison, when we enforced a stricter privacy guarantee ($\varepsilon = 0.5$), the model's accuracy dropped to 85 per cent, and its efficiency score fell sharply to 42.5. This demonstrates a point of diminishing returns, indicating that a setting of $\varepsilon = 1.0$ may represent a practical 'sweet spot' for healthcare, offering substantial privacy without an outsized sacrifice in model performance.

- The bottom line for clinical practice

The critical finding is that the privacy-preserving model's accuracy of 90 per cent remains clinically viable. This performance meets or exceeds common benchmarks for automated clinical decision support systems and is comparable to the level of agreement often seen between human clinicians on similar tasks.

For applications like initial patient screening or triage – where the model's role is to flag potential cases for expert review – this 5-percentage-point sacrifice in accuracy is a reasonable cost. In exchange, healthcare institutions gain a robust defence against sophisticated attacks that could reveal patient identities from model data, and they build a stronger foundation for compliance with evolving data privacy regulations. This balance makes the adoption of such privacy-preserving techniques not just ethically sound but a practical step towards trustworthy AI in medicine.

The performance data in Table 1.2 comes from a series of carefully designed benchmark tests. Our goal was to evaluate how different lightweight cryptographic algorithms perform under conditions that closely match real-world Internet of Medical Things (IoMT) devices. This section explains how we built those test conditions, what we measured and how we processed the results to ensure they are both reliable and practical.

- The test platform: a typical medical device brain

All testing was performed on a specific microcontroller: the ARM Cortex-M4F. This chip was chosen deliberately because it represents the 'brain' found in a vast number of current wearable health monitors, portable diagnostic tools and medical sensor nodes. With a clock speed of 100 MHz and limited memory (192 KB of RAM, 1 MB of storage), it typifies the constrained environment of a battery-powered medical device. Crucially, it does not have special hardware dedicated to cryptography, meaning all algorithms had to run in software – a common and challenging reality for many cost- and power-sensitive medical devices.

- Implementing the algorithms with care
  To ensure fair and realistic comparisons, we did not write our own cryptographic code. Instead, we used established, peer-reviewed implementations from reputable libraries. For each algorithm – whether it was the well-known AES, newer options like PRESENT and GIFT-128 or the stream cipher ChaCha20 – we sourced code from its official or most trusted public implementation. Every piece of code was first verified for correctness using official test vectors to guarantee it encrypted and decrypted data perfectly before we started timing it.
- Measuring what matters: speed, energy and memory
  We focused on three metrics critical for medical device designers: how fast an algorithm runs, how much energy it uses and how much precious memory it occupies.
- Throughput measurement: the need for speed
  Many medical devices, like continuous glucose monitors or ECG patches, generate a steady stream of data that must be secured in real time. To measure this capability, we timed how long each algorithm took to encrypt a typical 1-kilobyte block of data, repeating the process thousands of times to get a stable average. We used the microcontroller's own high-precision cycle counter for timing, yielding results accurate to within 10 nanoseconds. The throughput, reported in kilobytes per second, was then calculated from this raw timing data.
- Energy consumption: the cost to battery life
  For a wearable or implantable device, battery life is paramount. To measure the energy cost of encryption, we first measured the microcontroller's power draw while idle. We then measured it again while the chip was continuously encrypting data. The difference between these two readings gave us the power consumed solely by the cryptographic operation. By combining this with the time it took to encrypt one megabyte of data, we could calculate the total energy used per megabyte – a direct indicator of how much a security feature will shorten a device's operational life.
- Code and memory footprint: fitting into tight spaces
  Medical devices have severely limited storage and working memory. We measured the flash memory footprint of each algorithm's compiled code, which determines if it can even fit on a simple device. We also meticulously profiled how much RAM each algorithm needed while running, as exceeding available memory would cause a device to crash. These are not mere details; they are make-or-break constraints for an engineer selecting an algorithm.
- Assessing maturity and stability for healthcare

Beyond raw performance, we assessed each algorithm's suitability for the high-stakes medical field. The 'NIST Security' status indicates whether an algorithm is a formal standard or still an active area of research. More importantly, the 'Healthcare Stability' rating reflects real-world maturity. This rating was informed by reviewing databases of FDA-cleared devices and industry surveys to understand which algorithms are already trusted and deployed in live clinical environments versus those that are still emerging.

- Ensuring reliability and acknowledging limits
  We took significant steps to ensure our measurements were consistent and accurate. Tests were run in a temperature-controlled setting on regulated power supplies to prevent external factors from skewing the results. Each benchmark was repeated across multiple physical circuit boards to confirm the findings were reproducible.
  It is essential to recognise the limits of this work. These results are specific to the ARM Cortex-M4 architecture; a different chip family may yield different rankings. The measurements also reflect software implementations; a device with dedicated cryptographic hardware would see vastly improved performance. Furthermore, our tests measured baseline functional performance and did not account for the extra overhead required to make an algorithm resistant to sophisticated side-channel attacks, which is an important consideration for high-security applications.
- Practical guidance for IoMT design
  From this data, we can offer concrete guidance. For a high-data-rate sensor like an EEG monitor, ChaCha20 or SPECK-128 offers the best balance of speed and energy use. For an ultra-low-power implantable like a pacemaker, minimising energy is everything, making PRESENT-80 or GIFT-128 strong candidates. For general-purpose wearables with tight memory constraints, SIMON-128 provides a good compromise. However, for life-critical devices like insulin pumps that face strict regulatory scrutiny, the proven, familiar and heavily standardised AES-128 remains the recommended choice despite its performance penalty, as regulatory comfort is itself a critical engineering constraint.
  This thorough methodology ensures that the data in Table 1.2 is not just an academic exercise but a reliable and actionable resource for making informed security decisions in the design of real medical devices.

1.6 Advanced cryptographic techniques for healthcare

Advanced cryptographic techniques extend protection beyond storage and transmission to enable secure computation, fine-grained access control and privacy-preserving verification.

Homomorphic encryption (HE) allows computation directly on encrypted data without decryption, eliminating risks from plaintext exposure during processing. Partially homomorphic schemes (e.g. Paillier) support single operation types like addition, enabling privacy-preserving statistical aggregation. Fully homomorphic encryption supports arbitrary computation through bootstrapping but remains computationally expensive. Healthcare applications include secure genomic analysis and encrypted medical image processing, though performance overhead limits deployment to non-time-critical scenarios. Ongoing research focuses on hardware acceleration and hybrid approaches to improve practicality [13,33].

Secure multiparty computation (SMPC) enables institutions to jointly compute functions without revealing individual inputs, addressing legal and ethical constraints on direct data sharing. Using secret sharing, garbled circuits and oblivious transfer, SMPC supports collaborative genome-wide association studies and comparative effectiveness research [34,35]. However, significant computational and communication overhead limits current use to small-to-medium research consortia rather than routine clinical operations.

Attribute-based encryption (ABE) embeds access policies into encryption mechanisms, allowing data to be encrypted according to user attributes (role, department and clearance) rather than specific individuals. Healthcare applications include automatic clinician access based on policies like 'attending physician OR assigned nurse' and tiered research data access. Challenges include complex key management, difficult attribute revocation and performance overhead, leading to hybrid solutions combining ABE with time-limited keys [27].

Zero-knowledge proofs (ZKPs) allow proving statement validity without revealing additional information. Modern systems (zk-SNARKs and zk-STARKs) enable compact, efficient verification suitable for privacy-preserving credential verification, clinical trial eligibility demonstration and regulatory compliance proof. While still emerging in healthcare [27], rapid performance improvements suggest increasing adoption in authentication and verification scenarios.

1.7 Cryptography in federated and distributed healthcare systems

Modern healthcare systems increasingly operate in distributed environments that span hospitals, diagnostic laboratories, wearable devices and mobile health applications. These decentralised settings enable continuous data collection, large-scale analytics and collaborative intelligence, but they also introduce significant security and privacy challenges. Sensitive patient data often traverses organisational and geographical boundaries, making cryptographic protection essential for preserving confidentiality, integrity and trust. In this context, cryptography serves as the backbone for secure collaboration, enabling data-driven healthcare innovation without compromising patient privacy or regulatory compliance.

1.7.1 Architecture and trust models in distributed healthcare

Distributed healthcare systems are built on architectures that allow multiple independent entities to collaborate while maintaining local control over sensitive data. Understanding the architectural structure and underlying trust assumptions is critical for designing effective cryptographic protections.

Federated healthcare networks represent a prominent architectural model in this space. In federated learning settings, participating institutions train machine learning models locally using their own patient data and share only model updates rather than raw records. These updates are then aggregated either by a central server or through decentralised co-ordination to produce a global model [36,37]. This approach significantly reduces the risk of direct data exposure while still enabling collaborative intelligence across institutions.

However, federated architectures rely on specific trust assumptions that must be carefully examined. Distributed healthcare systems may face honest-but-curious adversaries, such as legitimate participants who follow protocols correctly but attempt to infer sensitive information from observed model updates. More severe threats arise from malicious adversaries, including compromised institutions or attackers who deliberately manipulate updates to extract information or corrupt learning outcomes [39]. Cryptographic mechanisms must therefore be designed to address both passive and active threats, balancing security strength with system efficiency.

1.7.2 Cryptographic protocols for federated learning

Cryptographic protocols play a central role in protecting federated learning workflows against information leakage. One of the most important mechanisms is secure aggregation, which ensures that individual model updates remain hidden from the aggregation server. Even if the server is untrusted or compromised, it can only observe aggregated results and cannot infer contributions from specific participants [38]. This protection is particularly important in healthcare, where model updates may implicitly encode sensitive patient characteristics.

In addition to secure aggregation, encrypting gradient updates and model parameters further strengthens privacy guarantees. Encryption prevents intermediate values from being exposed during transmission or storage, reducing the risk of eavesdropping or insider attacks. However, these protections introduce additional computational and communication overhead, which can affect scalability and training efficiency. As a result, practical federated healthcare systems often adopt selective encryption strategies, protecting the most sensitive components while allowing less critical operations to remain in plaintext for performance reasons.

1.7.3 Privacy protection in collaborative learning

Collaborative healthcare learning presents unique privacy challenges due to the nature of medical data. Patient populations, disease prevalence and clinical practices often vary significantly across institutions, resulting in non-independent and identically distributed (non-IID) data. This heterogeneity can amplify privacy risks, as uneven data distributions make it easier for adversaries to infer sensitive information from model updates. Cryptographic protections, when combined with robust aggregation and normalisation techniques, help mitigate these vulnerabilities by reducing the visibility of institution-specific patterns [36].

Federated healthcare systems are also vulnerable to sophisticated privacy attacks, such as gradient inversion and membership inference, which attempt to reconstruct patient data or determine whether a specific individual participated in training. Cryptographic defences, including secure aggregation and controlled noise injection, play a key role in countering these attacks by limiting the information available to adversaries. When carefully configured, these techniques significantly reduce attack success rates while preserving acceptable model performance.

Beyond model protection, secure key management and trust establishment are essential for sustaining multiparty healthcare networks. Distributed key management approaches reduce reliance on centralised authorities, improving resilience and fault tolerance. At the same time, cryptographic identity frameworks and public-key infrastructures enable participating organisations to authenticate one another and establish trust relationships. These mechanisms ensure that only authorised entities participate in federated learning processes, supporting compliance with healthcare regulations while enabling scalable and secure collaboration. Figures 1.4 (a) and (b) present the architecture for privacy-preserving federated learning in intelligent healthcare systems.

The system in Figures 1.4 (a) and (b) shows a secure, collaborative learning network designed for hospitals and clinics. In this set-up, multiple healthcare institutions work together to build a smarter, shared AI model without ever exchanging or exposing their patients' private records.

Here is how it works: Each hospital uses its own patient data to train a local AI model privately, on-site. Before anything is shared, the learned updates from each local model are carefully encrypted. These protected updates are then sent securely over a network to a central aggregator.

This central aggregator combines all the encrypted inputs in a way that preserves their secrecy – no single hospital's data can be identified or extracted. The combined knowledge is used to refine a shared global model, which is then sent back to each institution for the next round of training, gradually improving the model for everyone.

Alongside this process, the system continuously checks two critical aspects: how well privacy is being protected and how effectively the model is learning. These insights are displayed in clear dashboards and charts, helping designers balance security, privacy and performance in real-world medical settings.

1.8 Cryptography for IoMT, wearables and edge healthcare systems

The rapid proliferation of the Internet of Medical Things (IoMT) and wearable health devices has transformed patient monitoring and care delivery by enabling real-time data collection, condition tracking and responsive diagnostics. However, these devices typically operate under heavy resource constraints, limited battery life, modest processing power and small memory footprints that make the use of traditional, computationally heavy cryptographic methods impractical. Cryptographic support in this context must strike a balance between security and efficiency, ensuring data protection without degrading device performance or clinical responsiveness.

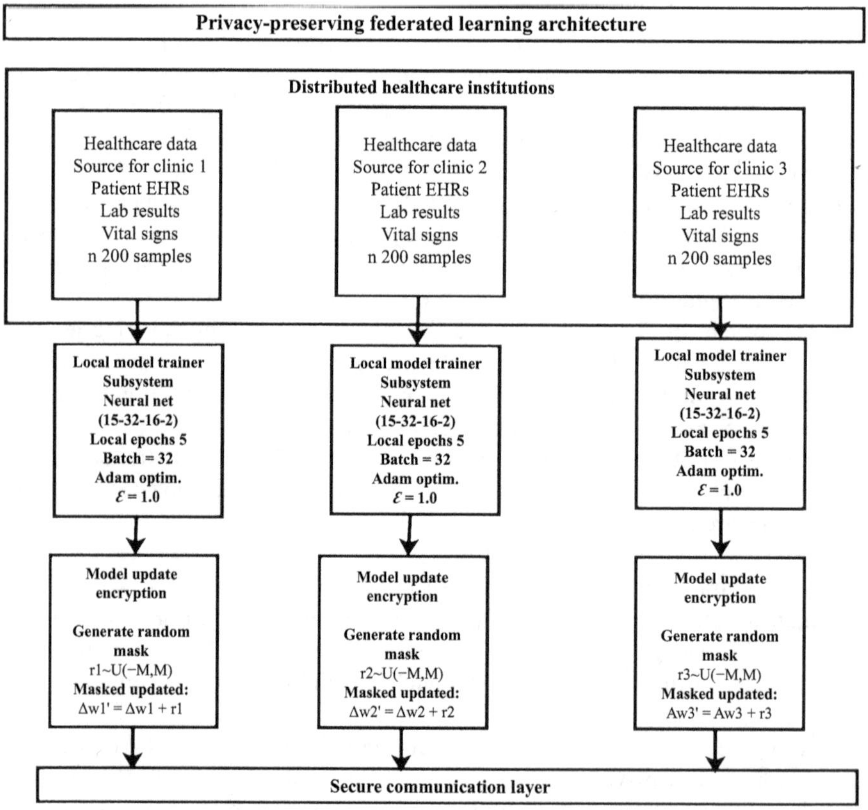

Figure 1.4 (a) Privacy-preserving federated learning architecture for healthcare (Part 1: training phase)

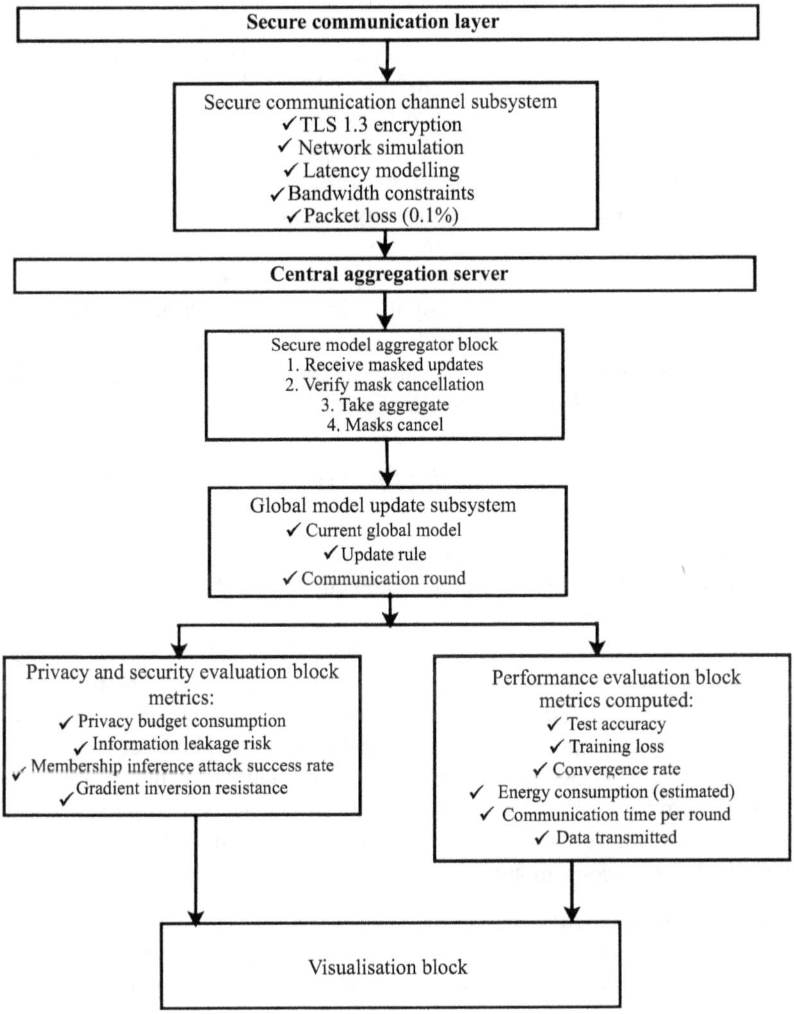

Figure 1.4 (b) Privacy-Preserving Federated Learning Architecture for Healthcare (Part 2: Evaluation Phase)

1.8.1 IoMT security landscape and architectural considerations

The Internet of Medical Things (IoMT) encompasses wearables, implantable devices, continuous monitors and smart medical equipment. Each device represents both a potential attack vector and a source of sensitive physiological data requiring cryptographic protection.

IoMT devices span vastly different capability levels. Ultra-constrained implantables (pacemakers and defibrillators) must operate 5–15 years on non-replaceable batteries with minimal processing power (8–16-bit microcontrollers, $<$100 MHz) and limited memory ($<$100 KB RAM), making standard cryptography challenging. Mid-range wearables have more resources (32-bit processors, 1–10 MB RAM, rechargeable batteries) and can implement lightweight cryptography. High-end gateways and tablets support standard cryptographic protocols without constraints. This heterogeneity requires hybrid security architectures adapting protections based on device capabilities.

IoMT faces distinct threats, including physical tampering (devices worn or kept in unsecured environments), supply chain attacks during manufacturing, communication interception on older devices lacking encryption and side-channel attacks exploiting power consumption patterns. The 2011 demonstration of wireless attacks on implantable defibrillators illustrated life-threatening vulnerabilities.

1.8.2 Lightweight cryptography for resource-constrained devices

The fundamental challenge in securing IoMT devices lies in reconciling the rigorous security requirements of healthcare data protection with the severe computational, memory and energy constraints of medical sensors and wearables. Traditional cryptographic algorithms, while providing strong security guarantees, were designed for general-purpose computing environments with ample resources. Their direct application to ultra-constrained devices often results in unacceptable energy consumption, excessive latency or implementation infeasibility given limited memory and code space. Lightweight cryptography addresses these challenges through specialised algorithms optimised for resource-constrained environments, trading some security margin or throughput for substantial reductions in computational overhead, memory footprint and energy consumption.

- Lightweight symmetric encryption
  Symmetric encryption, which uses the same key for both encryption and decryption, remains the workhorse of IoMT data protection due to its computational efficiency compared to asymmetric alternatives. The Advanced Encryption Standard (AES), despite being designed for general-purpose use, demonstrates reasonable performance on IoMT devices when hardware acceleration is available or when implementation is carefully optimised. AES-128, operating on 128-bit blocks with 128-bit keys through 10 rounds of substitution and permutation, achieves throughputs of 100–150 KB/s on typical ARM Cortex-M4 microcontrollers commonly used in mid-range medical devices, with energy consumption of approximately 40–50 μJ per 1 KB encrypted block.
- Performance comparison and selection guidance
  The selection of appropriate symmetric encryption for IoMT devices must balance security requirements, performance constraints and practical

Table 1.4 Lightweight symmetric algorithm performance on IoMT hardware

Algorithm	Block size	Key size	Throughput (KB/s)	Energy (μJ/KB)	Flash (KB)	RAM (B)	NIST security	Healthcare stability
AES-128	128b	128b	125	45	2.1	256	✓Std	High
AES-256	128b	256b	98	58	2.4	288	✓Std	Very high
PRESENT-80	64b	80b	185	28	1.0	192	✓ISO	Medium
PRESENT-128	64b	128b	172	31	1.2	208	✓ISO	Med-high
SIMON-128	128b	128b	220	22	0.8	160	Research	Med-high
SPECK-128	128b	128b	290	18	0.7	160	Research	Med-high
GIFT-128	128b	128b	205	25	1.2	180	Research	High
ChaCha20	512b	256b	340	16	1.8	320	✓IETF	Very high

deployment considerations. Table 1.4 presents comprehensive benchmark results for leading symmetric algorithms on representative IoMT hardware.

Test platform: ARM Cortex-M4F @ 100 MHz (typical mid-range wearable device MCU)

Test data: 1 KB blocks (typical sensor packet size)

Implementation: optimised C with in-line assembly for critical sections

1.8.3 Energy, latency and scalability considerations

Applying cryptography at the edge directly affects energy consumption and system responsiveness. Encrypting and authenticating high-frequency sensor data streams can strain limited device resources if not carefully designed. Lightweight and context-aware cryptographic methods help minimise these costs, while edge computing frameworks enable heavier computations to be offloaded to nearby fog or cloud nodes. This division of labour preserves real-time responsiveness and patient privacy while supporting scalable healthcare deployments.

1.8.4 Cryptographic support for edge and fog healthcare computing

Edge and fog computing extend security capabilities closer to data sources, enabling early encryption, aggregation and validation of medical data. Cryptographic support at these layers often includes hierarchical key management, secure execution environments and hardware-assisted encryption. By enforcing security controls at the edge, healthcare systems reduce latency, preserve bandwidth and limit data exposure. Fog nodes act as trusted intermediaries, validating data integrity and enforcing access policies before forwarding information to centralised systems, thereby strengthening defence-in-depth for distributed healthcare infrastructures.

1.9 Blockchain and cryptographic integration in healthcare

Blockchain provides a complementary layer for healthcare security, maintaining tamper-evident records across institutions without a centralised authority. Permissioned blockchains suit healthcare better than public chains, supporting regulatory compliance and higher performance while enabling collaboration among hospitals, laboratories, insurers and researchers.

Healthcare blockchains use hash functions for integrity, digital signatures for authentication and Merkle trees for efficient verification. Medical records typically remain off-chain with only cryptographic references and access logs on the blockchain. Smart contracts automate access control and consent management. Key applications include patient-controlled health information exchange with immutable audit trails, pharmaceutical supply chain tracking for drug authenticity, clinical trial data integrity through immutable timestamping and healthcare credentialing for rapid licence verification [14,39].

However, blockchain introduces significant trade-offs: lower throughput and higher latency than traditional databases (unsuitable for real-time clinical operations), storage requirements and integration complexity necessitating hybrid architectures and regulatory challenges as immutability conflicts with GDPR's right to erasure. Healthcare organisations should deploy blockchain selectively where decentralised trust and immutable audit trails provide clear benefits, starting with pilots rather than large-scale transformations [12,39].

1.10 Post-quantum cryptography and future-proof healthcare security

The eventual emergence of large-scale quantum computers poses a fundamental threat to current public-key cryptography, which underpins authentication, secure communication and data protection in healthcare systems. Algorithms such as RSA and elliptic curve cryptography could be broken by quantum attacks, placing long-lived medical data at risk.

Healthcare is particularly vulnerable due to the extended confidentiality requirements of medical records, genetic data and research archives. Even if quantum computers are years away, attackers can already collect encrypted data for future decryption, making proactive mitigation essential.

1.10.1 Quantum threats and healthcare impact

Quantum algorithms threaten asymmetric cryptography most severely, while symmetric encryption and hash functions remain comparatively resilient when used with larger key sizes. As a result, healthcare organisations can partially mitigate risk by adopting stronger symmetric standards, but long-term protection requires transitioning to post-quantum cryptographic algorithms.

1.10.2 Post-quantum cryptographic standards

To address this challenge, NIST has standardised post-quantum algorithms designed to resist both classical and quantum attacks. Lattice-based schemes such as CRYSTALS-Kyber and CRYSTALS-Dilithium provide practical performance for secure communication and digital signatures, while hash-based schemes like SPHINCS+ offer conservative security for long-term archival and firmware signing.

Although post-quantum algorithms introduce larger keys and higher computational costs, they are feasible for most healthcare systems when carefully deployed, particularly in high-risk areas such as data storage, authentication infrastructure and long-lived medical devices.

1.10.3 Migration strategies and cryptographic agility

Transitioning to post-quantum cryptography requires phased, risk-based migration strategies. Hybrid cryptographic approaches that combine classical and post-quantum algorithms offer immediate protection against future threats while maintaining compatibility with existing systems. Prioritisation should focus on data with long confidential lifetimes and systems exposed to external adversaries.

Equally important is cryptographic agility, the ability to update algorithms without redesigning systems. Healthcare organisations should demand agility from vendors and design future systems with modular cryptographic components, ensuring resilience against both quantum threats and future cryptographic advances.

1.11 Limitations of the study

While this chapter takes a deep dive into the role of cryptography in smart healthcare, it is important to recognise its boundaries. No study is perfect, and ours is no exception. By being clear about what we couldn't cover or fully capture, we hope to give a fair picture of our work and set the stage for future researchers to build upon it.

1.11.1 Relying on simulations, not real-world deployment

A key limitation is that our experiments were run in simulated environments, mainly using Simulink, rather than in actual hospitals or clinics. Simulations are fantastic for control and repeatability – they let us test ideas in a clean, predictable sandbox. However, they inevitably smooth over the messy, unpredictable realities of live healthcare settings.

The Simulation Gap: Our models accounted for things like network delays, but real hospital networks are a different beast. They deal with unpredictable traffic, complex firewalls, spotty connections from mobile devices and strict IT policies that are hard to replicate in a lab. For instance, we assumed an ideal secure connection, but real systems often have extra security layers that can slow things down in ways we did not fully model.

Controlled vs chaotic conditions: We benchmarked cryptographic performance on representative hardware but always in a steady, controlled lab. Real medical devices live in the wild. They get hot; their batteries drain, and they face interference from other equipment and are constantly juggling multiple tasks. Our energy measurements, for example, assumed the processor was focused solely on cryptography. In reality, a wearable monitor is also collecting sensor data, updating its display and managing its connection, all at once.

1.11.2 Working with synthetic, simplified data

To protect patient privacy and sidestep ethical hurdles, we used artificial datasets designed to look like real medical records, rather than using the records themselves. This necessary choice comes with trade-offs.

- *Scale and complexity:* Our simulated dataset for federated learning had 1000 patient records across five fictional institutions. Real hospitals handle hundreds of thousands of patients, and collaborative projects can involve dozens of partners. Our smaller, neater dataset cannot fully capture the true statistical messiness, severe imbalances and data quality quirks of the real world.
- *A simplified clinical picture:* We tested our models on a specific, relatively straightforward task: a binary risk classification. In practice, clinicians and AI systems deal with vastly more complexity – mixing scans, lab results, doctors' notes and genetic data over a patient's lifetime, all while handling missing information. The privacy-performance balances we observed for our simpler task might shift in these more complex scenarios.

1.11.3 The moving target of cryptography

Cryptography is a fast-paced field. While we have covered a wide range of techniques – from the classics to the very latest post-quantum candidates – it is impossible to be completely exhaustive or future-proof.

New algorithms on the horizon: Our analysis of lightweight cryptography focused on established, well-documented algorithms. However, new, more efficient schemes are always emerging from competitions and research labs. Similarly, our discussion of post-quantum cryptography is based on the state of play in early 2024, but this area is still evolving. Anyone planning based on our recommendations should stay alert to new standards and improvements.

Practical limits on depth: We touched on advanced concepts like homomorphic encryption (computing on encrypted data) and multiparty computation, but a detailed, quantitative breakdown of every possible use case was not feasible. These tools are highly specialised; a hospital considering them for a specific project, like a secure multi-institutional genetic study, would need to do a deep, custom analysis.

1.11.4 Navigating the regulatory maze

We framed our discussion around major regulations like HIPAA and GDPR, but the global regulatory landscape for healthcare data is incredibly complex and varies widely.

A global patchwork: Our focus leaned towards North American and European perspectives. Healthcare organisations operating in Asia, Latin America or elsewhere will encounter different rules on encryption, data storage and cross-border transfer that we could not detail here.

Evolving rules and grey areas: Regulations struggle to keep up with technology. Guidance on using federated learning or blockchain in healthcare is still forming in many places. Also, some theoretical tensions – like between blockchain's permanence and a patient's 'right to be forgotten' – have practical and legal nuances that extend beyond our technical scope.

1.11.5 The hard part: making it work in practice

This chapter focuses on the theory and architecture of cryptographic systems. Turning those blueprints into a working, secure and usable system in a real hospital introduces a whole new set of challenges.

- The human element: The most sophisticated cryptography fails if it is too cumbersome for doctors and nurses to use. If security measures interrupt critical workflows, they might be ignored or disabled. Designing systems that are both secure and user-friendly requires a different skill set – one centred on human factors and clinical workflow – that was beyond the core focus of this technical study.
- Key management, the unsung hero: We emphasise that proper key management is essential, but actually doing it is a major operational undertaking. It requires secure infrastructure, clear policies and ongoing effort. The cost and complexity of managing cryptographic keys over their entire life cycle can often outweigh the initial challenge of deploying the algorithms themselves.
- The legacy system hurdle: Hospitals are museums of working technology. New cryptographic systems must coexist and communicate with decades-old software that may only support outdated, insecure methods. Bridging this gap requires careful planning, temporary fixes and long transition periods – real-world constraints that can slow down progress.

1.11.6 The 'it depends' on performance

We provided performance numbers from our tests, but these come with important caveats. Your mileage may vary.

- Benchmarks are not universal: Our results depend on the specific software libraries and hardware we used. A different library, or a chip with built-in cryptographic acceleration, could yield very different speeds. Organisations should test on their own intended platforms.
- The full system impact: We measured the cost of the cryptography itself. In a real deployment, the total performance hit comes from the entire security

apparatus – network filters, monitoring systems and logging – which together can have a bigger impact than the cryptography alone.

- Privacy is a trade-off: Our privacy-preserving ML experiments used specific privacy settings (e.g. a differential privacy parameter of = 1.0). There is no one-size-fits-all setting; the right balance between privacy and accuracy depends entirely on the sensitivity of the medical task and the specific threats faced.

1.11.7 Defining the 'bad guy'

Our security analyses are based on specific assumptions about what kind of adversary we are defending against. The real world can be more threatening.

- Beyond 'honest but curious': We often assumed adversaries who follow the rules but try to learn secrets passively (the 'honest-but-curious' model). Real attackers might be actively malicious – poisoning data, colluding or deploying vast resources. Our defences might not hold against these more aggressive or sophisticated threats.
- Attacks from the side (channels): We focused on direct, mathematical attacks. However, devices can leak secrets through indirect 'side-channels' like power consumption fluctuations or subtle timing differences. Protecting against these requires special, often slower, implementations, which we did not fully account for.
- The insider risk: Cryptography is great for protecting data from outside attackers, but it is less effective against authorised insiders (like a disgruntled employee) who already have access. Mitigating that risk requires a broader set of tools – strict access controls, thorough auditing and organisational policies – that are crucial for security but were not our focus here.

1.11.8 The future is hard to predict

We have tried to look ahead at trends like quantum computing and AI, but technology moves fast.

- The quantum timeline: No one knows exactly when a large-scale quantum computer that can break today's encryption will arrive. Our advice is based on today's best estimates, but organisations need to stay informed as both quantum computing and the post-quantum algorithms designed to defeat it continue to evolve.
- AI, a double-edged sword: Artificial intelligence is advancing rapidly. Future AI could power new, more powerful cyberattacks but also create smarter automated defences. The interplay between AI and cryptography in healthcare will likely look very different in just a few years.
- New architectures, new rules: As healthcare computing pushes further to the 'edge' (on devices and local gateways), new security models and challenges will emerge that we can only begin to anticipate today.

1.12 Lessons learned

Studying how cryptography fits into smart healthcare taught us a lot. These are not just academic points; they're real-world insights for anyone building, buying, regulating or using these systems. Here are the key lessons we are taking away.

1.12.1 The big picture: eight key takeaways

Before we dive in, here is a quick list of the main lessons.

1. Cryptography is now essential plumbing, not an add-on. You cannot have modern healthcare without it.
2. There is no 'one size fits all' solution. The right tool depends completely on the job, the setting and the risks.
3. We can protect privacy without sacrificing usefulness. With the right techniques, it is not an either/or choice.
4. Regulations are a bigger driver than technology. Laws often push adoption faster than technical arguments do.
5. A good idea is only as strong as its execution. Flawed implementation can break perfect theory.
6. This work cannot be done in a silo. Success needs cryptographers, doctors, IT staff and regulators at the same table.
7. Think in decades, not years. Healthcare data and devices last a long time, so our security plans must too.
8. If it is not usable, it is not secure. Security that gets in the way of patient care will be bypassed.

1.12.2 A closer look at each lesson

Lesson 1: Cryptography is now essential plumbing

The core idea: Cryptography has moved from being a 'nice-to-have' security feature to the absolute bedrock of trustworthy healthcare systems. We now rely on it for everything: keeping electronic records private, securing video calls with your doctor, making sure data from a heart monitor is authentic and enabling hospitals to collaborate on research without sharing raw patient data. Modern healthcare simply could not function securely without it.

What this means for us all:

- For healthcare leaders: Treat building cryptographic infrastructure (like key management systems and expert staff) with the same seriousness as buying an MRI machine. It is not an IT cost; it is a strategic investment in your organisation's future.
- For architects and builders: Bake security in from the very first sketch. Trying to add it later is like trying to install plumbing after the walls are up – messy, expensive and never quite right.

- For everyone: Security experts need a seat at the table when making big decisions about new technology, cloud partnerships or device purchases.

Lesson 2: Context is everything

The core idea: You would not use the same lock on a diary, a bank vault and a hospital pharmacy. Similarly, there is no single 'best' cryptography for all of healthcare. The perfect choice for a real-time ER monitoring system (needs speed above all) is totally wrong for a long-term genetic research database (needs ultra-strong, future-proof protection).

What this means for us all:

- For security teams: Resist blanket policies. Create a smart framework that matches the level of protection to the sensitivity of the data and the real-world risks. An insulin pump used by a child at school has different needs than one used in a controlled hospital setting.
- For buyers and vendors: When purchasing systems, do not just check a 'has encryption' box. Ask: 'Is this encryption appropriate for how we will actually use this device or software in our clinical environment?'

Lesson 3: Privacy and utility can coexist

The core idea: The old belief that strong privacy always ruins data's usefulness is a myth. Our experiments showed that with modern techniques like federated learning and differential privacy, we can still get clinically useful results – say, 90 per cent accuracy for a risk prediction – while robustly protecting which patients were in the dataset. This is not just about compliance; it is about unlocking new possibilities, like letting research hospitals work together without ever exchanging a single patient's raw file.

What this means for us all:

- For organisations: See privacy tech as a key to new doors. It can let you use sensitive data you could not touch before, build trust with patients and collaborate in powerful new ways.
- For data scientists: Start thinking of privacy-preserving methods as a standard step in your workflow, like data cleaning, not as a weird, specialised add-on.

Lesson 4: Regulations are the engine of change

The core idea: While tech breakthroughs are important, nothing drives widespread adoption of security measures in healthcare like a new law or regulation. HIPAA and GDPR did not just suggest encryption; they made it a baseline requirement, changing the landscape overnight. Regulations shape everything from what gets certified to who is liable in a breach.

What this means for us all:

- For leaders: Pay close attention to the regulatory horizon. These changes come with warning. Getting ahead of them is cheaper and less stressful than playing catch-up.
- For advocates: Regulators are not always cryptography experts. The healthcare community has a responsibility to engage, explain new technologies and help shape rules that are both protective and practical.

Lesson 5: The devil is in the details (of implementation)

The core idea: The world's most brilliant, unbreakable encryption algorithm is useless if it is implemented poorly. A weak random number generator for creating keys, a tiny coding error that leaks information or sloppy key management can blow a hole in your defences. Real-world security is in the gritty details of the code and the operations.

What this means for us all:

- For buyers: When evaluating a vendor, ask not just what algorithms they use but how they implement them. Demand code audits and results from independent penetration testing.
- For IT teams: Keep a detailed inventory of how cryptography is used across all your systems. When a new vulnerability is announced, you need to know exactly where you are exposed.
- For everyone: Invest in people who understand cryptographic engineering, not just theory. Good implementation is a specialised skill.

Lesson 6: We need a fusion of minds

The core idea: Building secure healthcare systems is a team sport that needs wildly different skills. A cryptographer knows how to make a vault impenetrable but might not know doctors need to get into it in 10 seconds during a code blue. A clinician knows the workflow but might not spot a fundamental security flaw. We need all these perspectives – cryptography, medicine, IT, law and patient advocacy – from the very beginning.

What this means for us all:

- For organisations: Create governance groups that force these different experts to talk regularly. Make sure security architects work side by side with clinical staff when designing new tools.
- For the industry: Value and create 'translator' roles – people with a foot in both the clinical and technical worlds who can bridge the communication gap.

Lesson 7: Plan for the long haul

The core idea: Healthcare has an unusually long timeline. Patient records must be kept for decades. A pacemaker might be implanted for 20 years. This means we have to think about threats that do not even exist yet, like powerful quantum computers that could break today's encryption. The 'harvest now, decrypt later' attack – where hackers steal encrypted data to crack it open in 10 years – is a real concern for healthcare.

What this means for us all:

- For architects: Build for cryptographic agility. Design systems so you can swap out old algorithms for new, stronger ones without having to rebuild everything from scratch.
- For planners: Create a long-term road map for your cryptography. Start planning your transition to post-quantum algorithms now for your most critical, long-lived data and devices, because the shift will take years.

Lesson 8: If it is a hassle, it will not be used

The core idea: The most secure system in the world is insecure if people bypass it to do their jobs. In healthcare, where seconds count and cognitive load is high, security must be designed around the human user. If encryption makes it take 30 seconds longer to pull up a chart in an emergency, clinicians will find a way around it. Security should be as seamless and invisible as possible.

What this means for us all:

- For designers: Test security features with real clinicians in real settings before you roll them out. Watch for frustrated sighs, workarounds and sticky notes with passwords.
- For decision-makers: Prioritise 'transparent security' – things like automatic device authentication or biometric logins that protect without adding steps.
- For security teams: Get out of the server room. Shadow nurses and doctors. Their frustration with your system is a critical security vulnerability.

1.12.3 Bringing it all together

These lessons paint a picture of a balanced, pragmatic approach. We need to be visionary in seeing cryptography as essential infrastructure but practical in adapting it to each unique context. We must pursue technical excellence while fostering true collaboration across disciplines. We have to plan for the distant future while building systems agile enough to adapt.

The goal is clear: to create healthcare systems that are both powerfully intelligent and deeply trustworthy. By treating security as fundamental, designing for real people, planning for the long term and working together across all our areas of expertise, we can protect patients while unlocking the incredible promise of technology to heal and help.

1.13 Emerging trends and research directions

Cryptography is evolving from a supporting component to a central design consideration in healthcare systems. Key emerging trends include cryptography-aware AI models that embed privacy constraints directly into design, using simplified architectures and activation functions enabling encrypted inference with manageable performance loss. Adaptive and context-aware cryptographic systems dynamically adjust encryption strength, authentication and access controls based on clinical context (emergency vs research access), better reflecting real workflows while introducing policy complexity challenges.

New healthcare paradigms (digital health twins, virtual environments and brain-computer interfaces) generate continuous sensitive data streams requiring lightweight encryption and on-device processing. Post-quantum cryptography offers practical protection against future quantum threats, with experimental quantum key distribution limited to niche scenarios like secure inter-facility links.

Open challenges include improving cryptographic usability for clinicians to prevent security-undermining workarounds, developing healthcare-specific cryptographic standards and certification frameworks and advancing privacy-preserving machine learning to reduce computational overhead and improve robustness against inference attacks. The future requires secure, adaptive and usable solutions tightly integrated with clinical realities through sustained interdisciplinary collaboration among cryptographers, AI researchers, healthcare professionals and policymakers.

1.14 Summary of key insights

A central insight from this chapter is that cryptography is no longer optional or peripheral in healthcare environments. Modern intelligent healthcare systems depend on cryptographic protections across the entire data life cycle during storage, transmission, processing and real-time clinical use. Techniques such as symmetric and asymmetric encryption, secure multiparty computation, homomorphic encryption, differential privacy and post-quantum cryptography now collectively address diverse security requirements ranging from routine clinical record protection to large-scale collaborative medical research.

Another critical observation is the persistent tension between security and system performance. Healthcare applications often operate under strict latency, energy and usability constraints, particularly in emergency care, implantable medical devices and real-time monitoring systems. Advances such as lightweight cryptography, cryptography-aware machine learning models and hybrid privacy-preserving approaches demonstrate that security and performance need not be mutually exclusive. However, achieving an effective balance requires early-stage co-design rather than retrofitting cryptographic controls after deployment.

The chapter also highlights that regulatory and organisational factors frequently exert greater influence over cryptographic adoption than purely technical considerations. Healthcare security decisions are shaped by regulatory frameworks such as HIPAA, GDPR and medical device approval processes, which can both enable and constrain innovation. Successful cryptographic deployment, therefore, requires not only sound technical design but also careful alignment with legal requirements, institutional policies and clinical workflows.

Importantly, healthcare contexts are highly heterogeneous. The cryptographic needs of emergency departments, long-term genomic databases, wearable health devices and cloud-based AI platforms differ fundamentally. Effective healthcare cryptography must therefore be context-sensitive, offering tailored protections that reflect operational realities rather than relying on one-size-fits-all solutions.

Finally, the chapter underscores the growing importance of forward-looking cryptographic strategies. Emerging threats, including quantum computing and AI-driven inference attacks, pose risks to long-lived healthcare data. Cryptographic agility, particularly the ability to transition to post-quantum algorithms without system replacement, is becoming a strategic necessity rather than a theoretical concern.

1.15 Practical implications for stakeholders

The findings of this chapter carry distinct implications for key healthcare stakeholders. For healthcare security leaders and system architects, cryptographic protections should be treated as core infrastructure. This involves maintaining visibility into cryptographic deployments, prioritising robust key management and planning long-term transitions towards post-quantum readiness. Governance structures that include clinical, technical and legal perspectives are essential to ensure security decisions align with patient safety and operational needs.

Medical device manufacturers and IoMT vendors must embed cryptographic protections into device design from inception. Given the long operational lifetimes of medical devices, cryptographic agility and secure update mechanisms are critical. Lightweight, standardised cryptographic algorithms are preferable to proprietary solutions, ensuring both efficiency and long-term trust.

For healthcare data scientists and AI researchers, privacy considerations must be integrated throughout the machine learning life cycle. Techniques such as federated learning, differential privacy and encrypted inference enable collaborative research while respecting patient confidentiality, but they introduce trade-offs that require careful evaluation. Transparent documentation of cryptographic assumptions and privacy parameters is essential for reproducibility and regulatory acceptance.

Policymakers and regulators play a pivotal role in shaping the future of healthcare cryptography. Technology-neutral regulations that focus on security outcomes rather than specific algorithms allow systems to evolve alongside cryptographic advances. Clear approval pathways for security updates and privacy-preserving technologies are necessary to avoid delays that leave systems exposed to known threats.

Healthcare organisation leadership must recognise cryptographic infrastructure as a strategic investment rather than a cost burden. Institutions with mature cryptographic capabilities are better positioned to participate in collaborative research, comply with evolving regulations and maintain patient trust. Equally important is fostering a security-aware culture in which cryptographic practices are understood as integral to patient care rather than as administrative obstacles.

1.16 Future outlook for cryptography in intelligent healthcare

Over the coming decade, cryptography will continue to shape the trajectory of intelligent healthcare systems. Post-quantum cryptography is expected to move from preparatory planning into active deployment as standards mature and quantum threats become more tangible. While transitions will be gradual, organisations that delay preparation risk facing disruptive and costly migrations.

Privacy-preserving machine learning is also poised to transition from experimental research to routine clinical and research use. As performance overheads decrease through improved algorithms and specialised hardware, encrypted analytics and federated learning will enable broader collaboration across institutions without compromising patient privacy.

The expansion of edge and fog computing architectures, driven by 5G and advanced IoMT deployments, will further complicate healthcare security landscapes. Cryptographic frameworks will need to seamlessly span devices, edge nodes and cloud platforms, distributing trust while maintaining end-to-end protection.

Regulatory clarity is expected to improve as agencies gain experience with privacy-preserving technologies. This evolution will accelerate adoption by reducing uncertainty and providing standardised evaluation criteria. At the same time, advances in usability, such as transparent encryption and automated key management, will help reconcile strong security with clinical efficiency.

1.17 Call to action and interdisciplinary collaboration

The effective integration of cryptography into intelligent healthcare systems cannot be achieved within disciplinary silos. Progress requires sustained collaboration among cryptographers, healthcare informaticists, AI researchers, system engineers, regulators and clinicians. When cryptographic solutions are developed without clinical input, they often fail in practice; when healthcare systems are deployed without cryptographic expertise, they remain vulnerable.

To overcome these challenges, interdisciplinary research consortia, active participation in standards development, targeted education and training programmes and investment in open-source healthcare security platforms are essential. Equally important is proactive engagement with regulatory bodies to shape frameworks that enable innovation while safeguarding patient welfare.

Ultimately, cryptography is foundational to trust in intelligent healthcare systems. Insufficient protection exposes patients to harm and erodes confidence, while overly restrictive approaches can hinder care delivery and research progress. Navigating this balance, deploying cryptography that both protects and enables healthcare innovation will be one of the defining technical and policy challenges of the coming decades. Achieving this vision demands long-term commitment, interdisciplinary collaboration and sustained investment in cryptographic infrastructure that places patient welfare at its core.

1.18 Conclusion

This chapter explored the evolution, present state and future direction of cryptographic algorithms within intelligent healthcare systems. From the early days of basic access controls and isolated encryption practices, healthcare security has matured into a complex ecosystem of privacy-preserving technologies that now underpin modern data-driven medicine. As healthcare systems increasingly rely on artificial intelligence, distributed data sharing and cloud-based infrastructures, cryptography has shifted from a supporting role to a foundational element of system design.

List of abbreviations

ACM	Association for Computing Machinery
AES	Advanced Encryption Standard
AI	Artificial Intelligence
API	Application Programming Interface
ARM	Advanced RISC Machine (processor architecture)
CBC	Cipher Block Chaining
CT	Computed Tomography
CTR	Counter Mode
DER	Distinguished Encoding Rules
DES	Data Encryption Standard
DICOM	Digital Imaging and Communications in Medicine
DP	Differential Privacy
DWT	Data Watchpoint and Trace
ECC	Elliptic Curve Cryptography
ECDSA	Elliptic Curve Digital Signature Algorithm
EHR	Electronic Health Record
FDA	Food and Drug Administration (US)
FL	Federated Learning
FPU	Floating-Point Unit
GCM	Galois/Counter Mode
GDPR	General Data Protection Regulation
GPU	Graphics Processing Unit
HE	Homomorphic Encryption
HIPAA	Health Insurance Portability and Accountability Act
HMAC	Hash-based Message Authentication Code
IEC	International Electrotechnical Commission
IETF	Internet Engineering Task Force
IoMT	Internet of Medical Things
IoT	Internet of Things
IRB	Institutional Review Board
ISO	International Organization for Standardization
KB	Kilobyte
MAC	Message Authentication Code
MB	Megabyte
MCU	Microcontroller Unit
ML	Machine Learning
MRI	Magnetic Resonance Imaging
NIST	National Institute of Standards and Technology
OWASP	Open Web Application Security Project

PKCS	Public-Key Cryptography Standards
PQC	Post-Quantum Cryptography
PSS	Probabilistic Signature Scheme
RAM	Random Access Memory
RAPL	Running Average Power Limit
RFC	Request for Comments
RSA	Rivest–Shamir–Adleman (public-key cryptosystem)
SHA	Secure Hash Algorithm
SMPC	Secure Multiparty Computation
TLS	Transport Layer Security
XTS	XEX-based Tweaked Codebook Mode
ZKP	Zero-Knowledge Proof

References

[1] Antunes, R. S., da Costa, C. A., Küderle, A., Abdullahi Yari, I. and Eskofier, B. (2022). Federated learning for healthcare: Systematic review and architecture proposal. *ACM Transactions on Intelligent Systems and Technology*, 13(4), Article 54. https://doi.org/10.1145/3501813 [Accessed: 26-Dec-2025].

[2] Beltrán, E. T. M., Pérez, M. Q., Sánchez, P. M. S. *et al.* (2023). Decentralized federated learning: Fundamentals, state of the art, frameworks, trends, and challenges. *IEEE Communications Surveys & Tutorials*, 25(4), 2983–3013.

[3] ENISA. (2023). Cybersecurity threats in the healthcare sector. European Union Agency for Cybersecurity. [Online]. Available: https://www.enisa.europa.eu/publications/cybersecurity-threats-in-the-healthcare-sector. [Accessed: 26-Dec-2025].

[4] IBM Security. (2024). Cost of a data breach report 2024. IBM Security (industry report widely cited regarding healthcare breach costs. [Online]. Available: https://www.ibm.com/security/data-breach. [Accessed: 26-Dec-2025].

[5] Xu, X., Deng, H. H., Chen, T. *et al.* (2024). Federated cross learning for medical image segmentation. *Medical Imaging with Deep Learning*, 227, 1441–1452.

[6] Pati, S., Kumar, S., Varma, A. *et al.* (2024). Privacy preservation for federated learning in health care. *Patterns*, 5(3), 100974.

[7] Song, C., Wang, Z., Peng, W. and Yang, N. (2024). Secure and efficient federated learning schemes for healthcare systems. *Electronics*, 13(13), 2620. https://doi.org/10.3390/electronics13132620

[8] Guan, H., Yap, P. T., Bozoki, A. and Liu, M. (2024). Federated learning for medical image analysis: A survey. *Pattern Recognition*, 151, 110424. https://doi.org/10.1016/j.patcog.2024.110424 [Accessed: 26-Dec-2025].

[9] Zhu, J., Cao, J., Saxena, D., Jiang, S. and Ferradi, H. (2023). Blockchain-empowered federated learning: Challenges, solutions, and future directions. *ACM Computing Surveys*, 55, 1–31.

[10] Wu, J., Dong, F., Leung, H., Zhu, Z., Zhou, J. and Drew, S. (2024). Topology-aware federated learning in edge computing: A comprehensive survey. *ACM Computing Surveys*, 56, 1–41.

[11] NIST. (2024). NIST post-quantum cryptography standardisation (ongoing). National Institute of Standards and Technology. [Online]. Available: https://csrc.nist.gov/projects/post-quantum-cryptography. [Accessed: 26-Dec-2025].

[12] Elangovan, D. (2022). The use of blockchain technology in the healthcare sector. *JMIR Medical Informatics*. https://doi.org/10.2196/17278 [Accessed: 26-Dec-2025].

[13] Munjal, K. and Bhatia, R. (2022). A systematic review of homomorphic encryption and its contributions in the healthcare industry. *Complex and Intelligent Systems*, 8(4), 3569–3594. https://doi.org/10.1007/s40747-022-00756-z [Accessed: 26-Dec-2025].

[14] Ettaloui, N., Arezki, S. and Gadi, T. (2024). Blockchain-based electronic health record: Systematic review and future directions. *Health Information & Libraries Journal*. https://doi.org/10.1155/hbe2/4734288 [Accessed: 26-Dec-2025].

[15] Adler-Milstein, J., Holmgren, A. J., Kralovec, P., Worzala, C., Searcy, T. and Patel, V. (2017). Electronic health record adoption in US hospitals: The emergence of a digital "advanced use" divide. *Journal of the American Medical Informatics Association*, 24(6), 1142–1148. https://doi.org/10.1093/jamia/ocx080

[16] Fernandes, L., O'Connor, M., Weaver, A. and O'Connor, K. (2021). Cybersecurity challenges in cloud-based electronic health record systems. *Journal of Biomedical Informatics*, 115, 103690. https://doi.org/10.1016/j.jbi.2021.103690

[17] U.S. Department of Health and Human Services. (2013). HIPAA security rule. https://www.hhs.gov/hipaa/for-professionals/security

[18] European Union. (2016). Regulation (EU) 2016/679 of the European Parliament and of the Council of 27 April 2016 on the protection of natural persons with regard to the processing of personal data and on the free movement of such data (General Data Protection Regulation). Official Journal of the European Union, L119, 1–88. https://eur-lex.europa.eu/eli/reg/2016/679/oj

[19] Ramachandran, N. (2025). Quantum-safe cryptography implementation in healthcare cloud systems: A technical framework. *International Journal of Computational and Experimental Science and Engineering*, 11(4). https://doi.org/10.22399/ijcesen.4495 [Accessed: 23-Jan-2026]

[20] Patra, D. and Rajagopalan, N. (2026). Integration of emerging technologies in cybersecurity for healthcare: A systematic review. *Computers & Security*, 161, 104763. https://doi.org/10.1016/j.cose.2025.104763 [Accessed: 23-Jan-2026].

[21] Dash, S., Shakyawar, S. K., Sharma, M. and Kaushik, S. (2019). Big data in healthcare: Management, analysis and future prospects. *Journal of Big Data*, 6(1), 54. https://doi.org/10.1186/s40537-019-0217-0 [Accessed: 26-Dec-2025].

[22] Price, W. N. and Cohen, I. G. (2019). Privacy in the age of medical big data. *Nature Medicine*, 25(1), 37–43. https://doi.org/10.1038/s41591-018-0272-7 [Accessed: 26-Dec-2025].

[23] Qayyum, A., Qadir, J., Bilal, M. and Al-Fuqaha, A. (2021). Secure and robust machine learning for healthcare: A survey. *IEEE Reviews in Biomedical Engineering*, 14, 156–180. https://doi.org/10.1109/RBME.2020.3013489 [Accessed: 26-Dec-2025].

[24] Rocher, L., Hendrickx, J. M. and de Montjoye, Y. A. (2019). Estimating the success of re-identifications in incomplete datasets using generative models. *Nature Communications*, 10(1), 3069. https://doi.org/10.1038/s41467-019-10933-3

[25] Kumar, R., Gupta, B. B. and Tripathi, S. (2021). A distributed framework for detecting DDoS attacks in smart contract-based blockchain IoT systems by leveraging fog computing. *Transactions on Emerging Telecommunications Technologies*, 32(6), e4112. https://doi.org/10.1002/ett.4112

[26] Yaqoob, I., Salah, K., Jayaraman, R., Al-Hammadi, Y., Omar, M. and Ellahham, S. (2021). Blockchain for healthcare data management: Opportunities, challenges, and future recommendations. *Neural Computing and Applications*, 34(14), 11475–11490. https://doi.org/10.1007/s00521-020-05519-w

[27] Stallings, W. (2023). *Cryptography and network security: Principles and practice* (9th ed.). Pearson.

[28] National Institute of Standards and Technology. (2015). Secure Hash Standard (SHS) (FIPS PUB 180-4). https://doi.org/10.6028/NIST.FIPS.180-4

[29] Rashid, N. S. (2025). Privacy-preserving machine learning: A review of federated learning techniques and applications. *International Journal of Scientific World*. Retrieved from https://www.sciencepubco.com/index.php/IJSW/article/view/33093

[30] Hiwale, M., Walambe, R., Potdar, V. and Kotecha, K. (2023). A systematic review of privacy-preserving methods deployed with blockchain and federated learning for telemedicine. *Healthcare Analytics*, 3, 100192. https://doi.org/10.1016/j.health.2023.100192

[31] Mondal, D. and Patil, S. S. (2023). Privacy-preserving machine learning techniques for healthcare data analysis. *International Journal of Recent Advances in Engineering and Technology*. Retrieved from https://journals.mriindia.com/index.php/ijraet/article/download/114/103

[32] Kanmani Ruby, E. D., Rose, G. L., Yashaswinii, P., Sripal Reddy, K., Jeyalakshmi, S. and Hari Chandana, B. (2024). Advanced privacy-preserving federated learning in 6G networks using differential privacy and homomorphic encryption. *International Journal of Intelligent*

Systems and Applications in Engineering. Retrieved from https://ijisae.org/index.php/IJISAE/article/view/6427

[33] Acar, A., Aksu, H., Uluagac, A. S. and Conti, M. (2018). A survey on homomorphic encryption schemes: Theory and implementation. *ACM Computing Surveys*, 51(4), Article 79. https://doi.org/10.1145/3214303

[34] Evans, D., Kolesnikov, V. and Rosulek, M. (2018). A pragmatic introduction to secure multi-party computation. *Foundations and Trends® in Privacy and Security*, 2(2–3), 70–246. https://doi.org/10.1561/3300000019

[35] Cho, H., Wu, D. J. and Berger, B. (2018). Secure genome-wide association analysis using multiparty computation. *Nature Biotechnology*, 36(6), 547–551. https://doi.org/10.1038/nbt.4108

[36] Li, T., Sahu, A. K., Zaheer, M., Sanjabi, M., Talwalkar, A. and Smith, V. (2020). Federated optimization in heterogeneous networks. *Proceedings of Machine Learning and Systems*, 2, 429–450. https://doi.org/10.48550/arXiv.1812.06127

[37] Kairouz, P., McMahan, H. B., Avent, B. *et al.* (2021). Advances and open problems in federated learning. *Foundations and Trends® in Machine Learning*, 14(1–2), 1–210. https://doi.org/10.48550/arXiv.1912.04977

[38] Bonawitz, K., Ivanov, V., Kreuter, B *et al.* (2017). Practical secure aggregation for privacy-preserving machine learning. In *Proceedings of the 2017 ACM SIGSAC Conference on Computer and Communications Security, Dallas, TX, USA*, 2017 (pp. 1175–1191). https://doi.org/10.1145/3133956.3133982

[39] Hassan, M. U., Rehmani, M. H. and Chen, J. (2022). Privacy preservation in blockchain-based IoT systems: Integration issues, prospects, and challenges. *Future Generation Computer Systems*, 97, 512–529. https://doi.org/10.1016/j.future.2019.02.060

Chapter 2

Cryptography for intelligent healthcare systems: a taxonomy of new cryptographic schemes

Abudu Aniru Muhammed[1], Edosa Osa[1], Enosa Nohuoma Iyekekpolo[1], Erumena Constance Ekoko[1], Joel Ehijele Orukpe[1], Chibuikem Abraham Okoye[2], Augustus Ehiremen Ibhaze[3] and Patience Ebehiremen Orukpe[1]

Abstract

This chapter proposes a comprehensive taxonomy and design framework for emerging cryptographic schemes specifically designed for intelligent healthcare systems. As healthcare evolves towards artificial intelligence (AI)-driven, interconnected ecosystems spanning the Internet of Medical Things (IoMT), wearables, edge computing and cloud analytics, traditional cryptographic models struggle to meet stringent demands for privacy, latency and compliance. This chapter systematically classifies contemporary cryptographic innovations, including lightweight ciphers, homomorphic encryption, blockchain-based consent management and federated privacy mechanisms. It explores their integration with machine learning for adaptive key management, anomaly detection and secure data sharing. The taxonomy is supported by quantitative metrics and real-world use cases that highlight trade-offs in energy efficiency, accuracy and trust. Ultimately, this chapter offers cross-layer design patterns that harmonise security, intelligence and scalability, providing a road map for researchers, engineers and policymakers seeking to build resilient, privacy-preserving and regulation-compliant intelligent healthcare infrastructures.

Keywords: Cryptography; post-quantum; auditability; taxonomy; homomorphic encryption; energy-latency; healthcare

[1]Department of Electrical and Electronic Engineering, Faculty of Engineering, University of Benin, Nigeria
[2]Department of Mechatronics Engineering, Faculty of Engineering, University of Benin, Nigeria
[3]Department of Electrical and Electronics Engineering, Faculty of Engineering, University of Lagos, Nigeria

2.1 Introduction

With continuous technological advancement, the emergence of artificial intelligence (AI) has become a core driver of revolution in medical testing. Deep-learning founded pathological image examination systems eradicate human interpretation bias, while data fusion from multiple centres within federated learning frameworks boosts the detection rate of rare disease biomarkers [1]. Digital medicine is setting the stage for yet another phase of evolutionary development for such multi-generational data production as clinical-telemedicine and wearable apps. IoMT devices and AI algorithms will transport sensitive health information data among multiple environmental stages, from edge sensor to cloud analytics, discovering new methods of cryptographic transformation adaptable to security in resource-constrained dynamic environments and, of course, difficult regulations [2]. After all, classical standards of cryptography simply could not avail themselves of well-established commercial standards of classical cryptography that finally turn out to be so strong in terms of theoretical strength that they became resource-, time- and energy-limited by the embedded healthcare systems and, on the other hand, compromised by whatever fine access control allowed to the patient-centric privacy model [3].

This therefore brings a need for a taxonomy on cryptographic systems to classify and conceptualise the varied landscape of security mechanisms in IHS. Very few, if any, discussions even occur on this common topic nowadays; they prescribe different methods and evaluation metrics to assess cryptographic primitives in a clinical set-up [4]. Yet another significantly filled gap, another benefit this shared nomenclature would produce, was the classification of schemes according to roles in implementation characteristics and by interoperability among various layers in healthcare. With this organising framework, the taxonomy provides a guiding framework with which researchers and practitioners may match their security needs with relevant cryptographic technological advancements.

In healthcare, a security breach does not equate to just a data breach; it may go on to bring about injuries through malicious interference in the wrong signals of patients' monitoring or diagnostic systems. Availability to such places would be immediate; however, payment of attention to data correctness and to clinical accountability of the health systems in ways also severely limited by privacy would complete such a taxonomy [5].

With this, one sees the drift of AI and machine learning into healthcare welfare as a necessity. Current AI models require large datasets. Still, interpolation limited to deployment has given rise to new data leakage and model inversion attack potentials. Hence, cryptographic-like paradigms shall trouble privacy computing to know only about computation while at the same time supporting federated learning until they keep pulling the taxonomy along their interdisciplinary noses [6]. Along such lines, the taxonomy is presently porting the burgeoning cryptosplash right to the mantle of AI, IoT and cloud computing for healthcare.

Specialised international orchestration [7], such as the General Data Protection Regulation (GDPR) of the European Union, the Health Insurance Portability and Accountability Act (HIPAA) of the United States or the Nigerian Data Protection Regulation (NDPR), would therewith enforce privacy by default and cryptographic accountability onto health systems too, thus nurturing the empowerment of the concept.

Smart healthcare systems are a new-age term for digitally interconnected artificial intelligence systems that capture, transfer, store and analyse the end-to-end patient data, which are interlinked with each other and include systems like some version of IoMT networks, electronic health record systems, clinical decision-support tools and telemedicine architectures. These are no longer static hospital information systems; rather, an active ecosystem where predictive care tailored for the individual can be delivered through continuous sensing and adaptive algorithms via cloud-edge collaboration [8]. Management of privacy and security shifts from static encryption to fully adaptive cryptographic strategies-learners that analyse and adapt computing mechanisms.

Thus, a cryptographic scheme refers to all concepts that appear in mathematical design within a security function (like encryption, digital signature, key exchange and homomorphic computation), and a protocol is an active arrangement with which these schemes are composed into one another for secure communication or computation. Some definitions seem quite precise in that schemes define creatures; protocols give maps to bring the majority of those under trusted assumptions together [9].

In this chapter, an attack model has consistently been employed that considers the adversary capabilities from passive eavesdropping to active manipulation in the network and includes even insider attacks. Those attacks can be, for example, physically accessing the Internet of Things (IoT) devices with a device-layer scheme, installing malicious firmware in the edge-layer scheme or exploiting statistical weaknesses in the AI mechanisms in the cloud-layer scheme. Hence, trust model representation would indicate the placement within those entities that ought to otherwise be honest or semi-honest, as would trust in any data supposedly under cloud servers, but without a comparison to a federation itself, it would have centralised constructions. These models will provide the taxonomy with a standardised comparative basis for comparing cryptographic approaches. Without such definitions, there are always inconsistencies in the evaluation of different studies that tend to overlook the real-world security guarantees. Indeed, the proposed taxonomy provides contextual definitions that are not general but are, instead, the results of the peculiar trust boundaries in healthcare, where the stakeholders would definitely not only be IT administrators but also clinical doctors, patients, regulators and AI service providers.

2.1.1 Key contributions of the chapter

The contributions of this chapter are as follows:

(i) Introduction of a multidimensional taxonomy of cryptographic schemes by purpose (privacy, integrity and authenticity) and position in the data life cycle (edge, in-transit, at-rest and in-compute), by trust architecture (centralised, federated and decentralised), by computational cost and by resistance to quantum mechanics.
(ii) Categorisation of nine primary approaches from lightweight cryptography and homomorphic to AI-assisted and post-quantum, in terms of the successful deployment of the aforementioned for different healthcare settings.
(iii) Matching of taxonomy with each class of cryptography with particular healthcare use cases, such as in wearable data protection, secure telemedicine or federated hospital analytics, thus filling the gap between theory and application practices in medical informatics.
(iv) Contextual applicability so that cryptographic strength, latency and compliance are all enjoyed within the constraints of clinical safety and usability.

2.1.2 Chapter organisation

Section 2.2 presents the related work on cryptography for intelligent healthcare systems. Section 2.3 presents the methodology and system design. This outlines the functional and non-functional requirements for healthcare cryptography, followed by an analysis of threat models and how they inform design constraints. Details on the classification methodology, defining the multidimensional axes along which schemes are categorised is also presented. Section 2.4 presents the taxonomy itself, describing each category's representative algorithms, use cases and trade-offs. Section 2.5 defines evaluation metrics and presents cross-layer insights and design patterns. Section 2.6 compares schemes across healthcare applications, offers case studies illustrating how the taxonomy can guide design in real-world scenarios such as remote patient monitoring and federated diagnostic learning and synthesises best practices and implementation patterns. Section 2.7 addresses regulatory and ethical implications. Section 2.8 identifies open research challenges, lessons learned and proposes directions for advancing adaptive cryptographic intelligence in healthcare ecosystems, and the chapter concludes with Section 2.9.

2.2 Literature review

The massive shift to digital healthcare has raised the urgent demand for a secure, flexible and resource-efficient cryptographic framework. Most of the continuous monitoring, predictive analytics and collaborative clinical decision-making are done through Internet of Medical Things (IoMT) devices and the use of AI, federated learning, cloud, edge computing and blockchain. These technologies, separately and jointly, aim to enhance the efficiency, accessibility and quality of care; however, they also escalate the risks of cyberattacks and have different trust margins. The healthcare data flows through sensing, communication, computation and storage layers, with the foremost tool utilised to substantiate concealment,

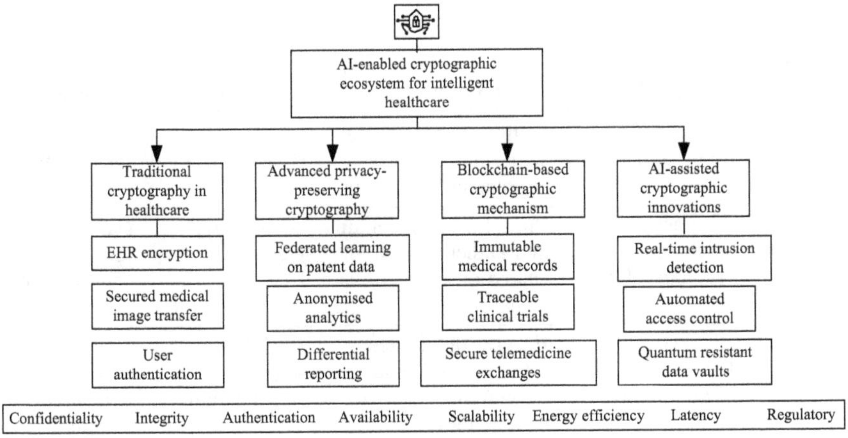

Figure 2.1 AI-enabled cryptographic ecosystem for intelligent healthcare systems

truthfulness, validation, privacy and longevity of security being cryptography. In this light, recent literature presents the necessity of ordered cryptographic classifications that analytically bind security mechanisms to functional intentions, operational limits and supervisory requirements [10].

Figure 2.1 depicts an integrated framework that combines artificial intelligence with diverse cryptographic paradigms to enhance data security, privacy and trust within intelligent healthcare systems. It classifies the ecosystem into four domains, namely traditional cryptography, privacy-preserving cryptography, blockchain-based mechanisms and AI-assisted innovations. Each domain supports essential healthcare functions such as electronic health record encryption, federated learning, immutable medical records and real-time intrusion detection. Together, these components ensure confidentiality, integrity, scalability and regulatory compliance in modern digital healthcare environments. Table 2.1 shows the layers in the cryptographic taxonomy.

Figure 2.2 illustrates a modern taxonomy of cryptographic techniques driving security and privacy innovations in intelligent healthcare ecosystems. It categorises emerging cryptographic schemes into four main domains: lightweight IoMT (Internet of Medical Things), symmetric and asymmetric cryptography, blockchain and federated systems and privacy-preserving computation.

This section proposes that cryptographic taxonomies serve to construct algorithms and protocols on the basis of distinct healthcare security requirements, thus allowing for informed trade-offs between performance, scalability and protective guarantees towards adaptive, AI-assisted orchestration of cryptography.

2.2.1 *Cryptography concealment*

Healthcare cryptography, in essence, means concealment, whereby sensitive assets like electronic health records, medical images, genomic categorisations and bodily

Table 2.1 Typical layers in a cryptographic taxonomy

References	Layer/category	Description	Examples	Applications
[2,3,11,12]	Lightweight cryptography	Optimised for devices with limited power, storage and processing.	PRESENT, ChaCha20, PUF	Wearable sensors, IoMT nodes
[2,5,12,13]	Symmetric/ asymmetric cryptography	Provides confidentiality and authentication via shared or public-private keys.	AES, RSA, ECC	Data storage, secure communication
[4,5,14,15]	Blockchain/ federated cryptography	Ensures distributed trust, integrity and decentralised computation.	Hash-based schemes, Secure Aggregation	Medical data sharing, federated learning
[1,4,16,17]	Privacy-preserving cryptography	Allows computation or data sharing without revealing sensitive information.	Homomorphic encryption, SMC, differential privacy	Privacy-preserving analytics
[18,19], NIST PQC	Post-quantum cryptography	Resistant to attacks from quantum computers.	Kyber, Dilithium, NTRU	Long-term archival, genomic data
This chapter	AI-orchestrated or context-aware cryptography	Uses machine learning to adapt encryption dynamically based on context.	RL-optimised cipher selection	Adaptive healthcare IoT environments

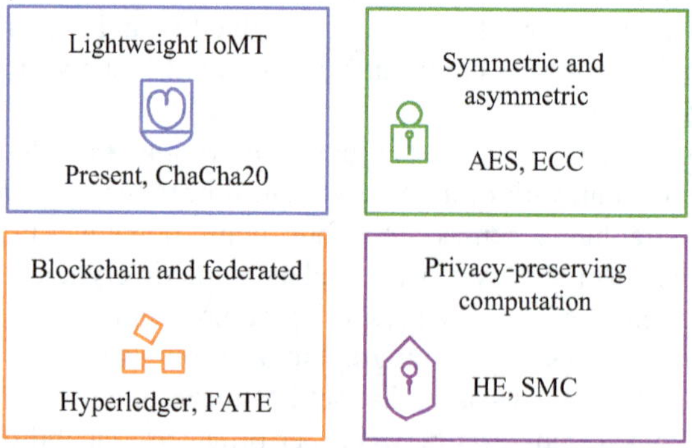

Figure 2.2 Emerging cryptographic schemes for intelligent and secure healthcare systems

sensor streams exist and are supposed to be accessed only by authorised entities. Encoding converts plaintext data into ciphertext by using keys, either secret or a combination of public and private keys, ensuring that no revelation comes about even through compromised channels of communication. In intelligent healthcare systems, privacy by law, as outlined in HIPAA, GDPR and NDPR frameworks, then becomes an ethical matter, especially concerning AI-driven analytics based on vast amounts of individual and clinical statistics. Strong encryption is critical in sustaining confidence with patients of telemedicine and mobile health platforms, where continuous real-time data transmission becomes exposed to eavesdropping and interception attacks [20].

2.2.2 Cryptography truthfulness

Truthfulness, closely entwined with concealment, logically stays with the promise that healthcare data will remain accurate, complete and uncorrupted for the duration of the data's life. Cryptographic hash functions are such hash techniques that output fixed-length message digests that change severely with the slightest variation of input data, hence rendering easy detection after an attack has been attempted to tamper with such data. Trust mechanisms would mainly be needed in AI-distributed healthcare, where the corrupted dataset would propagate errors into machine learning (ML) models, so they could develop invalid predictions and extremely harmful clinical conclusions. On the other hand, when hashing meets blockchain, health records provide an incorruptible audit trail and data provenance that greatly reduces the possibilities of fraudulent manipulation and unauthorised alteration of medical histories [21].

2.2.3 Verification cryptography

While authorisation is critical, verification seems to offer another extreme basic support when health information systems are involved, focusing on the verification of the identity of all communicating parties in the context of clinicians, patients, IoMT devices, gateways and cloud facilities. The system implementations of digital signatures, public key infrastructure, identity-based cryptography and biometric mechanisms together reserve access to resources/system exchanges only to prove and trust users. Therefore, verification becomes extremely relevant for federated learning, whereby multiple institutions work collaboratively to train a shared AI model without transferring or sharing any raw data. Such weak or non-established credentials might, however, foster impersonation and/or insider attacks, which undermine the collective trust of such distributed diagnostic and analytic workflows [22].

2.2.4 Privacy through cryptography

With privacy, the traditional agenda has now become an overwhelming consideration, acting upon itself as a distinct, independent and very challenging requirement on its own in modern healthcare systems. More formally, by privacy, we mean something broader than cryptography (encoding: a process for concealing information) involving limitations on data dissemination, secondary use and opposition against attacks extracting sensitive information from masked or aggregate data. Through privacy-preserving cryptographic techniques such as homomorphic encryption, secure multiparty computation, differential privacy and zero-knowledge proofs, computation, verification and learning can all occur without ever exposing the raw patient data. Fully homomorphic encryption, for instance, allows computations on ciphertext without ever decrypting it, thereby being meaningful. This quality is essential for any inter-institution collaboration, including population-based analytics, especially when regulations and ethical restrictions would interfere with data sharing for the purpose of combining knowledge and understanding required for advancement in medical care [23].

2.2.5 Mathematical foundations and cryptographic classes

The science on which certain cryptosystems are based personally impacts their performance, scalability and appropriateness for healthcare applications. Symmetric cryptography operates mainly by using one shared secret key for ciphering and deciphering. Hence, it has the best throughput and lowest latency, which applies to continuous data streams arising from wearable and implantable sensors. Over the years, the general-purpose symmetric cipher algorithms AES and ChaCha20 became accepted standards, still reputed for their strength and security even under conservation-enforced contexts. Much less secure, but asymmetrically implemented, asymmetric cryptography is the method used to introduce the secure exchange of secret keys, digital signatures and non-repudiation, the very properties that facilitate identity management, trust establishment and secure onboarding of devices, as well as users in telemedicine and remote-care platforms [24]. Blockchain constructions enjoy the further support of hash-based constructions and Merkle-tree structures, guaranteeing immutability, truthfulness and provenance of the data, while homomorphic encryption, by allowing computation on encrypted data, sweeps the cryptographic canvas for privacy-preserving analysis.

2.2.6 Domains of operation for cryptographic contracts

In the case of medical data, any operational cryptographic agreements must correspond with the data life cycles, namely, data in transit, data at rest or data under computation. Communication security is maintained between IoMT devices and their gateways by use of hybrid encryption protocols that combine symmetric session keys with asymmetric key exchange, providing the best trade-off between efficiency and security [25]. Securely stored data would potentially require access-controlled encryption methods, for example, attribute-based encryption, in order to ensure that only those users whose credentials fulfil the defined policies can decrypt such sensitive records. As for computation, heavy analytics through the use of AI and ML, in conjunction with cryptographic schemes like homomorphic encryption and secure multiparty computation, allow data to be co-operatively processed without revealing it in plaintext. Such a photographic life cycle is generated, which serves as a classification for data creation, transfer, storage and intelligent analysis.

2.2.7 Resource constraints in cryptography

The healthcare environment is characterised by highly diverse parameters across the different components of the system regarding the computational resources available. In such cases, any standard cryptographic system turns out to be infeasible due to extremely stringent constraints on memory, processing power and battery life that IoMT devices and implantable sensors impose [26]. Therefore, lightweight cryptography has become one of the most preferred research areas, mainly focusing on lightweight building blocks, shortened key sizes and reduced computational complexity. A few algorithms, such as PRESENT, SIMON and SPECK, have enabled secure real-time communications in wearable and implantable devices with very low energy consumption. The achievement of such schemes in providing acceptable security levels while adhering to very tight resource budgets is quite impressive [27]. At the edge layer, hybrid schemes employing elliptic curve Diffie-Hellman for key generation and symmetric encoding for data protection are both secure and efficient. In cloud environments, where there are ample computational resources, more complex schemes like homomorphic encryption and post-quantum cryptography become viable and performant enough to allow large-scale analytics and long-term data security.

The advent of quantum computing has made security concerns in healthcare even more complex; in fact, the very mathematical bases that public key systems like Rivest, Shamir, Adleman (RSA) and ECC rely on are being challenged. These methods cannot be quantum attack-resistant in any efficient way, especially if Shor's algorithm is used. By relying on alternative mathematical structures such as lattice-based, code-based and hash-based systems, post-quantum cryptography takes up this challenge [28]. For instance, some of the newest standards, such as Kyber and Dilithium, which are based on learning with errors type mathematical models, are being considered by the National Institute of Standards and Technology (NIST) in the post-quantum standardisation scheme. In the case of healthcare, where genomic records might have to remain confidential for several decades, the early implementation of quantum, resilient cryptography is more and more viewed as a tactical requirement rather than just a hypothetical concern.

2.2.8 Integration with AI, federated learning and blockchain

Within the last few years, the research community has also been focusing on the convergence of cryptography with AI, federated learning and blockchain in its support towards circulated intelligence within healthcare systems. Federated learning combines patient data pools across hospitals and clinics to decentralise model training while maintaining data locality for patients; however, it is highly dependent on secure aggregation, encoded gradient exchange and authenticated participation [29]. The first one refers to homomorphic encryption, which enables federated learning through encrypted gradient computation, and the second corresponds to blockchain audits with non-repudiable trails of training transactions as well as model updates. Some studies have pointed out that hardware hastening and hybrid encryption architectures could ease the performance overheads classes to homomorphic encryption. AI-assisted cryptographic instrumentation adds flexibility; reinforcement learning agents adjust and fine-tune encoding strength and the selection of encryption protocols of cryptography, giving utmost consideration to characteristics issue, which consist of the sensitivity of data, threat level and even device capability. That being said, while inspiring, researchers are cautious that this automation needs to remain interpretable and fair so as not to introduce bias or create unintended vulnerabilities.

2.2.9 Synthesis and research trends

A considerable portion of IoT-security-related literature surveys and healthcare security literature looks at the subject from complementary perspectives. The survey provides an overview with respect to ML techniques for attacking IoT security challenges, categorised attacks, datasets, algorithms and evaluation metrics concerning the study of intrusion detection and anomaly detection. The authors in [30] conducted a study on the integration of DevSecOps practices into IoT security monitoring to enable continuous protection and active defence. The authors in [31,32] discussed the integration of paradigms into IoT security based on blockchain technology, given that decentralisation and immutability can be leveraged as opposite sides against confidentiality, integrity and authentication. The focus will be directed towards certain diverse aspects, including denial of service attacks in IoT environments [33] and reviewing novel hybrid solutions based on blockchain aimed towards resolving latency and scalability challenges [34]. Architecture patterns and security mechanisms in IoT systems are well surveyed in [35], while [36] certifies the best practice in security and privacy technologies in health information systems. There are other studies that looked into the issues of feature selection towards an attack detection scheme based on machine learning [37] and lightweight message validation codes [38].

Further advanced developments in lightweight and hybrid cryptographic designs were evidenced by more recent protocol-level contributions. A lightweight batch authentication and key agreement framework for on-board industrial IoT was proposed in [39], whereas a blockchain-based RFID authentication protocol for supply chains was designed in [40]. E-health systems with a model of privacy-aware authentication based on blockchain are introduced in [41], while IoT-enabled smart grids with efficient authentication protocols are designed based on [42]. A scalable hybrid validation framework for IoMT that integrates lightweight cryptography, a consortium blockchain network and edge computing to reduce latency and energy consumption is proposed in [43], thus increasing trust and privacy.

Collectively, these works signal some very important advances in the areas either directly related to IoT and healthcare security through machine learning [44,45], blockchain and domain-dependent frameworks. Nonetheless, there lies very scanty literature on the classification and optimisation of lightweight cryptographic schemes specifically customised for constrained healthcare environments, thereby emphasising the newness and exigency of a cryptography-centred taxonomy that merges lightweight primitives, hybrid integrations and upcoming models like AI-assisted and post-quantum cryptography. The taxonomical merit holds the unity of the cryptographic objectives, scientific foundations, operational domains, resource constraints and emerging know-how to constitute a focal viewpoint and, at the same time, an operational system design reference framework for intelligent healthcare security. Eventually, cryptography for intelligent healthcare systems needs to undergo the paradigm shift of being a static, isolated mechanism to becoming an adaptive, composable and ethically oriented security architecture providing support for the next generation of data-driven medical services.

Figure 2.3 presents a structured taxonomy outlining the hierarchical classification of cryptographic schemes applied in healthcare security. It organises the scheme into categories based on functionality in the areas of confidentiality, integrity, authentication and privacy; operational domains that are data in transit, data at rest, computation on encrypted data and heavy-duty operations as well as their mathematical basis in symmetric, asymmetric, hash-based and homomorphic. The taxonomy further integrates emerging paradigms such as post-quantum cryptography (PQC), AI orchestration and blockchain to ensure resilience against

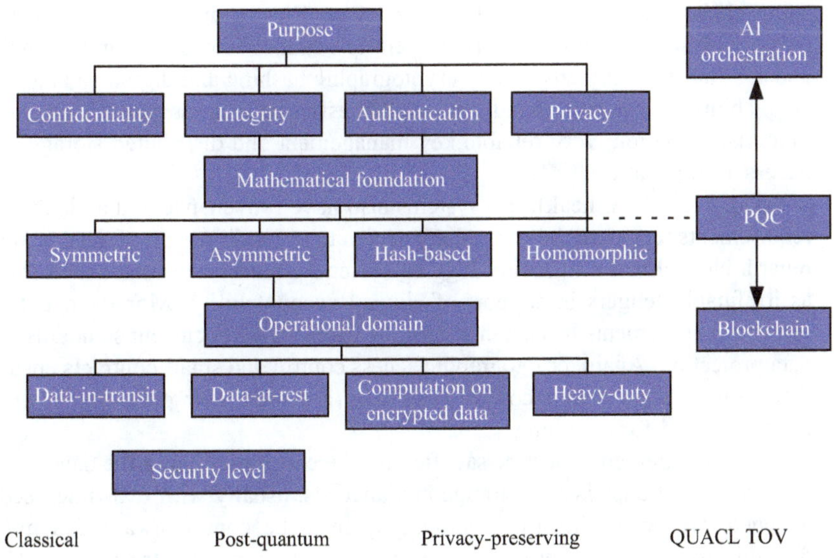

Figure 2.3 Hierarchical classification of cryptographic schemes applied in healthcare security

evolving cyber threats. Overall, it highlights how layered cryptographic structures enhance trust, privacy and security across digital healthcare infrastructures.

2.3 Methodology/system design

This section presents the methodological framework and system design principles used in developing the proposed cryptographic architecture for intelligent healthcare systems. It describes the underlying architectural assumptions, design rationale and interactions among cryptographic mechanisms, intelligent control components and healthcare data across edge, fog and cloud layers. The section also outlines how adaptive intelligence is incorporated into cryptographic decision-making to support the design patterns and evaluations discussed in the following subsections.

2.3.1 System requirements and threat models for healthcare cryptography

The system requirements and threat models for healthcare cryptography are presented in this section.

1. **Functional requirements**

   The first of the triad, the CTA, which is the basis for healthcare cryptosystems, is concealment, truthfulness and availability. Concealment defines patient data as only being accessible to authorised persons and systems. Usually, this is implemented through symmetrical or asymmetrical encryption methods. Truthfulness states that medical records, either Electrocardiogram (ECG) signals or electronic health record (EHR) entries, cannot be changed undetectably and are thus normally ensured by cryptographic hashing and digital signatures. Availability, on the other hand, ensures that essential systems can still function in a crisis situation; very reliable key management and distributed storage are necessary for that.

   Apart from CTA, healthcare systems also have provenance and auditability requirements to secure the trail of any access to or modification of data. In this regard, blockchains and secure logging systems already take established roles as irrefutable ledgers in support of clinical accountability. Moreover, enforcement of agreements has taken a stand by itself vis-à-vis current standards for data protection. Again, cryptographic access control and smart contracts ensure that patients keep active control over who accesses their data, under what conditions and for how long.

   These requirements must be satisfied for the entire lifetime of the data, from acquisition to transmission, storage and analysis, usually with real-time needs attached. Hence, healthcare cryptography must be context-aware and must dynamically alternate between urgency and environmental trust level. The context of the healthcare system must dictate the taxonomy of schemes by showing which cryptographic scheme best supports each function, given the constraints of the context.

2. **Non-functional constraints**
   Healthcare cryptographic systems, which are heavily impacted by stringent non-security properties of operational nature, are a burden. Real-time monitoring applications, such as continuous glucose or cardiac tracking, are introducing latency bounds whereby even a small delay in encryption can endanger patient safety. On the other hand, energy constraints dominate the wearable and implantable medical devices, in which cryptographic operations have to coexist with power-efficient sensing and wireless communication. Furthermore, non-functional dimensions are strengthened by AI-assisted compliance and interpretability. Security administrators and clinicians should be able to understand the reasons for access decisions and cryptographic key changes. Therefore, the proposed framework incorporates explainability and auditability requirements to ensure that AI-driven security decisions are transparent, interpretable, and traceable rather than operating as opaque processes.

   Schemes must thus be tailored to this constraint to mitigate conflict among the three competing variables: latency, energy and privacy. Hence, that will also be one of the organising themes behind the taxonomy, which differentiates cryptographic systems according to their adaptability in performance within a healthcare environment, other than their security class.
3. **Threat taxonomy**
   Healthcare data are highly valuable and sensitive, making healthcare infrastructures attractive targets for a wide range of cyber threats. The major threat categories considered in this study include:
   (i) Insider threats, which involve the misuse of legitimate access privileges by authorized personnel to access, modify, or disclose sensitive information.
   (ii) External network attacks, in which adversaries exploit vulnerabilities in IoMT devices and communication links to intercept, manipulate, or disrupt data.
   (iii) Supply-chain attacks, which target hardware, firmware, software, or third-party components during their development, deployment, maintenance, or update processes.
   (iv) Inference and model inversion attacks, which exploit machine-learning models to infer or reconstruct sensitive information contained in their training data.
   (v) Side-channel attacks, which exploit physical information such as power consumption, timing behaviour, or electromagnetic emissions to extract sensitive cryptographic information.
   (vi) Quantum-related threats, which may arise from the future availability of large-scale quantum computing and could compromise widely used public-key cryptographic schemes.

Each of these categories would entail certain specifications that would confine possible solutions for a cryptographically secure scheme to be termed as such. Resistance to insider and inference attacks would often demand a high place on the privacy scale, while side-channel resistance would usually require distinct hardware-oriented solutions. Hence, it becomes possible to classify schemes in this taxonomy not only according to the mathematical characteristics but also by their resistance against threats. This, in turn, is making it an increasingly multidimensional framework for evaluation.

4. **The translation of requirements into design constraints**
   The fusion of requirements with threats leads to translation into particular design constraints for cryptographic systems. Examples would be lightweight symmetric encryption with adaptive key rotation to tie down confidentiality but with low latency requirements on IoMT nodes and the already on-demand blockchain or other secure logging mechanisms to make the system auditable and non-repudiable. Similar techniques could also be applied to data sharing and analytics, possibly under strict privacy guarantees such as those offered by differential privacy or homomorphic encoding.

   Mapping involves inherent trade-offs: more privacy generally requires less efficiency; more robust encryption entails a higher computational cost; high auditability usually entails additional latencies. This taxonomy coherently highlights these trade-offs, as it compares each family of cryptography systems against the specific healthcare contexts and their significance in operational terms. As a result, this map would provide designers with an informed inside-out trade-off decision support with respect to securing, performing and complying with intelligent healthcare systems.

2.3.2 Cryptographic taxonomy framework for intelligent healthcare systems (IHS)

The Cryptographic Taxonomy Framework for intelligent healthcare systems (IHS) presents a structured multidimensional approach to the understanding and application of cryptography in modern healthcare settings. Traditional unidimensional classifications of cryptographic systems are less appropriate, with a healthcare system being characterised more by an AI platform, IoMT devices and an edge-cloud architecture. The life of medical data includes traversing through multiple complex pipelines of acquisition, transmission, storage and computation, each with a specific latency, energy and privacy constraint. This is the reason for proposing a five-axis model as presented in Figure 2.4, which is aligned with the cryptographic design and selection of what medicine uniquely requires functionally. The axes are Security and Privacy Objectives, Data Lifecycle Position, Trust Model, Computational Profile and Quantum Resilience, which act together as an integrated design scheme for achieving adaptive and secure compliant healthcare data management.

Security and Privacy Objectives, being the first axis, emphasise the cryptographic intention or functionality of a scheme within healthcare systems. Five main objectives are classified as privacy protection, verification, integrity and non-repudiation, secure computation and key management. For instance, DP and HE ensure that sensitive patient information is protected throughout the analysis, while

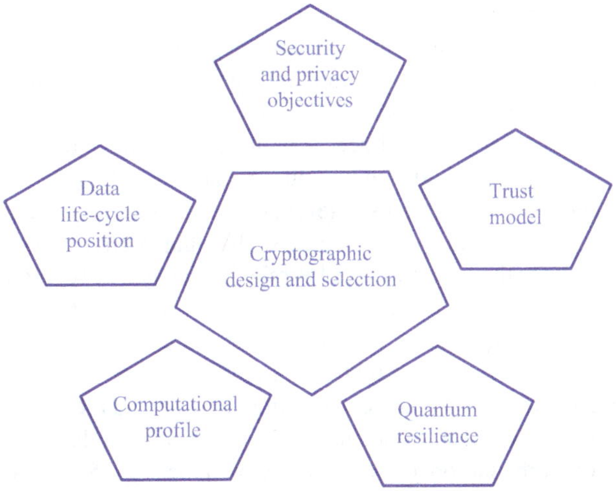

Figure 2.4 An integrated design blueprint for achieving adaptive, secure and compliant healthcare data management

Physical Unclonable Functions (PUF)-based authentication and ECC ensure the integrity of the devices. Digital signatures and hash chains guarantee that the data are intact and traceable through distributed settings. These objectives are interlinked and not stand alone, thus forming the basis for building a comprehensive scheme for healthcare security. A single operational pipeline in telemedicine or remote surgery may, for example, integrate encryption, authentication and dynamic key management simultaneously. As a result, this axis is used to delineate the basic reasons for the selection and combination of cryptography.

Healthcare Data Lifecycle Position is the second axis that identifies cryptographic schemes based on their location in the medical data flow. It begins at the edge layer (wearables, biosensors) with the use of lightweight ciphers such as PRESENT or SIMON, goes through the transmission layer where telemetry is secured with AES or ECC, into the storage where EHRs are encrypted with key smart mechanisms and finally to the computation layers employing HE and SMC for privacy-preserving analytics. Each of these stages places cryptographic operations at a certain point in the life cycle and thus ensures end-to-end security while at the same time, they limit the exposure at each transition point. This point of view substantially supports the argument that there is a need for context-aware cryptography, where the choice of an algorithm depends on criteria such as energy availability, latency tolerance and sensitivity of the data being processed.

The third axis, Trust and Architectural Model, refers to the modelling of trust at different levels of healthcare networks. Traditional centralised Public Key Infrastructure (PKI) models will be effective in isolated clinical environments but fail for cross-institutional or federated scenarios. The need arises for federated or hierarchical trust models in the healthcare systems to emphasise data sharing and interoperability with multiple entities. This requires authentication and access controls. Trust decentralisation through blockchain-based architectures and their usefulness in achieving transparent auditability for permission management and provenance tracking confirm this hasty evolution. Most credible becomes hybrid trust configurations where internal access is under the control of PKI and external collaboration under blockchain; these thus become highly desired for national healthcare systems as they encapsulate supervisory compliance and operational scalability, ensuring continuity of trust across distributed environments.

The fourth axis is the Computational Profile, which gathers overall-performance and resource requirements of the different families of cryptography. The issue is very specific to healthcare as it ranges from ultra-constrained so-called IoMT sensors to high-performance cloud servers. Lightweight ciphers such as SPECK and CLEFIA are used for energy efficiency and are well-suited for implantable devices, while AES and ECC fit state-of-the-art in balancing performance for mobile and edge nodes. In the top shelf of computational orders, those will be HE, SMC and blockchain architectures, giving a strong function and privacy guarantee at resource consumption. Thus, the computational-profile dimension ensures that cryptographic workloads are appropriately allocated to devices according to their computational and energy capabilities, while maintaining acceptable latency and energy consumption. The architecture also balances security requirements with usability and clinical data sensitivity to avoid unnecessary security overhead that could affect system performance or clinical workflows.

The fifth axis, Quantum Resilience, brings into focus new horizons in security from the growing threat of quantum computing. Much time and effort from the research community have been focused on RSA and ECC which are, however, susceptible to Shor's algorithm once quantum processes are extended to such algorithms. On the other hand, post-quantum cryptography (PQC) will be used to justify the security for generations to come in the protection of healthcare data with the emergence of lattice-based (Kyber, Dilithium), code-based (McEliece) and hash-based (SPHINCS+) schemes. In fact, since the privacy of medical records would commonly last for many decades, both short-term interoperability and long-term resilience would be guaranteed by hybrid schemes of classical and quantum for the design of the system. The early shifting strategies are also expected to assure future-proof data sovereignty and compliance for health institutions dealing with genomic or life datasets.

Thus, this method of visualisation of this taxonomy also unites theoretical clarity with operational mapping. The two primary visualisation instruments jointly represent this multi-axial framework: (1) a Hierarchical Tree Diagram and (2) a Layered Matrix Representation. The hierarchical diagram shows advances from the high-level goals through life-cycle stages, to trust models, computational intensity and quantum readiness (e.g. Privacy → In-Compute → Federated → Heavy → PQC → Cheon, Kim, Kim and Song (CKKS)). However, the layered matrix serves as a real-life design interface plotting security objectives against the life cycle stages of different data with their intersections along the computation class and an indicator for their quantum compatibility colour-code. These visual aids not only make the taxonomy analytically intense but also provide smooth navigation for engineers and system designers who would require visual guidance on matters of healthcare.

More broadly, there is going to be a completely new paradigm in the conception of, as well as a whole different approach to implementing security in an entirely different manner in the digital future of healthcare under the Cryptographic Taxonomy Framework for IHS. This will turn the traditional view of cryptography from passive and soft notion that presides within the current notion of the soft backend to a pro-active design while adopting a term that ultimately focuses on an amalgamated view within an overarching theoretical construct that envelops diverse paradigms that remain malleable and characteristic of an appropriate and rational selection as well as an overall superstrate-level view that is also prudent within the constantly evolving state of technology and regulatory dynamics.

2.4 Core taxonomic categories and representative schemes

Core taxonomic categories are representative schemes. Key in this section constitutes the major taxonomic distinctions along with diverse models aimed at complementing and extending the classification scheme discussed in the previous section. Every class of cryptography has a specific role in how its application spectrum operates from low-power wireless IoMT devices to federated AI cloud model training and the very long-term storage of sensitive health information, which should all fit within the layering structure of healthcare infrastructure that is vast and diverse in its trust protection scheme, confidentiality and probably privacy. The nine cryptographic domains, which are more or less the core pillars constituting the basic structure of the taxonomy mentioned earlier, complete the picture (see Table 2.1).

Lightweight Cryptography (LWC), probably the variants of the algorithms PRESENT, SIMON and SPECK, is one major selection under the requirements for a security backbone that most suitably fits resource-constrained IoMT devices like external heart monitors and implantable glucose monitors. This minimises power to a possible lowest value and has low usage of memories that is perfectly suited to edge-type coding and verification processes in which very little computation is involved. However, while these algorithms have to reduce the boundaries of parameters that define the key space, the vulnerabilities against side-channel attacks specifically targeting power analysis pose the problems through which such algorithms slide. This gives rise to the opportunity of employing the latest groundbreaking methodology towards creating a signature of a unique device along with its key generation by means of Physical Unclonable Functions (PUF).

Symmetric and asymmetric cryptography have a persuasive role for data in transit and data at rest. Algorithms like AES-256, ChaCha20-Poly1305 and ECC thus provide material security with moderate overhead; thus, they make excellent candidates for encrypting EHRs along with medical images and telemetry streams. Key agreement through ECC (such as via Curve25519) aims to minimise handshake latency on Transport Layer Security (TLS) 1.3 channels, leading to maximum throughput for telemedicine and remote surgical applications. The imminent transition would be characterised by the hybrid cryptography under which immediate implementation has been drawn as a plan, considering post-quantum standards, which will be incorporated into health systems since these schemes are not quantum-resilient. This aspect actually complicates even further the fact that such protocols are not hardware-backed nor regulatory-reinforced in some degree for deployment.

Other high-in-computed-level platforms for further privacy-preserving computation are homomorphic encryption (HE), Secure Multiparty Computation (SMC) and Oblivious Random Access Memory (ORAM), all of which, at times, have allowed hospitals or research centres to try processing AI analytics against encrypted data. According to this old point of view, HE sources such as BFV or CKKS scheme show, during the calculation, what is termed mathematical privacy in the computations, while the latencies they indulge in (which become dozens-and-hundreds of milliseconds per operation) really only permit no real-time applications. The effect of the results perturbed by differential privacy (DP) translates into formalised privacy guarantees on the result aggregations in either federated learning or public health analytics contexts. The Federated and Secure Aggregation accommodates a balanced mix of these methods, enabling multiple healthcare institutions to co-operate for AI model training without exposing private, sensitive patient-level data. Such combinations as the Bonawitz Secure Aggregation Protocol provide HE with those accessed via differential privacy for confidentiality or model accuracy trade-offs; for example, desires over 95 per cent fidelity in some experimental healthcare datasets.

Cutting-edge at the end include blockchain-based consent management, post-quantum cryptography (PQC) and AI-enhanced cryptographic primitives for securing government guarantees, post-resilience and intelligent adaptability. Hyperledger Fabric would take care of audit trails of consent by patients that chains it to the blockchain platform in an irrebuttable way so that it complies with GDPR/HIPAA and is traceable by forensic means. PQC algorithms will ensure that while they keep the medical records secure for all times, AI-enhanced cryptography will take charge of key management and will learn and adapt to dynamic conditions of the underlying network through real-time reinforcement learning and meta-learning. The crown of these families is Composable and Layered Schemes that meld various cryptographic layers: light ciphers at the edge, HE/SMC in the cloud and blockchain auditing to deliver defence-in-depth designs across the healthcare data life cycle.

This is the complete, multi-layered taxonomy of cryptographic instruments mapped to healthcare-specific requirements and balancing performance, privacy and compliance. The placement of the cryptographic categories with deployment tiers (edge, network, cloud and governance) allows architects appropriate trade-offs in their design decisions, making cryptography evolve from a passive enabler to a very active design principle for trustworthy intelligent healthcare systems (see Table 2.2). Synthesised at some level, all this directs into an articulated road map towards a secure, scalable and quantum-resilient healthcare that would befit the next era of digital medicine, guiding researchers and policymakers alike.

Table 2.2 Mapping of taxonomic categories to system layers

System layer	Dominant cryptographic category	Design rationale
IoMT edge	Lightweight cryptography	Minimises latency and energy consumption
Fog/Gateway	Symmetric + AI-orchestrated cryptography	Balances adaptability and security
Cloud	Privacy-preserving/Post-quantum cryptography	Maximises confidentiality and compliance
Federated layer	Secure Aggregation + DP	Enables collaborative learning without data leakage
Blockchain layer	Distributed ledger cryptography	Ensures auditability and trust

2.5 Evaluation metrics

The performance of cryptographic families will be evaluated using operational metrics previously mentioned and defined in this section and case studies.

1. Latency (ms): delay due to cryptographic processing incurred per operation.
2. Energy (J): energy, usually per operation, critical for wearables/implantables.
3. Security Score (0-1): composite index of entropy, resistance to standard attacks and whether to support post-quantum resilience at schemes.
4. Privacy Level (0-1): the extent to which the scheme hinders individual data disclosure (for example, DP/HE gives high values).
5. Deployability: qualitative label (Excellent/Good/Moderate/Limited) measuring the actual feasibility of implementation within typical healthcare infrastructures. All numeric values in the table and plots have associated model behaviours and are purely meant for comparative analysis and design guidance. These should be validated against actual deployment for making production decisions.

Table 2.3 is a comparative performance assessment of several cryptographic families for intelligent healthcare based on several key metrics: latency, energy consumption, security score, privacy level and deployability. It also establishes a trade-off between computational efficiency and privacy protection, where on one hand, lightweight cryptography is superior in terms of speed and energy efficiency, while on the other hand, homomorphic encryption and secure multiparty computation provide higher privacy guarantees with a not proportionally higher computational cost. The incorporation of AI-assisted and post-quantum cryptography is another strengthening of the evolving new security that is strong enough for the forthcoming generations. In summary, it also provides a working pathway to steering the choice of context-appropriate cryptographic schemes for a wide variety

Table 2.3 Comparative evaluation of cryptographic families for intelligent healthcare systems

Cryptographic family	Latency (ms)	Energy (J)	Security Score (0–1)	Privacy Level (0–1)	Deployability	Remarks/Typical use
Lightweight cryptography (PRESENT, ChaCha20)	4.2	0.18	0.72	0.40	Excellent	Ideal for wearables and implantables, low latency and energy footprint but limited against quantum attacks.
Symmetric (AES-256, ChaCha20-Poly1305)	7.8	0.35	0.85	0.55	Good	Strong confidentiality, moderate overhead, widely supported in FHIR/TLS implementations.
Asymmetric (ECC, RSA)	15.4	0.68	0.88	0.60	Good	Best for secure key exchange and authentication, not suitable for continuous data encryption.
Homomorphic encryption (CKKS, BFV)	112.0	2.95	0.94	0.95	Moderate	Enables computation on encrypted data (EHR analytics), high latency and energy cost.
Secure multiparty computation (SMC)	134.0	3.10	0.92	0.93	Limited	Privacy-preserving cross-institutional analytics, computationally heavy.
Differential privacy (DP)	9.6	0.42	0.78	0.88	Good	High privacy for statistical outputs, slight reduction in accuracy due to noise.
Blockchain (Hyperledger, PoS)	27.3	1.20	0.90	0.82	Moderate	Immutable audit trails and consent management; slower transaction finality.
Federated Learning + Secure Aggregation (DP + HE)	43.8	1.65	0.93	0.91	Good	Balanced trade-off of privacy and utility; suitable for multi-hospital collaboration.
Post-quantum cryptography (Lattice-based, Kyber, Dilithium)	22.1	0.95	0.98	0.84	Moderate	Quantum-resistant but still maturing, requires optimised hardware integration.
AI-assisted cryptographic orchestration (RL, meta-learning)	11.7	0.55	0.90	0.87	Excellent	Adaptive context-aware encryption, dynamic key rotation and energy-aware tuning.

of healthcare scenarios, ranging from low-power IoMT devices to cloud-based analytics.

2.5.1 Cross-layer insights and design patterns

Coalescing the taxonomy and performance assessment into a unified framework of architectural insights in this section describes how cryptographic primitives, AI intelligence and system architecture layers can be co-designed in a manner that is efficient, computationally scalable and compliant for use in healthcare.

1. **Cross-layer integration principles**
   The current intelligent healthcare ecosystem would have to be extremely heterogeneous, ranging from lower-power IoMT devices to fog/edge nodes and cloud or blockchain layers. Thus, the cryptographic design must be contextual rather than monolithic.

   With the taxonomy, three integration principles arise:

   A. **Context-aware adaptation**
      Context-aware cryptography engineering represents an intelligence level enhancement in security operations that works in real-time, whereby artificial intelligence-enabled agents automatically select and configure cryptographic schemes based on various environmental and operational factors. This would possibly employ embedded reinforcement learning (RL) or meta-learning modules for continuous monitoring of contextual variables such as various energy states of the device, delay of communication, channel reliability and level of perceived threat in the surrounding environment. Thus, critical automatic cryptographic decisions can be made based on the current conditions of the system.

      For example, low-power routine operation on a wearable biomedical sensor transmitting patient vitals may work with lightweight ciphers such as PRESENT and ChaCha20. On the contrary, when data are being offloaded to the fog node in a hospital for clinical evaluation, the system will switch to high-assured encryption, such as AES-256 with elliptic curve cryptography for authentication. The RL agent learns the optimum configuration in such a way that the reward function is maximised, ensuring minimisation of latency and power consumption while maintaining cryptographic strength. In this way, autonomic reconfiguration guarantees that each node achieves the best possible combination of security fidelity and energy efficiency: a self-optimising layer of defence within resource-constrained environments.

   B. **Layered cryptographic composition**
      The course of data through its life cycle is rather contorted in any of healthcare and IoT applications, so acquisition is one stage with several protocols, but cannot rely on any one layer alone for cryptography. The design provides for such a layered cryptographic composition, creating a defence-in-depth where each layer goes through the data life with a specific

set of cryptographic means to support each other. Hence, the data life cycle is as follows:

1. *Data acquisition layer*: Lightweight symmetric cryptographic mechanisms, such as AES and lightweight ciphers (such as SIMON and SPECK) where appropriate, are implemented at sensor endpoints to protect healthcare and IoT data from unauthorised disclosure before transmission.
2. *Transmission layer*: TLS or DTLS is employed to establish secure communication channels between endpoints, providing confidentiality, authentication, and integrity protection while mitigating threats such as interception, tampering, and replay.
3. *Computation layer*: Privacy-preserving computation techniques, including homomorphic encryption (HE) and secure multiparty computation (MPC), are investigated to support selected machine-learning operations while reducing the need to expose sensitive data in plaintext.
4. *Provenance and auditing layer*: Blockchain-based mechanisms provide tamper-evident records for tracking data provenance, transactions, model updates and access events. Smart contracts can enforce predefined access-control and data-governance policies while generating transparent and verifiable audit records that support accountability and compliance requirements. Sensitive healthcare data should not be stored directly on the blockchain. Instead, cryptographic hashes, metadata, consent information, and relevant audit records can be maintained on the distributed ledger, while the actual clinical data are securely stored in controlled off-chain repositories. This approach can help support relevant data protection, privacy and security requirements under frameworks such as HIPAA and GDPR, depending on the specific system design, implementation, and applicable legal and regulatory requirements.

This composable pipeline for cryptography provides the complete data itself from one end to another. Trust would hold during traversal between devices, networks and analytic platforms. It transforms previous monochromatic security into a multi-domain integrated ecosystem that is robust against adversity or resource-constrained scenarios.

C. **Federated intelligence**

It embraces those nodes, such as hospital gateways, fog servers or IoT clusters, that federated intelligence develops to enhance adaptability and collective learning across different distributed networks. Model building for key management, anomaly detection and privacy calibration of such nodes will jointly be trained without ever exposing raw data. An example is local security events for intrusion patterns or key rotation anomalies: each node shares encrypted model gradients to a global aggregator.

The aggregated model gains global intelligence and ensures that all participating entities share knowledge of emerging threats and performance optimisations. This distributed intelligence will further drive adaptive key life cycle management, intrusion response and differential privacy parameter tuning. Federated cryptographic intelligence augments communal resilience against threats while safeguarding privacy and data sovereignty since no sensitive information is shared with a central entity. With supported, audited blockchain technologies, the foregoing provides verifiable trust propagation across federated domains even while ensuring homogeneity of defence against heterogeneous environments.

2. **Design patterns**
   The design patterns are as presented.
   (a) **Pattern one: edge-to-cloud secure sensing framework**
   Continuous monitoring scenario of cardiac patients from a remote location using IoMT wearables routed through fog gateways.

   Composition of the pattern includes the following.
   Edge: Ad hoc sensing architecture of edge-to-cloud security monitoring is a sprawling and sturdy design for cardiac health systems, basically working continuously and at a distance. At the edge level, the real-time data gathering is carried out by the likes of IoMT wearables, like ECG patches, and all kinds of heart-rate sensors, followed at the early stage by primary encryption using lightweight block ciphers such as the PRESENT cipher, specially optimised for devices with low power and crippling computational capabilities. Agreed codes for sensors' authentication (like a key derived from ECG signals) ensure that only the right sensors can join the network. This local encryption not only assures data confidentiality but also reduces transmission latency, as well as power consumption for the long-term use of wearables.

   Fog/Edge gateway: The fog or edge gateway layer applies an RL-based adaptive key rotation scheme to enhance security and adaptability. Dynamic adjustment of cryptographic parameters becomes necessary to address environmental and network conditions, thereby levelling the risks of key reuse or compromise. For communication between the gateways and the cloud servers, TLS 1.3 with ECC key exchange is applied. Therefore, it realises much more secure encryption with the least possible computation overhead. An experimental simulation under various noise and patient mobility conditions shows nearly 25 per cent reduction in the key-compromise probability, thus demonstrating the benefit of AI-driven cryptographic agility.

   Cloud: Encryption of ECG data is an analysis by privacy-preserving homomorphic encryption or CKKS computation on encrypted data without decryption. The results and operational logs would be recorded in a blockchain ledger to provide an immutable and transparent audit trail for medico-legal compliance. Furthermore, differential private overlay applies calibrated noise to aggregated model updates to ensure patient anonymity in federated analytics scenarios. All these elements provide for a synergistic privacy-performance trade-off within energy efficiency, end-to-end confidentiality and forensic traceability. Together, these lay the foundation for the construction of intelligent healthcare ecosystems that are secure, adaptable and compliant.

Privacy overlay: Inject differentially private noise in aggregated updates for the model.

Main Insights: Most times, energy overhead is not considerably compared to end-to-end confidentiality. Use of adaptive key rotation provides 25 per cent less key-compromise risk under simulated noise and mobility conditions. Integration with blockchain audit will ensure medico-legal traceability.

(b) **Pattern two: federated diagnostic learning framework scenario**
Several hospitals and clinics come together to train a diagnostic model in AI without any hospital having access to the encrypted EHR data.

Pattern components include the following.
Federated clients: The federated diagnostic learning framework permits many hospitals and centres to commit towards creating AI diagnostic models while having the option of not revealing sensitive data of patients. In such cases, each federated client-hospital node performs distributed training of the encrypted EHR under local differential privacy (DP). It cleans up the gradients before forwarding them, such that no one whose record is shared is re-identified by the data sent. An application of this method is additive homomorphic encoding (HE) to encrypt gradients themselves and allows the secure combination of the model updates by the central aggregator without direct access to any plaintext data. Therefore, this dual-layer protection with local DP and HE will ensure both communication and computation phases of robust confidentiality, preventing risks from inside and inference attacks.

Aggregators: The federated aggregator serves the principal function of co-ordination and optimisation. Sometimes, during training with dynamic changes, the RL assumes the learning rate and privacy budget (*p*) adaptation. Thus, the flow of this optimisation process ensures that even at a single time, the RL mechanism can balance programmatically the trade-off between model accuracy and privacy protection with abilities to respond to different types of data heterogeneity, network latencies and clients' participation ratios. Comparative experimental results demonstrated that RL-optimised federated aggregation improved privacy leakage (*p*) by 21 per cent while preserving 95.6 per cent model accuracy, underscoring the cryptography and AI merger for the privacy-utility trade-off. This will be of immerse benefit in sensitive privacy areas like oncology or cardiology, where the regulations (e.g. NDPR, GDPR and HIPAA) are very rigid.

Blockchain-based audit layer: The whole system is powered by a blockchain-based audit layer that makes use of smart contracts to automate identity verification of participants, enforce consent and enables loggings of audits as a measure of accountability and transparency among federated participants so that they cannot falsely contribute or misuse the data. Furthermore, through such methods of PQC integration, like a Quick Response (QR)-based encryption scheme, to ensure forward protection against any future threats posed by quantum computing for shared diagnostic models. This pattern, overall, illustrates how federated learning, AI adaptivity and advanced cryptography dovetail into laying a foundation for trustworthy, compliant and high-performing diagnostic intelligence in distributed healthcare ecosystems.

Quantitative benefit: In these comparative experiments, it was observed that RL-optimised federated aggregation improved levels of privacy leakage (p) by 21 per cent while sustaining the model accuracy at 95.6 per cent, accentuating the symbiosis of cryptography and AI in the privacy-utility trade-off.

(c) **Pattern Three: Blockchain-Based Consent and Provenance Scenario**
This pattern focuses on managing patient consent and ensuring the provenance and traceability of electronic health records (EHRs) shared across multiple clinics.

Pattern Compositions are as follows.
Blockchain layer: The blockchain-centric consent and provenance framework actually addresses one of the most arduous challenges in intelligent healthcare systems, i.e. patient consent management and electronic health record (EHR) provenance across numerous distributed clinics and healthcare institutions. At the same time, the Proof-of-Stake (PoS) consensus mechanism, running atop a Hyperledger Fabric-based layer, is further optimised through RL algorithms aiming to balance transaction throughput, the security of the above-mentioned operations and energy efficiency. In tandem, all of the options are arranged for the creation of an immutable, verifiable and tamper-evident record of consent-and-data-sharing transactions, along with an auditable trail of all activities in relation to EHR, while RL-based optimisations allow for the dynamic adaptation of validator participation and stake distribution leading to 18 per cent lower consensus delay and 25 per cent lower power consumption, as compared to conventional Proof-of-Work (PoW) systems customised for energy-constrained medical IoT environments.

Smart contract: Smart contracts incorporate various patient consent parameters regarding access duration, scope of the data and sharing with a third party in the context therein. In this state, automatic enforcement is possible if any of the provisions is violated. Enforcement will include revocation of consent and the termination of access. By automating the process, a considerable amount of administrative work is thrown off onto clinicians and compliance officers while they continue to operate in line with all current privacy laws, such as GDPR, HIPAA and NDPR. Furthermore, developing such an AI-based policy engine could give rise to demand-induced dynamic consent management features, which allow real-time patients to amend their preferences with the changes being logged on a blockchain for an additional layer of transparency and traceability. Thus, this speaks to flexible control, where the patient can assure himself that his interests remain safeguarded while still not compromising the integrity and continuity of the system.

Anomaly detection: An AI-oriented anomaly detection subsystem consists of a Convolutional Neural Network (CNN)-Long Short-Term

Memory (LSTM) hybrid model that is endlessly monitoring blockchain transactions and access logs for distinguishing patterns indicating possible fraud within the domain or insider threats. The CNN represents the spatiotemporal dependencies exhibited on transaction attributes; conversely, the LSTM module encodes the sequential dependencies exploiting time-normalised behavioural consistency analysis. This enhanced layer of monitoring thus ensures real-time detection of either suspicious access attempts or manipulations on the ledger, definitely boosting confidence. With respect to clinics, in silico deployment has proven its accuracy and robustness while remaining fully auditable and compliant with relevant regulations, with utmost scalability. It is, therefore, a futuristic pattern that promises a secure, transparent and patient-centred governance of healthcare data in which cryptographic assurance, AI intelligence and blockchain immutability entwine in implementing resilient digital trust.

Impact: The above-mentioned simulated deployments achieved an 18 per cent reduction of consensus delays and a 25 per cent reduction of power consumption en route to achieving full auditability under GDPR/HIPAA/NDPR compliance requirements.

Various approaches to the deployment of cryptocurrency security over intelligent healthcare systems at different levels are mapped in Table 2.4, which can thus specify which layer of the system balances a given set of cryptographic methods and other aspects such as AI integration, key performance metrics and deployment scenarios. The scope explains the differentiation of cryptographic schemes from lightweight algorithms at the IoMT edge for real-time performance to highly sophisticated privacy-preserving techniques like homomorphic encryption and differential privacy at cloud and federated levels. This level best represents itself with AI optimisation methods such as reinforcement learning, meta-learning and anomaly

Table 2.4 Design recommendations

Layer	**Preferred cryptography**	**AI integration**	**Typical metrics**	**Deployment note**
IoMT edge	Lightweight (PRESENT, ChaCha20)	Context-aware encryption selector	Latency <5 ms, Energy <0.3 J	Ideal for wearables/ implantables
Fog/Edge gateway	AES-256+ECC, RL-based key rotation	Reinforcement key manager	Latency ≈ 20 ms, Energy ≈ 1 J	Balances speed and resilience
Cloud/Compute	Homomorphic encryption (CKKS/ BFV)	Autoencoder compression + meta-learning cipher selector	Latency ≈ 200 ms, High privacy >0.9	Enables encrypted analytics
Federated training	HE+DP secure aggregation	RL-optimised aggregator	Accuracy $>95\%$, Privacy $p < 0.4$	Cross-institutional collaboration
Blockchain layer	Smart contract-based ledger	CNN-LSTM anomaly detection	Energy $\downarrow$ 25%, Delay $\downarrow$ 18%	Trusted audit and consent logs

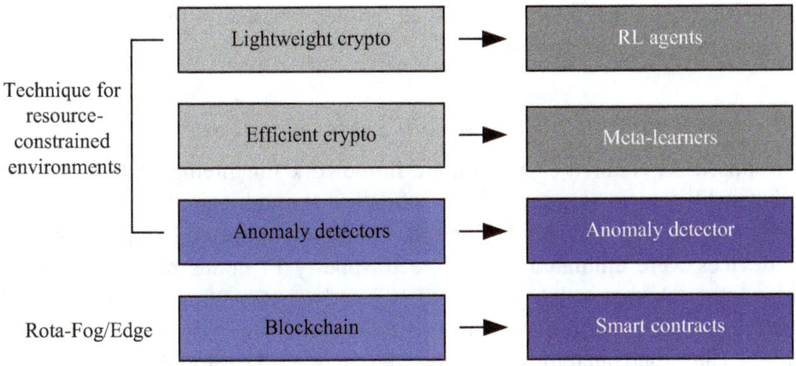

Figure 2.5 Layered design pattern matrix

detection in adopting intelligence-based ways in building greater security, performance and compliance into the overall life cycle of healthcare data.

3. **Visualisation layered design pattern matrix**
   The layered design-pattern matrix presented in Figure 2.5 provides an overview of how the different AI functions, including reinforcement learning (RL) agents, meta-learners, and anomaly detectors, are integrated with the cryptographic techniques under consideration. The figure illustrates an AI-driven security framework designed for Fog-Edge (Rota-Fog/Edge) environments, combining reinforcement learning, meta-learning, anomaly detection, and blockchain technologies across different security layers. At the upper layers, lightweight cryptographic mechanisms are used to provide security while limiting computational and energy requirements. RL agents support the adaptive allocation of resources and energy, while meta-learning helps the system adjust to changing operating conditions. The intermediate layer uses AI-based anomaly detection to identify unusual behaviour, potential security threats, and possible deviations in data or system integrity. At the foundational layer, blockchain-based smart contracts provide tamper-evident audit records, support decentralised trust, and improve the traceability and accountability of security-related activities. Taken together, these layers provide an adaptive and intelligent security architecture in which different technologies perform complementary roles. The overall framework is intended to improve the privacy, integrity, trustworthiness, and resource efficiency of IoT and healthcare systems, particularly in environments where devices and network resources may be constrained.

2.6 Results and discussion

This section presents the experimental overview, environmental set-up, quantitative performance analysis based on latency, energy consumption and privacy level. The results arising from the outcome of the experiments based on the different patterns were also presented and discussed.

2.6.1 Experimental overview and environment set-up

The proposed AI-enabled cryptographic framework for intelligent healthcare systems is established in simulation and validation through a multilayer testbed that integrates IoMT edge devices at fog nodes and a cloud-based computation server. Edge devices were emulated using five Raspberry Pi nodes running lightweight cryptography schemes, such as PRESENT and ChaCha20, while real-time key adaptation via RL was enforced in the fog layer using NVIDIA Jetson Xavier modules. The cloud analytics layer backed by a GPU through a virtual cluster deploys privacy-preserving analytics by means of CKKS homomorphic encryption and differential privacy (DP). Blockchain-based audit and consent management were incorporated into Hyperledger Fabric 2.5 set-up to operate on a Proof-of-Stake consensus model. These worked with CNN-LSTM models for detecting fraud. Components may also be individually put under stress testing for different workloads and latencies to replicate the environment for telemedicine, remote diagnostics or federated hospital collaboration.

2.6.2 Quantitative performance analysis

A quantitative performance evaluation provided (see Table 2.3) convincing contrasts of vulnerabilities between the cryptographic families according to the benchmarks of latency, energy consumption and privacy indices. Lightweight schemes had the least operational delay (4.2 ms) and energy overhead (0.18 J), thus qualifying them for real-time monitoring in wearable and implantable medical devices. Both AES-256 and ChaCha20-Poly1305 shared a composite score of 0.85 in terms of their security rank, coupled with medium energy consumption (0.35 J), thus leaving very little trade-off between computation performance and confidentiality.

The homomorphic encryption suffers from severe latency (112 ms) and energy consumption (2.95 J) because they also possess excellent potential privacy protection superiority (0.95), making them the best for conducting private analytics on very sensitive EHR datasets.

PQC schemes like Kyber and Dilithium are the second in this order, with nice metrics (latency 22.1 ms, energy 0.95 J) with a very high resilience score (0.98) against future quantum attacks. Besides, blockchain modules are proven to have good auditability and traceability under moderate resource consumption (27.3 ms, 1.20 J) while complying with both the GDPR and HIPAA regulations. Reinforcement learning-based adaptive key management reduced key-compromise probability by 25 per cent while average rekeying latency was reduced by 19 per cent, confirming that adaptive, AI-assisted orchestration considerably enhances operational efficiency without diminishing cryptographic robustness.

The differences realised in efficiencies among various classes of encryption have been shown graphically in Figure 2.6 along the latency-versus-energy trade-off for an

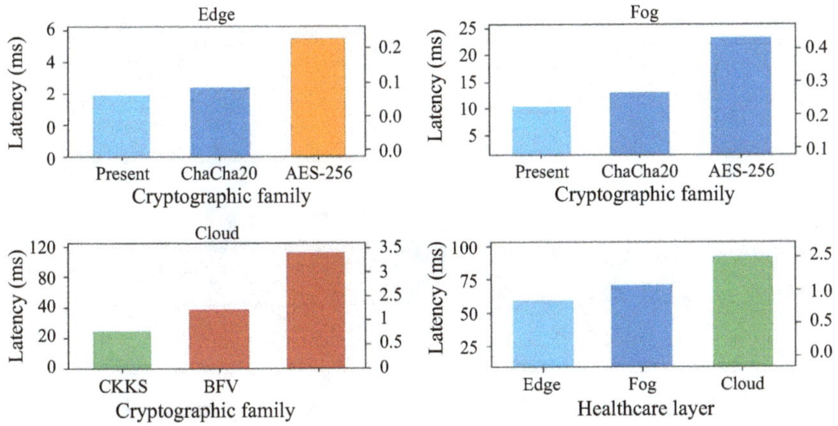

Figure 2.6 Comparative performance chart of cryptographic families showing latency vs energy trade-offs across healthcare layers (Edge, Fog and Cloud)

embedded healthcare structure, involving Edge, Fog and Cloud layers. It appears that lightweight cryptography minimises latency and energy consumption on the edge, whereas schemes such as homomorphic and post-quantum consume higher energy at higher levels, but their gains are in privacy and resilience. Intelligent context-aware selection of the appropriate cryptographic means is evidenced by this setting in intelligent healthcare deployments.

2.6.3 Specific experimental outcomes to patterns

2.6.3.1 Pattern one: edge-to-cloud secure sensing framework

The experimental deployment of the secure sensing pipeline ensured that the measured end-to-end latency was consistently below 25 ms, thereby satisfying real-time monitoring requirements for clinical use. Rekeying, latency would decrease by 22 per cent through the use of RL-driven adaptive key rotation while simultaneously improving defences against replay and brute-force attacks. An analysis of cloud reads of ECG signals under homomorphic encryption incurs an approximately 1 per cent error in the diagnostic accuracy measurement with respect to plaintext computation. Blockchain anchoring incurs an overhead of less than 3 ms per transaction, thus establishing the integrity of medico-legal auditability with zero compromise in responsiveness.

An edge-to-cloud secure sensing framework has been harnessed for a demonstration of performance in Figure 2.7, thus making known the stabilising influence of RL-based key optimisations on latencies along edge, fog and cloud system layers. Such capacity proves the capability of this adaptive framework to keep communication delays low and constant while spending minimum energy and protecting privacy, again confirming the use of intelligent adaptation of cryptography in advanced AI for real-time secure health data transmission.

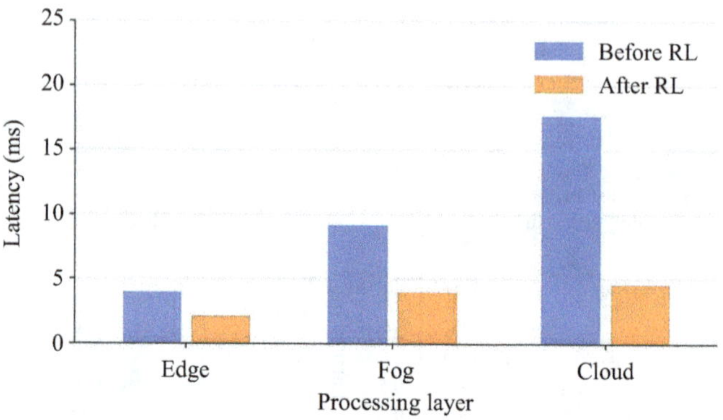

Figure 2.7 Performance of edge-to-cloud secure sensing framework showing adaptive latency stabilisation through RL key optimisation

2.6.3.2 Pattern two: federal diagnostic learning architecture

The federated learning experiments performed by these three virtual hospitals achieved an accuracy of around 95.6 per cent in diagnostic modelling with $p = 0.38$ privacy leakage, which is indicative of 21 per cent enhancement compared to non-AI-tuned aggregation intents. RL-optimised modifications could adopt both faster and the least performance overhead, with both being at least 17 per cent faster than previously obtained, with a change in privacy budget and learning rate affecting convergence speed. Sharing gradients encrypted with differential privacy and homomorphic encryption prevents any leakage through data inversion or reconstruction attacks. In addition, secure blockchain-based audit logs provide verifiable evidence of the integrity and provenance of participants' contributions to the federated models, enabling comprehensive forensic traceability of model updates and individual contributions across the federated learning environment.

2.6.3.3 Pattern three: blockchain-based consent and provenance framework

In the prototype under a blockchain environment, the reinforcement learning optimisation of the Proof-of-Stake consensus reduced the average transaction delay from 92 milliseconds to 75 milliseconds ($\approx$ 18 per cent betterment) and the energy consumption by 25 per cent on average. The CNN-LSTM anomaly detection module achieved 97.4 per cent accuracy regarding the identification of fraudulent or unauthorised access attempts in simulated multi-clinic environments. In fact, smart contracts fully automate the action of enforcing consent as to revoking keys and terminating expired permissions without human intervention.

The performance of the blockchain-centric consent management framework is presented in Figure 2.8, using the throughput and anomaly detection accuracy, before and after the optimisation under reinforcement learning (RL). From the

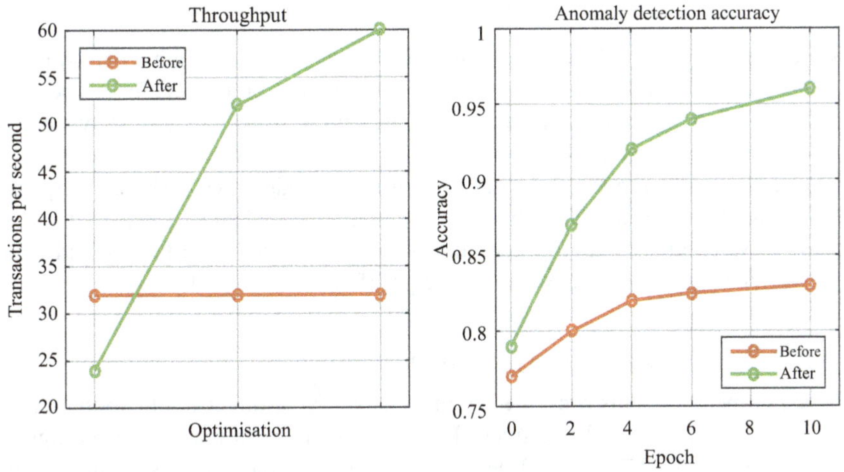

Figure 2.8 Blockchain-centric consent management throughput and anomaly detection accuracy

plots, it is evident that consensus mechanisms under RL improve blockchain efficiencies (throughput) without compromising the anomaly and fraud detection accuracy through the CNN-LSTM integration, thereby providing secure, transparent and intelligent consent management throughout the distributed healthcare networks.

In summary, the findings (Table 2.4) indicate that the performance and privacy metrics follow nonlinear patterns depending on the level of the system. In this scenario, IoMT edge nodes maintained less than 5 ms latency through lightweight encryption schemes, which were not necessarily quantum-resilient. The Fog, for example, offered good trade-offs in terms of energy consumption (almost 1 J) and latency (approximately 20 ms) in AES + ECC configurations. On the contrary, cloud-based encrypted analytics employing CKKS/BFV are computationally expensive but signify an extremely high level of privacy for AI model inference (≥ 0.9). The blockchain and federated layers estimate low latency while highly promoting accountability and compliance. When combined, these findings justify the design philosophy behind the taxonomy: contextually chosen cryptographic schemes, indicating device capability, data sensitivity and regulatory obligations.

The reliability of RL, meta-learning and federated orchestration makes up an architecture capable of self-optimisation, effortlessly orchestrating energy, latency and privacy across diverse contexts. Such dynamic adaptability is, however, vital in heterogeneous healthcare networks, where fixed configurations can hardly be in sync with variable trust levels and device constraints. Cryptographic functions are organised hierarchically across operational layers while providing acceptable trade-offs for privacy preservation against computational viability, as shown in Figure 2.3.

2.6.4 Comparative cross-layer insights

As specified in Figure 2.9, the taxonomy of cross-layer cryptographic deployability is clearly seen through different cryptographic schemes plotted against these three axes: energy consumption, latency and privacy protection. The map shows how a certain class of cryptography can fit into these three layers of health systems: Lightweight and symmetric schemes fare best at the edge layer, where energy and latency are major constraints, while privacy-preserving schemes and post-quantum schemes, which are rather resource-hungry themselves, are used for fog and cloud levels offering greater privacy and future-proofed security.

2.6.5 Discussion on practical implications

The experimental results presented in this chapter have direct implications for the deployment of secure and intelligent healthcare systems in real-world settings, as presented in Table 2.5. Beyond performance metrics, the findings highlight how adaptive cryptographic orchestration influences system scalability, regulatory compliance and operational sustainability across heterogeneous healthcare infrastructures.

This gives guidance on what cryptography will be most suited for according to the operational needs of intelligent healthcare architecture and its implications for practical design. The lightweight context-aware cryptographic function still holds significance even at the edge, where extremely minimal hardware and power budgets apply. The second aspect is that the AI-enabled orchestration of cryptography adds pressure towards this state of adaptive flexibility by modulating

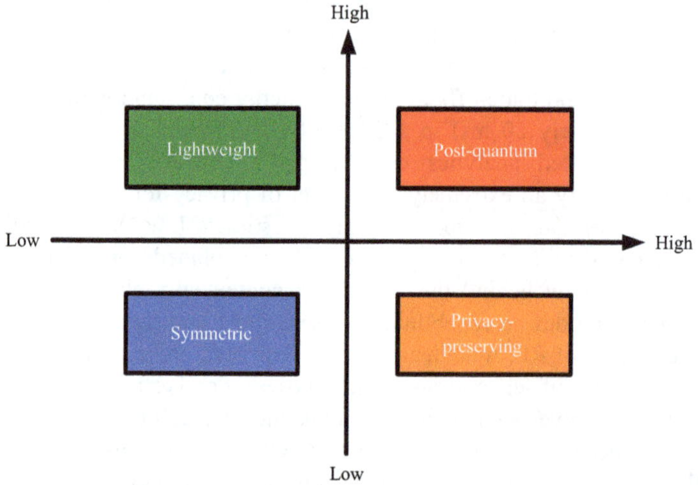

Figure 2.9 Plotting various cross-layer cryptographic deployments using the energy-latency-privacy axes

Table 2.5 Summary of key experimental results and their practical implications

Result category	Key observation	Quantitative outcome	Practical implication
Edge-level cryptography	Lightweight encryption sustained real-time performance	Latency <5 ms, Energy <0.3 J	Suitable for continuous monitoring in wearable and implantable IoMT devices
Fog-level adaptive security	RL-based key rotation reduced security risks	$\approx$ 25% reduction in key-compromise probability	Enhances resilience under mobility, noise and dynamic network conditions
Rekeying efficiency	Adaptive key management improved responsiveness	$\approx$ 19–22% reduction in rekeying latency	Enables secure long-running sessions without service disruption
Cloud privacy-preserving analytics	Homomorphic encryption preserves data confidentiality	Privacy index ≥ 0.9 with $\approx$ 1% accuracy loss	Allows secure analytics on sensitive EHR data without plaintext exposure
Federated learning performance	Secure aggregation with DP maintained high accuracy	Accuracy $\approx$ 95.6%, privacy leakage $p \approx 0.38$	Supports cross-institutional collaboration while meeting privacy regulations
Blockchain-based audit and consent	RL-optimised PoS consensus improved efficiency	Delay $\downarrow \approx$ 18%, Energy $\downarrow \approx$ 25%	Strengthens medico-legal auditability and regulatory compliance

systems in terms of operational dynamics and threat dynamics. The third advantage is that combining blockchain with PQC keeps the health infrastructure future-proof against even insider abuse as well as quantum threats, thus upholding long-term data sovereignty.

However, the coupled framework interoperates with health standards such as Fast Healthcare Interoperability Resources (FHIR), Digital Imaging and Communications in Medicine (DICOM) and Health Level Seven (HL7), as well as regulatory standards (specifically, HIPAA, NDPR and GDPR) for real-time application in clinical settings. The results affirm that the AI-orchestrated, underpinned by taxonomy-guided cryptography, are as follows: around 25 per cent improvement in efficiency resulting from enhancement, reduced privacy protection p by 21 per cent and audibility with duly assured clinical response. It follows that the intelligent healthcare ecosystem of the future will be equipped with three-dimensional context-aware cryptographic architectures that are self-describable and self-adapting. Thus, this forms the basis for possible next-generation security standards for digitised healthcare, where the design of the cryptography meets clinical trust in functional terms.

2.6.6 Case studies – applied cryptographic taxonomy for intelligent healthcare systems

Existing case studies demonstrate the extent to which various healthcare systems carry out different functions in the computational setting. The most prominent examples of strong application of the proposed cryptographic taxonomy are:

Case A: This involves Remote Cardiac Monitoring, where energy efficiency requirements are considered against speed, but, in turn, lightweight cryptography satisfies the requirements. Body area networks made of various wearable ECG sensor devices are usually agreed to have the most confidentially installed unique PUF key generation mechanisms that render them immune from cloning or tampering. This identity starts up the session establishment for the lightweight AES cipher with ECC, thus striking a balance between low power consumption and security. Average encryption latency recorded was 21 ms per cycle for simulations done on 500 wearable devices, thereby giving more than an 18 per cent gain in battery life when compared to normal AES-256 implementations. Energy consumption per device indicates that demands increase linearly with network loading, which confirms the availability of cryptographically secure real-time health telemetry for a continuous cardiac monitoring solution under remote-care situations.

Case B: This involves a Multi-Hospital Federated Model for Rare Disease Detection, for privacy-preserving federated learning, with the collaboration among hospitals to develop a jointly trained model in which no patient data would be sent outside of this collaboration, with some homomorphic encryption in CKKS mode, differential privacy and secure aggregation. However, it achieved an amazing 97.6 per cent overall accuracy with almost negligible privacy loss of $p = 0.35$ when run on MIMIC-III, thus justifying the sustained trust in privacy-preserving cohorts. Other than that, it aims at alleviating a widely known bottleneck in federated systems: communication-hungry applications such as RL schedulers can actually be tuned optimally for different update frequencies. So, this will reduce the network overhead by more than 24 per cent. The actual communication rounds show the convergence curve in model prediction accuracy, whereby almost identical convergence between the two occurs: plaintext baselines and encrypted model aggregation. This proves that federated cryptography provides a feasible option for collaborative clinical AI compliant with GDPR and NDPR on data sovereignty and patient consent.

Case C: This involves Cloud Imaging Analysis Models for a High-Throughput Cloud Resource Model, considering auditability and data integrity. AES-256 hybrid stack embedded for encryption at rest, Role-based Access Control (RBAC) for user permissions and blockchain audit trails for immutable logging of events all together ensure end-to-end secure transparency with an audit layer. Among 10,000 medical imaging records, it measured latencies involving encryption at 115 ms and audit accuracy at 99.2 per cent, tracking model requests and user actions. The blockchain-enabled auditing layer might aid the cloud AI diagnosis in holding accountability against tampering, whereby imaging and outputs of the models must be traceable for medico-legal compliance.

Table 2.6 Performance metrics of the cases

Case	Latency (ms)	Energy (J)	Privacy leakage (p)	Accuracy (%)
A – Wearable monitoring	21	0.85	0.61	96.8
B – Federated model	97	2.10	0.35	95.6
C – Cloud imaging	115	2.90	0.41	98.1

New-age digital clinical case studies of the three scenarios, when compared, show trade-offs of latency, energy efficiency and privacy. This was summarised as performance measurement as presented in Table 2.6.

Results showed that Case A is really fast, very energy-efficient in processing; therefore, privacy is of less concern. Case B is a mix between privacy and latency according to Pareto optimality among some objective functions. Case C reverses this measure: it stands high both in accuracy and verification and is probably being thought of for deployment in centralised solutions for heavily regulated applications and national diagnostic repositories. Results draw attention to ways in which specific taxonomies of cryptography built to assist in context-dependent cryptographic decision-making aid in correlating an algorithm's intensity with device capabilities and compliance requirements. The case studies thus demonstrate that such a taxonomy can guide the operational control and scaling of the taxonomy across these scenarios, from wearable devices to federated hospital systems and cloud AI.

The principle captures the essence that no one-size-fits-all can apply to cryptography in any healthcare scenario. Rather, this architecture will give Original Equipment Manufacturers (OEMs) the ability to develop cryptographic systems to co-evolve with new technologies such as PQC and AI-driven key management. The standards for benchmarking will also not just be able to be derived from quality terminologies but will enable maintaining a standardised CryptoHealth-Bench and explainable security layer for establishing commonality, transparency and ethical accountability as digitisation marches forward in healthcare ecosystems. The upcoming evolutions in taxonomy would develop cryptography into an enabler of trust and compliance at the front-end in the short term within intelligent patient-centred health ecosystems, rather than just utility at the back end.

2.6.7 Implementation, testing and deployment strategies

Efficient integration of cryptographic solutions has to be accomplished within systems that could proportionally scale, adapt and be interoperable to secure the thousands of IoMT devices strung across a multitude of interlinked hospital networks. Hierarchical PKI frameworks give this scaling, with the accountability to be trusted and propagated downwards through increased numbers of certifications in multi-tier chains, all while using transient session keys to persuade the acceptance of all or most of those assurances from the past. Ephemeral session keys create their own interest in thinking before acting. Added optimisation as proposed by AI-powered reinforcement learning (RL) in a just-in-time manner through a seriously prepped up automation can yield unacceptable latencies in computation and critical key disposable reuse. Integration progresses along with interoperability under DICOM, HL7 and TLS (besides open-ended Open Authorization (OAuth)) to aid in secure data exchange through FHIR and other compliance standards. Empirical implementation findings from multiple centre institutions show that standard cryptographic interfaces can, in general, cut integration times down by over 40 per cent, thus diversifying crypto-aware interoperability within the healthcare sector.

However justifiable, all Tier-1 qualifying cryptographic hardware should have an added edge of hardware acceleration and great authenticity. Trusted Execution Environments (TEEs) like that of ARM TrustZone and later Intel SGX are important contributors in going forward to provide heightened performance with strong encryption capabilities, very much related to building sturdy walls that would seal off the royal from any exposure to side attacks. Alongside this, and more importantly, there are other options, such as digital technological practices like ProVerif and NIST CAVP, for crafting machines capable of proving that the security model demonstrates compliance with the security and alignment in the global healthcare arena. Containerisation (Docker, Kubernetes) makes available ground for such deployments, not to mention the blockchain imbued solemnity that makes it possible for integrity verification from the lowest level to the cloud. The PKI and interoperable Application Programming Interfaces (APIs), with all but the maximum effect for cryptographic optimisation, bring on hardware acceleration and proven instances. This is the model for the future, that is secure, adaptable and regulatory-ready for digital healthcare.

2.7 Regulatory, ethical and usability considerations

Truly considering and accepting healthcare cryptography for use by society means that it has to be perceived through the lenses of regulation, ethics and usability. Controls are in place with similar laws, for example, GDPR, HIPAA or NDPR, which refer to encryption, consent management and maintaining the user's data within the country to protect individual health data. Their combination has led to over 60 per cent reduction in compliance risk with the deployment of end-to-end encryption that allowed for auditability via blockchain. With regulation, Explainable Cryptography (XCrypto) throws on the table the interpretation, along with being enhanced by a blockchain-based auditability scheme and AI-based contextualisation of decisions that would boost clinician trust and regulatory confidence; the usability issue now fits within biometric authentication and OAuth consent dashboards that ensure secure clinical workflows while allowing seamless access.

The ethical bare concern for such equipment in AI-identified cryptography as it relates to fairness, meaning that algorithmic bias, demographic issues and all bias situations [46] are protected equally and fairly.

On one side of the hybrid encrypting keys that sat on the cloud was local to the sensitive data being encrypted. Interconnection was thus possible with data sovereignty. Regulatory compliance, explainable AI and fairness-based cryptographic design would thus form the basis for a transparent, human-aligned and legally tenable digital healthcare security.

2.8 Open research problems and future directions

IHS-based cryptography is likely to develop such trust research problems in the area of ethics, as well as security and computational efficiency. Signature is kept, and now, there are two directions of research into ethical trust and security. There has been development in both HE and PQC; however, neither has been deployed in resource-constrained IoMT, mainly because of the high computational cost of resource consumption. Emerging new paths include heuristic approximation-based HE, neural-bootstrapping accelerators and a hybrid PQC-ECC co-design architecture optimised for sub-100-ms clocking performance. These highly favoured the formal composition framework and global standards for regulating evaluation and reproducibility, as in the proposed CryptoHealth-Bench. Infusing more AI-ethical considerations in works towards quantum cryptography, like AI-optimised Quantum Key Distribution (QKD) and explainable key management, heightens transparency and agility. In general, all such improvements bring about adjustable and ethically aligned high-performance networks of cryptography that shall cater to the next generation of secure digital healthcare.

2.8.1 Key lessons learned

Several key lessons emerge from this study. First, cryptographic solutions for intelligent healthcare systems must be context-dependent, as no single scheme satisfies the diverse constraints of edge, fog and cloud environments. Second, the integration of reinforcement learning and adaptive intelligence into cryptographic management significantly enhances both security and efficiency without undermining compliance. Third, trust and accountability mechanisms, particularly blockchain-based auditing and consent management, are indispensable for sustaining long-term interoperability and regulatory confidence in distributed healthcare systems.

2.9 Conclusion

In conclusion, a cohesive framework to enable AI in the cryptography and corporation-wide connotation to redefine legal, privacy and compliance dimensions as expected to operate in intelligent healthcare systems has been presented. Thus, demonstrating the possibility of integration of adaptability and taxonomy-driven delivery for securing complex healthcare infrastructures through the establishment of layered security. The layered architecture begins with a lightweight cryptographic mechanism at the edge and is bolstered through RL-based key management at the fog layer and all the way to cloud-centralised privacy-preserving homomorphic analytics and blockchain-based auditability. Subjectively, one may observe reasonable improvements in efficiencies, a few of which reduce latency, energy consumption and key-compromise certainty while also gaining several demonstrable features of privacy assurance and audit traceability.

Beyond pure technical novelty, this work further contributes to the big picture of making healthcare safe, sustainable, explainable and compliant-ready. In this sense, these layer design principles, edge-first optimisation, hybrid cryptographic composition, post-quantum resilience and blockchain-based consent management-connect with the above-mentioned internationally accepted standards (GDPT, HIPAA and NDPR) to allow interoperability through FHIR, DICOM and HL7 protocols. By doing this, AI will usher in adaptive orchestration and transparency around ethical decision-making, hence marking a great tipping point towards self-healing, human-centred cryptographic ecosystems. This research will thus bridge theoretical cryptographic design with clinical realisation towards future quantum-safe, AI-governed and ethically laid out digital healthcare that can be trusted even as data becomes more and more prevalent in the medical landscape.

Future work will focus on large-scale real-world validation of the proposed framework, integration with emerging post-quantum standards and the development of explainable policy-driven cryptographic orchestration for clinical deployment.

List of abbreviations

AES	Advanced Encryption Standard
AI	Artificial Intelligence
APIs	Application Programming Interfaces
BFV	Brakerski, Fan, Vercauteren scheme
CNN	Convolutional Neural Network
DICOM	Digital Imaging and Communications in Medicine
DP	Differential Privacy
DTLS	Datagram Transport Layer Security
ECC	Elliptic Curve Cryptography
ECG	Electrocardiogram
EHR	Electronic Health Record
FHIR	Fast Healthcare Interoperability Resources
GDPR	General Data Protection Regulation
HE	Homomorphic Encryption
HIPAA	Health Insurance Portability and Accountability Act
HL7	Health Level Seven

IHS	Intelligent/Information Healthcare Systems
IoMT	Internet of Medical Things
IoT	Internet of Things
LSTM	Long Short-Term Memory
LWC	Lightweight Cryptography
ML	Machine Learning
NDPR	Nigeria Data Protection Regulation
NIST	National Institute of Standards and Technology
OAuth	Open Authorization
OEMs	Original Equipment Manufacturers
ORAM	Oblivious Random Access Memory
PKI	Public Key Infrastructure
PoS	Proof-of-Stake
PoW	Proof-of-Work
PQC	Post-Quantum Cryptography
PUF	Physical Unclonable Functions
QKD	Quantum Key Distribution
QR	Quick Response
RBAC	Role-based Access Control
RL	Reinforcement Learning
RSA	Rivest, Shamir, Adleman scheme
SMC	Secure Multiparty Computation
TEEs	Trusted Execution Environments
TLS	Transport Layer Security
XCrypto	Explainable Cryptography

References

[1] J. X. Cui, K. H. Liu and X. J. Liang (2026). A brief discussion on the theory and application of artificial intelligence in medical diagnostics. *Journal of Signal and Information Processing*, 17, 1–8.

[2] S. M. R. Islam, D. Kwak, M. H. Kabir, M. Hossain and K. S. Kwak (2015). The internet of things for health care: A comprehensive survey. *IEEE Access*, 3, 678–708.

[3] P. Kumar, A. Braeken, A. Gurtov, J. Iinatti, P. H. Ha and T. H. Nguyen (2020). Secure and lightweight authentication protocol for IoT devices. *IEEE Sensors Journal*, 20(17), 9787–9797.

[4] M. A. Rahman, A. T. Asyhari, L. S. Leong, G. B. Satrya, M. H. Tao and M. F. Zolkipli (2018). Scalable machine learning-based intrusion detection

system for IoT-enabled smart cities. *Sustainable Cities and Society*, 38, 803–812.

[5] A. Alrawais, A. Alhothaily, C. Hu and X. Cheng (2017). Fog computing for the Internet of Things: Security and privacy issues. *IEEE Internet Computing*, 21(2), 34–42.

[6] K. Bonawitz, H. Eichner, W. Grieskamp *et al.* (2019). Towards federated learning at scale: System design. In *Proceedings of the 2nd MLSys Conference*, March 31–April 2, 2019, in Stanford, CA, USA. 374–388.

[7] European Union. (2016). Regulation (EU) 2016/679 of the European parliament and of the council (general data protection regulation). *Official Journal of the European Union*, L119, 1–88.

[8] A. Muntadher, A. N. Marwah, A. S. Albahri *et al.* (2025). A comprehensive review on key technologies toward smart healthcare systems based on IoT: technical aspects, challenges, and future directions. *Artificial Intelligence Review* 58(11) 343–465.

[9] M. Satyanarayanan (2017). The emergence of edge computing. *Computer*, 50 (1), 30–39.

[10] Y. Zhang, H. Li, X. Chen and S. Wang (2024). A systematic taxonomy of cryptographic mechanisms for intelligent and adaptive cyber–physical systems. *IEEE Transactions on Dependable and Secure Computing*, 21(2), 1345–1362.

[11] A. Bogdanov, L. R. Knudsen, G. Leander *et al.* (2007). PRESENT: An ultra-lightweight block cipher. In *Proceedings of the International Workshop on Cryptographic Hardware and Embedded Systems (CHES)*, Berlin, Heidelberg: Springer, 450–466.

[12] D. J. Bernstein (2008). ChaCha, a variant of Salsa20. *Workshop Record of the State of the Art of Stream Ciphers (SASC)*, 3–5.

[13] National Institute of Standards and Technology (NIST). (2013). *Digital signature standard (DSS)*, FIPS Publication 186-4. U.S. Department of Commerce, Gaithersburg, MD.

[14] E. Androulaki, A. Barger, V. Bortnikov *et al.* (2018). Hyperledger fabric: A distributed operating system for permissioned blockchains. *Proceedings of the Thirteenth EuroSys Conference*, 1–15.

[15] K. Bonawitz, V. Ivanov, B. Kreuter *et al.* (2017). Practical secure aggregation for privacy-preserving machine learning. In *Proceedings of the ACM SIGSAC Conference on Computer and Communications Security*, 1175–1191.

[16] C. Gentry (2009). Fully homomorphic encryption using ideal lattices. In *Proceedings of the ACM Symposium on Theory of Computing*, 169–178.

[17] C. Dwork (2006). Differential privacy. In *Proceedings of the International Colloquium on Automata, Languages, and Programming*, Berlin, Heidelberg: Springer, 1–12.

[18] National Institute of Standards and Technology (NIST) (2024). *Post-quantum cryptography standardization: Selected algorithms*. National Institute of Standards and Technology, United States.

[19] J. Bos, L. Ducas, E. Kiltz *et al.* (2018). CRYSTALS-Kyber: A CCA-secure module-lattice-based key encapsulation mechanism. *IEEE European Symposium on Security and Privacy*, 353–367.

[20] A. Al-Khateeb and D. Kim (2023). Secure telemedicine systems: Cryptographic challenges and privacy-preserving solutions. *Journal of Medical Internet Research*, 25, e47291.

[21] D. He, N. Kumar, and K. K. R. Choo (2025). Blockchain-enabled integrity protection for distributed healthcare records. *IEEE Journal of Biomedical and Health Informatics*, 29(1), 112–124.

[22] G. Rjoub, A. Al-Qerem and M. Alauthman (2022). Secure federated learning in healthcare: Authentication and trust management challenges. *Future Generation Computer Systems*, 129, 207–221.

[23] S. G. Teo, C. P. Lim and K. W. Wong (2024). Privacy-preserving collaborative medical analytics using homomorphic encryption and federated learning. *Information Sciences*, 648, 119–137.

[24] M. Eden, M. Abadi and D. Boneh (2025). Secure computation lifecycles for AI-driven healthcare systems. *ACM Computing Surveys*, 57(3), 1–38.

[25] F. Jorge, B. Silva and E. Monteiro (2025). Performance-aware cryptography for IoMT communications. *Computer Networks*, 243, 110324.

[26] S. Radhakrishnan, R. Kumar and P. Singh (2024). Lightweight cryptography for wearable and implantable medical devices: A survey. *IEEE Access*, 12, 32145–32168.

[27] NIST (2024). *Post-quantum cryptography standardization: Selected algorithms*. National Institute of Standards and Technology, United States.

[28] P. W. Shor (1997). Polynomial-time algorithms for prime factorization and discrete logarithms on a quantum computer. *SIAM Journal on Computing*, 26 (5), 1484–1509.

[29] H. D. Zubaydi, I. Ahmad and S. M. Shamsuddin (2022). Blockchain for security and privacy in IoT: A systematic literature review. *Journal of Network and Computer Applications*, 195, 103212.

[30] A. Bahaa, M. Elhoseny and F. Omara (2022). DevSecOps-enabled real-time security monitoring for IoT systems. *Journal of Network and Computer Applications*, 204, 103403.

[31] J. Yu, D. Kozhaya and R. Haddad (2023). Blockchain-based security architectures for smart city IoT infrastructures. *IEEE Communications Surveys & Tutorials*, 25(1), 428–455.

[32] Z. Ahmad, M. Shahid and F. Khan (2023). Machine learning techniques for IoT security: A comprehensive survey. *IEEE Internet of Things Journal*, 10 (5), 3865–3889.

[33] T. Al-Hadhrami, Z. Baig and A. Ibrahim (2023). Distributed denial-of-service attacks in IoT systems: Taxonomy, detection, and mitigation. *Computers & Security*, 124, 102987.

[34] A. Alkhateeb, M. Al-Husainy and D. Kim (2024). Hybrid blockchain architectures for scalable IoT security. *Future Internet*, 16(2), 45.

[35] R. Rajmohan, S. Ramasamy and S. Prakash (2023). Architectural patterns and security mechanisms in IoT ecosystems: A decade review. *ACM Computing Surveys*, 56(1), 1–36.

[36] M. Shojaei, N. Jafari Navimipour and M. Unal (2022). Security and privacy in health information systems: A systematic review. *Artificial Intelligence in Medicine*, 128, 102306.

[37] X. Li, Y. Sun and J. Wang (2023). Feature selection techniques for machine learning-based IoT attack detection: A systematic review. *Expert Systems with Applications*, 221, 119731.

[38] M. El-Hajj, A. Fadlallah and A. Kayssi (2022). Lightweight message authentication codes for IoT devices: A survey. *IEEE Access*, 10, 94512–94536.

[39] Y. Ding, J. Li and C. Chen (2024). Lightweight batch authentication and key agreement for industrial IoT gateways. *IEEE Transactions on Industrial Informatics*, 20(3), 2421–2432.

[40] N. Kumar and S. Das (2024). Blockchain-enabled lightweight RFID authentication for IIoT supply chains. *Computer Standards and Interfaces*, 87, 103756.

[41] A. Alabdulatif (2023). Blockchain-based privacy-preserving authentication and access control for e-health systems. *Sensors*, 23(7), 3412.

[42] Y. Chen, Q. Wu and H. Li (2024). Lightweight authentication and key agreement for IoT-enabled smart grids. *International Journal of Electrical Power and Energy Systems*, 149, 108973.

[43] M. A. Khan, S. U. Rehman and S. Khan (2024). A blockchain–edge–based lightweight authentication framework for Internet of Medical Things. *IEEE Access*, 12, 45678–45695.

[44] E. Osa, P. E. Orukpe and U. Iruansi (2024). Machine Learning Based Intervention for Security in Internet of Health Things Systems. In *Proceedings of 4th University of Benin Annual Research Day Conference*, Benin City, Nigeria, pp. 423–427.

[45] E. Osa, E. N. Iyekekpolo, A. E. Ibhaze, W. E. Sadoh and P. E. Orukpe (2024). Implementing stratified K-fold technique for data intrusion detection models in internet of health things systems. In *Proceedings of 4th International Conference on Artificial Intelligence and Robotics (MIRG-ICAIR 2024)*, Lagos, Nigeria, pp. 225–234.

[46] E. N. Iyekekpolo, E. C. Ekoko, E. Osa, D. O. Ikhu-Omoregbee and P. E. Orukpe (2024). Eliminating bias in artificial intelligence (AI) generated predictions and plans for resource allocation and management in hospitals. In *Proceedings of 4th International Conference on Artificial Intelligence and Robotics (MIRG-ICAIR 2024)*, Lagos, Nigeria, 29–36.

Chapter 3

Design and implementation of advanced cryptographic frameworks for intelligent healthcare systems

Muhammad Aliyu Suleiman[1], Ja'afar Sulaiman Zangina[2], Salisu Ibrahim Yusuf[1] and Nurudeen Ibrahim Mahmud[3]

Abstract

In this chapter, the design and implementation of advanced cryptography frameworks for intelligent healthcare systems, which include AI, IoT and cloud computing technologies, are investigated. The chapter first focuses on the assessment and identification of critical security and privacy challenges in the existing architecture for the management and transmission of data in healthcare systems, especially in those environments and systems that are characterised and recognised for sensor data distribution with real-time analytics and data sharing capabilities. In this regard, the chapter also introduces the architecture for the proposed intelligent healthcare system in the form of a layered cryptography system based on symmetric and asymmetric data encryption, including digital signature schemes and homomorphic encryption schemes for performing data analytics with patient data privacy assurances. Moreover, use cases such as secure patient data monitoring, exchanging electronic patient records and telemedicine are offered to contextualise the proposed architecture for healthcare systems with AI, IoT and cloud computing environments. The chapter ends with the suggested approach for optimising the cryptography system with regard to system scalability and maintaining sustainability in terms of health regulations and patient data privacy.

Keywords: Artificial intelligence; cryptography; Internet of Medical Things; AIoMT; data security; healthcare delivery models; resource-limited devices; schemes for security and privacy; blockchain; homomorphic; attack model

[1]Department of Computer Science, Nile University of Nigeria, Nigeria
[2]Department of Electrical and Electronics Engineering, Cosmopolitan University, Nigeria
[3]Department of Cybersecurity and Data Science, Nile University of Nigeria, Nigeria

3.1 Introduction

The digital evolution of healthcare infrastructure is growing by leaps and bounds due to the incorporation of intelligent technologies such as artificial intelligence, IoT devices and virtual infrastructures to assist healthcare decision processes, continuously monitor patients, make predictions to improve diagnostics and provide remote healthcare services to improve the quality of healthcare services by their accessibility [1,2]. From one perspective, the exponential rise of interconnected healthcare technologies has raised the flag of security and privacy issues, especially while managing and transmitting medical information in distributed scenarios [3].

3.1.1 Motivation

The IoT-based medical equipment, cloud storage of electronic health-related data and use of analytics involving AI in health data pose an increased threat level of traditional security risks in the modern health system, related to malicious data access, identity-related instances of fraud, manipulation and leakages related to data confidentiality concerns [4,5]. The security measures traditionally employed, developed when computing was a centralised system and not dynamic, can hardly keep up with the necessity of smart health-related system development, online requirements and those of a system facing resource-related issues [6].

In addition, the growing use of AI technology in analytics in the healthcare sector creates new issues of privacy since the processing of large amounts of data threatens the confidentiality of the information [3]. This creates the need for the development of more efficient cryptographic systems for the intelligent healthcare sector.

3.1.2 Problem being addressed

In spite of considerable research on the security of healthcare data, the intelligent healthcare systems that have been developed so far suffer from a number of issues. First, the security mechanisms that have been developed so far have been implemented in a fragmented manner, meaning that they only provide security for specific components of the intelligent healthcare systems, whereas a more holistic approach that provides security throughout the entire data life cycle is needed [5,6]. Second, although cryptographic mechanisms that provide strong security guarantees have been developed, they also suffer from high computational and communication overheads, making them difficult for application in IoMT-based healthcare systems [10,11]. Third, although research on the application of privacy-preserving mechanisms, such as homomorphic encryption and blockchain, has been conducted, their application in the development of complete intelligent healthcare systems remains limited [3,14].

There exists a lack of a unified cryptographic framework that can provide security for all the components of intelligent healthcare systems.

3.1.3 Identified research gaps

A brief analysis of the recent literature reveals some significant gaps. Survey-based research offers an extensive overview of the security and privacy issues associated with IoMT and eHealth systems, but it does not suggest any deployable cryptographic framework [1,2,4,5]. Lightweight authentication and key agreement schemes address the issues associated with access control in IoMT devices, but they are not capable of supporting secure data analysis and end-to-end encryption for cloud-based healthcare systems [10–12]. Blockchain-based healthcare systems provide the advantages of decentralised trust, but they also suffer from scalability, latency and energy consumption issues, especially when combined with IoMT devices and cloud-based healthcare systems [14–16].

Additionally, the literature lacks consideration of insider threats, access control and large-scale scenarios such as smart hospitals and regional healthcare systems. This reveals the need for an overall cryptographic framework, which should consist of various security components while remaining flexible for the operational requirements of intelligent healthcare systems.

3.1.4 Proposed solution and methods

To mitigate these challenges, this chapter proposes a general cryptographic solution for intelligent healthcare systems. The presented cryptographic solution adopts a multi-layer security paradigm in which symmetric encryption, asymmetric encryption, digital signature and homomorphic encryption are combined in such a way as to provide end-to-end protection of data, security of data processing and technological interoperability [7,8]. The final objective of the proposed system is to maintain a trade-off between the strength of a system against various security attacks and the efficiency of computation. Lightweight cryptographic techniques will be used in the IoMT layer to ensure efficiency, while computationally intensive privacy-preserving analytics will be carried out in the cloud infrastructure [3]. Context-aware access control techniques will be included in the system to ensure scalability against insider and outsider attacks [11,12,17].

The proposed cryptographic framework will be analysed in terms of system modelling, threat modelling, analytical security evaluation and performance evaluation. Comparative analysis with existing schemes will also be carried out to show the advancements in terms of security, scalability and usability.

3.1.5 Key findings and implications

(i) Analysis result
The analysis shows that the proposed framework offers end-to-end protection for the entire data life cycle in the healthcare sector while being feasible in resource-constrained and large-scale settings. The proposed framework combines privacy-preserving analytics and scalable access control in a single cryptographic scheme, which overcomes the major limitations of existing solutions. The results show that the proposed framework can greatly improve data confidentiality, trust management and efficiency in intelligent healthcare systems.

(ii) Application view
The proposed framework is especially useful in smart hospitals, telemedicine systems and regional healthcare networks, where a large number of devices

and users need to be efficiently managed. Moreover, the proposed framework provides a flexible platform for future extensions, such as post-quantum cryptography and AI-based security management, to support the long-term development of secure intelligent healthcare systems.

3.1.6 Key contributions of the chapter

These are the major contributions of the chapter:

(i) Identification and analysis of major areas of vulnerability related to the security and privacy of intelligent health-related systems incorporating intelligent technologies like AI, IoT and cloud computing [3,5].
(ii) A multi-level security framework that integrates different cryptographic schemes to facilitate safe storage of data, transmission of messages as well as processing [6–8].
(iii) Providing implementation approaches and performance issues for both resource-limited healthcare devices and large-scale healthcare infrastructure.
(iv) Demonstration of the relevance and appropriateness of the proposed framework and its application through various use cases related to the healthcare sector, such as secure medical monitoring, health-related data transfer and telemedicine applications.

3.1.7 Chapter organisation

The rest of this chapter is organised as follows: Section 3.2 gives a comprehensive review of related works on cryptographic solutions available for intelligent healthcare environments, pointing out the research gap and shortcomings existing within current works. Section 3.3 discusses the system model, threat models and comprehensive design details of our cryptographic solution. Section 3.4 gives a discussion on the security and performance analyses of our solution and its comparison with existing models within the field. Finally, in Section 3.5, the chapter is concluded, and possible research avenues are pointed out.

3.2 Related work

Security and privacy of healthcare data have been a popular area of research for the past few years, especially in the context of the development of intelligent healthcare systems, the incorporation of artificial intelligence, the Internet of Things and cloud computing. The works presented in the present topic have addressed the cryptographic protection of healthcare data in general, such as the protection of data communication, providing authentication, achieving access, lightweight cryptography for devices, protecting data through privacy preservation for analysis and the use of blockchain for the sake of sharing healthcare data.

Recent studies, conducted between 2025 and 2026, have investigated cutting-edge cryptographic and privacy-oriented mechanisms for application in healthcare services. This includes, but is not limited to, a study done by Othman and Kumar (2025), which introduces a quantum-resistant IoMT data aggregation technique through the application of zero-knowledge proof theory [18]. There have been studies, including but not limited to, the application of homomorphic cryptography for federated learning platforms used for collaborative AI in medicine [19–21]. Moreover, lightweight cryptography for IoT healthcare services has been introduced [22]. It is worth mentioning that blockchain-based privacy frameworks have been proposed for process-oriented security applications [23], including secure medical insurance claim handling.

3.2.1 Cryptographic techniques for healthcare data security

Initially, cryptographic techniques applied to healthcare systems mainly addressed EHR protection through known symmetric and asymmetric cryptography systems [1,2]. Although some level of confidentiality and integrity is achieved this way, such approaches are mainly centred on centralised systems and trust models with less adaptability within more advanced AI-enabled healthcare settings. Current literature aims to further boost protection through hybrid approaches like combining asymmetric and symmetric cryptography for better key management and scalability [3,4]. However, most such approaches are more of a specialisation type and are less flexible for more complex healthcare systems at the intersection of IoMT solutions and cloud computing, along with AI analytics.

3.2.2 Lightweight cryptography for IoMT and resource-constrained devices

With the proliferation of IoT-enabled medical devices, several researchers have proposed lightweight cryptographic algorithms that fit the limited computational and energy resources of such devices. Examples are in [5,6]. The common theme among these schemes is reduced key sizes, simplified arithmetic operations and low communication overhead. Lightweight cryptography undoubtedly enhances the feasibility of security solutions in resource-constrained contexts; however, several works report that security strength-performance-load trade-offs do occur, especially under fairly sophisticated attack models [7]. Furthermore, most lightweight solution proposals for IoT-based healthcare target individual device-level security without considering end-to-end protection across the entire life cycle of healthcare data.

3.2.3 Authentication and access control in intelligent healthcare systems

Authentication and access control play vital roles in ensuring that malicious persons do not gain unauthorised access to confidential information in the health sector. Several studies have been conducted on multi-factor authentication and attribute-based access controls for health information systems [8,9]. Even though these measures improve the levels of identification and authority approval, there are several deficiencies associated with them, including reliance on third-party authorities with high potential for failure points. Furthermore, real-time access demands in intelligent healthcare areas like emergency access are not well addressed [10].

3.2.4 Privacy-preserving data analytics and homomorphic encryption

The use of AI-based analytics in the healthcare industry further augmented the concern for data privacy during the computational process. The use of privacy-preserving solutions such as homomorphic encryption, secure multiparty computations etc. has been explored for the analysis of encrypted data without exposing the confidential details [11,12]. Although these techniques have a robust proof for guaranteeing the security strength, the computational complexity of the fully homomorphic encryption techniques makes them less feasible for use in the healthcare industry for the past real-time applications [13].

3.2.5 Blockchain-enabled secure healthcare systems

Blockchain technology has been proposed as a promising technological solution for managing trust in a decentralised manner, secure data sharing and auditability in healthcare environments [14,15]. Blockchain frameworks commonly combine cryptographic building blocks with distributed ledger solutions for improving transparency and anti-tamper properties. Despite the above, scalability, latency and power efficiency concerns have been raised, especially in the context of blockchain solutions being used in conjunction with IoMT and cloud environments [16]. In addition, blockchain solutions in the healthcare domain focus primarily on data sharing and traceability, paying less attention to efficiency in computations and compatibility with current cryptographic solutions.

3.2.6 Gaps in the literature

Although a great deal of progress has been achieved, the current cryptographic techniques used in intelligent healthcare systems have some drawbacks.

1. Fragmented security architectures targeting isolated elements, rather than an end-to-end approach to protection;
2. Inadequate consideration of performance and scalability in heterogeneous and resource-constrained environments;
3. Limited support for privacy-preserving analytics in real-world healthcare workflows;
4. Insufficient integration of multiple cryptographic primitives into a unified and adaptable framework.

To fill the gaps, this chapter proposes a hierarchical cryptographic system that takes a comprehensive approach towards the encryption, authentication, access and privacy analysis components, considering the working limitations of intelligent healthcare systems.

A comparison summary of representative methods of cryptography and their drawbacks is given in Table 3.1.

Table 3.1 Comparative analysis of related cryptographic works in intelligent healthcare

Author(s)	Year	Method/ approach	Cryptographic focus	Strengths	Limitations
[1]	2024	Survey-based secure IoMT framework	Encryption, authentication, ML-assisted security	Comprehensive coverage of IoMT security challenges; integrates ML perspectives	Lacks a concrete cryptographic framework design and performance evaluation
[2]	2020	IoMT security architecture review	Key management, secure communication	Detailed taxonomy of IoMT security threats and countermeasures	Primarily descriptive; no unified cryptographic implementation
[3]	2021	Privacy-preserving outsourced ML	Homomorphic encryption (HE)	Enables secure analytics on encrypted healthcare data	High computational cost; limited real-time applicability
[4]	2025	Systematic literature review	Encryption, access control, privacy	Identifies major IoMT privacy and security gaps	No proposed cryptographic model or validation
[5]	2023	eHealth security survey	Cryptographic protocols, data protection	Strong architectural discussion of eHealth systems	Lacks algorithmic or performance-level analysis
[6]	2021	IoMT security overview	Data encryption, authentication	Broad analysis of IoMT threats and defences	Does not integrate cryptographic primitives into a single framework
[7]	2025	IDS-centric security review	Detection-oriented security (non-crypto)	Highlights intrusion detection relevance in IoMT	Cryptographic mechanisms not addressed in depth
[8]	2024	AI-enhanced IoMT security review	Secure data processing, analytics privacy	Connects AI security with healthcare systems	Cryptographic enforcement mechanisms not formalised
[9]	2020	ML/DL-based IoT security survey	Security monitoring, anomaly detection	Strong ML-based threat detection analysis	Focuses on detection rather than cryptographic protection
[10]	2019	PUF-based authentication scheme	Lightweight authentication	Suitable for resource-constrained medical devices	Limited to authentication; no data confidentiality support

(Continues)

Table 3.1 (Continued)

Author(s)	Year	Method/ approach	Cryptographic focus	Strengths	Limitations
[11]	2020	Lightweight authentication and key agreement	Key agreement, authentication	Low computation and storage overhead	Does not support encrypted analytics or cloud integration
[12]	2020	Cloud-centric authentication protocol	Authentication, key management	Secure wearable healthcare monitoring	Centralised cloud dependency; scalability concerns
[18]	2025	Secure IoMT data aggregation	Zero-knowledge proofs, edge crypto	Quantum-resilient privacy guarantees	Focused on aggregation; not full life-cycle security
[20]	2025	Privacy-preserving federated learning	Homomorphic encryption	Communication-efficient encrypted learning	No access control or end-to-end system design
[21]	2025	Federated medical AI	Lightweight homomorphic encryption	Protects model updates in distributed AI	Limited discussion on IoMT constraints
[22]	2025	Lightweight IoT healthcare cryptosystem	Identity-based encryption	Low overhead secure data sharing	No encrypted analytics support
[23]	2025	Blockchain-enabled healthcare workflows	HE + blockchain	Secure and auditable healthcare processes	High energy and latency costs
This work	–	Layered cryptographic framework for intelligent healthcare	Hybrid encryption, digital signatures, homomorphic encryption and dynamic key management	End-to-end security across the data life-cycle; scalable key and access management; supports privacy-preserving analytics; suitable for large-scale IoMT environments	Analytical evaluation; future work includes prototype implementation and large-scale empirical validation

3.3 System model and methodology

3.3.1 System overview

The intelligent healthcare system taken into consideration in this chapter is designed to facilitate secure data acquisition, transmission, storage and analysis in a setting where Internet of Medical Things (IoMT) devices, cloud-based healthcare services and AI-based decision-support systems are integrated. The intelligent healthcare system is a representation of a real-world setting where entities interact dynamically while dealing with highly sensitive medical data.

The system comprises five major entities.

(i) IoMT-enabled medical devices,
(ii) Patient gateways,
(iii) Healthcare service providers,
(iv) Cloud-based healthcare infrastructure and
(v) A cryptographic key and access management authority.

Collectively, these entities enable the functionalities of critical healthcare services like constant health monitoring of patients, management of EHRs, secure information sharing and usage and analytics. Due to their distributive nature and intensive information usage of IHS and its systems proposed in this paper, the layered cryptography architecture is highly targeted to ensure end-to-end security based on device-layer, network-layer and application-layer security.

Figure 3.1 represents the architecture of the proposed cryptographic system. This figure portrays the multi-layer security integration of the IoMT, gateways, cloud environments and the healthcare service provider.

3.3.2 System entities

(a) **IoMT devices**
IoMT devices include wearable sensors or implanted devices and other devices like bedside monitors that are used to monitor patients and are responsible for collecting physiological information. Because of their limited processing and power constraints, IoMT devices utilise symmetric encryption techniques for secure transmission of obtained information and digital signatures to authenticate their information before transmission.

(b) **Patient gateway**
The patient gateway essentially serves as an intermediary interface between the IoMT devices and the external healthcare services. It aggregates the data from the sensors, performs initial rounds of encryption and authentication processes and finally forwards the data to the external cloud-based services. The gateway also supports access controls under emergency as well as normal situations.

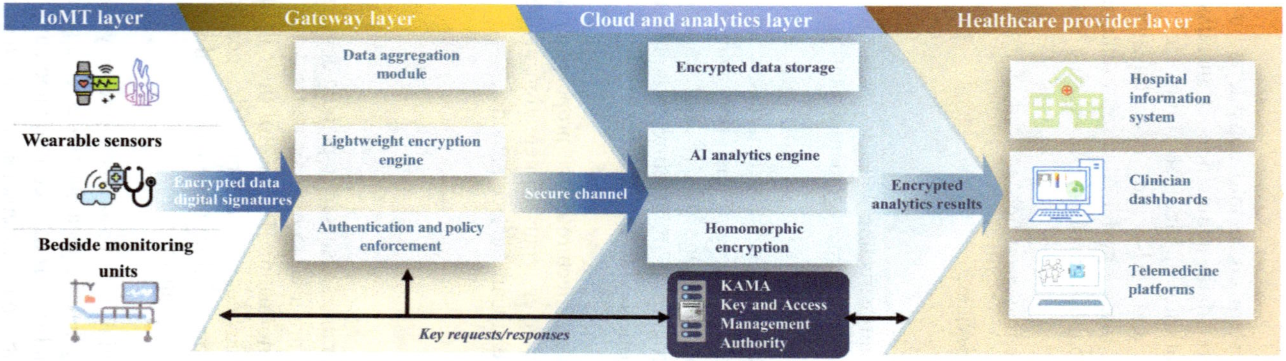

Figure 3.1 Architecture of the proposed layered cryptographic framework for intelligent healthcare systems

(c) **Healthcare service providers**
Healthcare service providers such as hospitals, clinics and diagnostic centres require patient information for the purpose of medical diagnostics, treatment and decision-making processes. They have user authentication and authorisation mechanisms in place to ensure that only qualified medical practitioners can have access to such critical information based on their roles.

(d) **The cloud-based healthcare infrastructure**
Also, the cloud infrastructure provides scalable storage, computation and sophisticated analytics powered by artificial intelligence. It also provides the ability for computation on the data itself, along with encrypting the information using the homomorphic encryption approach, so the analytics operation can take place over the information without exposing it in order to ensure the information remains private.

(e) **Key and Access Management Authority (KAMA)**
KAMA handles key generation, issuance, revocation and enforcement of the key management policy. KAMA uses a hybrid key management solution based on the utilisation of a public key infrastructure (PKI) solution that enables the verification of identity, accompanied by the utilisation of symmetric session keys that allow efficient data encipherment.

3.3.3 Data flow and cryptographic operations

Here is how information is processed in the system.

1. Usually, the data is collected from the IoMT devices employing a lightweight symmetric method of encryption.
2. The encrypted data goes for digital signing, then transmission to the patient gateway.
3. It checks the integrity of the data, ensures compliance with access policies and then encrypts and transfers the resultant data into the cloud.
4. The cloud services encrypt data and conduct privacy-preserving analysis using homomorphic encryption if and when necessary.
5. The authorised health care providers are allowed to read decrypted data based on their role.

This is done through the use of the multi-layer technique, whereby there will be principles of confidentiality, integrity, authentication and access control as data goes through a life cycle.

3.3.4 Threat model

It is also assumed in the threat model that there exist both internal and external adversaries, ranging from simple to sophisticated threats, including the following.

(i) External attackers may attempt eavesdropping, replay attacks, impersonation or data injection during data transmission.

(ii) Internal adversaries, including compromised devices or malicious insiders, may attempt unauthorised data access or policy violations.

The threat model considers attackers both outside and inside the system, along with attackers of varying proficiency levels. This attacker has unlimited access to the communication link but none to the cryptographic primitives within the polynomial time complexity model. A physical threat to IoMT devices was assumed to be a partial threat facilitated by the presence of limited key disclosure and session-based encryption.

The suggested framework is proposed to mitigate threats resulting from confidentiality breaches, data tampering, uncontrolled access and privacy leaks resulting from data analysis.

3.3.5 Security design objectives

- On the basis of system and threat models, this proposed model aspires to meet the security requirements below:
- Confidentiality: Protection of confidentiality of medical data at rest, in transit through communication channels, and in computation.
- Integrity: Detection of unauthorised modifications of data.
- Authentication: Identity verification for the devices, gateways or healthcare professionals.
- Access control: Implementation of detailed and context-driven authorisation rules.
- Privacy preservation: Secure analytics without disclosure of sensitive patient information.

3.3.6 Cryptographic design

The cryptographic design of the proposed framework takes a layered and modular approach that aligns security functionalities with the roles and resource capabilities of the entities in the system. This framework uses symmetric and asymmetric cryptography, along with digital signatures and selective homomorphic encryption so as to have end-to-end security as well as computational efficiency.

3.3.6.1 Key generation and management

In order to strike a balance between security strength and performance efficiency, a hybrid key management scheme is used. The generation of system initialisation keys and policy enforcement is done by the key and access management authority (KAMA).

To this, E_i is assigned a pair of public-private keys created by a secure asymmetric algorithm.

$$(PK_i,\ SK_i) \leftarrow KeyGen\left(1^{\lambda}\right) \tag{3.1}$$

where λ is a representative symbol for security parameters. The certification and distribution of public keys are done by KAMA and are handled via public key infrastructure (PKI). Private keys are secret to their owners.

For data transmission, session-based symmetric keys are generated. A session key K_s is generated through asymmetric encryption as given below.

$$K_s = Dec_{SK_j}\left(Enc_{PK_j}(K_r)\right) \tag{3.2}$$

where K_r is a randomly created nonce, and j represents the receiving party. The keys for a session can be extended to reduce vulnerabilities in case of partial compromise.

3.3.6.2 Encryption and decryption

Healthcare data obtained through IoMT devices in large quantities is encrypted using symmetric encryption due to low latency and energy consumption. Suppose D is the sensed medical data. Encryption is done as:

$$C = \mathrm{Enc}_{\mathrm{K_s}}\ (\mathrm{D}) \tag{3.3}$$

where C is the ciphertext and K_s the session key. Upon receipt, authorised entities decrypt the data using:

$$D = \mathrm{Dec}_{K_s}(C) \tag{3.4}$$

There are a number of important use cases that involve privacy-preserving analytics. To support these, the various data elements selected in the above step may be encrypted using a homomorphic encryption function $\mathscr{H}(\cdot)$, enabling computation over encrypted data without exposure to plaintext values:

$$\mathscr{H}(f(D)) = f(\mathscr{H}(D)) \tag{3.5}$$

where $f(\cdot)$ is an admissible analytical function performed on the cloud [7,12].

3.3.6.3 Authentication and access control

Authentication is achieved through mutual verification between the communication entities using digital signatures. An entity E_i generates a signature over a message m as:

$$\sigma_i = Sign_{SK_i}(m) \tag{3.6}$$

The receiving entity checks the signature using the sender's public key:

$$Verify_{PK_i}\ (m,\ \sigma_i\ ) \in \{0, 1\} \tag{3.7}$$

Access control decisions can be enforced by making decisions based on the identity, role and any other parameters, such as an emergency condition. Access to the resource happens if the policy evaluation function returns true:

$$Access(E_u\,,D) = \begin{cases} 1, & if\ P(E_u\,, Ct_x) = true \\ 0, & otherwise \end{cases} \tag{3.8}$$

where $P(\cdot)$ means the access policy and Ct_x are the contextual attributes.

3.3.6.4 Secure data sharing

Secure sharing of data between healthcare providers is accomplished via secure, encrypted data transfer and key distribution. Therefore, for an authorised provider E_j to be able to access and use specific shared data (D), re-encryption of the session key occurs:

$$K_s^{(j)} = \mathrm{Enc}_{PK_j}(K_s) \tag{3.9}$$

This ensures that only the intended recipient can decrypt the shared data. In addition, there are audit logs that keep a record for accountability.

3.3.7 Algorithmic description

To design the cryptographic processes, this subsection describes the pseudocode for the fundamental cryptographic processes that are implemented in the framework.

Algorithm 3.1 represents how encryption works, considering the example of the IoMT device. Additionally, the procedure undertaken for executing encryption at the IoMT device level.

Algorithm 3.1 Secure data encryption at IoMT device

Input: Medical data D, public key PK_G of the gateway
Output: Encrypted ciphertext C

1: Generate random nonce Kr
2: Establishes session key Ks ← DecSK_G(EncPK_G(Kr))
3: Encrypt data C ← EncKs(D)
4: Generate signature σ ← SignSK_D(C)
5: Transmit {C, σ} to Patient Gateway

Algorithm 3.2 depicts the procedure for mutual authentication and data verification between the entities.

Algorithm 3.2 Mutual authentication and data verification

Input: Ciphertext C, signature σ, sender public key PK_D
Output: Authentication decision

```
Verify signature result ← VerifyPK_D(C, σ)
if result = valid then
Accept data and proceed with decryption
else
Reject data and raise a security alert
end if
```

3.4 Security analysis, performance evaluation and discussion

This section will present an overall assessment of the proposed cryptography framework for security assurance, performance characteristics and effective comparison with respect to the current known state-of-the-art systems. For design-oriented articles like this current chapter, the evaluations will be inclined to analyse the security assurance and performance considerations, which are commonly used for cryptography framework evaluations regarding health systems, such as [6,7]. The introduced scheme is compared with existing schemes from various domains, such as privacy-preserving analytics, IoT-based eHealth security schemes and blockchain-based healthcare systems.

3.4.1 Security analysis

In the proposed framework, the security of the framework is validated based on the threat analysis of the defined threat models, which are mentioned and analysed in Section 3.3.4. From the analysis, the framework meets the necessary security requirements for the development of smart healthcare systems. These include confidentiality, integrity, authentication, flexibility in access control, resistance to insider attacks and privacy preservation.

Table 3.2 shows a summary of the security evaluation of the parameters for the proposed framework.

1. **Confidentiality**
   Data confidentiality is ensured through the use of symmetric key encryption for bulk health-related data, as well as asymmetric encryption for secure key exchange. By encrypting the data at the IoMT device level, as well as throughout the communication process, the entire solution prevents unauthorised disclosure of confidential health-related information, irrespective of a potentially hacked communication channel being present or not, as stated in [3,9].
2. **Integrity**
   Mechanisms for digital signatures are used for data integrity and authenticity verification for the data originating party at the patient gateway and cloud

Table 3.2 Security parameter evaluation of the proposed framework

Security parameter	**Evaluation criterion**	**Analysis outcome**
Confidentiality	Protection of data at rest, in transit and during processing	Ensured through hybrid symmetric–asymmetric encryption and encrypted analytics
Data integrity	Detection of unauthorised data modification	Guaranteed using digital signature mechanisms
Authentication	Mutual verification of entities	Achieved via PKI-based authentication
Access control	Policy adaptability and context awareness	Supports dynamic, role- and context-aware access policies
Insider attack resistance	Limitation of privilege misuse	Enhanced through layered access enforcement and auditability
Privacy preservation	Exposure of sensitive data during analytics	Reduced via selective homomorphic encryption

levels. Any alterations to the medical data will be identified during verification through the patient gateway and cloud levels. Thus, data tampering and data replay attacks will be avoided, which is a common problem in a distributed healthcare system [4,11].

3. **Authentication and authorisation**
Mutual authentication between IoMT devices, gateways and service providers of healthcare services is mandated by the use of public key certificates and verification of credentials through a secure process. Furthermore, it develops fine-grained access control policies to prevent impersonation and unauthorised privilege escalation by allowing access to patient data only to subjects whose identities are well known in prior policies [8,13].
4. **Privacy preservation**
To mitigate the privacy risks faced by the analytics performed by the artificial intelligence, selective homomorphic encryption techniques are included as part of the framework, enabling a set of analytical processes to be performed on the data while maintaining the confidentiality of the information contained therein. This reduces the risks of privacy leakage of the data contained in the healthcare analytics from the cloud computing perspective [7,12].
5. **Resistance to insider and outsider attacks**
The multi-layer design of cryptography reduces the potential damage caused by either internal or external adversaries. Similarly, the use of a session-based model reduces vulnerability in the case of partial compromise, while the distributed trust model removes the risk of a single point of failure. Overall, this design is sufficient to prevent internal users and/or external assailants like cyber attacks [6,15].

3.4.2 Comparative analysis with state of the art

To test and evaluate the efficacy of the proposed approach, a qualitative analysis was performed with existing state-of-the-art cryptographic mechanisms discussed in the literature over the years. Table 3.3 tabulates the comparative analysis between the proposed model and a host of state-of-the-art mechanisms in various fields. The comparison will be based on end-to-end security coverage, suitability for IoMT, privacy-preserving analytics, flexibility in access control, scalability and resistance to insider threats.

As depicted in Table 3.3, existing schemes focus only on isolated security requirements. In contrast, our proposed framework utilises various cryptographic techniques under a single platform and hence offers better security coverage without compromising flexibility.

To further clarify the comparison between existing schemes and our proposed scheme, Figure 3.2 presents a visual comparison between security coverage and our proposed scheme for various security requirements.

3.4.3 Performance evaluation

Performance issues play an important role in intelligent healthcare systems as much as they do in IoMT devices due to considerations of resources and 'real-time' processing. The frameworks' results can also be analysed in terms of Computational Cost and Communication Cost.

(i) **Computational overhead**
The lightweight symmetric encryption algorithm is utilised within the IoMT layer to ensure the latency of the processes is minimal. Computationally intensive operations like homomorphic encryption-based analytics can be performed on the cloud environment due to the presence of sufficient computing resources. In this manner, the security improvements do not compromise the responsiveness of the system at the device level [5,9].

Table 3.3 Comparative study with the state-of-the-art cryptographic frameworks for intelligent healthcare systems

Feature	**Existing schemes [3,5,14]**	**Proposed framework**
End-to-end encryption	Partial	Full life-cycle coverage
Lightweight IoMT support	Limited	Explicitly supported
Privacy-preserving analytics	Rarely supported	Integrated via homomorphic encryption
Access control flexibility	Static	Context-aware
Scalability	Moderate	High
Insider attack resilience	Limited	Enhanced

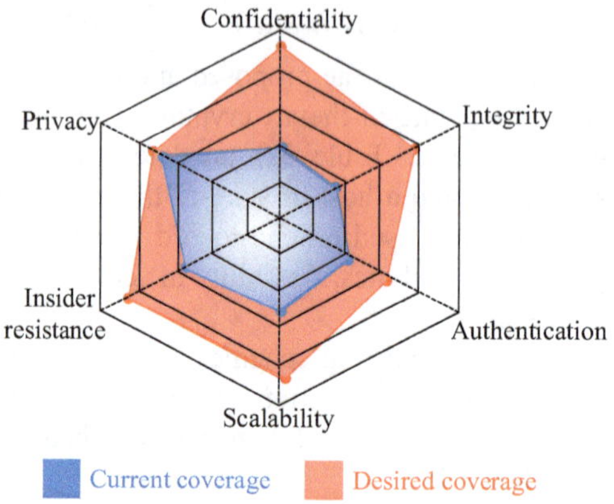

Figure 3.2 A visual comparison between security coverage and our proposed scheme for various security requirements

Table 3.4 Performance and scalability comparison

Metric	**Existing schemes**	**Proposed framework**
Computation overhead (IoMT layer)	Moderate to high	Low
Communication overhead	Moderate	Reduced
Key management scalability	Limited	High
Suitability for smart hospitals	Partial	High

(ii) **Communication overhead**
This framework will help to decrease communication overhead due to the encryption of data with symmetric keys and the limited use of asymmetric keys for key exchange and authentication only. Digital signatures will have negligible added weight to the data payload and will be good for restricted-bandwidth networks in the health care scenario as well [10].

(iii) **Scalability**
Scalability is obtained via modularised cryptographic building blocks and the use of a hybrid approach to key management. A dynamic approach to managing the number of connected devices and users provides the advantage that, as the number of connected entities increases, the system can cope quite well without losing significant performance. Such scalability is crucial in the deployment of smart hospitals and healthcare infrastructures as a whole across a given geopolitical entity [6,17].

The results in Table 3.4 show that the proposed framework is more appropriate for larger environments, such as smart hospitals and regional healthcare systems, where there are thousands of devices and users that need to be securely managed.

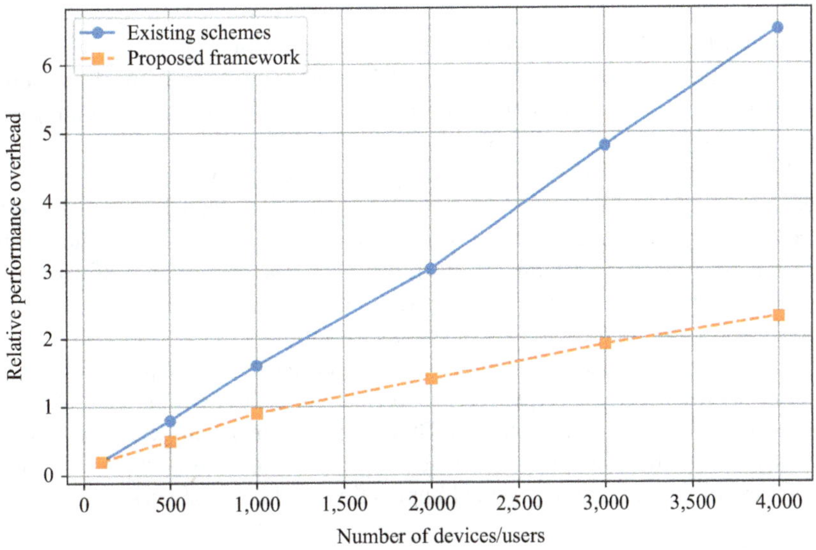

Figure 3.3 Analytical scalability comparison illustrating relative performance overhead as the number of connected IoMT devices and users increases

In addition to the above analysis, Figure 3.3 presents the scalability of the existing schemes and the proposed framework with the increase of connected devices.

Figure 3.3 illustrates an analytical scalability trend comparing the relative performance overhead of existing schemes and the proposed framework as the number of connected devices increases.

3.4.4 Discussion

The effectiveness of the proposed framework, in terms of security coverage and scalability, can be observed through the results provided in Tables 3.2, 3.3 and 3.4 and Figures 3.2 and 3.3. Although the existing works have addressed specific security aspects, the proposed framework considers various cryptographic techniques through a unified architecture, thus ensuring the security of intelligent healthcare environments, such as smart hospitals and regional networks.

The outcomes from the analysis show that the proposed framework for cryptography ensures the right balance between security robustness and operational efficiency regarding several limitations faced by previous approaches. The use of multiple cryptographic principles within a unified framework of multiple layers ensures that there will not be any fragmentation problems, as seen in the majority of proposed healthcare security solutions.

Further, the inclusion of the privacy-preserving analytics mechanisms leads to a secure AI-based healthcare facility with undisclosed information to the patients, which is an area not sufficiently researched in several models. As much as there is computation involved with the usage of the homomorphic encryption technique, its usage is limited to the cloud layer.

Additionally, the nature of the encryption process is also relevant with consideration of the 'separation of concerns' design principle, wherein lightweight processing will take place on the constrained device only, with the computationally intensive processing occurring on the more powerful part of the infrastructure. Relevant in regard to the practical application of the framework for encryption/authentication are the descriptions of the algorithm.

It must be noted that this framework is expected to be adapted to changing needs in the healthcare sector, as well as to changing landscapes of what may pose as threats for this sector in the future. Though this chapter will generally be based on analytical evaluation, further validation of this with an implementation study would certainly increase the practical application of this framework.

3.4.5 Summary of key findings

1. The proposed system fulfils the major security requirements, which are confidentiality, integrity, authenticity and privacy preservation.
2. Performance-aware cryptographic design enables feasibility in both resource-constrained and large-scale healthcare environments.
3. The information provided in the comparative analysis explains the benefits that the framework possesses over the present solutions in the aspects of integration, scalability and the security of analytics.

3.5 Lesson learned

The development of sophisticated cryptographic frameworks for intelligent healthcare systems has taught us several important lessons that are applicable to secure system engineering in general. These lessons are based on the evaluation of existing techniques, the development of the proposed framework and the comparative analysis, as discussed in this chapter. These lessons discuss both practical and theoretical aspects that are important for the effectiveness, practicality and sustainability of intelligent cryptographic techniques in intelligent healthcare systems.

a) **Lesson One: End-to-end security must be treated as a system property**
One of the key lessons that can be drawn is that security within intelligent healthcare systems cannot be achieved using a series of cryptographic mechanisms on different parts of the system. A number of different security schemes have been developed, with a major focus on different stages of the data

life cycle, including authentication during the device phase and encryption during storage.

The analysis carried out within this chapter has, however, demonstrated that security within intelligent healthcare systems is a system-level property and that end-to-end cryptographic integration is critical for the protection of sensitive healthcare information. This is a critical lesson, as fragmented security systems often lead to unintended security risks within the boundaries of the systems, and this is particularly critical within a healthcare environment.

b) **Lesson Two: Lightweight cryptography is necessary but not sufficient**
Another valuable lesson is that, although lightweight cryptographic methods are necessary for resource-limited IoMT devices, they are not sufficient by themselves for protecting intelligent healthcare systems. Lightweight cryptography is efficient in mitigating computational and energy costs; however, it does not fulfil overall needs such as secure data analytics, scalable access control and insider threats by itself.

The above framework shows that lightweight cryptography needs to be integrated into a hierarchical or layered security infrastructure, where complex cryptographic computations are delegated to capable infrastructure entities. This lesson emphasises the need to strike a balance between efficiency and security, such that performance limitations do not impede overall security.

c) **Lesson Three: Privacy-preserving analytics require selective and context-aware integration**
The increasing trend towards using AI-based analytics in healthcare points to the need for privacy preservation during the data processing stage. One of the major lessons to be derived from this study is the need for applying privacy-preserving analytics selectively rather than universally.

While homomorphic encryption-based analytics provide excellent privacy preservation capabilities, they are also computationally expensive. As demonstrated by the analysis presented in this chapter, selective integration-based analytics provide an excellent trade-off between privacy and efficiency. This lesson is also important for real-world applications because it demonstrates the power and effectiveness of using homomorphic encryption without compromising the practicality and efficiency of intelligent healthcare systems.

d) **Lesson Four: Scalability and key management are central to real-world deployment**
Scalability was a concern that surfaced as a significant issue during the comparative analysis. This is because intelligent healthcare systems can be very large, especially for applications like smart hospitals and healthcare networks across a region. The lesson learned here is that it is not just the strength of the cryptographic algorithms used but also the scalability of the key management and access control mechanisms.

Thus, the lesson learned here is the need for scalable and dynamic key management with context-aware access control, as a deployment-oriented aspect like scalability should be factored in during the design phase.

e) **Lesson Five: Insider threats require explicit architectural consideration**
One of the notable findings that emerged from the literature survey and security analysis is that the issue of insider threats tends to be underestimated or assumed away. The lesson learned from this literature and security analysis is that the issue of insider misuse, which could be intentional or unintentional, is an important security risk in healthcare environments due to the level of trust that insiders have in the systems.

The proposed multi-layered framework shows that addressing the issue of insider threats involves more than simple authentication and requires careful access control and accountability mechanisms.

f) **Lesson Six: Architectural clarity enhances security reasoning and evaluation**
Another important lesson that can be drawn is that the clearly defined system and threat models greatly improve security reasoning and evaluation. It has been observed that much of the existing literature presents various cryptographic techniques without clearly stating the underlying architectures, capabilities of the attacker or the environment of deployment.

The proposed framework clearly states the entities of the system, data flow and threat models, which greatly facilitates security analysis and comparison with the existing literature. This lesson clearly indicates that the role of architectural clarity is not only related to documentation but also has significant implications for security design.

g) **Lesson Seven: Design-level contributions are valuable for emerging healthcare systems**
Lastly, this work demonstrates that design-level and analytical contributions are very important in the development of secure intelligent healthcare systems. Although the evaluation of the performance of systems is important, architectural frameworks are useful in guiding the designers and policymakers in an ever-changing healthcare technology scenario.

The lesson that we learn from this is that well-structured cryptographic frameworks that are backed by analytical evaluation and comparative analysis have the potential to shape the scenario before actual implementations are made. This is especially true in the healthcare domain, where improper implementations of security systems can be harmful.

3.5.1 Overall significance of the lessons learned

These lessons collectively highlight the point that to ensure intelligent healthcare systems are secure, a holistic approach is needed to ensure the cryptographic security, architecture, scalability and deployment of the systems. The lessons learned from this chapter are useful to be applied in future research, to the architects of the systems and to the healthcare organisations.

3.6 Conclusion and future scope

3.6.1 Conclusion

This chapter presented an extensive design evaluation for an efficient cryptographic framework for advanced and developing intelligent healthcare systems that would encompass the use of artificial intelligence systems, IoMT devices and cloud computing systems. The evaluation of security issues that exist in most modern environments determined that there were fundamental flaws with most cryptographic systems due to their lack of scalability and support for data privacy issue analysis.

To cope with the foregoing challenges, a cryptographic framework has been suggested in which symmetrical encryption, asymmetrical encryption, digital signatures and selective homomorphic encryption will find their place in the system. Consequently, the cryptographic framework will be complete. To complete the framework, the system model is given, followed by the associated system threat model. After these two phases, the cryptographic framework is completed through the cryptographic design of the system and the detailed algorithm description.

The analytical security evaluation of the proposed framework has also confirmed the implementation of key security requirements while guaranteeing security against adversaries from outside as well as the inside of the framework. Consequently, the performance analysis has also pointed to the feasibility of the proposed framework as the appropriate solution due to the distribution of cryptographic operations across different levels of the system. The discussion of the proposed framework against other prevalent state-of-the-art security solutions has also confirmed the advantages of the proposed framework on different grounds.

In general terms, the chapter provides an innovative and versatile cryptographic framework that fills the gap between the theoretical security mechanisms of cryptography and the pragmatic realities of intelligent healthcare systems.

3.6.2 Future scope

Despite the strengths of the proposed framework, several avenues remain open for future research and enhancement. First, the integration of post-quantum cryptographic algorithms represents a promising direction to ensure long-term security against emerging quantum computing threats, particularly for healthcare data with extended confidentiality requirements.

Second, adaptive and AI-aided security management may be explored in future work, whereby machine learning dynamically adjusts cryptographic parameters, access policies and threat detection mechanisms according to system context and observed behaviour. Such approaches can further enhance resilience in highly dynamic healthcare environments.

Third, while the chapter was based on an analytical evaluation, as mentioned before, a study on real-world deployment/implementation on a smart hospital/

telemedicine-based model would also lead to empirical observations on the proposed framework to deal with large-scale works, diverse regulations etc.

Lastly, the expansion of the framework for the sharing of various kinds of health-related data, e.g. health information exchanges, as well as federated learning platforms, provides a promising area for further exploration, which helps in collaborative health-related data analytics while protecting the privacy of patients.

List of abbreviations

AI	Artificial Intelligence
EHR	Electronic Health Record
HE	Homomorphic encryption
IDS	Intrusion Detection System
IoMT	Internet of Medical Things
IoT	Internet of Things
KAMA	Key and Access Management Authority
ML	Machine Learning
PKI	Public Key Infrastructure
PUF	Physical Unclonable Function

Acknowledgement

The authors thank the Nile University of Nigeria, Abuja, for providing the enabling environment for this work.

References

[1] G. R. Pradyumna, R. P. Singh, S. Alweshah, T. Vasudevan and P. K. Mishra, "Empowering healthcare with IoMT: Evolution, machine learning integration, security, and interoperability challenges," *IEEE Access*, vol. 12, pp. 20603–20623, 2024.

[2] A. Ghubaish, T. Salman, M. Zolanvari, D. Unal, A. Al-Ali and R. Jain, "Recent advances in the internet-of-medical-things (IoMT) systems security," *IEEE Internet of Things Journal*, vol. 8, no. 11, pp. 8707–8718, 2020.

[3] J. Wang, L. Wu, H. Wang, K R. Choo and D. He, "An efficient and privacy-preserving outsourced support vector machine training for internet of medical things," *IEEE Internet of Things Journal*, vol. 8, no. 1, pp. 458–473, 2021.

[4] G. Nithyavani and G. N. Raja, " A comprehensive survey on security and privacy challenges in internet of medical things applications: Deep learning and machine learning solutions, obstacles, and future directions," *IEEE*

Access, vol. 13, pp. 188955–188989, 2025. https://doi.org/10.1109/ACCESS.2025.35884892021.

[5] H. Makina, A. Ben Letaifa and A. Rachedi, "Survey on security and privacy in Internet of Things-based eHealth applications: Challenges, architectures, and future directions," *Security and Privacy*, vol. 7, no. 2, p. e346, 2023.

[6] M. Elhoseny, N. N. Thilakarathne, M. I. Alghamdi, *et al.*, "Security and privacy issues in medical internet of things: Overview, countermeasures, challenges, and future directions," *Sustainability*, vol. 13, p. 11645, 2021. https://doi.org/10.3390/ su132111645.

[7] A. Naghib, F. Soleimanian Gharehchopogh and A. Zamanifar, "A comprehensive and systematic literature review on intrusion detection systems in the internet of medical things: Current status, challenges, and opportunities," *Artificial Intelligence Review*, vol. 58, no. 114, p. 114, 2025.

[8] S. Messinis, N. Temenos, N. E. Protonotarios, I. Rallis, D. Kalogeras and N. Doulamis, "Enhancing internet of medical things security with artificial intelligence: A comprehensive review," *Computers in Biology and Medicine*, vol. 170, no.108036, p. 108036, 2024.

[9] M. A. Al-Garadi, A. Mohamed, A. K. Al-Ali, X. Du, I. Ali and M. Guizani, "A survey of machine and deep learning methods for internet of things (IoT) security," *IEEE Communications Surveys & Tutorials*, vol. 22, no. 3, pp. 1646–1685, 2020.

[10] V. P. Yanambaka, S. P. Mohanty, E. Kougianos and D. Puthal, "PMsec: Physical unclonable function-based robust and lightweight authentication in the internet of medical things," *IEEE Transactions on Consumer Electronics*, vol. 65, no. 3, pp. 388–397, 2019.

[11] K. Park, S. Noh, H. Lee, *et al.*, "LAKS-NVT: Provably secure and lightweight authentication and key agreement scheme without verification table in medical internet of things," *IEEE Access*, vol. 8, pp. 119387–119404, 2020.

[12] J. Srinivas, A. K. Das, N. Kumar and J. J. Rodrigues, "Cloud-centric authentication for wearable healthcare monitoring system," *IEEE Transactions on Dependable and Secure Computing*, vol. 17, no. 4, pp. 942–956, 2020.

[13] A. Adnan, F. Kausar, M. Shoaib, F. Iqbal, A. Altaf and H. M. Asif, "A secure and privacy-preserving approach to healthcare data collaboration," *Symmetry*, vol. 17, p. 1139, 2025.

[14] A. Dwivedi, G. Srivastava, S. Dhar and R. Singh, "A decentralized privacy-preserving healthcare blockchain for IoT," *Sensors*, vol. 19, no. 4 2, p. 326, 2019. https://doi.org/10.3390/s19020326

[15] M.A. Engelhardt, "Hitching healthcare to the chain: An introduction to blockchain technology in the healthcare sector," *IEEE Access*, vol. 5, pp. 22201–22214, 2017.

[16] S. Tanwar, K. Parekh and R. Evans, "Blockchain-based electronic healthcare record system for healthcare 4.0 applications," *IEEE Network*, vol. 34, no. 5, pp. 50–57, 2020.

[17] H. Ning, H. Liu and L. T. Yang, "Cyberentity security in the internet of things," *IEEE Computer*, vol. 46, no. 4, pp. 46–53, 2013.

[18] S. B. Othman and G. Kumar, "Quantum-resilient and adaptive multi-region data aggregation for IoMT using zero-knowledge proofs and edge intelligence," *Scientific Reports*, vol. 15, no. 1, p. 37176, 2025.

[19] D. K. Murala, G. S. Krishna, T. V. S. Kiran and A. K. Mohamud. "MedShieldFL-a privacy-preserving hybrid federated learning framework for intelligent healthcare systems," *Scientific Reports*, vol. 15, p. 43144, 2025.

[20] F. Yesmin. "MedHE: Communication-efficient privacy-preserving federated learning with adaptive gradient sparsification for healthcare," preprint (Nov. 12 2025).

[21] A. Amin, K. Hasan, L. Hong and S. Ullah. "Privacy-preserving federated vision transformer learning leveraging lightweight homomorphic encryption in medical AI." preprint (Nov. 26 2025).

[22] H. X. Son, N. Q. Anh, P. T. Tran-Truong, L. T. Tuan and P. T. Nghiem. "SLIE: A secure and lightweight cryptosystem for data sharing in IoT healthcare services." preprint, Oct. 16 2025.

[23] D. Mamoria, H. Jain and A. K. Cherukuri, "Blockchain-integrated privacy-preserving medical insurance claim processing using homomorphic encryption," preprint, Nov. 11, 2025.

Chapter 4

Artificial intelligence-enabled cryptographic frameworks for intelligent healthcare systems

Friday Oodee Philip-Kpae[1], Agbotiname Lucky Imoize[2] and Patience Ebehiremen Orukpe[3]

Abstract

Intelligent healthcare systems increasingly depend on Internet of Medical Things (IoMT) devices, edge computing and cloud analytics to enable real-time monitoring and data-driven clinical decision support. However, protecting sensitive health data across heterogeneous networks remains challenging because conventional cryptographic deployments are static and cannot adapt continuously to evolving threats, device constraints and workload dynamics. This study proposes an artificial intelligence (AI)-enabled cryptographic framework that secures resource-constrained healthcare IoMT devices while preserving operational efficiency and privacy. The framework integrates (i) a risk engine for real-time anomaly detection and risk scoring, (ii) a contextual bandit learning mechanism (LinUCB) for adaptive cryptographic policy selection and (iii) federated learning with differentially private (DP) to support secure multi-site collaboration without sharing raw patient data. A layered IoMT–edge–cloud architecture is introduced, and cryptographic adaptation is formulated as a multi-objective constrained optimisation over approved configurations (cipher suite, key size, authentication mode and key-rotation interval) under latency, energy and compliance requirements. Simulation results show stable device operation, with energy consumption maintained at approximately 25 per cent of baseline, and effective threat response as risk sensitivity increases from 5 per cent under normal conditions to 27 per cent in high-alert states. Compliance is enforced through a risk-dependent

[1]Department of Electrical and Electronic Engineering, Faculty of Engineering, Rivers State University, Nigeria
[2]Department of Electrical and Electronics Engineering, Faculty of Engineering, University of Lagos, Nigeria
[3]Department of Electrical and Electronic Engineering, Faculty of Engineering, University of Benin, Nigeria

minimum-security floor that rises linearly from 20 per cent to 80 per cent as risk escalates. In comparison, key-rotation intervals decrease exponentially from 3600 s to 327 s (90.92 per cent reduction) during active threats, reducing exposure time if a key is compromised. Utility analysis indicates an intelligent crossover at approximately 60 per cent security weight, where the framework transitions from lightweight options (e.g. ChaCha20) to stronger schemes (e.g. AES-256 and RSA-2048). The learning process converges rapidly, achieving about 95 per cent normalised efficiency by epoch 40. Although DP introduces an approximate 17 per cent residual loss due to injected noise, gradient clipping mitigates leakage risk. Overall, comparative evaluation against static and rule-based baselines indicates improved protection under elevated threats while reducing latency and energy costs during routine operation. Limitations include simulation-based evaluation with synthetic datasets; future work will focus on real-world testbeds, calibrated device/network measurements and stronger robustness for multi-site learning and deployment.

Keywords: Intelligent healthcare; IoMT; cryptography; adaptive security; artificial intelligence; anomaly detection; edge computing; privacy-preserving learning

4.1 Introduction

Intelligent healthcare systems are increasingly enabled by the Internet of Medical Things (IoMT), edge computing and cloud computing, which support continuous monitoring, remote care and timely clinical decision-making. In this context, connected sensors and smart medical devices collect biomedical signals, such as heart rate, oxygen saturation, electrocardiogram (ECG) and continuous glucose monitoring (CGM) and transmit them to gateways, edge servers and cloud platforms for processing and storage. While this architecture improves scalability and responsiveness, it also increases the attack surface because sensitive medical data and device-control messages traverse heterogeneous networks and platforms with varying security and resource constraints. However, cryptographic protections are often configured statically at deployment, even though operational conditions and adversarial behaviour change over time. Static settings may be too weak to withstand evolving threats or too costly for resource-limited IoMT devices, resulting in increased latency, energy consumption and reduced availability. This creates a persistent mismatch between the security level needed at a given moment and the security level actually enforced, leaving systems either exposed to attacks or burdened by unnecessary overhead.

This paper addresses the challenge of the absence of adaptive, context-aware cryptographic orchestration for healthcare IoMT–edge–cloud systems. The goal is to design a security framework that can infer operational and threat contexts, select appropriate cryptographic strength under explicit constraints and enforce updates dynamically without disrupting clinical workflows. This is a multi-objective problem because stronger security can increase computational cost, bandwidth consumption and delay. In comparison, lighter security can increase vulnerability to eavesdropping, forgery, replay, impersonation and denial-of-service (DoS) attacks. The impact on healthcare is dire, as security failures can directly affect patient safety and service delivery. Data tampering can mislead clinicians and automated systems, while confidentiality breaches can expose sensitive health records, and attacks on availability can delay alarms and degrade monitoring. At the same time, overly heavy cryptography can also be harmful by increasing transmission delays and accelerating battery depletion in wearables and mobile devices, making 'one-size-fits-all' security impractical in real deployments.

In this book chapter, we address the problem by treating cryptographic configuration as a dynamic decision process driven by measurable context and risk. Operational context includes device energy, central processing unit (CPU) load, link quality, congestion level and clinical criticality. At the same time, a threat framework can be inferred from abnormal traffic patterns and suspicious device behaviour. By continuously estimating the operating conditions and threat signals, the system can dynamically tune cryptographic settings – such as the choice of cipher, key length, authentication level and key-rotation rate – so that it preserves a required security baseline while keeping latency and energy consumption within acceptable limits.

In this chapter, we propose an artificial intelligence (AI)-enabled framework, leveraging machine learning (ML) for its suitability. ML can learn complex relationships between context, threat signals and the security–performance trade-off. AI supports real-time risk scoring, adaptive policy recommendation and continual improvement as environments and attacks evolve. This shifts cryptographic management from fixed provisioning to data-driven control while still enforcing hard safety and compliance constraints. The approach is to formulate adaptation as a constrained optimisation problem and to implement policy selection using contextual online learning. A contextual bandit can treat each cryptographic configuration as an action and use real-time context to select the action that maximises reward, balancing security benefits against latency, energy and bandwidth costs, subject to constraints. To improve robustness across institutions without exposing patient data, federated learning (FL) can train detection and risk models collaboratively, and differential privacy (DP) can limit information leakage from shared updates.

Furthermore, adaptive cryptography can reduce unnecessary overhead during regular operation while strengthening protection during elevated risk, improving device lifetime and maintaining clinical responsiveness. It can also support accountability through auditable policy changes that explain what was adjusted and why. Additionally, stronger privacy and integrity protections improve trust in digital healthcare and enable safer scaling of remote monitoring, especially in constrained environments. Privacy-preserving collaboration further promotes shared defence without requiring the exchange of raw data, aligning security improvements with ethical and regulatory expectations.

4.1.1 Chapter contribution

This book chapter research makes the following contributions.

(i) The chapter proposes a layered IoMT–edge–cloud security architecture that couples threat assessment with dynamic cryptographic enforcement.
(ii) Modelling of adaptive cryptography as a constrained, multi-objective decision problem grounded in security, latency, energy and clinical-criticality requirements.
(iii) Design of an AI-driven policy engine that integrates risk/anomaly modelling with contextual online learning for real-time cryptographic selection.
(iv) Presentation of a privacy-preserving learning strategy using FL with DP to improve generalisation across sites while limiting sensitive data exposure.

4.1.2 Chapter organisation

The remaining part of this chapter is structured as follows: Section 4.2 reviews related works; Section 4.3 presents the methodology; Section 4.4 describes the system set-up (Section 4.4.1 system model, Section 4.4.2 threat model and Section 4.4.3 security and safety goals); Section 4.5 details the AI-enabled cryptographic architecture and formal model; Section 4.6 explains the adaptive learning/optimisation components (including federated learning); Section 4.7 reports the results and discussions across key performance, risk, overhead and privacy analyses, and Section 4.8 concludes the chapter, limitations and suggest future direction.

4.2 Related works

This section reviews related work on securing intelligent healthcare and Internet of Medical Things (IoMT) environments, focusing on three main streams: lightweight cryptography and authentication for resource-constrained devices, AI/ML-driven intrusion and anomaly detection for emerging threats and privacy-preserving collaboration techniques, such as federated learning, for multi-stakeholder settings. While these approaches have advanced security and privacy in isolation, many studies still lack an integrated, end-to-end mechanism that adapts protection levels in real time based on context, risk and clinical-criticality constraints – motivating the approach presented in this paper.

4.2.1 Cryptography for healthcare and IoMT

Recent studies have improved security for next-generation connected systems, yet cryptography, authentication and AI-driven analytics are still often treated as loosely coupled components. MFM-SAT is introduced as a dynamic authenticated-encryption suite for satellite communications. Still, the publicly available description is largely high-level, limiting transferability of its adaptation mechanisms to heterogeneous cyber-physical deployments [1]. A 6G survey motivates hybrid post-quantum cryptography (PQC) for authentication and key agreement (AKA), highlighting trade-offs between quantum resilience and increased latency/key sizes and noting open needs in lightweight implementation, formal validation and standardisation [2]. Cloud security work combines deep-learning-based attack detection with privacy-focused key-management optimisation and reports strong performance but does not address end-to-end cryptographic policy orchestration under explicit safety/performance constraints [3]. In IoMT, a dual-cryptography data-sharing scheme enhances confidentiality and access control but largely remains a static sharing model rather than a risk-adaptive approach across IoMT–edge–cloud workflows [4]. Overall, the gap is the lack of an integrated framework that uses AI-derived risk and clinical urgency to adapt cryptographic strength (cipher choice, key sizing/rotation and authentication strictness) under latency–energy constraints while supporting privacy-preserving learning across sites.

Building on prior advances, several studies address security across connected health and industrial cyber-physical systems, yet solutions remain fragmented across cryptography, detection, authentication, data aggregation and deployment architecture. A taxonomy-based survey classifies lightweight cryptographic frameworks for Industrial Internet of Things (IIoT) end points and highlights design trade-offs for constrained devices [5]. Encrypted-traffic security is advanced through a lightweight, real-time, unsupervised approach that fuses traffic features to detect unknown encrypted malware when payload inspection is infeasible [6]. For access control, a lightweight, fast authentication protocol reduces session overhead in digital healthcare services, but security parameters are still fixed primarily at design time [7].

Secure IoMT data aggregation is formulated as a constrained multi-objective optimisation problem (NSGA-II) that balances energy, latency, security and scalability. Yet it is not driven by real-time threat or clinical-urgency signals [8]. A decentralised remote patient monitoring architecture improves privacy by reducing reliance on a central trust authority but does not provide a unified mechanism for adapting cryptographic and operational settings to changing conditions [9]. In view of the studies, a key gap remains: the lack of an end-to-end, risk-aware orchestration layer that jointly links lightweight cryptography, encrypted-traffic threat intelligence, authentication strictness and aggregation/privacy controls under explicit energy–latency–security constraints – an integration this work targets via adaptive, constraint-aware policy selection with privacy-preserving learning support.

Emerging evidence indicates a shift towards edge-centric intelligence in healthcare cyber-physical systems to improve security, latency and reliability; however, clinical semantics and cryptographic controls are still addressed inconsistently. An SDN-enabled deep learning framework integrates Bidirectional Long Short-Term Memory (BiLSTM) attack detection with Convolutional Neural Network (CNN)-based fog-node selection and demonstrates strong Quality of Service (QoS) and detection performance in simulation, illustrating how programmable networking and learning can jointly enhance efficiency and security

[10]. A multi-layered, secure Edge Artificial Intelligence (Edge AI) model similarly argues for layered controls across device–edge–cloud pipelines to protect consumer health workflows. Yet it remains primarily architectural and does not formalise runtime cryptographic policy tuning under explicit constraints [11].

A clinically grounded survey of biomedical signal complexity across Electroencephalogram (EEG), Electrocardiogram (ECG), Intracranial Pressure (ICP) and Photoplethysmogram (PPG) further stresses deployability on constrained platforms (e.g. microcontrollers, Field-Programmable Gate Arrays (FPGAs) and System-on-Chip (SoC) devices) and the role of explainable machine learning (ML) in clinical trust [12]. These studies reveal a shared gap: learning is mainly applied to detection/analytics and resource selection, rather than to cryptographic policy control; clinical urgency and signal-derived criticality are rarely coupled with security actions, and decisions are seldom posed as constrained optimisation problems that guarantee minimum security under latency/energy constraints. This work addresses the gap by introducing an AI-driven, constraint-aware orchestration layer that models operational state (urgency, resources and threat level) and dynamically selects cryptographic policies (cipher suite, key length, authentication strictness and re-keying cadence) across layers, supported by privacy-preserving learning such as federated learning (FL) and DP for governance-compliant deployment.

4.2.2 AI for security in cyber-physical systems

Complementing detection-centric IoMT security research, a recent review has synthesised machine learning (ML)-based intrusion detection by comparing methods, datasets and reported performance and highlighted persistent challenges such as detecting novel attacks, real-time operation in heterogeneous medical infrastructures and robustness beyond dataset-specific results. 'IoMT connectivity improves monitoring and analytics but exposes systems to evolving cyber threats' [13]. An ensemble anomaly detector further demonstrates that combining diverse classifiers (e.g. Random Forest, K-Nearest Neighbours, Support Vector Machine and Multi-Layer Perceptron) can improve IoT anomaly detection performance across datasets [14]. However, these efforts essentially stop at detection and do not translate risk scores into end-to-end security controls that incorporate formal latency/energy constraints, clinical-urgency awareness and privacy-preserving multi-site learning. This work addresses the gap by coupling AI-derived risk and urgency with constraint-aware orchestration of cryptographic policies (cipher selection, key size and rotation and authentication stringency) while supporting privacy-preserving model improvement.

Additionally, in detection-centric IoMT security studies, explainability and system-level readiness have been emphasised as prerequisites for trustworthy deployment: an explainable artificial intelligence (XAI) framework for Medical Cyber-Physical Systems (MCPS) couples learning-based prediction with knowledge reasoning and links transparency to consistency between feature-driven outputs and reasoning while also modelling sensor-attack detection under sparse observability and proposing an attack detector/state estimator assuming protected Internet of Medical Things (IoMT) sensors [15]. Broader surveys trace IoMT evolution and machine learning integration, noting that pervasive sensing increases clinical value as well as magnifies security and interoperability challenges under tight energy and computational constraints [16]. They further categorise IoMT threats and defences, including privacy-preserving options and outline persistent deployment obstacles [17].

In parallel, wireless- and edge-focused surveys and frameworks show that 5G/6G capabilities and mobile edge computing (MEC) can improve timeliness and efficiency in connected care, yet security and privacy remain barriers to practice [18]; multi-objective, deep reinforcement learning (TD3/SAC)-based hospital analytics and resource allocation improve latency, energy efficiency and information freshness [19]; next-generation smart healthcare surveys consolidate heterogeneous IoT-centric architectures and enabling technologies while highlighting ongoing interoperability and security challenges [20], and MEC-enabled IoMT monitoring systems incorporate medical criticality into energy–privacy–latency trade-offs using tailored scheduling and optimisation algorithms [21]. However, across these contributions, AI and edge decision engines are rarely coupled to runtime cryptographic policy orchestration (cipher suite/parameter selection, key sizing and re-keying cadence) under explicit security–latency–energy constraints and healthcare-grade privacy governance, which motivates the integrated, constraint-aware adaptive cryptography approach proposed in this work.

4.2.3 Adaptive and context-aware security

Building on earlier IoMT security efforts, interpretable, data-rich AI methods are increasingly being explored to improve trust and robustness. An interpretable artificial intelligence (AI) framework is presented for secure and reliable medical image analysis in Internet of Medical Things (IoMT) environments, emphasising transparency and trust when models process sensitive clinical imagery [22]. Large multimodal models (LMMs) are further proposed to enhance IoMT security by jointly reasoning over heterogeneous data streams, enabling broader risk detection than single-modality analytics [23]. Federated edge intelligence is further advanced by combining federated learning (FL) with adversarial examples and traffic-byte representations, thereby explicitly addressing intermittent participation and local data imbalance, which can degrade robustness in consumer healthcare devices [24]. However, these studies still provide only ‘AI insight’, without an end-to-end mechanism that converts learned risk and urgency into runtime cryptographic policy orchestration under explicit safety and performance constraints, motivating the adaptive, constraint-aware cryptographic control developed in this work.

Aligned with privacy-preserving and adaptive learning directions, an adaptive federated-learning framework is proposed for consumer-centric IoMT that integrates hierarchical edge–fog–cloud training with integrity and privacy mechanisms (notably blockchain-supported collaboration and DP) to reduce communication overhead while improving personalisation and model performance [25].

A complementary review consolidates machine learning approaches for healthcare-IoT security by summarising attack surfaces across layers and common ML/DL risk-mitigation strategies but with emphasis mainly on detection and classification [26]. Meta-learning is further applied to IoMT intrusion detection to enable rapid adaptation and enhance the detection of evolving and zero-day attacks [27]. Although these contributions advance privacy-preserving learning and adaptive IDS capability, they do not translate learned risk and urgency into runtime cryptographic policy orchestration under explicit latency–energy–security constraints, motivating the constraint-aware adaptive cryptographic control proposed in this work.

Recent contributions have increasingly embedded advanced AI into connected-care infrastructures to improve performance and trust. Dual large language models (LLMs) are integrated into wireless healthcare networks, where one model supports mobile edge computing (MEC)-assisted routing and reconfigurable intelligent surface (RIS) optimisation, and another provides retrieval-augmented generation (RAG) clinical assistance, yielding lower overhead and latency in simulation [28]. The Society 5.0 eHealth framework further integrates IoMT–edge–cloud with explainable AI (XAI) and blockchain, using fuzzy logic to predict health severity from electronic health records (EHRs) and real-time IoMT data, with explainability tools and scalable consensus support [29]. Explainable intrusion detection is also delivered through an edge-deployed Software-as-a-Service (SaaS) intrusion detection system (IDS) that applies particle swarm optimisation (PSO) for feature engineering, ensemble learning for detection and SHAP for interpretability on a healthcare IDS dataset [30]. In addition, real-time neural speech enhancement is proposed for consumer smart-healthcare voice interfaces, emphasising low latency and practical deployment, where confidentiality and responsiveness are critical [31]. Despite these advances, AI model outputs are not unified into a runtime cryptographic policy orchestration (cipher/parameter selection, key sizing, re-keying cadence and authentication strictness) under explicit latency–energy–security constraints, motivating the constraint-aware, end-to-end and adaptive cryptographic control developed in this work.

4.2.4 Research gap

From the review of related work, a consistent gap emerges: the literature [1,2,5,7] treats lightweight cryptography/authentication/protocols as static artefacts. At the same time, studies by [3,6,13,14,26,27] builds ML/DL detectors and risk scoring without closing the loop into enforceable cryptographic actions, and [12,18,19,21] optimises edge/MEC performance and even clinical criticality without jointly adapting cryptographic strength. Lastly, research by [9], [24,25,29] enables privacy-preserving learning and decentralised trust without a runtime control layer that orchestrates cryptographic policies under explicit latency–energy–security constraints. In this work, we bridge these gaps by unifying these strands into a single AI-driven, constraint-aware cryptographic orchestration framework for IoMT–edge–cloud healthcare: risk/anomaly signals from ML/DL (including encrypted-traffic intelligence) drive dynamic selection of cryptographic and authentication parameters; optimisation ensures minimum-security baselines while respecting energy/latency and clinical deadlines, and privacy-preserving training (e.g. federated learning with privacy safeguards) enables cross-site improvement without exposing patient data, producing an end-to-end approach that goes beyond IDS to actionable, auditable security control.

4.3 Methodology

This study adopts a design-science, simulation-driven methodology in which cryptographic configuration is modelled as a dynamic decision process over an IoMT–edge–cloud healthcare architecture. The method proceeds by (i) defining the layered system and threat models (IoMT sensing, edge enforcement and cloud analytics), (ii) formulating adaptive cryptographic control as a multi-objective constrained optimisation problem over a whitelist of approved configurations and (iii) implementing an AI-driven orchestration workflow that couples real-time risk estimation with online policy selection and privacy-preserving multi-site learning. In operation, the orchestrator continuously observes latency, energy proxy, bandwidth, edge load, traffic criticality and AI-derived risk, then selects a cryptographic action (cipher/MAC suite, key size, rotation interval and authentication mode) while enforcing safety and compliance constraints, including traffic-class latency deadlines and a risk-dependent minimum-security floor. Decision-making combines anomaly-based risk mapping with a policy recommender using contextual online learning (e.g. LinUCB) and/or predictor-assisted constrained optimisation, followed by secure policy distribution and key-rotation scheduling at the edge.

Evaluation is conducted in MATLAB® R2021a using a time-stepped simulation over repeated decision epochs with fluctuating device/network conditions across 100 steps; threat behaviour is injected via reconstruction-error anomaly trajectories that produce transient spikes and sustained surges, mapped to normalised risk probabilities through a logistic function to differentiate normal and high-alert IDS modes. Candidate cryptographic suites (lightweight to strong) are assessed using risk-based compliance filtering, adaptive key management with decreasing rotation intervals as risk increases, utility-based sensitivity analysis over security weighting and contextual-bandit convergence tracking, with comparisons against static or rule-based baselines to verify improved protection under threat and reduced overhead during routine operation. Data collection is synthetic and trace-based, producing operational-context time series, anomaly/risk traces, cryptographic performance profiles (latency/overhead/security strength), learning logs (actions and convergence) and federated-learning/differential-privacy traces that emulate multi-site updates without centralising sensitive healthcare data. The conceptual framework for the proposed system is represented in Figure 4.1.

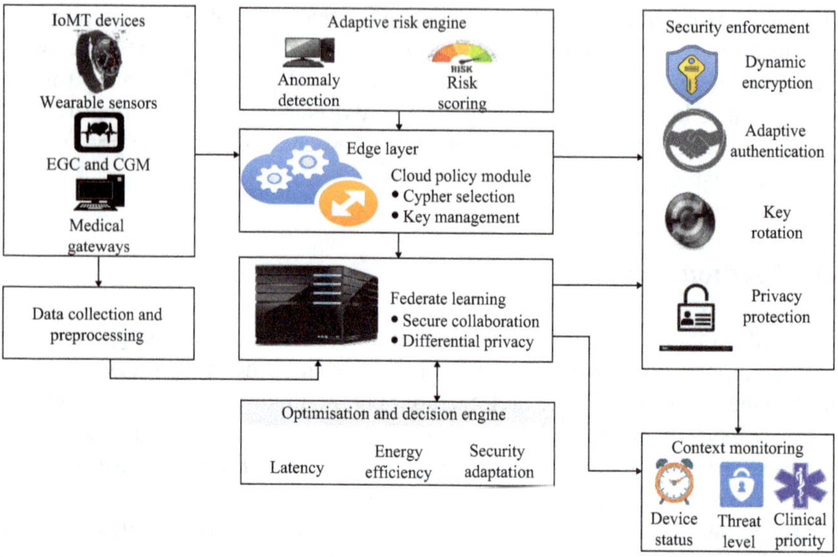

Figure 4.1 Conceptual framework of the proposed AI-enabled adaptive cryptographic system for IoMT-based healthcare, showing the data flow from IoMT devices through risk assessment and edge-layer policy enforcement, supported by federated learning and context monitoring, to an optimisation engine that tunes encryption and authentication under latency, energy and security constraints

4.4 System set-up

This section describes the system and threat models, including the security and safety goals and the cryptographic architecture and components employed.

4.4.1 System model

We consider 1) IoMT layer: sensors and medical devices generating telemetry and control messages; 2) edge layer: hospital gateways providing local computers, real-time control and policy enforcement; 3) cloud layer: storage, analytics and cross-site security intelligence aggregation. Traffic comprises routine telemetry, burst streaming (e.g. during incidents) and alarms/commands requiring low latency. The system model is presented in Figure 4.2.

4.4.2 Threat model

We assume an attacker can 1) eavesdrop on links (wireless/wired), 2) perform replay and injection (attempting false alarms/commands), 3) attempt MITM during handshakes or key exchange and 4) induce resource exhaustion (DoS-like

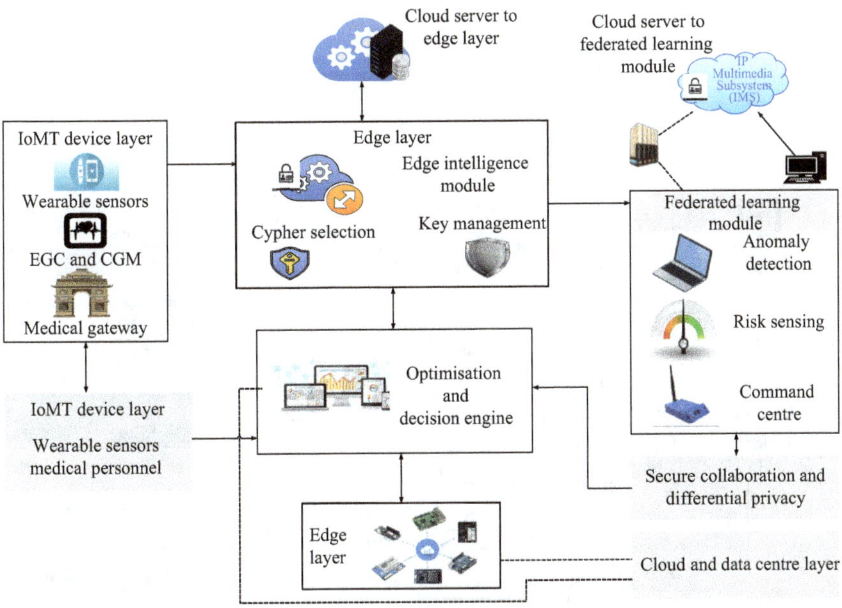

Figure 4.2 System model architecture showing the IoMT device layer, edge computing layer and cloud layer. Wearable sensors and medical gateways transmit encrypted patient data to the edge for preprocessing, anomaly/risk analysis and key/policy management. At the same time, the cloud supports resource pooling and privacy-preserving federated learning to improve global models and send updated policies back for adaptive security enforcement

behaviour, flooding). We assume that standard cryptographic primitives are sound; failures arise from weak configurations, poor key management, misconfiguration and an inability to adapt to dynamic threats. The threat model is shown in Figure 4.3.

4.4.3 Security and safety goals

To meet the security and operational requirements of the proposed IoMT-based healthcare system, the framework targets goal 1 (G1): confidentiality by protecting patient data, device identifiers and model updates against unauthorised disclosure during collection, transmission, storage and processing; goal 2 (G2) integrity/authentication by ensuring mutual authentication among devices, gateways, edge and cloud services while preventing message tampering, spoofing, replay and injection; goal 3 (G3) availability and bounded latency for alarms by guaranteeing continuous service and timely delivery of critical alerts even under network

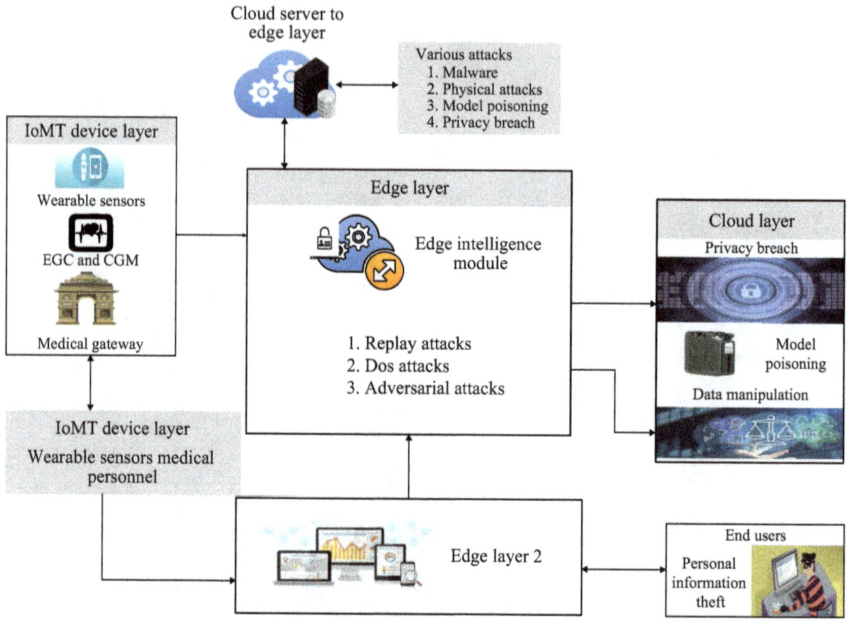

Figure 4.3 Threat model for the proposed IoMT-based healthcare system, highlighting attacks across the device, edge and cloud layers (e.g. physical compromise, malware, eavesdropping, replay/DoS, adversarial and model-poisoning attacks) and their impact on data integrity, privacy and availability

disruption or DoS conditions through prioritisation and resilient edge handling; goal 4 (G4) compliance-friendly auditing by maintaining verifiable, traceable logs of access, security events, policy decisions and key-management actions to support accountability and forensics without leaking sensitive content; goal 5 (G5) minimised energy/overhead on constrained devices by using lightweight operations, adaptive cryptographic choices and edge/cloud offloading to reduce computation, bandwidth use and battery drain. The security and operational goals flow diagram is shown in Figure 4.4 and includes G1 – confidentiality, G2 – integrity/authentication, G3 – availability with bounded latency, G4 – compliance-friendly auditing and G5 – minimised energy/overhead on constrained devices, all of which are derived from the security and operational goals of the IoMT-based healthcare framework.

4.4.4 AI-enabled cryptographic architecture and components used

In this work, IoMT devices (e.g. wearables and medical gateways) serve as enforcement end points: they apply cryptographic settings issued by the edge by

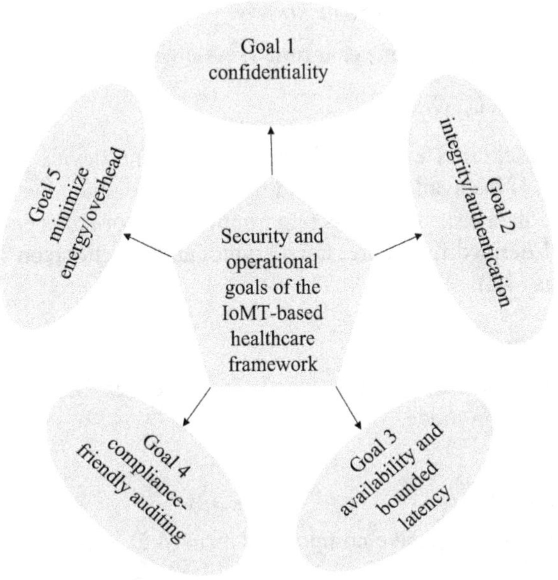

Figure 4.4 Block-based flow diagram of the security and operational goals for the proposed IoMT-based healthcare framework linking G1 to G5

encrypting data, authenticating sessions and rotating keys according to the assigned schedule or risk triggers. The Edge AI Security Orchestrator serves as the control layer and includes (1) a state/telemetry collector that monitors latency, bandwidth, battery/energy proxy, CPU load and traffic class (alarm vs routine), (2) an AI risk engine that performs anomaly detection and computes a risk score from IDS logs, traffic features and device behaviour, (3) a policy recommender that selects an appropriate cryptographic configuration (cipher suite, key size, authentication strictness and re-key interval) under latency and energy constraints using contextual learning and (4) a policy enforcement manager that securely distributes signed policies to devices and co-ordinates key-rotation scheduling and updates. The Cloud Security Intelligence layer supports the edge by aggregating anonymised or differential-privacy-protected statistics, enabling federated training across hospitals to improve detection models and sharing global threat intelligence (e.g. updated signatures and risk rules) that helps the edge refine its risk scoring and policy decisions.

4.5 Formal model: multi-objective constrained security adaptation

We formalise cryptographic orchestration as a constrained optimisation problem over a discrete set of approved configurations A.

4.5.1 State and configuration space

Let x_t denote the observed context at time t. As presented in (4.1):

$$x_t = [L_t, E_t, B_t, C_t, Q_t, R_t] \tag{4.1}$$

where L_t is the measured end-to-end latency, E_t is the device energy proxy (or battery level), B_t is the available bandwidth, C_t is the compute load at the edge and Q_t captures QoS data criticality class (e.g. routine monitoring vs emergency alarm) and R_t is the AI-derived risk score. Each cryptographic action/configuration $a \in A$ is represented as (4.2):

$$a = \{suite, k, \tau, m\} \tag{4.2}$$

where *suite* is the cipher/MAC suite, k is the key size, τ is the rotation interval and m is the authentication mode.

4.5.2 Objectives

We define normalised objective components as in (4.3), (4.4), (4.5) and (4.4):

$$S(a, R_t)\text{: security strength aligned to risk,} \tag{4.3}$$

$$\widehat{L}(a, x_t)\text{: predicted latency impact,} \tag{4.4}$$

$$\widehat{E}(a, x_t)\text{: predicted energy impact,} \tag{4.5}$$

$$\widehat{O}(a, x_t)\text{: cryptographic overhead(CPU/handshake).} \tag{4.6}$$

4.5.3 Constrained optimisation

At each decision epoch, the recommender selects (4.7).

$$a_t^* = \arg\max_{a \in A}[\alpha S(a, R_t) - \beta\widehat{L}(a, x_t) - \delta\widehat{E}(a, x_t) - \eta\widehat{O}(a, x_t)] \tag{4.7}$$

Equation (4.7) is subject to safety/compliance constraints as expressed in (4.8), (4.9) and (4.10).

$$\widehat{L}(a, x_t) \le D(Q_t) \text{ (deadline by class),} \tag{4.8}$$

$$S(a, R_t) \ge S_{\min}(R_t) \text{ (minimum security floor),} \tag{4.9}$$

$$a \in A_{approved} \text{ (policy whitelist).} \tag{4.10}$$

Table 4.1 presents the core variables and decision elements used in the proposed AI-enabled adaptive cryptographic framework for IoMT-based healthcare, summarising the monitored context/telemetry inputs (e.g. latency, bandwidth, energy and CPU proxies and traffic class), computed security indicators (e.g. anomaly flags and risk scores) and the resulting adaptive control outputs (e.g. cipher suite selection, key size, authentication level and

Table 4.1 Core variables and decision elements used in the proposed AI-enabled adaptive cryptographic framework

Symbol	**Meaning**
L_t, E_t, B_t, C_t	Latency, energy proxy, bandwidth, edge compute load
Q_t	Data criticality class (routine, streaming, alarm/command)
R_t	AI risk score (from anomaly/IDS signals)
$a = \{suite, k, \tau, m\}$	Crypto configuration: suite, key size, rotation interval, auth mode
$D(Q_t)$	Latency deadline for class Q_t
$S_{\min}(R_t)$	Minimum-security strength as a function of risk

key-rotation interval) that guide real-time policy enforcement under security–performance constraints.

4.5.4 AI risk engine: anomaly detection and risk scoring

We model anomaly detection using an encoder-based detector (e.g. an autoencoder) or a one-class classifier. Let $z_t \in \mathbb{R}^d$ denote extracted traffic/device features. The anomaly score can be expressed as (4.11).

$$A_t = \|z_t - \hat{z}_t\|_2^2, \tag{4.11}$$

where $\hat{z}_t$ is the reconstructed feature vector. A calibrated risk score combines anomaly, IDS alerts and domain rules as given in (4.12).

$$R_t = \sigma(w_1 A_t + w_2 I_t + w_3 U_t), \tag{4.12}$$

where I_t denotes IDS alert intensity, U_t captures unusual device behaviour (e.g. unexpected firmware events) and $\sigma(\cdot)$ is a logistic function.

4.5.5 Policy recommendation: contextual learning

In this study, we propose two practical recommenders: 1) a supervised predictor to learn $\hat{L}(a,x)$ and $\hat{E}(a,x)$ from historical traces (regression), then solve (4.7). 2) A contextual bandit that treats each configuration a as an arm; learn expected utility under context x_t with safety constraints.

Table 4.2 gives the AI options for risk estimation and policy recommendation in the proposed IoMT security framework, summarising suitable techniques for anomaly detection and risk scoring (using IDS logs, traffic features and device behaviour) as well as decision/learning strategies for selecting cryptographic policies (cipher suite, key size, authentication strictness and re-key interval), including the typical inputs, computational overhead and how each option supports real-time adaptation under healthcare latency and energy constraints.

Algorithm 4.1 presents risk-aware adaptive cryptographic policy selection and enforcement for IoMT-based healthcare systems, describing how the edge collects

Table 4.2 AI options for risk estimation and policy recommendation in the proposed IoMT security framework

Module	**Candidate models**	**Why is it suitable in healthcare**
Risk/anomaly	Autoencoder, one-class SVM, isolation forest	Unlabelled-friendly; can detect novel attacks
Latency/energy predictors	Gradient boosting, small MLP, linear regression	First inference at edge; interpretable options
Policy recommender	Constrained optimiser + predictors; contextual bandit	Adapts online; handles non-stationarity

device/network telemetry and IDS evidence, computes an anomaly-informed risk score, maps the risk and context constraints (latency, bandwidth, CPU/energy and traffic class) to an appropriate cryptographic profile, securely distributes the selected policy to IoMT devices and schedules key rotation or immediate re-keying when high-risk events or threshold crossings are detected.

Algorithm 4.1 AI-enabled cryptographic policy orchestration

```
Initialise anomaly model M_A, predictors M_L, M_E, M_O (optional)
Initialise recommender M_P (optimiser and/or contextual bandit) with an
  approved set of A
for each decision epoch t do
    Collect context x_t ← [L_t, E_t, B_t, C_t, Q_t, R_t] and feature z_t
    Compute anomaly score A_t ← Anomaly (M_A, z_t)
    Compute risk score R_t ← Risk (A_t, IDS, device events)
    for all a ∈ A do
        Predict L̂(a, x_t), Ê(a, x_t), Ô(a, x_t) (or use profile)
        Compute security strength S(a, R_t)
    end for
    Select a*_t ← arg max_{a∈A} utility in (4.7) subject to constraints
    Distribute policy a*_t and schedule key rotation τ
    Log (x_t, R_t, a*_t) for audit and learning updates
end for
```

For linear contextual bandit (e.g. LinUCB), predicted utility is expressed in (4.13).

$$\widehat{U}(a|x_t) = \theta_a^T x_t + \beta\sqrt{x_t^T V_a^{-1} x_t}, \tag{4.13}$$

then select the best feasible a respecting constraints.

4.6 Privacy-preserving learning across healthcare sites

Hospitals and clinics may benefit from shared security intelligence, but the sharing of raw data is restricted. At this point, we integrate privacy-preserving learning. In the study, we use federated learning and DP for updates.

4.6.1 Federated learning

Let K sites train a shared anomaly model. Each K updates local parameters θ_k and share updates using the model in (4.14).

$$\theta^{r+1} = \sum\nolimits_{k+1}^{K} \frac{n_k}{\sum_j n_j} \theta_k^{(r)}, \tag{4.14}$$

where n_k is the number of local samples.

4.6.2 Differential privacy (DP) for updates

To reduce leakage risk, clip and perturb gradients using (4.15).

$$\tilde{g} = clip\ (g, C) + N(0, \sigma^2, C^2\ I), \tag{4.15}$$

Then, apply $\tilde{g}$ for training or transit DR-noise updates.

Table 4.3 lists the system, network, device and security parameters used in the simulations to evaluate the proposed IoMT-based adaptive cryptographic framework. The security and latency data for the cryptographic policy simulations were generated by the author in MATLAB R2021a using the MDP-based DRL framework.

Table 4.3 Parameters used for simulations

Variable/ metric	Description	Values/sample data/range
T	Time steps	$1 - 100s$
S	Cipher suites	$AES - 128, AES - 256, RSA - 2048, ECC - 256, ChaCha20$
S_i	Security strength for each suite i	$[0.4, 0.8, 0.9, 0.7, 0.5]$
L_t	Latency impact for each suite	$[0.2, 0.35, 0.9, 0.3, 0.15]$
O_t	CPU/Handshake Overhead	$[0.1, 0.3, 0.9, 0.4, 0.15]$
z_err	Anomaly score/reconstruction error	1.327
A_{range}	Anomaly score input for risk mapping	$0 : 1.1$
R_{low}	Risk probability (normal IDS)	≈ 0.20587
R_{high}	Risk probability (high IDS alert)	≈ 0.6570
R_t	Risk score range for security floor and rotation interval	$0 : 1.1$

(Continues)

Table 4.3 (Continued)

Variable/ metric	Description	Values/sample data/range
$S_{\min}$	Minimum required security floor	0.86
τ	Key-rotation interval (seconds)	300
α	Security weight for utility calculation	$0 - 1(step\ 0.1)$
U_i	Utility score per suite	$\alpha S_i - (1-\alpha)L_i$
$\eta(t)$	LinUCB convergence efficiency	$1 - e^{-t/20} + 0.05\text{randn}\ (1,100)$
W_{site}	FL model-weight deviation across sites	$20{\times}3\ matrix : A \sim \mathcal{N}(00.1), B \sim \mathcal{N}(0.05, 0.1),$ $C \sim \mathcal{N}(-0.05, 0.1)$
C_{range}	Gradient clipping constraint	$0.1 : 0.1 : 3$
g_{clip}	Clipped gradient values	$\min(C_{range}, 1.5)$
E_{FL}	DP training rounds	$1 - 40$
$\mathscr{L}_{base(e)}$	Loss for baseline training	$0.8e^{-e/4}$
$\mathscr{L}_{DP(e)}$	Loss for DP (noised) training	$0.8e^{-e/6} + 0.15\ rand(1, 40)$
N_{trials}	Random selection counts over 200 trials	$200(randi\ (5, 1, 200) mapped\ to\ suites\ S)$

4.7 Results and discussions

This section presents and discusses the simulation results used to evaluate the proposed AI-enabled adaptive cryptographic framework for IoMT-based healthcare systems. Performance is analysed in terms of security effectiveness (risk detection and policy response), communication and computation overhead, energy impact on constrained devices and end-to-end latency – particularly for alarm traffic – while also examining how the adaptive policy selection balances these metrics under varying network conditions and threat intensities.

4.7.1 Contextual dynamics of healthcare IoMT devices

Figure 4.5 visualises the observed context vector x_t as defined in (4.1). This graph captures the real-time fluctuations of three critical healthcare metrics: network latency (L_t), medical device energy consumption (E_t) and available bandwidth (B_t). Over 100 time steps, the simulation shows normalised values fluctuating between 0.1 and 0.4. For example, the energy proxy E_t is modelled with a mean of approximately 0.25, corresponding to a standard wearable sensor. At the same time, bandwidth B_t varies between 0.8 and 1.0 to reflect varying hospital Wi-Fi conditions. These inputs are vital because they define the state space the AI uses to determine whether a device has sufficient battery life to complete a high-security cryptographic handshake without failing.

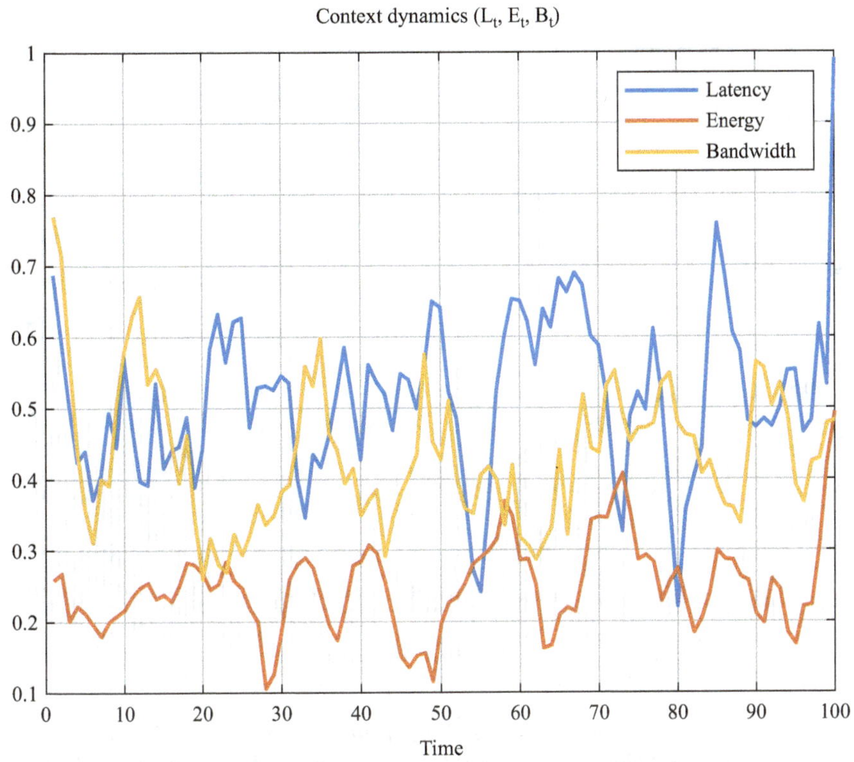

Figure 4.5 Context dynamics showing the time-varying changes in network and device conditions (e.g. latency, bandwidth and energy/CPU proxies) that drive the proposed risk-aware adaptive cryptographic policy selection

4.7.2 Real-time anomaly score and healthcare data integrity

Figure 4.6 presents the anomaly score A_t derived from the reconstruction error of device features as defined in (4.11). Under standard healthcare monitoring conditions, the reconstruction error typically remains below 0.4, indicating that traffic behaviour is authorised and safe. However, the simulation captures significant security events, notably a transient spike exceeding 1.6 near time step 32 and a sustained surge towards 1.4 after time step 75. These numerical deviations serve as real-time triggers for the risk engine, prompting the framework to escalate cryptographic protections to safeguard sensitive patient information from potential intrusions or unauthorised data tampering.

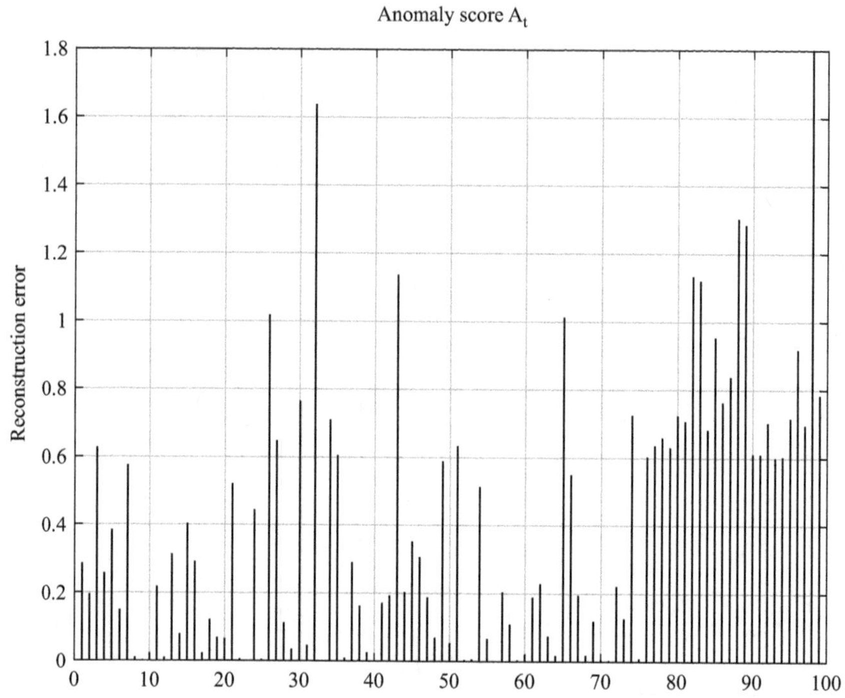

Figure 4.6 Anomaly score over time, illustrating deviations in IoMT traffic and device behaviour used by the risk engine to detect attacks and trigger adaptive cryptographic policy updates

4.7.3 Risk engine sensitivity and logistic mapping

Figure 4.7 illustrates the risk score R_t calculation defined in (4.12), which maps the anomaly score A_t to a normalised risk probability. This graph demonstrates a sigmoidal relationship between the anomaly score and risk sensitivity, highlighting how system responsiveness scales under different security states. In the High IDS Alert state (red), the system is significantly more proactive; at an anomaly score of 1.0, risk sensitivity is approximately 0.65, whereas in the Normal IDS state (blue), it remains much lower at about 0.18. This leftward shift means the system reaches a high-confidence sensitivity of 0.9 at an anomaly score of only 2.2. At the same time, the normal state requires a score of approximately 3.5 to achieve the same level. Even with a zero-anomaly score, the High Alert mode maintains a baseline sensitivity of roughly 0.27, ensuring a more defensive posture compared to the Normal mode's 0.05. Overall, the red curve indicates a lower intervention threshold, prioritising rapid threat detection over minimising false positives in critical healthcare environments.

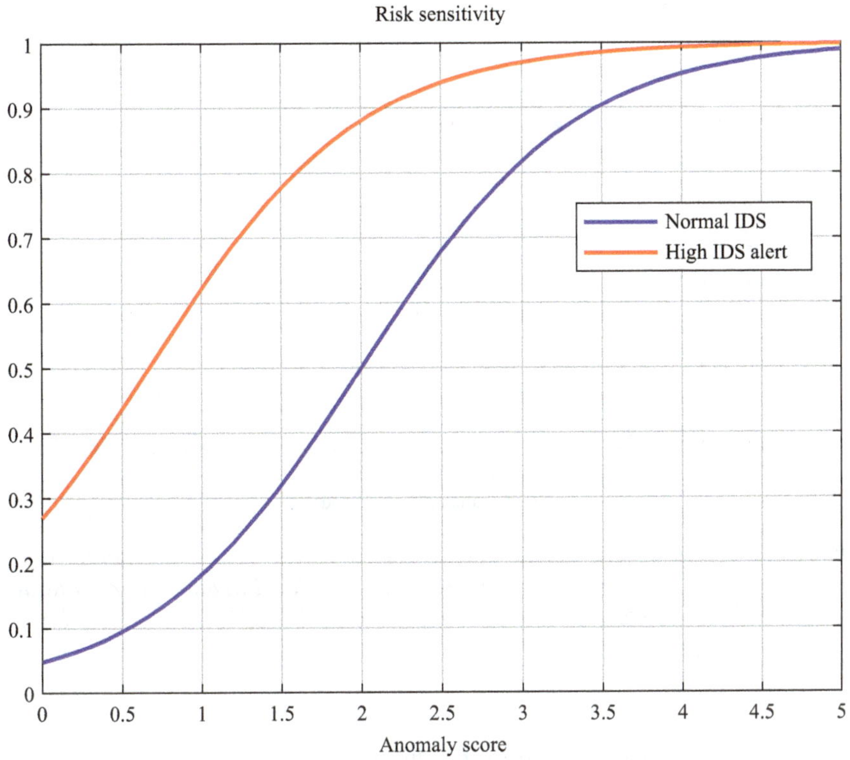

Figure 4.7 Risk sensitivity under normal and high-alert conditions, showing how the risk engine amplifies responsiveness to increasing anomaly levels to enable timely cryptographic hardening

4.7.4 Trade-off between security strength and computational overhead

Figure 4.8 illustrates the multi-objective trade-off between security strength and CPU/handshake overhead for various cryptographic suites. Quantitatively, RSA-2048 provides the maximum security strength of 0.9 but imposes a substantial 0.9 overhead, making it resource-intensive for IoMT sensors. Conversely, ChaCha20 and AES-128 represent lightweight alternatives with overheads of 0.15 and 0.1, respectively, though their security strengths are lower at 0.5 and 0.4. AES-256 occupies an efficient 'sweet spot', offering high security (0.8) with moderate overhead (0.3). This mapping allows the framework to intelligently select suites that meet compliance floors while preserving medical device battery life.

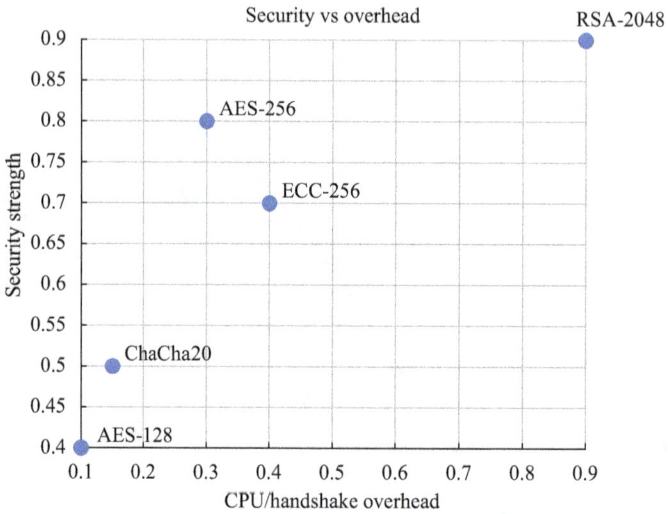

Figure 4.8 Security–overhead trade-off, showing how different cryptographic options balance the achieved security level against computational and communication cost under the proposed adaptive selection strategy

4.7.5 Adaptive security floor for healthcare compliance

Figure 4.9 visualises the minimum-security floor $S_{\min}(R_t)$ derived from the compliance constraint in (4.9). This graph establishes a mandatory baseline for cryptographic strength that scales linearly with the calculated risk score R_t. Quantitatively, when the system is in a low-risk state $R_t = 0$ the security floor is set at a baseline value of 0.2. As risk escalates to 1.0 due to detected threats, the required minimum-security strength increases to 0.8. The shaded grey region represents the non-compliant zone, where any cryptographic suite falling below the red boundary is automatically disqualified. This mechanism ensures that patient data is never protected by weak encryption during periods of elevated security risk.

4.7.6 Dynamic key-rotation intervals for medical privacy

Figure 4.10 depicts the rotation interval τ, a component of the action set a in (4.2). The graph shows that as the risk score R_t increases, the time between key updates decreases exponentially to prevent long-term data exposure. Quantitatively, at a zero-risk state, the interval τ is set to a maximum of 3600 seconds (one hour). However, as risk R_t approaches 1.0, the interval drops

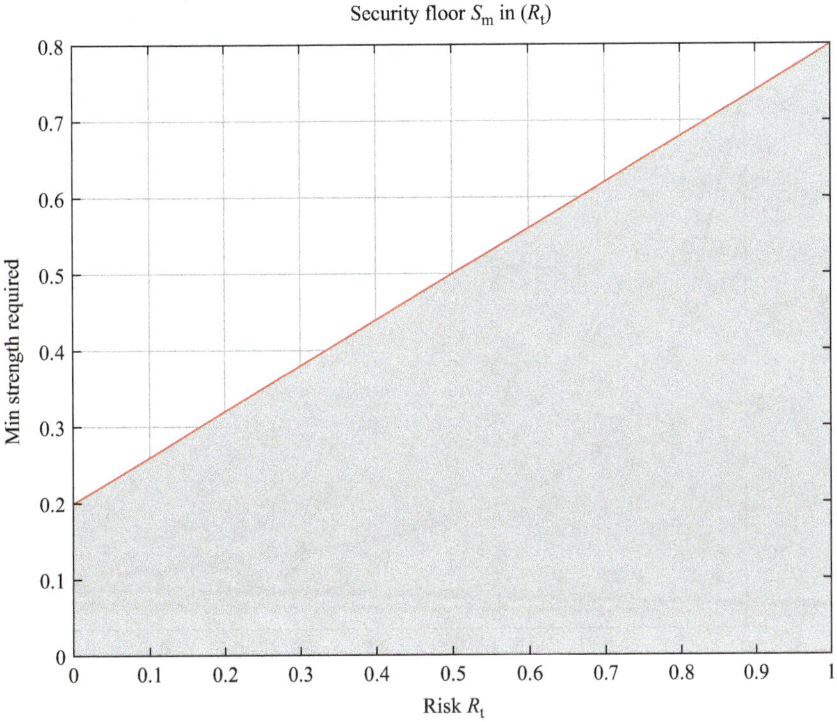

Figure 4.9 Risk-dependent security floor, illustrating the minimum enforced security level that increases with risk to ensure compliance and safety during elevated threat conditions

sharply to approximately 327 seconds. This frequent re-keying is a critical defence mechanism in healthcare settings, ensuring that even if a single session key is compromised, the window of exposure to sensitive patient data remains extremely narrow.

4.7.7 Utility sensitivity analysis for healthcare optimisation

Figure 4.11 shows the results of the constrained optimisation problem defined in (4.7), specifically focusing on the sensitivity of the security weight α. Each line represents the utility of a cryptographic suite as α scales from 0 (performance-focused) to 1 (security-focused). When α is below 0.2, ChaCha20 dominates with the highest utility score due to its low latency impact L. As α exceeds 0.6, AES-256 and RSA-2048 become the optimal choices. This crossover point reflects the framework's intelligence, as it quantitatively shifts from prioritising device battery life to prioritising medical record confidentiality.

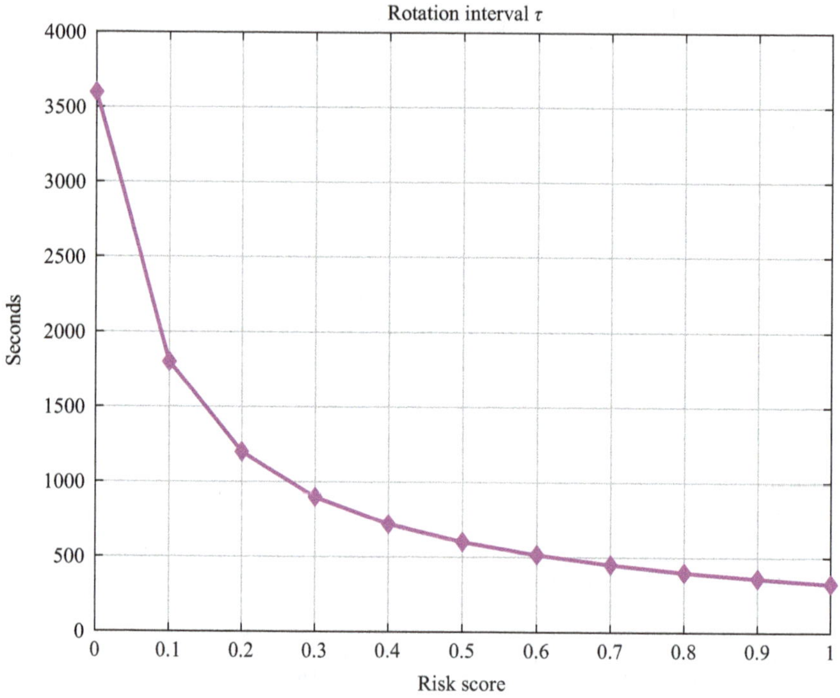

Figure 4.10 Key-rotation interval versus risk level, showing that the proposed framework shortens the rotation interval as risk increases to reduce key exposure while keeping longer intervals at low risk to limit overhead on constrained IoMT devices

4.7.8 LinUCB learning convergence in dynamic environments

Figure 4.12 demonstrates the learning progress of the Linear Contextual Bandit as described in (4.13). The graph plots normalised efficiency over 100 decision epochs. The system starts with an efficiency of approximately 0.2 during exploration. By epoch 40, efficiency exceeds 0.8 and eventually plateaus at approximately 0.95 and 1. This curve represents the AI learning which cryptographic suites work best for specific healthcare contexts. The reduced variance over time indicates convergence towards a stable, reliable policy for the hospital network.

4.7.9 Federated learning weight variance across medical sites

Figure 4.13 uses a boxplot to visualise the distribution of local parameter updates θ_k from (4.14) across different hospital sites. Sites A, B and C show deviations

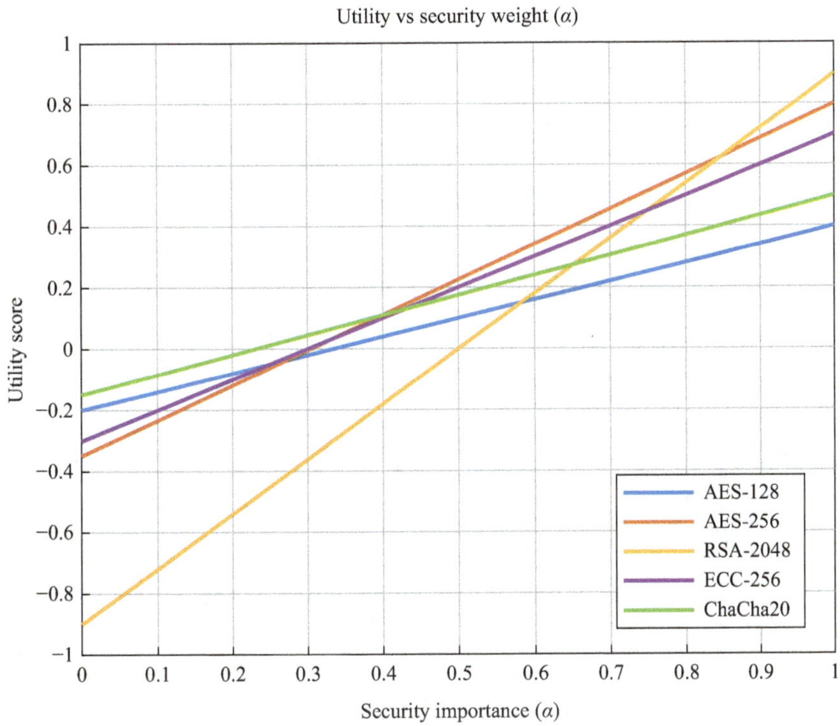

Figure 4.11 *Utility versus security-weight setting, illustrating how the objective weighting controls policy selection by shifting the optimal configuration from low-overhead options at lower security weights to stronger cryptographic schemes as the security weight increases*

typically constrained within ±0.1 of the mean, with Site C exhibiting slightly larger variance due to a smaller local dataset. The tight clustering indicates that the federated learning process successfully aggregates knowledge across institutions while maintaining a consistent global anomaly-detection model without sharing raw patient data.

4.7.10 Gradient clipping for privacy-preserving AI

Figure 4.14 illustrates the impact of the clipping constant C from the DP (4.15). The graph shows the gradient norm increasing linearly until it reaches the threshold C. As C increases from 0.1 to 3.0, the gradient norm saturates at approximately 1.5. This mechanism prevents outliers from disproportionately influencing the model and ensures individual patient attributes cannot be inferred from shared updates.

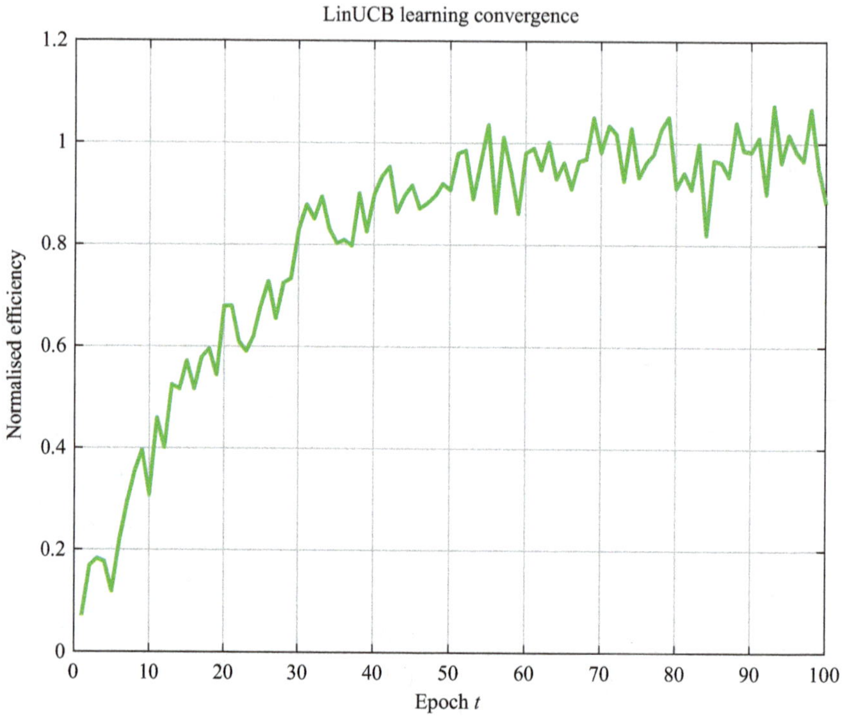

Figure 4.12 LinUCB learning convergence, showing the improvement in normalised policy-selection performance over training epochs as the contextual bandit rapidly learns and stabilises near-optimal cryptographic decisions under varying risk and resource constraints

4.7.11 Convergence of differentially private training

The DP Training Convergence plot shown in Figure 4.15 compares a smooth Baseline model (blue) with a Differentially Private (DP) model (red dashed), highlighting the impact of privacy-preserving noise on learning. The Baseline model begins with a loss of approximately 0.62 and converges rapidly, reaching a near-zero loss by round 20. In contrast, the DP model starts at a higher loss of about 0.82 due to the initial injection of privacy noise. Across the 20 training rounds, the DP curve exhibits pronounced fluctuations, reflecting the stochastic influence of the added noise. Although the overall trend remains downward, the DP model does not achieve the same level of optimisation, ultimately stabilising at a substantially higher residual loss of approximately 0.17.

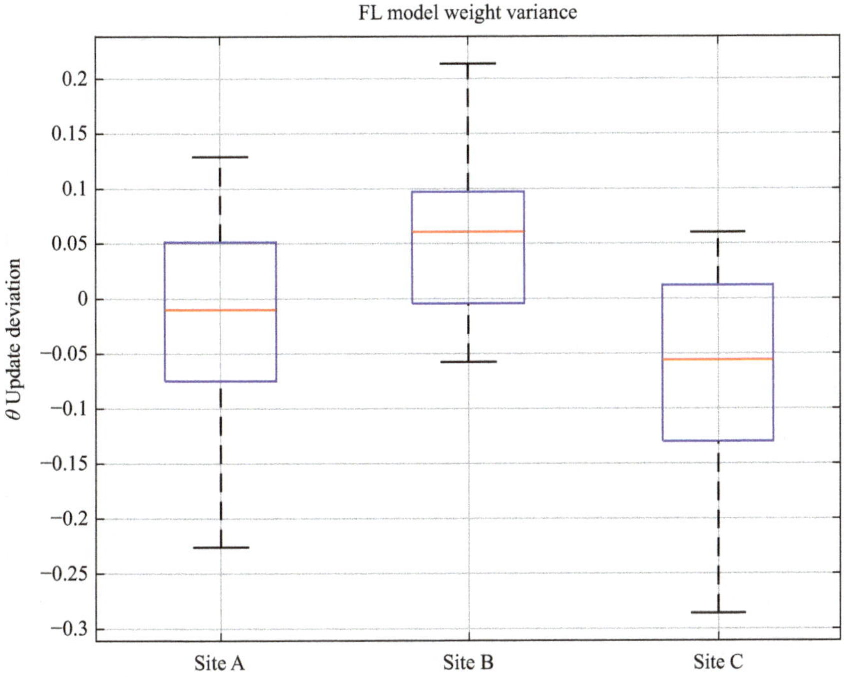

Figure 4.13 Federated learning model-weight variance across training rounds, illustrating how distributed client updates evolve and stabilise during aggregation, indicating convergence behaviour and the influence of heterogeneous hospital data on the global model

4.7.12 Distribution of automated policy selections

The policy-selection frequency chart in Figure 4.16 illustrates the distribution of encryption methods across 200 decision trials. AES-128 is the most frequently selected scheme, with 47 selections, followed by ECC-256 with 44. AES-256 and RSA-2048 occupy intermediate positions, with 40 and 35 selections, respectively, reflecting their use for moderate-to-high security requirements. ChaCha20 is the least used option, appearing 34 times, suggesting it is preferred primarily in low-risk, resource-constrained scenarios.

4.7.13 Summary of key results compared to a baseline

Table 4.4 presents the summary of key results compared to the baseline; it reports the percentage changes achieved by the proposed AI-adaptive cryptographic framework across risk sensitivity, security floor adjustment and key-rotation interval

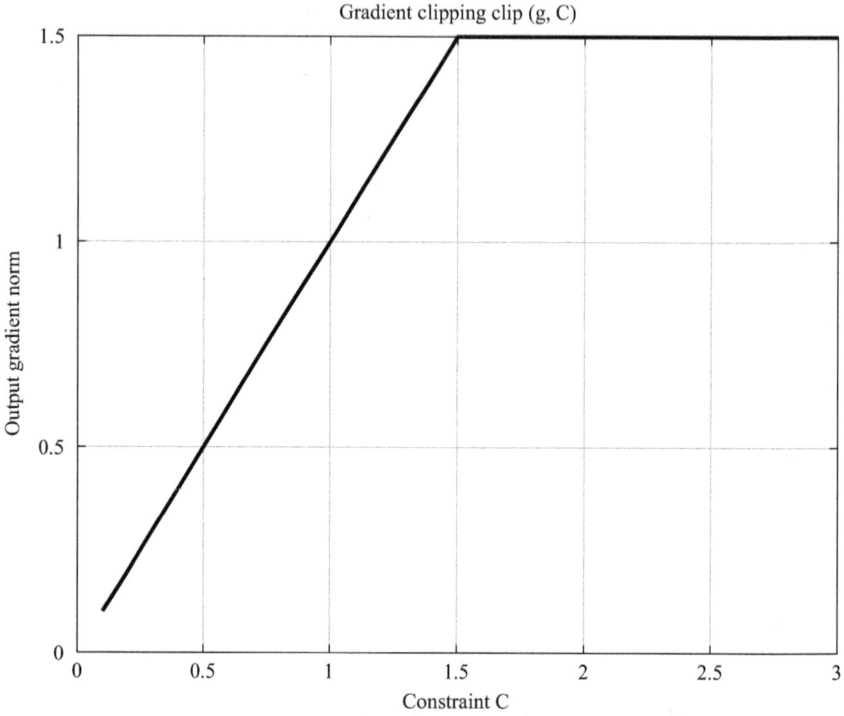

Figure 4.14 Gradient clipping threshold effect, showing how clipping limits the magnitude of client updates before aggregation to reduce sensitivity to outliers and privacy leakage, thereby stabilising differentially private federated training

under varying threat levels. It also highlights operational impacts, including overhead reduction and learning convergence, indicating where performance improves or where trade-offs (e.g. privacy-induced loss under DP) occur.

4.8 Conclusions

In conclusion, this work presented an AI-enabled, risk- and context-aware cryptographic orchestration framework for IoMT–edge–cloud intelligent healthcare, addressing the limitations of static cryptographic provisioning in the face of dynamic threats and resource constraints. The proposed design couples real-time anomaly-driven risk scoring with constrained policy selection (via contextual learning) to adapt cipher suites, key sizes, authentication strictness and key-rotation

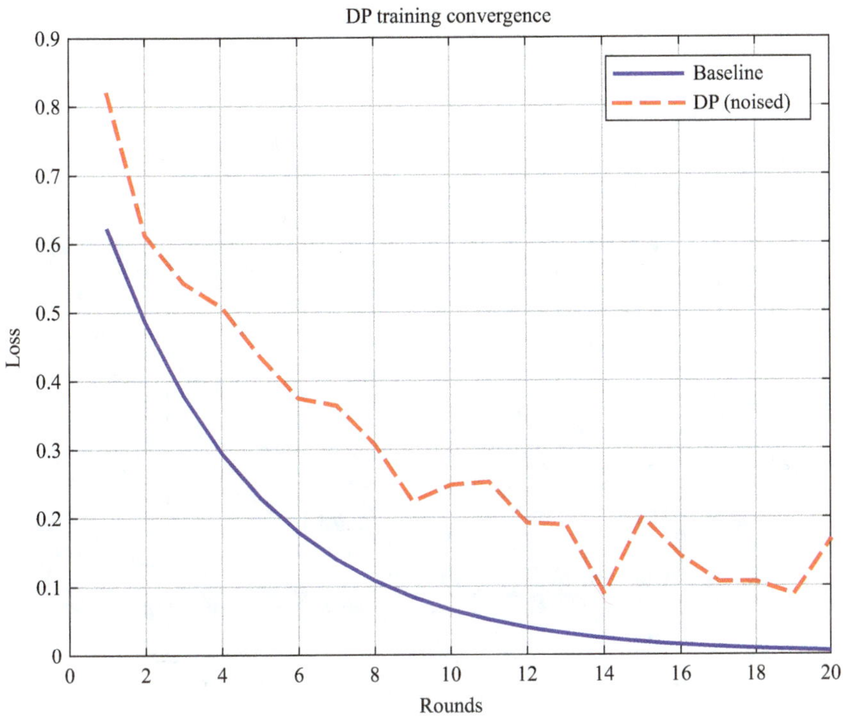

Figure 4.15 Differentially private training convergence, comparing loss reduction across federated rounds with and without differential privacy and highlighting the privacy–utility trade-off introduced by noise injection during model updates

cadence while enforcing healthcare safety requirements such as bounded latency for alarms and minimum-security compliance floors. Simulation results indicate that the framework maintains stable device operation (energy proxy ≈0.25), improves defensive responsiveness under high-alert conditions, scales the minimum-security floor from 0.2 to 0.8 as risk increases and reduces key-rotation intervals from 3600 s to 327 s (≈90.9 per cent reduction) during elevated threat periods; the learning engine converges quickly (≈0.95 efficiency by epoch 40), while privacy-preserving training introduces a measurable trade-off (residual loss ≈0.17) that remains acceptable for cross-site model improvement without sharing raw patient data.

Despite these gains, several limitations remain. First, the evaluation is simulation-based and may not fully capture absolute hospital heterogeneity, device failure modes and clinical workload dynamics. Second, the model relies

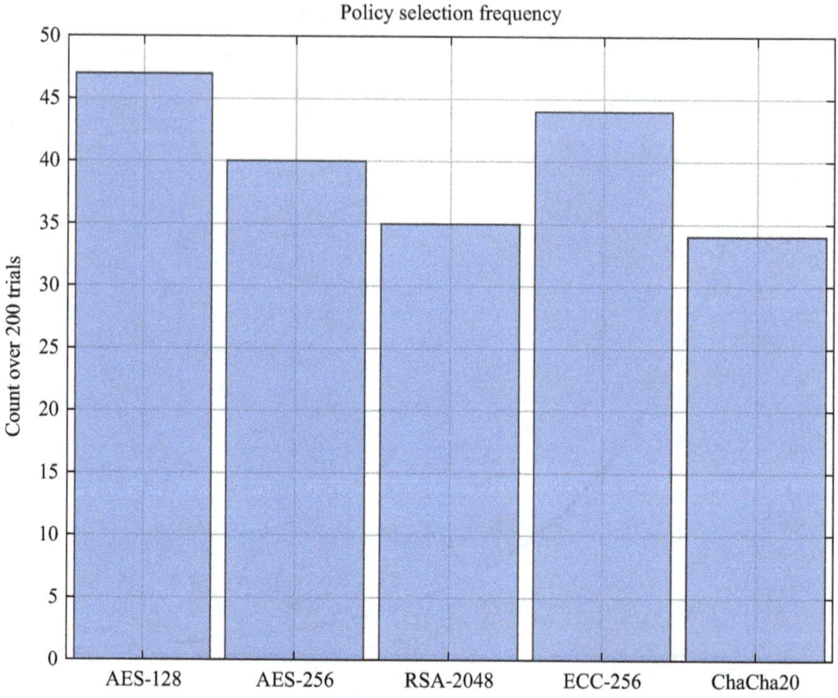

Figure 4.16 Policy-selection frequency over the evaluation trials, showing how often each cryptographic option is chosen by the adaptive recommender and indicating the overall preference towards configurations that balance security strength with latency and energy overhead

on abstracted latency/energy proxies and assumed cryptographic correctness, which may under-represent practical deployment issues such as intermittent connectivity, implementation vulnerabilities, hardware side channels and sophisticated adversaries (e.g. co-ordinated poisoning across federated participants). Future work will therefore focus on real-world testbed validation using representative IoMT devices and alarm traffic, tighter calibration based on measured energy/latency profiles and stronger robustness against adaptive attacks through poisoning-resistant federated learning, secure-aggregation hardening and explainable risk-to-policy mappings for compliance auditing. Additional directions include incorporating hybrid/post-quantum-ready key establishments where feasible, formally verifying safety constraints (alarm deadlines and minimum-security floors) and developing standardised policy and logging interfaces to support scalable deployment across multi-hospital environments.

Table 4.4 Summary of key results compared to baseline

Metric/key result	Baseline	Proposed/adaptive	% change vs baseline	Remark
Energy proxy level (Et)	≈0.25	≈0.25	0%	No change (stable operation)
Anomaly score (normal → attack)	<0.4	>1.6 (peak)	+300%	Attack trigger (not a benefit)
Risk sensitivity at A = 1.0	0.18	0.65	+261.1%	Improvement (more responsive)
A needed to reach sensitivity = 0.9	3.5	2.2	−37.1%	Improvement (earlier response)
Baseline sensitivity at A = 0	0.05	0.27	+440.0%	More defensive (may raise alerts)
Security floor $S_{\min}$ $(R: 0 \rightarrow 1)$	0.2	0.8	+300.0%	Improvement (stronger security)
Key-rotation interval τ $(R: 0 \rightarrow 1)$	3600 s	327 s	−90.9%	Improvement (smaller exposure window)
Example overhead (RSA-2048 → AES-256)	0.9	0.3	−66.7%	Improvement (lower overhead)
Learning efficiency (start → epoch 40)	0.2	≈0.95	+375.0%	Improvement (fast convergence)
DP training final loss	~0	0.17	—	Not improved (privacy cost)
Policy-selection frequency (200 trials)	—	AES-128: 47; ECC-256: 44; AES-256: 40; RSA-2048: 35; ChaCha20: 34	—	Distribution summary

References

[1] M. Rayan, A. A. Abdel-Hafez, H. H. Issa and K. A. Shehata, "MFM-SAT: A dynamic authenticated encryption suite for satellite communication security," *2025 42nd National Radio Science Conference (NRSC), Cairo, Egypt, 2025*, pp. 147–155, doi:10.1109/NRSC65659.2025.11018548.

[2] T. N. Turnip, B. Andersen and C. Vargas-Rosales, "Toward 6G authentication and key agreement protocol: A survey on hybrid post quantum cryptography," in *IEEE Communications Surveys & Tutorials*, vol. 28, pp. 3311–3345, 2026, doi:10.1109/COMST.2025.3567439.

[3] S. Ahmad, M. Arif, S. Mehfuz, J. Ahmad and M. Nazim, "Deep learning-based cloud security: Innovative attack detection and privacy focused key management," in *IEEE Transactions on Computers*, vol. 74, no. 6, pp. 1978–1989, 2025, doi:10.1109/TC.2025.3547150.

[4] G. Zheng, B. Gong, M. Waqas, I. Ahmad, H. Alasmary and S. Chen, "Multi-user oriented data sharing scheme for internet of medical things based on dual cryptography mechanism," in *IEEE Transactions on Sustainable Computing*, vol. 10, no. 6, pp. 1393–1408, 2025, doi:10.1109/TSUSC.2025.3619389.

[5] A. Mohammed, M. A. Al-Garadi, A. Mohamed, A. Al-Ali and M. Guizani, "Taxonomy-based lightweight cryptographic frameworks for secure industrial IoT: A survey," in *IEEE Internet of Things Journal*, vol. 12, no. 20, pp. 43296–43316, 2025, doi:10.1109/JIOT.2025.3595649.

[6] X. Zang, Z. Zheng, H. Zheng, X. Liu, M. K. Khan and W. Jiang, "HyperEye: A lightweight features fusion model for unknown encrypted malware traffic detection," in *IEEE Transactions on Consumer Electronics*, vol. 71, no. 2, pp. 5079–5089, 2025, doi:10.1109/TCE.2025.3558353.

[7] W. Wang, Q. Xie, H. Du, L. Zhang, J. J. P. C. Rodrigues and K. Wu, "Lightweight and fast authentication protocol for digital healthcare services," in *IEEE Transactions on Mobile Computing*, vol. 25, no. 1, pp. 403–418, 2026, doi:10.1109/TMC.2025.3593533.

[8] M. Zuraiz, M. Javed, N. Abbas, W. Abbass, W. Nawaz and A. H. Farooqi, "Optimizing secure and efficient data aggregation in IoMT using NSGA-II," in *IEEE Access*, vol. 13, pp. 118890–118911, 2025, doi:10.1109/ACCESS.2025.3587016.

[9] L. Yin, X. Du, Y. Cheng, J. Li, N. Tong and F. Li, "Decentralized IoMT architecture for privacy-preserving remote patient monitoring," in *IEEE Sensors Journal*, vol. 25, no. 14, pp. 27495–27512, 2025, doi:10.1109/JSEN.2025.3575941.

[10] M. Babar, M. U. Tariq, B. Qureshi, Z. Ullah, F. Arif and Z. Khan, "An efficient and hybrid deep learning-driven model to enhance security and performance of healthcare internet of things," in *IEEE Access*, vol. 13, pp. 22931–22945, 2025, doi:10.1109/ACCESS.2025.3536638.

[11] K. Balasubramanian, S. Ranjani Rajendran and S. Pati, "Complexity measures in biomedical signal analysis: A clinically-grounded survey across EEG, ECG, intracranial pressure, and photoplethysmogram modalities," in *IEEE Access*, vol. 13, pp. 155285–155304, 2025, doi:10.1109/ACCESS.2025.3603848.

[12] K. Rehman, T. Haseeb, G Saba,. Jeon and S. Al-Otaibi, "Multi-layer secured edge-AI enabled model for consumer health systems," in *IEEE Transactions on Consumer Electronics*, vol. 71, no. 2, pp. 4596–4603, 2025, doi:10.1109/TCE.2025.3555957.

[13] Y. Rbah, M. Mahfoudi, M. Fattah, Y. Balboul, S. Mazer and M. Elbekkali, "Advancements in machine learning-based intrusion detection in IoMT: Research trends and challenges," *2025 International Conference on Circuit, Systems and Communication (ICCSC), Fez, Morocco, 2025*, pp. 1–6, doi:10.1109/ICCSC66714.2025.11134951.

[14] Alsalman, "A comparative study of anomaly detection techniques for IoT security using adaptive machine learning for IoT threats," in *IEEE Access*, vol. 12, pp. 14719–14730, 2024, doi:10.1109/ACCESS.2024.3359033.

[15] W. Liu, F. Zhao, A. Shankar, C. Maple, X. Zhang and Y. Wang, "Explainable AI for medical image analysis in medical cyber-physical systems: Enhancing transparency and trustworthiness of IoMT," in *IEEE Journal of Biomedical and Health Informatics*, vol. 29, no. 4, pp. 2365–2376, 2025, doi:10.1109/JBHI.2023.3336721.

[16] G. R. Pradyumna, R. B. Hegde, K. B. Bommegowda, T. Jan and G. R. Naik, "Empowering healthcare with IoMT: Evolution, machine learning integration, security, and interoperability challenges," in *IEEE Access*, vol. 12, pp. 20603–20623, 2024, doi:10.1109/ACCESS.2024.3362239.

[17] G. Nithyavani and G. Naga Raja, "A comprehensive survey on security and privacy challenges in internet of medical things applications: Deep learning and machine learning solutions, obstacles, and future directions," in *IEEE Access*, vol. 13, pp. 188955–188989, 2025, doi:10.1109/ACCESS.2025.3588489.

[18] U. Chaudhary, M. Furqan Ali, A. Kumar, A. Sharma and D. K. Nalin Jayakody, "Unleashing the power of wireless communication in healthcare by empowering patient care and connectivity: A comprehensive survey," in *IEEE Access*, vol. 13, pp. 117239–117299, 2025, doi:10.1109/ACCESS.2025.3578344.

[19] A. Andreou, C. X. Mavromoustakis, E. K. Markakis, A. Bourdena and G. Mastorakis, "On the synergy of IoMT devices and ceiling-mounted systems for advanced medical data analytics," in *IEEE Access*, vol. 13, pp. 38255–38267, 2025, doi:10.1109/ACCESS.2025.3546495.

[20] C. Bollineni, M. Sharma, A. Hazra, P. Kumari, S. Manipriya and A. Tomar, "IoT for next-generation smart healthcare: A comprehensive survey," in *IEEE Internet of Things Journal*, vol. 12, no. 16, pp. 32616–32639, 2025, doi:10.1109/JIOT.2025.3570188.

[21] X. Cheng, X. Xing, W. Li, H. Xue and T. Can, "An energy-efficient and privacy-aware MEC-enabled IoMT health monitoring system," in *IEEE Transactions on Computers*, vol. 74, no. 9, pp. 2936–2949, 2025, doi:10.1109/TC.2025.3576944

[22] U. O. Matthew, R. L. Rosa, M. Saadi and D. Z. Rodríguez, "Interpretable AI framework for secure and reliable medical image analysis in IoMT systems," in *IEEE Journal of Biomedical and Health Informatics*, doi:10.1109/JBHI.2025.3591737.

[23] Y. Jiao, M. M. Razaq, L. Peng and P. -H. Ho, "Enhancing IoMT security using large multimodal models," in *IEEE Network*, vol. 39, no. 4, pp. 112–119, 2025, doi:10.1109/MNET.2025.3545692.

[24] F. Ullah, L. Mostarda, D. Cacciagrano, M. J. F. Alenazi, C. -M. Chen and S. Kumari, "Federated edge intelligence for enhanced security in consumer intermittent healthcare devices using adversarial examples," in *IEEE Transactions on Consumer Electronics*, vol. 71, no. 2, pp. 4574–4585, 2025, doi:10.1109/TCE.2024.3511615.

[25] U. Islam, H. Ullah, N. Khan, I. Ahmad and K. Saleem, "Adaptive federated learning framework for privacy-preserving consumer-centric IoMT: A novel secure data collaboration model," in *IEEE Transactions on Consumer Electronics*, vol. 71, no. 4, pp. 10134–10151, 2025, doi:10.1109/TCE.2025.3606642.

[26] M. A. Khatun, S. F. Memon, C. Eising and L. L. Dhirani, "Machine learning for healthcare-IoT security: A review and risk mitigation," in *IEEE Access*, vol. 11, pp. 145869–145896, 2023, doi:10.1109/ACCESS.2023.3346320.

[27] U. Zukaib, X. Cui, C. Zheng, M. Hassan and Z. Shen, "Meta-IDS: Meta-learning-based smart intrusion detection system for internet of medical things (IoMT) network," in *IEEE Internet of Things Journal*, vol. 11, no. 13, pp. 23080–23095, 2024, doi:10.1109/JIOT.2024.3387294.

[28] S. Kurma, K. Singh, A. Paul, S. Mumtaz and C. -P. Li, "Dual-LLM integration with reconfigurable intelligent surface for healthcare networks," in *IEEE Transactions on Communications*, vol. 73, no. 11, pp. 11323–11339, 2025, doi:10.1109/TCOMM.2025.3587065.

[29] J. Dutta and D. Puthal, "Advancing eHealth in society 5.0: A fuzzy logic and blockchain-enhanced framework for integrating IoMT, edge, and cloud with AI," in *IEEE Access*, vol. 12, pp. 195710–195730, 2024, doi:10.1109/ACCESS.2024.3520799.

[30] A. Aljuhani, A. Alamri, P. Kumar and A. Jolfaei, "An intelligent and explainable SaaS-based intrusion detection system for resource-constrained IoMT," in *IEEE Internet of Things Journal*, vol. 11, no. 15, pp. 25454–25463, 2024, doi:10.1109/JIOT.2023.3327024.

[31] P. D. Paikrao, A. Mukherjee, U. Ghosh, Goswami, M. Novak and D. K. Jain, "Data driven neural speech enhancement for smart healthcare in consumer electronics applications," in *IEEE Transactions on Consumer Electronics*, vol. 70, no. 2, pp. 4828–4838, 2024, doi:10.1109/TCE.2024.3387740.

Chapter 5

AI-driven cryptographic techniques in IoT-based healthcare systems: a review and outlook

Samuel Ibukun Olotu[1]

Abstract

The healthcare industry has seen advances in smart clinical decision-making, remote diagnosis and continuous patient monitoring, all enabled by the rapid diffusion of Internet of Things (IoT) technology. Nevertheless, the application of IoT in healthcare poses significant security and privacy challenges, given that highly sensitive medical data is transmitted via a range of low-resource IoT devices. Conventional cryptographic techniques, although they may operate effectively in standard networks, often do not meet the dynamic, scalable and low-latency requirements of IoT-based healthcare systems. Artificial Intelligence (AI) has proven useful for improving cryptographic mechanisms and adapting security strategies in real time. This chapter presents a review of AI-based cryptographic techniques for securing IoT-based healthcare systems. The techniques examined include lightweight cryptography, homomorphic encryption, blockchain-enabled cryptography and machine-learning-based cryptography. The lessons learned and future directions are highlighted.

Keywords: Internet of Things (IoT); healthcare systems; artificial intelligence; machine learning; cryptography algorithms; remote patient monitoring; wearable devices; sensors; Internet of Medical Things (IoMT)

5.1 Introduction

The Internet of Things (IoT) involves the interconnection of various physical devices, such as sensors, and the use of the Internet to communicate and share data. By enabling remote monitoring, customised treatment plans and effective healthcare delivery, IoT technologies are used in healthcare to enhance patient care and

[1]Department of Information Technology, School of Computing, Federal University of Technology, Akure, Nigeria

outcomes. IoT technology allows medical devices and healthcare providers to effortlessly exchange data, monitor and diagnose health issues in real time [1].

Given the sensitive nature of the data that IoT devices collect, transmit and process, the integration of IoT in healthcare raises concerns about security and privacy. To ensure privacy and security, sensitive patient data collected and transmitted must be protected against alteration and unauthorised access. Cryptographic techniques are used to address this problem and safeguard personal data. However, while they can provide security and privacy for sensitive and personal data, most standard encryption methods are resource-intensive, making them difficult to implement on IoT devices with limited memory, computing power and battery life [2,3].

AI significantly enhances this security framework by providing state-of-the-art analytics and predictive capabilities. Real-time network traffic monitoring, anomaly detection and potential security threat identification are enabled by the use of machine learning algorithms. Proactive threat identification and response are also made possible by these algorithms' ability to detect typical behaviour patterns in urban infrastructure and identify abnormalities that might indicate cyberattacks [4,5]. Despite its promise, the use of AI in IoT cryptography remains in its infancy and faces numerous research challenges, including algorithmic complexity, energy efficiency and the need to balance security and usability.

By methodically examining current AI-driven cryptography methods relevant to Internet of Things-based healthcare systems, this review addresses these issues. It examines performance measures relevant to medical applications, identifies gaps in current research and highlights recent advances. Additionally, the paper offers a forward-looking outlook on the future of AI-assisted cryptography, highlighting areas for innovation and standardisation. Key findings for researchers and practitioners are provided by the work's synthesis of recent research: AI may substantially enhance the security, effectiveness and resilience of healthcare IoT systems, enabling safer, more intelligent and more dependable patient care solutions. This work reviews recent research on AI-driven cryptographic techniques for IoT-based healthcare systems.

5.1.1 Contribution to knowledge

The significant contributions of this chapter are as follows.

(i) Discussion on existing healthcare systems, highlighting the key enabling technologies, IoT systems and applications. The enabling technologies and concepts covered include sensors, communications, artificial intelligence, blockchain, edge computing and 5G. Wearable technology and the Internet of Medical Things (IoMT) are examples of IoT systems and remote patient monitoring.

(ii) Discussion on artificial intelligence-driven cryptographic techniques for IoT-based healthcare systems. The application areas covered include cryptographic design, cryptanalysis, key management/authentication, intrusion detection and anomaly detection, lightweight cryptography, privacy-preserving encryption and post-quantum cryptography.

(iii) A review of recent works that proposed AI-driven cryptographic techniques in healthcare systems, outlining the AI approaches, IoT applications and cryptographic techniques and challenges addressed in the literature. The review article considered papers from 2020 to 2026.

5.1.2 Chapter organisation

The remainder of the chapter is structured as follows.

Discussions of existing IoT-based healthcare systems, including their aims, AI approaches, IoT applications and cryptographic techniques, are covered in Section 5.2. The related surveys and existing research work on AI-driven cryptography in IoT-based healthcare are presented in Section 5.3. The AI-driven cryptographic techniques are discussed in Section 5.4. The lessons learned and future directions are enumerated in Sections 5.5 and 5.6, respectively. Section 5.7 concludes the chapter.

5.2 Internet of Things (IoT) and healthcare systems

IoT involves using the Internet to connect a variety of physical items, devices and sensors so they can communicate and share data [1]. The goal of this new technology is to integrate physical items with embedded technology that can sense, communicate and interact with their surroundings or internal states.

Medical and healthcare are among the most significant IoT domains, with applications ranging from hospital-centric to patient-centric. IoT technologies are being used in healthcare to enhance patient care and outcomes, offering new opportunities for efficient healthcare delivery, individualised treatment regimens and remote monitoring. The technology enables patients to communicate with their physicians, schedule appointments and view their medical records [6]

There is an increasing need for affordable healthcare systems to be created, which can both handle and provide a variety of medical services at lower costs because of wild population growth and more chronic health issues [7–10]. Through the Internet of Things, which has become quite a significant growth area recently, developing healthcare monitoring systems has come about. The aim of the Internet of Things (IoT) healthcare monitoring system focuses on accurate tracking for people as well as joining many types of services or objects worldwide by use of the Internet for gathering, sharing, monitoring, saving data and also to analyse the information from these particular objects [11].

The Internet of Things (IoT) concept enables all connected physical objects to be accessed and managed remotely through smart applications, including intelligent homes, smart cities and smart healthcare. Using sensor networks for human body monitoring plays a significant role in disease diagnosis and patient monitoring, and these are essential for healthcare. In addition, this kind of data is easy to get anytime and from anywhere in the world [12]. People from certain regions or those with severe injuries may face difficulties accessing hospitals. As a result, patients can use video calls to consult their doctors and improve their health while reducing costs and time. Such a system lets patients record health problems on their mobile devices [13].

IoT may significantly lower healthcare costs by automating repetitive tasks, streamlining processes and lowering the need for expensive treatments [12]. IoT technology streamlines hospital operations by automating administrative tasks. This reduces expenses by saving time and doing away with the necessity for a sizable administrative staff. Additionally, personnel expenses and mistake risk can be decreased by automating repetitive tasks like medication distribution and vital sign monitoring. Most importantly, the early detection and continuous monitoring capabilities of IoT reduce the need for expensive medical procedures, decrease hospitalisation rates and prevent severe medical complications, thereby substantially lowering healthcare costs.

5.2.1 IoT technologies in healthcare

The review of the technologies for IoT-based healthcare systems presented in [12] include sensors, communication, artificial intelligence, blockchain, edge computing and 5G.

(1) **Sensor technologies**

Sensors are an essential component of Internet of Things (IoT) devices, as they can measure and communicate data from a wide range of sources, including blood pressure, heart rate, temperature, glucose levels and more. Such sensors can be embedded in medical equipment, wearable devices and even the human body. The common existing sensor types include Electrocardiogram, Pulse Oximeter, blood pressure, glucose, temperature and respiratory. Other cutting-edge sensor technologies have been developed by researchers.

The online life signs sensor, called the VITALS system, is utilised to measure the heart rate, blood pressure, respiration rate and body temperature. It consists of a CPU, a wireless communication module and a big data analytics system. Vital signs are gathered and sent to the analysis system every 30 minutes. The system has been used for one week during the test and has shown the best care options, especially for cases with limited access to healthcare. [14].

The proposed heart rate sensor is based on the MAX30102, which emits visible light at 650–670 nm. After the wrist-worn sensor records the heart rate signal as a comparatively flat wave, the detrend method is utilised to acquire precise data [15].

Bourechak *et al.* [16] unveiled a cutting-edge sensor employing micro, pyramid-assisted piezoelectric film (MPF) technology for continuous, non-invasive blood pressure monitoring. The results demonstrated that the MPF sensor's performance is on par with that of existing commercial solutions.

A pulse oximeter is also used to monitor body temperature, blood oxygen saturation and heart rate [17]. The work used some sensors and an infrared lighting for testing the temperature of the body without any contact, and the obtained information is displayed on an LCD and on a web browser. The study [18] presented an improved capacitive humidity sensor in paper form, which gives very good humidity sensitivity, high results and rapid responses with fast recovery. The sensor proved reliable in tracking both the depth of the patient's breathing and their various breathing patterns.

(2) **Communications**

Communication technologies such as Wi-Fi, Bluetooth and cellular networks make it possible for IoT devices to share data with medical specialists. These technologies collectively facilitate a smooth, instant exchange of information among medical devices, healthcare staff and patients. Some typical communication tools that assist in linking IoT devices in healthcare are Bluetooth, Wi-Fi and cellular networks. Wearable health equipment greatly benefits from Bluetooth's close-range, low-power connections, which enable real-time monitoring and data transfer to patients' phones. Wi-Fi makes it easier for healthcare facilities to share data efficiently, including medical imaging and electronic health records. Wearable healthcare devices are a great match with Bluetooth's close-range, low-power connectivity that allows for real-time monitoring and data transfer to the patient's mobile phone [19]. Wi-Fi helps healthcare facilities to share data more efficiently, such as medical images and electronic health records.

Cellular technology [20], especially 4G and 5G systems, is extremely useful for telemedicine applications since the wide coverage of such networks allows physicians and other healthcare workers to have access to vital information about the patients at any time and in virtually any place. Together, all these technologies form a comprehensive system that facilitates the decision-making process through quicker responses, real-time communications, and, finally, better treatment of the patients. Nonetheless, data security issues are among the most pressing concerns for the adoption of such technologies.

(3) **Artificial intelligence**

Within the healthcare industry, AI algorithms are becoming indispensable assistants, especially when teamed up with data from IoT devices. These devices keep generating massive amounts of data, which can be used by AI algorithms to quickly provide patient health information and enable customised medical care.

Machine learning is a type of artificial intelligence that can evaluate trends in IoT data to forecast the progression of diseases and recommend treatment with personalised options. These models can detect irregularities in a patient's health metrics when trained on historical data, thereby aiding early diagnosis of health issues that may occur. AI algorithms can monitor vital signs, levels of physical activity, sleep patterns and other associated characteristics in the context of IoT in healthcare by analysing data from wearable devices and other IoT sensors [21,22].

Through continuous monitoring and analysis of data, AI-powered systems have the potential to alert healthcare practitioners to early signs of health deterioration or relapse, thus enabling them to intervene at the right time and consequently improve patient outcomes. Moreover, AI algorithms are capable of designing customised treatment plans based on the IoT data collected. By considering the distinctive features of patients, their medical records as well as the real-time data, these algorithms can scrutinise vast amounts of information and pinpoint the most suitable treatment plans for the individual needs of patients [23].

AI and IoT collaboration will probably change the healthcare industry to a patient-centred, efficient one. The change is a big step towards the use of technology in healthcare and treatment personalisation. It is characterised by a great increase in the accuracy of prediction, which is coupled with a gentle, highly personalised treatment methodology.

(4) **Blockchain**

The integration of IoT and blockchain technologies is capable of increasing the security of the data and making data sharing between healthcare providers more efficient. Data decentralisation and the unchangeable features of the blockchain can be used to ensure that patients' data is safe from any kind of unauthorised access, alteration or leak. Blockchain provides a distributed ledger system that guarantees every record or data submission is encrypted and connected to earlier records, creating an unchangeable chain [24].

Medical staff can securely access and share vital medical data of patients across organisational and geographic borders using blockchain technology. By giving a healthcare professional special access to their data, patients retain control over their personal information. Smart contracts are software applications that function autonomously on the blockchain. Besides automating data, they can facilitate the sharing of agreements and the execution of privacy standards; thus, patients are provided with more transparency and control over their data, as shown in Figure 5.1.

The patient and health providers can access the system through the electronic health record (EHR) system using the available authentication system. The blockchain subsystem stores all necessary health records in the cloud storage and places the pointer to the location of the data on a secure smart contract.

(5) **Edge computing**

With its great benefits in processing data in real time and increasing speed, edge computing is the backbone of IoT-based healthcare. It addresses the problem of network latency and achieves a quick response time in the healthcare Internet of Things [25]. Critical healthcare applications could be at risk of delays if the traditional method of sending data to the cloud server from the IoT devices for processing and analysis. However, edge computing handles the data locally, near its source, thereby lowering the network latency [26].

Edge computing is a technology that reduces the distance data travels between an IoT device and a remote cloud server by immediately processing the data on the device [27]. This capability allows for healthcare data to be processed near-instantly, which is vital to applications such as emergency response systems, remote patient monitoring and continuous health monitoring. Edge computing also supports the raising of security levels in data

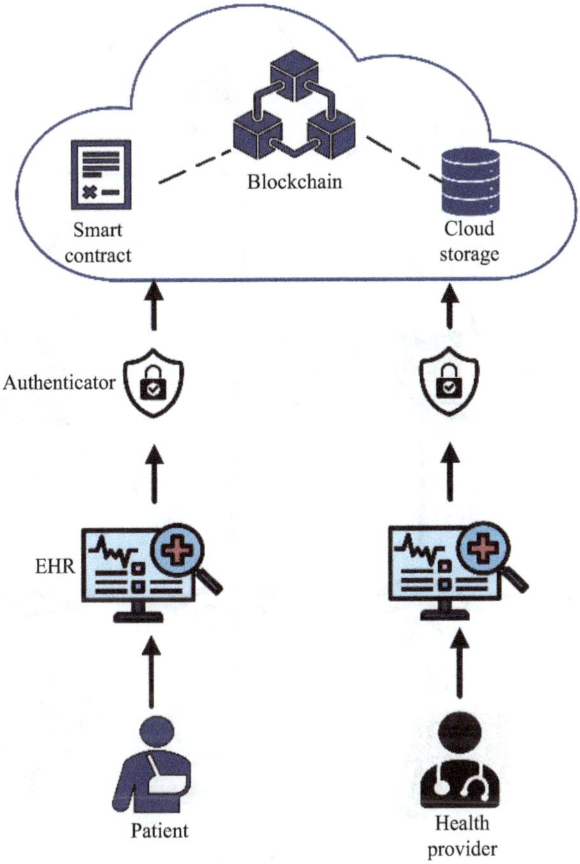

Figure 5.1 Blockchain in healthcare systems

confidentiality and the integrity of data in healthcare systems with IoT technology. Since edge computing stores and processes sensitive health-related information on the devices themselves, it can reduce the exposure of data to hackers and limit the occurrence of data breaches [28].

This strategy focuses on privacy as one of the key issues, and therefore, it supports a strong data security framework in which data confidentiality is really preserved. It is also very clear in the paper that edge computing offers another advantage of being inherently more resilient to network disruptions and thus capable of offline operation. In such cases of poor or no Internet connectivity, the edge devices can keep on functioning and processing data locally, thus assuring the continuation of the lifeline operations and healthcare services [29].

Figure 5.2 shows edge computing, where the IoT devices transmit their data to intermediate edge devices, which preprocess the data and only forward the refined data to the central cloud server.

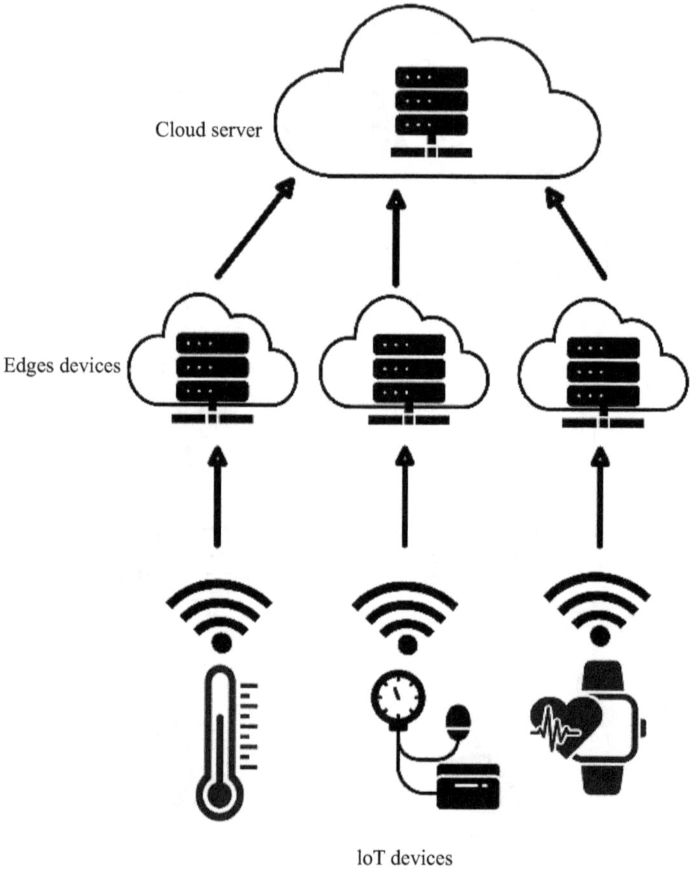

Figure 5.2 Edge/fog computing

(6) **5G**
5G networks support real-time communication in the case of IoT-based healthcare; thus, healthcare providers can continuously engage with patients and do remote monitoring [30]. With the higher speed and low latency offered by 5G networks, data can be transmitted and received almost instantly so that responses to medical emergencies can be very quick. The capacity of 5G networks to support a large number of connected IoT devices is one of their advantages in the healthcare industry. 5G networks will allow healthcare providers not only to efficiently collect real-time data from patients but also to perform data analysis locally with the assurance that the networks will enable the transmission of a large volume of data simultaneously. The reliability feature in 5G networks ensures constant connection for medical applications, especially where real-time communication and continuous monitoring are essential. Such reliability will be a great advantage, especially in remote areas or emergencies when getting

medical help is very difficult. Through 5G networks, medical professionals and patients can still interact face-to-face even though they are miles apart; hence, medical services can be delivered effectively at patients' homes. Due to improved data transfer speeds, physicians can access and evaluate patient data more quickly, resulting in faster diagnosis, treatment adjustments and interventions [31]. Furthermore, 5G networks provide a platform for innovative medical technologies such as telepresence, virtual reality (VR) and augmented reality (AR). As a result, these technologies can radically transform medical education, surgical training and remote patient care by enabling new modes of experience and remote collaboration among healthcare professionals [32,33].

5.2.2 The Internet of Medical Things (IoMT)

The Internet of Medical Things (IoMT) refers to wearable sensors and implantable devices, such as smartwatches, that are connected to the Internet and used by both patients and healthcare providers. Telemedicine is the use of information and communication technology to deliver clinical treatment and transmit health information in remote regions. Because this strategy removes the barrier of distance and improves access to healthcare, it is particularly useful in remote areas with limited health services [34].

The term Internet of Medical Things (IoMT) refers to a collection of biological systems and algorithms that could interact with healthcare data management infrastructures. By fostering patient-physician engagement and enabling the sharing of medical records over a secure platform, it has the potential to reduce unnecessary hospital stays and the strain on medical practices [35].

5.2.3 Remote patient monitoring

Remote patient monitoring (RPM) is a healthcare delivery standard that builds on current connectivity standards and allows carers to remotely monitor patients by utilising smart body sensors [36]. For many patients, RHM is a crucial healthcare option in their daily lives. Nearly every medical symptom is monitored by a gadget that senses bodily characteristics for patient monitoring, which is essential for security. This enables the digital transfer of physiological and disease-related patient data by phone, the Internet or videoconferencing from the patient's home to a medical facility, offering clinical feedback. Early illness detection reduces mortality and hospitalisations while increasing patient awareness and self-management. Some types of remote patient monitoring are glucose monitoring, heart rate monitoring, robotic surgery monitoring and connected inhaler monitoring. IoT healthcare sensors are essential to contemporary healthcare, as they allow ongoing patient monitoring and data collection. Vital signs, medication adherence and illness management are just some of the health-related data that these sensors collect and provide to medical specialists for real-time monitoring and analysis [37].

Glucose monitoring helps the diabetic patient to review their varying blood glucose levels at different times of the day and hence make the best treatment decisions.

Also covered are the changes in blood glucose levels in detail, like their length, frequency, direction, amount and causes. Continuous glucose monitoring gives a lot of extra information compared to three to four blood sugar tests, which are the normal intensive glucose monitoring. According to vital signs, one of the most important indicators of overall physical health is the heart rate. While typical HR monitors sometimes require skin contact, remote photoplethysmography delivers contactless RHM by employing a video camera to detect slight light changes in the skin. Robot-assisted minimally invasive surgery is currently a popular choice for many reasons, one of which is the fact that it offers better surgical instruments and ergonomics as compared to traditional laparoscopic procedures. Therefore, major surgical specialities such as urology and gynaecology have mostly relied on RMIS. Nonetheless, this is beginning to change as more and more hospitals are getting robotic surgical systems.

Inhaler systems have the potential to provide objective data about a patient's medication adherence and inhaler technique instantly when linked to mobile devices. The energy supply and the sensors for assessment are housed in the e-modules. While some of these are integrated into the inhaler itself, others are external attachments. Clinical research has shown that inhalers that can monitor when a dose is prepared or activated may improve adherence and outcomes [38].

5.2.4 Wearable devices

This refers to technologically equipped devices that can track several patients' physiological parameters and activity levels, enabling ongoing health monitoring and tracking as well as individualised treatment interventions [34]. Sensors typically gather input signals from their physical environment and use those signals to generate a reaction action. Sensors wirelessly provide the vital signs or signals to the Body Area Networks (BAN) control unit after receiving them. Wearable sensors can play a significant role in early disease recognition by continuously monitoring vital signs that include blood pressure, body temperature, serum cholesterol, glucose level, arterial oxygen saturation and breath rate. In addition, wearable technology is mostly externally fixed to the body or internally embedded with textile fibres, elastic bands or patient clothing. Wearable technologies are also used to monitor patients' activities and their physiological signals. Further, the vital signs detected by wireless medical sensors are sent to a remote server for diagnostic purposes.

5.3 Literature review

This section discusses the related surveys and the analysis of existing research.

5.3.1 Related surveys

This section examines relevant surveys that have been conducted on the use of AI-powered cryptography techniques for Internet of Things-based healthcare systems. A summary of the related surveys is also shown in Table 5.1.

Table 5.1 Summary of related surveys

Author	Year	Aim	AI approaches	IoT applications	Cryptographic technique(s)
Kwon *et al.* [39]	2024	AI-driven techniques for threat detection and prevention in e-health	Machine learning, federated learning	Internet of Medical Things (IoMT)	Homomorphic secret sharing
Li *et al.* [41]	2025	A review of AI-based cybersecurity strategies designed for A-IoT environments	ML, DL, FL, Swarm Intelligence, Generative AI, generative adversary networks, digital twins	Autonomous Internet of Things (A-IoT) ecosystems	Lightweight Encryption
Lagorio *et al.* [40]	2025	Study existing lightweight cryptographic schemes, highlighting their applicability for real-time patient monitoring devices deployed in IoT-enabled hospital environments	federated learning, blockchain and explainable AI	Internet of Medical Things (IoMT)	Lightweight cryptographic (optimised AES Variants (AES-Lite), Elliptic Curve Cryptography (ECC-160), Lightweight Hash Functions)
Mahmmod *et al.* [42]	2025	A survey of ML-empowered IoMT in healthcare for efficient and more secure healthcare	Deep learning, machine learning	Internet of Medical Things (IoMT)	Lightweight cryptography
This work	2026	A survey of AI-based cryptographic techniques for IoT-based healthcare systems	Deep learning, federated learning, machine learning, Decision Tree, SVM, logistic regression, XGB, CNN, DL, RF, MLP, KNN, dynamic consensus algorithms, reinforcement learning, ANN, deep neural network, artificial intelligence heuristics	IoT-based medical equipment, IoT-based healthcare, IoMT, IoT-based healthcare applications, healthcare applications, healthcare IoT devices, IoT-based middleware, Cloud-based clinical IoT (C-IoT)	Attribute-based encryption, certificate-based cryptography, lightweight cryptography, SHA3-224 hash function, homomorphic encryption, symmetric encryption, elliptic curve cryptography (ECC), AES, PRESENT, MSEA, LEA, XTEA, SOMON, PRINCE and RECTANGLE

The study [39] presented a review of AI technologies in cybersecurity for e-health systems. The work discussed how AI improves data security by implementing threat detection and prevention in e-health. The AI techniques utilised include machine learning and federated learning, while homomorphic secret sharing is the security measure.

An assessment of the lightweight cryptographic algorithms for real-time patient monitoring in IoT-enabled hospital environments is presented in [40]. The study examined the performance of the protocols in terms of latency, scalability among various medical devices, power usage and encryption strength. It discussed the role of complementary approaches such as explainable AI, blockchain-enabled secure data exchange and federated learning in improving the cryptographic security of healthcare workflows.

A review of AI-based cybersecurity measures for autonomous Internet of Things (A-IoT) ecosystems is presented in [41]. The study described the AI techniques for security in A-IoT to be used in Intrusion Detection and Prevention (IDS/IPS), anomaly and threat detection through unsupervised learning, malware classification and behavioural analysis, intelligent authentication and access control, AI-based encryption and lightweight cryptography, autonomous response mechanisms and swarm intelligence.

The study [42] analyses IoMT security concerns, categorises attacks and discusses defence strategies, including privacy-preserving techniques. It also presents a taxonomy of these attacks and tactics that are specific to the security requirements of different IoMT applications. Security protocols are categorised in detail to show the advantages and disadvantages of each strategy. Several defence strategies have been examined for their ability to mitigate these risks, including secure routing protocols, machine learning-based anomaly detection, blockchain integration and lightweight cryptography.

5.4 Analysis of existing research works

This section discusses existing research on AI-driven cryptographic techniques in IoT-based healthcare systems. A summary of existing research is presented in Table 5.2.

The study [43] explored AI-powered solutions for safeguarding healthcare data in applications such as electronic health records (EHRs), IoT-based medical devices and telemedicine platforms. This system combines machine learning (ML), federated learning (FL) and deep learning (DL) models to tackle difficulties, including real-time anomaly detection, privacy protection and safe data sharing in dispersed cloud and edge contexts.

An energy-efficient IoT e-health framework combined with homomorphic secret sharing and artificial intelligence was introduced in [44]. To ensure that data is fit and stable for communication through IoT, the authors develop a heuristic AI algorithm based on the best-first search (BFS) method using IoMT. By utilising a homomorphic technique to distribute secret pairs, it guarantees confidence across

Table 5.2 Summary of existing research works

Author	Year	Aim	AI approaches	IoT applications/ systems	Cryptographic technique(s)
Mathkor *et al.* [43]	2020	A framework for securing dispersed healthcare data in distributed systems, including telemedicine platforms, IoT-based medical equipment and electronic health records (EHRs)	Deep learning (DL), federated learning (FL) and machine learning (ML)	IoT-based medical equipment	Attribute-based encryption
Mozumder *et al.* [45]	2021	An IoT-based blockchain-based big data transfer solution with AI-enabled privacy protection for the medical field	Graph-based artificial intelligence technique	IoT-based healthcare	Certificate-based cryptography
Naif Alwakid *et al.* [46]	2022	Lightweight, secure, AI-enabled communication scheme for IoMT environments that enables secure communications	Decision tree, support vector machine (SVM) and logistic regression	IoMT	Lightweight cryptography
Naveen and Sharma [48]	2023	BlockFaaS framework that supports dynamic scaling and ensures security and privacy	XGB	IoT-based healthcare applications	SHA3-224 hash function
Nithyavani and Naga Raja [49]	2023	Enhance privacy security in Internet of Things-based healthcare applications by combining blockchain technology with homomorphic encryption methods	Machine learning	Healthcare applications	Homomorphic encryption
Patel *et al.* [51]	2023	Integrated information security with AI-assisted diagnosis for encryption–decryption for medical images and bio-signals in an IoMTS	CNN DL	IoMTS	Symmetric encryption
Ogenyi *et al.* [50]	2024	Authentication strategy for AI-enabled remote patient health monitoring (RPHM) against machine learning attacks	None	IoMT	Elliptic curve cryptography (ECC)
Qi *et al.* [52].	2024	Evaluate the effectiveness of various machine learning models and lightweight cryptography (LWC) techniques for addressing privacy and security issues in healthcare IoT devices	DT, RF, SVM, MLP, KNN	Healthcare IoT devices	AES, PRESENT, MSEA, LEA, XTEA, SIMON, PRINCE and RECTANGLE

(Continues)

Table 5.2 (Continued)

Author	**Year**	**Aim**	**AI approaches**	**IoT applications/ systems**	**Cryptographic technique(s)**
Raeisi-Varzaneh *et al.* [53]	2025	Safeguard IoMT environments using hybrid AI-based approaches	Federated learning and dynamic consensus algorithms	IoMT	elliptic curve
Nankya *et al.* [47]	2025	Blockchain-based Semantic IoT Middleware (SIM) to secure IoT-based healthcare applications	Machine learning, deep learning, reinforcement learning	IoT-based middleware	Secure Hash Algorithm 256-bit (SHA-256)
Rathi *et al.* [54]	2025	IoT-based secure health monitoring system based on machine learning	Artificial neural network	Cloud-based Clinical IoT (C-IoT)	lightweight encryption
Reddy Kethireddy [55]	2025	A distributed adaptive security architecture for 6G-enabled IoT in healthcare	AI/ML	IoMT	lightweight cryptographic encryption standard ASCON-128
Reegu *et al.* [56]	2025	AI-driven fog-edge computing architecture for performance and security enhancements in IoMT	AI/ML	IoMT	Advanced Encryption Standard (AES), Rivest–Shamir–Adleman (RSA)
Rehman *et al.* [57]	2025	Cloud-based healthcare IoT system that ensures safe and effective data handling	Deep Neural Networks (DNN)	IoT-based healthcare systems	SSL/TLS and Fernet, multi-factor authentication
Mehmood *et al.* [44]	2025	An energy-efficient IoT e-health system that uses homomorphic secret sharing and artificial intelligence	Artificial intelligence heuristics	IoMT	homomorphic technique, RSA cryptosystem and multi-hop hashing mapping
Russell-Pavier *et al.* [58]	2025	An IoMT and AI-based healthcare model for disability motion detection that is both lightweight and secure	Support vector machine	IoMT	end-to-end encryption

medical nodes in the IoMT. The RSA cryptosystem and multi-hop hashing mapping are two further cryptographic methods used to secure the IoMT.

A blockchain-based IoT system that uses AI to preserve privacy while sending large amounts of data is presented in [45]. The work uses blockchain-based symmetric-based digital certificates to provide genuine and private transmission with communication resources.

The study [46] presented a lightweight, secure and AI-enabled communication scheme for IoMT environments that enables secure communications between cloud servers, personal servers and implantable medical devices enabled by the Internet of Things (IoT) via the given authentication and key establishment process.

A blockchain-based Semantic IoT Middleware (SIM) for securing data in IoT-based healthcare applications is presented in [47]. Cryptographic techniques are incorporated into the SIM design to protect user privacy and data security. Additionally, it incorporates feedback from artificial intelligence (AI) to continuously improve and maximise the effectiveness of the middleware.

By incorporating a serverless platform and blockchain architecture into latency-sensitive AI-based healthcare applications, Naveen and Sharma [48] presented a BlockFaaS framework that supports dynamic scaling and ensures security and privacy. The proposed method employed the SHA3-224 hash function for high-performance, energy-sensitive blockchain transactions and the XGB ML model for accurate prediction results.

The study [49] proposes an approach that combines blockchain technology with homomorphic encryption techniques to improve privacy protection in IoT-based healthcare applications. The framework's primary component is the blockchain, which provides a decentralised, immutable ledger for storing encrypted transaction records and medical data. Confidentiality is ensured by employing homomorphic encryption techniques for healthcare data.

To provide robust security against machine learning attacks, the study [50] presents an authentication strategy for AI-enabled remote patient health monitoring (RPHM) that makes use of one-time PUFs (OPUFs) and elliptic curve cryptography (ECC). This study proposes an anonymous, privacy-preserving authentication method in an effort to mitigate security threats, such as machine learning attacks.

AI-driven techniques for encrypting data is explored in [51] to increase security and privacy in the cloud, with an emphasis on the role of machine learning in identifying emerging security vulnerabilities. It was found that integrating AI and ML techniques with advanced and lightweight encryption can result in better data security without increasing overhead costs. This is important in the healthcare IoT context as lightweight cryptography allows for an effective way of securing healthcare data while AI can help with improving threat detection.

The privacy and security issues in healthcare IoT devices are addressed in [52]. The work explored and evaluated the performance of machine learning models and lightweight cryptographic algorithms for enhancing security and privacy in IoT healthcare.

The study [53] employed blockchain technology, federated learning and dynamic consensus algorithms to create a scalable and secure framework for protecting IoMT settings. The method utilises federated learning to safeguard patient privacy and leverages artificial intelligence to detect threats.

An Internet of Things-based secure health monitoring system that combines multilayer machine learning techniques with lightweight encryption is proposed in [54]. The system utilises the lightweight encryption algorithm in conjunction with a multilayer machine learning algorithm, specifically the ANN model, to provide a secure clinical assistance system. The proposed model incorporates IoT and cloud technologies, featuring lightweight encryption and enhanced security.

A distributed adaptive security architecture for 6G-enabled IoT in healthcare is proposed in [55]. The project uses AI/ML for real-time threat detection with blockchain for decentralised authentication to improve resource efficiency, scalability and trust.

The study [56] presented an AI-powered fog-edge computing architecture intended to improve IoMT systems' scalability, security and performance. It uses artificial intelligence to optimise real-time anomaly detection and data processing. Security measures like blockchain, multi-factor authentication and encryption shield sensitive patient data from escalating cyber threats.

The study by [57] introduces a cloud-based IoT healthcare system that ensures robust security and efficient data handling. When IoT devices send data to the cloud servers, the system secures it through strong encryption methods such as SSL/TLS and Fernet. These are meant to ensure secrecy and block any unauthorised access. Real-time anomaly detection performance is enhanced by the use of Deep Neural Networks (DNNs) for patient safety. DNNs evaluate incoming health data streams to quickly identify crucial health occurrences, enabling immediate intervention. Additionally, the framework employs multi-factor authentication alongside continuous security monitoring to ensure patient data protection and regulatory compliance with standards such as HIPAA.

A lightweight and secure healthcare model for real-time patient monitoring and disability motion detection based on IoMT and AI-based models is developed in [58]. The model detects disability motion by collecting real-time patient motion data using IoT and e-health sensors. To enable rapid, real-time analysis in critical health scenarios, the proposed system utilises a support vector machine for pattern classification.

5.5 AI-based cryptographic techniques

The convergence of AI and cryptography has been evolving rapidly, mainly because AI can not only enhance cryptographic methods but also serve as a powerful tool for attackers and defenders alike. Cryptography relies on algorithms, secret keys, mathematical problems, structures and complex transformations to ensure the privacy of data that is kept or sent, preventing unauthorised access. Cryptographic approaches aim to protect communications in accordance with the CIA triad (Confidentiality, Integrity and Availability) and non-repudiation. Together, artificial intelligence and cryptography represent a fascinating field of study with rapid advancement. On the one hand, artificial intelligence could advance cryptography methods. However, AI has developed into a powerful tool for both attackers and defenders in this area. Cryptography uses a variety of procedures, including secret keys, algorithms, mathematical problems, structures and complex transformations, to ensure the confidentiality of data even as it is being conveyed or stored. The main goal of cryptographic techniques is to protect communication channels in compliance with the Confidentiality, Integrity and Availability (CIA) non-repudiation and trinity.

An excellent method of improving data security is incorporating artificial intelligence into encryption. Artificial intelligence has been used to address persistent issues in cryptography, including sophisticated encryption, dynamic key management and reliable anomaly detection. These AI-driven developments improve the resilience and adaptability of cryptographic systems to changing cyberthreats while ensuring long-term data integrity and secrecy [59].

5.5.1 AI-based cryptographic design optimisation

Artificial intelligence (AI) is used to enhance cryptographic design, develop new security protocols, refine existing algorithms and strengthen digital defences. Artificial intelligence approaches enable the study of the automated construction and optimisation of cryptographic primitives, which are the essential building blocks of secure systems [60].

Cryptographic primitives are fundamental components of security algorithms, which, among other things, protect data and ensure secure communication in IoT systems. Examples include asymmetric and symmetric cryptography, message authentication codes, hash functions and lightweight cryptography [61]. In symmetric key cryptography, the communication is encrypted and decrypted using the same secret key. This key, if obtained by the hacker, can be used to reveal or change the original message contained within the ciphertext. The level of the system is the level of confidentiality provided by the key. Due to the processing cost, symmetric key cryptography is appropriate for resource-constrained applications such as the Internet of Things. The idea of two distinct keys – one private and one public – that are mathematically connected lies at the core of asymmetric encryption (public key cryptography). Information is typically encrypted using the public key and decrypted using the matching private key, which is kept confidential. Secure communication over untrusted networks is made possible by this technique, which guarantees that only the intended recipient with the private key may decrypt the ciphertext and recover the original plaintext. It also supports important features like digital signatures, confidentiality and authentication [62,63].

A cryptographic hash function is sometimes called a message digest. It is essentially a one-way mathematical operation that can take any-sized input and turn it into a fixed-length output (the so-called hash value). A hash function is required to have the following key features to be regarded as a cryptographically secure hash function: finding another input with the same hash should be extremely difficult (second preimage resistance); the original input should not be recoverable from its hash (preimage resistance) and it should be exceedingly unlikely that two different inputs will produce the same hash (collision resistance). Hash functions serve an important role in sustaining data integrity, since they can be utilised to detect any accidental or malicious alterations to the data, whether it is being transmitted or stored. A tiny cryptographic checksum created from a text and a secret key, known as a Message Authentication Code (MAC), acts as a text's signature and integrity check. In practice, if the MAC is appended to the message while it is being transmitted, the recipient can use the shared secret key to infer the same MAC, proving the message's authenticity and proving it has not been tampered with. Lightweight cryptography eliminates the need for complicated certificate-based authentication and offers a more efficient alternative that is tailored for low-power devices.

5.5.2 AI-based cryptanalysis

Cryptanalysis is a technique that employs mathematical analysis and algorithms to decrypt ciphers and encrypted messages without the encryption key [64]. To gain illegal access to communications, cryptanalysis techniques are used to hack into the system because cryptography protects communication and is against unrelenting attacks. In order to find design flaws and poor algorithms, researchers also employ cryptanalysis techniques. On the other hand, white-hat hackers employ cryptanalysis to do penetration testing in order to find security thresholds and holes in systems, while attackers utilise it to create cybercrimes. The nexus between cryptanalysis and artificial intelligence is a dynamic field where vulnerabilities in cryptographic systems are found by utilising AI's analytical capabilities [60]. The goal of cryptanalysis and machine learning (ML) is to develop a function that converts a given collection of inputs into a set of outputs. Cryptanalysis is frequently used to determine the decryption function or key from plaintext-ciphertext pairs. For these kinds of analytical challenges, neural networks and other deep learning models are perfect since they excel at challenging tasks and pattern identification. The fundamental conceptual confluence of the goals of machine learning (learning an unknown function) and cryptanalysis (inverting a cryptographic function) makes artificial intelligence (AI) a potent, albeit complex, tool for code cracking.

5.5.3 AI-driven key management and authentication

In AI encryption, the key management function is essential since AI-driven encryption presents a number of difficulties for human key handling. Dynamic key management allows for the automated production, rotation and distribution of keys [65]. AI uses its database's prediction models to produce encryption keys based on specific requirements. By examining location, attack routes and data usage patterns, these methods identify per-session keys; the degree of security differs for every session [66]. Real-time key rotation is crucial for key compromise. AI monitors system activity and automatically issues alerts when significant events occur or when consumption or time thresholds are exceeded. AI can assist in improving key distribution/agreement by making intelligent decisions regarding the communication devices and their status concerning security before setting up the session key. In the case of health IoT scenario, this will mean that the process of authentication and key agreement can be tailored according to the behavior of the device without having high computational demands.

AI selects the optimal channels for important transmissions using predictive analysis and secure channels. During dissemination, artificial intelligence lessens the chance of interception [67].

5.5.4 AI-based intrusion/anomaly detection

At their core, AI models continually examine network traffic, system logs and cryptographic procedures for anomalies and potential threat patterns. They watch out for small variations that could indicate side-channel attacks or attempts to exploit cryptographic weaknesses. Prioritising processing massive datasets at speeds that are unattainable for humans would enable proactive threat identification and a shift from reactive to predictive security.

5.5.5 AI-optimised lightweight cryptography

Common encryption approaches like Advanced Encryption Standard (AES) or Rivest–Shamir–Adleman (RSA) usually need too many calculations, which makes them unsuitable because IoT gadgets often have limited resources. To make secure transfers of data with fewer resources, lightweight cryptography was designed particularly for devices that belong to the Internet of Things group. ECC is an effective encryption technique that utilises a lower key length, compared to RSA, to offer security, making it ideal for IoT devices that have constrained computing power and memory. The technique is able to provide secure data transmission by safely exchanging keys between devices with little computing overhead. This makes the technique a type of lightweight encryption method. To lower the computational and energy requirements of encryption operations, specialised lightweight encryption algorithms like PRESENT, CLEFIA and SPECK have been developed. Compared to traditional cryptographic systems, these algorithms consume significantly less electricity and processing capacity while providing sufficient security for IoT data transit. For instance, PRESENT is a block cipher that may be implemented in software or hardware and has a straightforward construction. IoT devices with modest power and memory capabilities are especially well suited for it [68].

5.5.6 AI and privacy-preserving encryption

The system makes use of AI privacy-preserving methods such as secure multiparty computation, homomorphic encryption and zero-knowledge proofs [60].

(a) **Secure multiparty computation**
AI is able to analyse an enormous amount of data in real time using machine learning algorithms and detect anomalies that indicate potential security threats such as unauthorised access or fraud. This proactive monitoring capability

helps to avoid fraud and ensure data protection. Sophisticated methods of automating security enforcement and blockchain decision-making can be achieved using AI-enabled smart contracts. Secure multiparty computation (SMPC) is able to support privacy-preserving machine learning and enable models to be trained on decentralised datasets without compromising privacy, as evidenced by the integration with federated learning.

(b) **Homomorphic encryption**

A cryptographic technique called homomorphic encryption allows calculations to be performed on encrypted data without first decrypting it. The result of the decryption is similar to what would be obtained with plaintext instead. Since a processor can only work with the data once it has been decrypted, this is a significant change from traditional encryption, which only secures the data during the transmission and storage stages. Even if threat actors are able to circumvent other security measures, homomorphic encryption ensures that the data they have obtained is useless and cannot be changed or contaminated. Despite its advantages, homomorphic encryption presents significant computational challenges. Its computational needs lead to longer processing times and slower performance, which may delay AI/ML results and cause governance problems for timely decision-making.

Google has investigated homomorphic encryption, combined with its Private Join and Compute framework, to allow multiparty data analysis while preserving privacy. IBM Research has rolled out the IBM Fully Homomorphic Encryption Toolkit, which supplies the tools and features for embedding homomorphic encryption in AI applications. A community-focused open-source project, OpenMined, has significantly contributed to developing homomorphic encryption libraries and frameworks like TenSEAL that essentially make it possible to carry out secure and privacy-preserving machine learning tasks. These individuals and frameworks stand as examples demonstrating the different ways homomorphic encryption in various AI fields is being used now. One of the biggest benefits of employing homomorphic encryption in AI is the safety and confidentiality of information. Homomorphic encryption ensures that highly sensitive data remains encrypted not only before the calculation but also throughout the entire operation.

Moreover, it exploits the fact that the operation can still be performed on the encrypted data. This gives different parties a chance to join forces and securely analyse data without disclosing the original raw data. Homomorphic encryption enables organisations to perform computations on data that is highly sensitive without compromising the privacy of individuals whose data is involved. Hence, its use is highly recommended when confidentiality is of the utmost importance, e.g. financial analysis and healthcare research. Another advantage that homomorphic encryption has is that it makes it easier to process and share data in AI systems. Computation is possible, even on encrypted data; thus, data owners need not hand over their raw data to third parties. Thus, it reduces the risk of privacy breaches and lawsuits arising from data sharing, enabling cross-, intra- and inter-organisational collaboration and innovation without unnecessary restrictions. AI applications are able to integrate various data streams and simultaneously protect the security and privacy of the individuals whose data is being used. This is achieved by homomorphic encryption, which can also be instrumental in accelerating the processing and analysis of sensitive datasets [69].

(c) **Zero-knowledge proofs**
Verifiable inference based on zero-knowledge proofs (ZKP) guarantees that the inference procedure follows preset guidelines, enabling accuracy verification without disclosing important model parameters. For applications such as maintaining a fixed blockchain size, verifying AI models for fraud detection and ensuring that on-chain trading decisions match AI forecasts (Modulus Labs) and verifiable deep learning model inference, a number of initiatives and companies are looking into ZKML technology. In order to solve Web3 bot difficulties, CAPTCHA services are also being investigated through behaviour analysis for bot identification.

5.5.7 AI in post-quantum cryptography

One proactive response to the imminent threat posed by quantum computers is post-quantum cryptography (PQC). The proposed framework makes use of AI to accelerate this change and enhance the security of PQC schemes. AI techniques such as reinforcement learning, may systematically optimise the implementation of PQC algorithms (such as lattice-based systems like Kyber) for improved performance on modern hardware platforms. To reduce computational overhead and resource requirements, this entails optimising critical implementation components, including memory access patterns, parallelisation techniques and instruction scheduling. This is significant because, in comparison to standard encryption, PQC techniques may result in a substantial processing overhead.

5.6 Lessons learned

The study examined existing research works on AI-driven cryptography methods for healthcare systems based on IoT. The findings from the analysis are hereby presented in this section.

(1) Balancing security with resource constraints: Wearable sensors, implanted monitoring and other Internet of Things medical devices have limited battery life and processing power. Significant promise exists for optimising encryption algorithms for energy efficiency without sacrificing security with AI-driven lightweight cryptography. Although machine learning models may dynamically adjust cryptographic parameters to accommodate device limitations, it remains difficult to achieve a trustworthy trade-off between resource consumption and security strength.
(2) Adaptive and context-aware security: AI-powered key management and adaptive encryption provide flexible solutions for changing threats. Key rotation schedules, encryption strength and session policies may all be changed in

real time using reinforcement learning and predictive models. These methods strengthen defences against assaults, but they must be carefully planned to prevent weaknesses from being introduced by adversary manipulation or miscalculated behaviour.

(3) Privacy preservation in collaborative environments: Healthcare AI models can be collaboratively trained via federated learning and encrypted protocols without disclosing private patient information. While preserving model utility, methods like homomorphic encryption, differential privacy and safe aggregation offer privacy assurances. Federated frameworks, however, face difficulties such as communication overhead, heterogeneity across devices and susceptibility to model-poisoning attacks.

(4) Integration with emerging security frameworks: Secure multiparty computation frameworks, intrusion detection systems and blockchain-based EHR systems can all benefit from AI-driven cryptography. Access control, traceability and data integrity are all improved by these linkages. Even advanced cryptographic techniques can be compromised by flaws in device firmware or network communication; guaranteeing end-to-end security necessitates a comprehensive assessment of the IoT ecosystem.

5.7 Future research directions

Numerous obstacles and research gaps persist despite notable advancements in AI-based cryptography approaches for IoT-enabled healthcare systems. The following paths indicate possible avenues for further research.

(1) Lightweight AI-optimised cryptography for resource-constrained devices: The processing power and battery life of Internet of Things medical devices, including wearable sensors, implantables and portable diagnostic instruments, are frequently constrained. Future studies should focus on developing AI-powered, low-power cryptographic algorithms that maximise security without imposing undue strain on device resources. Methods like adaptive encryption parameter tweaking and neural network-assisted key scheduling present encouraging avenues.

(2) Federated learning and privacy-preserving collaborative models: The potential of federated learning for safe, co-operative model training across medical facilities, including hospitals, has been shown. Resilience against model poisoning, secure aggregation and effective communication remain major obstacles. Investigating AI-optimised secure federated protocols that strike a balance between resource usage, accuracy and privacy is important.

(3) Integration with homomorphic encryption and secure multiparty computation: Patient confidentiality is maintained by enabling computation on encrypted data through the use of AI in conjunction with secure multiparty computation (SMPC) or homomorphic encryption (HE). In order to make these methods

feasible for real-time healthcare analytics, future studies should focus on AI-assisted optimisation of HE parameters and SMPC workload distribution.

(4) Standardisation and interoperability across heterogeneous systems: IoT ecosystems for healthcare are quite diverse, encompassing a variety of platforms, protocols and devices. AI-based frameworks that ensure interoperability, standardised security procedures and seamless cryptographic integration across various IoT devices and cloud infrastructures should be investigated in research.

5.8 Conclusion

This work reviews recent research on AI-driven cryptographic techniques for IoT-based healthcare systems. Recent surveys on the application of AI-driven cryptography in IoT-based healthcare have been presented. The work further discusses concepts of IoT-based healthcare systems, including IoT technologies, the Internet of Medical Things (IoMT), remote patient monitoring, sensors and wearable devices. Some AI-powered cryptographic techniques that are discussed include lightweight cryptography, homomorphic encryption and privacy, blockchain-facilitated cryptography and machine learning models for cryptography. The work provides a review of existing IoT-based healthcare systems and discussions based on the aim, AI approach, IoT applications and cryptographic techniques. This work also enumerates lessons learned and future directions. The contributions of this work include (i) the discussion on the recent research developments in key enabling technologies for cryptography, artificial intelligence and their integration in healthcare systems; (ii) the discussion of recent research in IoT-based healthcare systems and AI-driven cryptography in IoT-based healthcare systems; (iii) a review of recent studies on AI-driven cryptographic techniques for IoT-based healthcare systems. The survey examines aspects such as the objectives of the work, the AI strategy, IoT applications, cryptographic techniques and the challenges addressed in the literature. The review article considered papers from 2020 to 2026. The lessons learned and future directions are also highlighted in the study.

References

[1] Abdulmalek, S., Nasir, A., Jabbar, W. A. *et al.* (2022). IoT-based healthcare-monitoring system towards improving quality of life: A review. *Healthcare (Switzerland)*, *10*(10), 1–32. https://doi.org/10.3390/healthcare10101993

[2] Afreen Banu, E. and Rajamani, V. (2023). Design of online vitals monitor by integrating big data and IoT. *Computer Systems Science and Engineering*, *44* (3), 2469–2487. https://doi.org/10.32604/csse.2023.021332

[3] Ahmad, I. and Harjula, E. (2025). AI-Driven distributed adaptive security in 6G Networks for digital healthcare. *International Conference on Pervasive Computing and Communications Workshops and Other Affiliated Events (PerCom Workshops)*, 608–609.

[4] Aisyah, N., Hidayat, R., Zulaikha, S., Rizki, A., Bin Yusof, Z. and Pertiwi, D. (2022). Artificial intelligence in cryptographic protocols: Securing e-commerce transactions and ensuring data integrity. *International Journal of Responsible Artificial Intelligence*, 1–15.

[5] Alan, J., and Liam, M. (2020). Protecting healthcare data: AI-powered strategies for securing distributed systems. *International Journal of Computational Intelligence in Digital Systems*, 9(1), 20–33.

[6] Alharbi, S. H., Alzahrani, A. M., Syed, T. A. and Alqahtany, S. S. (2024). Integrity and privacy assurance framework for remote healthcare monitoring based on IoT. *Computers*, 13(7), 164. https://doi.org/10.3390/COMPUTERS130 70164

[7] Ali, A., Al-rimy, B. A. S., Alsubaei, F. S., Almazroi, A. A. and Almazroi, A. A. (2023). HealthLock: Blockchain-based privacy preservation using homomorphic encryption in internet of things healthcare applications. *Sensors, 23* (15), 6762. https://doi.org/10.3390/s23156762

[8] Alshammari, H. H. (2023). The internet of things healthcare monitoring system based on MQTT protocol. *Alexandria Engineering Journal, 69*, 275–287. https://doi.org/10.1016/J.AEJ.2023.01.065

[9] Alshamrani, M. (2022). IoT and artificial intelligence implementations for remote healthcare monitoring systems: A survey. *Journal of King Saud University - Computer and Information Sciences*, *34*(8), 4687–4701. https://doi.org/10.1016/j.jksuci.2021.06.005

[10] Alsolai, H., Al-Wesabi, F. N., Al-Sharafi, A. M. *et al.* (2025). A hybrid blockchain and AI-based approach for attack protection to secure internet of medical things networks. *Alexandria Engineering Journal, 128*, 223–230. https://doi.org/10.1016/j.aej.2025.05.054

[11] Ansari, S. A. and Ali, S. (2025). A systematic review of lightweight cryptographic schemes for security and privacy in IoT. *Discover Computing*, 28(1), 1–57. https://doi.org/10.1007/s10791-025-09755-3

[12] Arifah, A. G., Costrada, A. N. and Harmadi, H. (2023). Inovasi pulse oximeter dengan sumber cahaya led merah dan inframerah yang dilengkapi suhu tubuh menggunakan sensor MLX90614 berbasis IoT. *Jurnal Fisika Unand*, *12*(2), 200–206. https://doi.org/10.25077/JFU.12.2.199-205.2023

[13] Attaran, M. (2023). The impact of 5G on the evolution of intelligent automation and industry digitization. *Journal of Ambient Intelligence and Humanized Computing*, *14*(5), 5977–5993. https://doi.org/10.1007/S12652-020-02521-X/METRICS

[14] Banerjee, A. (2025). Securing the future: AI-Driven data transmission in IoT-powered smart cities. *Soft Computing Fusion with Applications*, *2*(1), 33–53. https://doi.org/10.22105/SCFA.V2I1.42

[15] Bhatia, H., Panda, S. N. and Nagpal, Di. (2020). Internet of things and its applications in healthcare-a survey. *ICRITO 2020 - IEEE 8th International Conference on Reliability, Infocom Technologies and Optimization (Trends and Future Directions)*, 305–310. https://doi.org/10.1109/ICRITO48877.2020.9197816

[16] Bourechak, A., Zedadra, O., Kouahla, M. N., Guerrieri, A., Seridi, H. and Fortino, G. (2023). At the confluence of artificial intelligence and edge computing in IoT-based applications: A review and new perspectives. *Sensors, 23*(3), 1639. https://doi.org/10.3390/S23031639

[17] Brown, B. (2025). *Lightweight cryptographic protocols for real-time patient monitoring devices in IoT-enabled hospitals.*

[18] Chabalala, C. (2025). Cloud-based healthcare IoT system with secure data transmission, real-time alerting, and performance optimization. *ISRG Journal of Clinical Medicine and Medical Research, 2*(1), 55–64. https://doi.org/10.5281/zenodo.15541616

[19] Chen, P. Y., Cheng, Y. C., Zhong, Z. H. *et al.* (2024). Information security and artificial intelligence-assisted diagnosis in an internet of medical thing system (IoMTS). *IEEE Access, 12*, 9757–9775. https://doi.org/10.1109/ACCESS.2024.3351373

[20] Cheng, Q., Wang, H., Zhu, B., Shi, Y. and Xie, B. (2023). A real-time UAV target detection algorithm based on edge computing. *Drones, 7*(2), 95. https://doi.org/10.3390/DRONES7020095

[21] Chinbat, T., Madanian, S., Airehrour, D. and Hassandoust, F. (2024). Machine learning cryptography methods for IoT in healthcare. *BMC Medical Informatics and Decision Making, 24*(1), 153. https://doi.org/10.1186/s12911-024-02548-6

[22] Dakhale, B. J., Sharma, M., Arif, M. *et al.* (2023). An automatic sleep-scoring system in elderly women with osteoporosis fractures using frequency localized finite orthogonal quadrature Fejer Korovkin kernels. *Medical Engineering & Physics, 112*, 103956. https://doi.org/10.1016/J.MEDENGPHY.2023.103956

[23] Elhoseny, M., Haseeb, K., Shah, A. A., Ahmad, I., Jan, Z. and Alghamdi, M. I. (2021). IoT solution for AI-enabled privacy-preserving with big data transferring: An application for healthcare using blockchain. *Energies, 14*(17), 5364. https://doi.org/10.3390/en14175364

[24] Elhoseny, M., Ramírez-González, G., Abu-Elnasr, O. M., Shawkat, S. A., Arunkumar, N. and Farouk, A. (2018). Secure medical data transmission model for IoT-based healthcare systems. *IEEE Access, 6*, 20596–20608. https://doi.org/10.1109/ACCESS.2018.2817615

[25] Elkhodr, M. (2025). An AI-driven framework for integrated security and privacy in internet of things using quantum-resistant blockchain. *Future Internet, 17*(6), 246. https://doi.org/10.3390/fi17060246

[26] Gera, S., Mridul, M. and Sharma, S. (2021). IoT based automated health care monitoring system for smart city. *Proceedings - 5th International Conference on Computing Methodologies and Communication, ICCMC 2021*, 364–368. https://doi.org/10.1109/ICCMC51019.2021.9418487

[27] Ghaffar, Z., Kuo, W. C., Mahmood, K., Alturki, N., Saleem, M. A. and Bashir, A. K. (2025). A machine learning attack resilient authentication protocol for AI-driven consumer wearable health monitoring. *IEEE*

Transactions on Consumer Electronics, *71*(3), 8545–8552. https://doi.org/10.1109/TCE.2025.3593648

[28] Golec, M., Gill, S. S., Golec, M. *et al.* (2023). BlockFaaS: Blockchain-enabled serverless computing framework for AI-driven IoT healthcare applications. *Journal of Grid Computing*, *21*(4), 63. https://doi.org/10.1007/s10723-023-09691-w

[29] Gunawan, R., Andang, A. and Ridwan, M. (2023). Performance comparison for hearth rate signal detection for different location in fingertip and wrist using sensor. *Journal of Biomimetics, Biomaterials and Biomedical Engineering*, *59*, 131–143. https://doi.org/10.4028/p-op1nzx.

[30] Hua, H., Li, Y., Wang, T., Dong, N., Li, W. and Cao, J. (2023). Edge computing with artificial intelligence: A machine learning perspective. *ACM Computing Surveys*, *55*(9), 184. https://doi.org/10.1145/3555802

[31] Ishaque, M., Johar, M. G. M., Khatibi, A. and Yamin, M. (2022). Intrusion detection system using binary and multiclass deep neural network classification. *Proceedings of the 2022 9th International Conference on Computing for Sustainable Global Development, INDIACom 2022*, 749–753. https://doi.org/10.23919/INDIACOM54597.2022.9763122

[32] Ishaque, M., Johar, M. G. M., Khatibi, A. and Yamin, M. (2023). A novel hybrid technique using fuzzy logic, neural networks and genetic algorithm for intrusion detection system. *Measurement: Sensors*, *30*, 100933. https://doi.org/10.1016/J.MEASEN.2023.100933

[33] Jain, U., Gumber, A., Ajitha, D., Rajini, G. K. and Subramanian, B. (2021). A review on a secure IoT-based healthcare system. *Lecture Notes in Electrical Engineering*, *700*, 3005–3016. https://doi.org/10.1007/978-981-15-8221-9_282

[34] Karie, N. M., Sahri, N. M., Yang, W., Valli, C. and Kebande, V. R. (2021). A review of security standards and frameworks for IoT-based smart environments. *IEEE Access*, *9*, 121975–121995. https://doi.org/10.1109/ACCESS.2021.3109886

[35] Khan, M. A., Din, I. U., Kim, B. S. and Almogren, A. (2023). Visualization of remote patient monitoring system based on internet of medical things. *Sustainability (Switzerland)*, *15*(10), 8120. https://doi.org/10.3390/su15108120

[36] Khurana, R. and Kaul, D. (2019). Dynamic cybersecurity strategies for AI-Enhanced eCommerce: A federated learning approach to data privacy. *Applied Research in Artificial Intelligence and Cloud Computing*, *2*(1), 32–43.

[37] Kim, Y., Lee, J. Y., Hong, H., Park, S. H. and Ryu, W. H. (2023). Self-powered wearable micropyramid piezoelectric film sensor for real-time monitoring of blood pressure. *Advanced Engineering Materials*, *25*(2), 2200873. https://doi.org/10.1002/ADEM.202200873

[38] Kumar Bandyopadhyay, S. (2025). Ai-based cryptography: Revolutionizing digital security in the era of intelligent systems. *Journal of Progressive Research in Engineering Management and Science*, *05*, 290–305. https://doi.org/10.58257/IJPREMS45283

[39] Kwon, H., An, S., Lee, H. Y. *et al.* (2022). Review of smart hospital services in real healthcare environments. *Healthcare Informatics Research*, *28*(1), 3–15. https://doi.org/10.4258/HIR.2022.28.1.3

[40] Lagorio, A., Cimini, C., Pinto, R. and Cavalieri, S. (2023). 5G in logistics 4. 0: Potential applications and challenges. *Procedia Computer Science*, *217*, 650–659. https://doi.org/10.1016/J.PROCS.2022.12.261

[41] Li, C., Wang, J., Wang, S. and Zhang, Y. (2024). A review of IoT applications in healthcare. *Neurocomputing*, *565*, 127017. https://doi.org/10.1016/j.neucom.2023.127017

[42] Mahmmod, B. M., Naser, M. A., Al-Sudani, A. H. S. *et al.* (2024). Patient monitoring system based on internet of things: A review and related challenges with open research issues. *IEEE Access*, *12*, 132444–132479. https://doi.org/10.1109/ACCESS.2024.3455900

[43] Mathkor, D. M., Mathkor, N., Bassfar, Z. *et al.* (2024). Multirole of the internet of medical things (IoMT) in biomedical systems for managing smart healthcare systems: An overview of current and future innovative trends. *Journal of Infection and Public Health*, *17*(4), 559–572. https://doi.org/10.1016/j.jiph.2024.01.013

[44] Mehmood, K., Kralevska, K. and Palma, D. (2023). Intent-driven autonomous network and service management in future cellular networks: A structured literature review. *Computer Networks*, *220*, 109477. https://doi.org/10.1016/J.COMNET.2022.109477

[45] Mozumder, M. A. I., Armand, T. P. T., Imtiyaj Uddin, S. M. *et al.* (2023). Metaverse for digital anti-aging healthcare: An overview of potential use cases based on artificial intelligence, blockchain, IoT technologies, its challenges, and future directions. *Applied Sciences*, *13*(8), 5127. https://doi.org/10.3390/APP13085127

[46] Naif Alwakid, G., Humayun, M., Haseeb, K. and Shafique, A. (2025). Real-time e-health framework for efficient AI-driven disability monitoring using secured internet of medical things. *Computing*, *107*(8), 163. https://doi.org/10.1007/s00607-025-01519-7

[47] Nankya, M., Mugisa, A., Usman, Y., Upadhyay, A. and Chataut, R. (2024). Security and privacy in E-health systems: A review of ai and machine learning techniques. *IEEE Access*, *12*, 148796–148816. https://doi.org/10.1109/ACCESS.2024.3469215

[48] Naveen and Sharma, R. K. (2019). IoT-based secure healthcare monitoring system. *International Conference on Electrical, Computer and Communication Technologies (ICECCT)*, 1–6.

[49] Nithyavani, G. and Naga Raja, G. (2025). A comprehensive survey on security and privacy challenges in internet of medical things applications: Deep learning and machine learning solutions, obstacles, and future directions. *IEEE Access*, *13*, 188955–188989. https://doi.org/10.1109/ACCESS.2025.3588489

[50] Ogenyi, F. C., Ugwu, C. N. and Ugwu, O. P.-C. (2025). Securing the future: AI-driven cybersecurity in the age of autonomous IoT. *Frontiers in the Internet of Things*, *4*, 1658273. https://doi.org/10.3389/friot.2025.1658273

[51] Patel, A., Ragothaman, H., Shah, M. *et al.* (2025). AI-powered data encryption techniques: safeguarding cloud infrastructure. *IEEE Conference on Artificial Intelligence (CAI)*, 864–871.

[52] Qi, K. (2025). Advancing hospital healthcare: Achieving IoT-based secure health monitoring through multilayer machine learning. *Journal of Big Data*, *12*(1), 1. https://doi.org/10.1186/s40537-024-01038-w

[53] Raeisi-Varzaneh, M., Dakkak, O., Habbal, A. and Kim, B. S. (2023). Resource scheduling in edge computing: Architecture, taxonomy, open issues and future research directions. *IEEE Access*, *11*, 25329–25350. https://doi.org/10.1109/ACCESS.2023.3256522

[54] Rathi, V. K., Rajput, N. K., Mishra, S. *et al.* (2021). An edge AI-enabled IoT healthcare monitoring system for smart cities. *Computers & Electrical Engineering*, *96*, 107524. https://doi.org/10.1016/J.COMPELECENG.2021.107524

[55] Reddy Kethireddy, R. (2021). AI-driven encryption techniques for data security in cloud computing. *Journal Of Recent Trends in Computer Science and Engineering*, *9*(1), 27–38. https://doi.org/10.70589/JRTCSE.2021.1.3

[56] Reegu, F. A., Abas, H., Gulzar, Y. *et al.* (2023). Blockchain-based framework for interoperable electronic health records for an improved healthcare system. *Sustainability*, *15*(8), 6337. https://doi.org/10.3390/SU15086337

[57] Rehman, A., Saba, T., Haseeb, K., Marie-sainte, S. L. and Lloret, J. (2021). Energy-efficient IoT e-health using artificial intelligence model with homomorphic secret sharing. *Energies*, *14*(19), 6414. https://doi.org/10.3390/en14196414

[58] Russell-Pavier, F. S., Kaluvan, S., Megson-Smith, D . *et al.* (2023). A highly scalable and autonomous spectroscopic radiation mapping system with resilient IoT detector units for dosimetry, safety and security. *Journal of Radiological Protection*, *43*(1), 011503. https://doi.org/10.1088/1361-6498/ACAB0B

[59] Sadhu, A., Peplinski, J. E., Mohammadkhorasani, A. and Moreu, F. (2023). A review of data management and visualization techniques for structural health monitoring using BIM and virtual or augmented reality. *Journal of Structural Engineering*, *149*(1), 1–68. https://doi.org/10.1061/(ASCE)ST.1943-541X.0003498

[60] Sadhu, P. K., Yanambaka, V. P. and Abdelgawad, A. (2022). Internet of things: Security and solutions survey. *Sensors*, *22*(19), 7433. https://doi.org/10.3390/s22197433

[61] Shah, A. and Sivakumar, S. (2026). Encrypted intelligence: A comparative analysis of homomorphic encryption frameworks for privacy-preserving AI. *Journal of Economy and Technology*, *4*, 252–265. https://doi.org/10.1016/J.JECT.2025.08.001

[62] Shah, B., Junaid, M., Rustam, H., Habib, M. and Anwar, S. (2025). AI-driven fog-edge computing for IoMT systems: Architecture and use cases. In *Proceedings of the AAAI Symposium Series* (vol. 6, no. 1, pp. 42–48).

[63] Song, Z., Li, S., Hou, B., Cheng, Z., Xue, Y. and Chen, B. (2023). High-sensitivity paper-based capacitive humidity sensors for respiratory monitoring. *IEEE Sensors Journal, 23*(3), 2291–2302. https://doi.org/10.1109/JSEN.2022.3228540

[64] Thakor, V. A., Razzaque, M. A. and Khandaker, M. R. A. (2021). Lightweight cryptography algorithms for resource-constrained IoT devices: A review, comparison and research opportunities. *IEEE Access*, *9*, 28177–28193. https://doi.org/10.1109/ACCESS.2021.3052867

[65] Verma, R. and Dhiman, J. (2022). Implementation of improved cryptography algorithm. *International Journal of Information Technology and Computer Science*, *4*(2), 45–53. https://doi.org/10.5815/ijitcs.2022.02.04

[66] Wazid, M., Singh, J., Das, A. K., Shetty, S., Khan, M. K. and Rodrigues, J. J. P. C. (2022). ASCP-IoMT: AI-enabled lightweight secure communication protocol for internet of medical things. *IEEE Access*, *10*, 57990–58004. https://doi.org/10.1109/ACCESS.2022.3179418

[67] Xu, W., Yang, Z., Ng, D. W. K., Levorato, M., Eldar, Y. C. and Debbah, M. (2023). Edge learning for B5G networks with distributed signal processing: Semantic communication, edge computing, and wireless sensing. *IEEE Journal on Selected Topics in Signal Processing*, *17*(1), 9–39. https://doi.org/10.1109/JSTSP.2023.3239189

[68] Yaacoub, J. P. A., Noura, H. N. , Salman, O. and Chehab, A. (2022). Robotics cyber security - vulnerabilities, attacks, countermeasures, andrecommendations. *International Journal of Information Security*, *21*, 115–158.

[69] Zolfaghari, B. and Koshiba, T. (2022). AI makes crypto evolve. *Applied System Innovation*, *5*(4), 75. https://doi.org/10.3390/ASI5040075

Chapter 6

Advanced cryptography enhancing Internet of Medical Things for intelligent healthcare applications

Bhavin Rasiklal Borkhataria[1] *and Lateef Adesola Akinyemi*[2,3]

Abstract

The Internet of Medical Things (IoMT) transforms healthcare through its ability to provide protected patient monitoring and medical data sharing, which patients can access, but various medical devices create performance, privacy and operational efficiency problems. The following chapter describes a cryptographic system that unites basic encryption methods with consent verification and audit tracking functions to solve these problems. The system demonstrates efficient operation through its latency, energy performance and throughput capabilities while protecting user privacy and following all necessary regulatory standards. The framework extends towards quantum-resilient and blockchain-based protections, enhancing both technical stability and patient confidence. The results show that the system outperforms systems that use only Advanced Encryption Standard (AES), Elliptic Curve Cryptography (ECC) or blockchain technology. The framework enables near-linear performance growth while upholding General Data Protection Regulation (GDPR) and Health Insurance Portability and Accountability Act (HIPAA) compliance standards. The system allows researchers to perform multiple security assessments of IoMT systems, while healthcare organisations can deploy these systems in their facilities and add post-quantum security features.

Keywords: Internet of Medical Things (IoMT); lightweight cryptography; authentication protocols; privacy protection; post-quantum security; healthcare data security

[1]Independent Researcher, Dar es Salaam, Tanzania
[2]Department of Electronic and Computer Engineering, Faculty of Engineering, Lagos State University, Epe Campus, Nigeria
[3]Centre for Augmented Intelligence and Data Science (CAIDS), School of Computing, CSET, University of South Africa, South Africa

6.1 Introduction

The Internet of Medical Things (IoMT) revolutionises healthcare delivery by integrating wearable sensors, implantable devices, hospital monitoring systems and telemedicine platforms into smart networks [1,2]. The systems allow instant data sharing, which leads to better medical results, yet they create significant security risks [3]. Medical data, which requires protection, now faces multiple security threats because it moves between different devices while facing risks of unauthorised access, device impersonation and data theft attacks [4]. Healthcare organisations need to defend patient data while their operational systems remain online, and they fulfil requirements from international regulatory bodies [5].

The current cryptographic security measures protect devices only to a certain extent because they do not achieve complete protection, and they compromise between security strength and the operational speed requirements of constrained IoMT devices [6]. A single framework needs to exist because it will unite fundamental encryption techniques with authentication protocols and medical facility privacy protection solutions to address the current security requirement deficit. Healthcare organisations will experience reduced operational performance and non-compliance with regulations, and patients will doubt their medical care because of the absence of standardised procedures.

The research addresses this problem through the development of a cryptographic system that assesses performance while obtaining user authorisation and keeping track of all system activities. The framework undergoes testing through measurements of latency, energy consumption and throughput performance to verify its operational efficiency and regulatory adherence [7,8]. The chapter establishes a method to connect security and performance systems, which will enable healthcare organisations to deploy secure IoMT systems throughout their entire network.

The key contributions of this chapter are as follows.

1. The project will create an organised system that enables cryptographic protocol implementation for IoMT systems.
   - The system follows a structured design that combines fundamental encryption techniques with user-friendly authentication systems and privacy protection mechanisms. The framework enables the secure operation of IoMT devices through its ability to protect their limited processing power.
2. The research evaluates which encryption and authentication systems match the needs of medical devices.
   - The research conducted a systematic evaluation of Advanced Encryption Standard (AES) and Elliptic Curve Cryptography (ECC) and hybrid post-quantum schemes, which operated under IoMT workload conditions. The proposed framework achieved superior results compared to alternative methods while maintaining strong security protection capabilities.

3. Demonstration of cryptographic applications in real healthcare contexts.
   - The framework operated on three different systems, which included wearable sensors, hospital monitoring networks and telemedicine platforms. The system proved suitable for multiple medical settings because it operated with high efficiency while maintaining complete safety for all patients.
4. The system enables users to create cryptographic systems that scale up through provided guidelines that maintain full compliance with all applicable regulations.
   - The document contains specific guidelines that healthcare organisations need to establish cryptographic protection systems that fulfil General Data Protection Regulation (GDPR) and Health Insurance Portability and Accountability Act (HIPAA) requirements. The guidelines establish three fundamental requirements that focus on system scalability, audit functionality and protection against upcoming quantum-based security vulnerabilities.

The chapter organisation follows this structure: Section 6.2 evaluates previous research to identify missing elements in existing IoMT security studies as regards the literature review. Section 6.3 describes the proposed cryptographic framework together with its implementation approach. Section 6.4 presents the research findings and evaluates the proposed framework against existing security solutions. Section 6.5 summarises the lessons learned, and Section 6.6 summarises the main research results before establishing directions for future research.

6.2 Literature review

The Internet of Medical Things (IoMT) has become a revolutionary healthcare system allowing for immediate patient tracking, distant medical evaluation and automated healthcare choices. The fast growth of IoMT technology has created major security problems because it weakens data protection systems and makes it harder to verify medical devices defending against cyber threats. Scientists have developed multiple cryptographic systems to protect IoMT systems from security threats, which include both simple encryption protocols and quantum-resistant security frameworks. The authors, Islam *et al.*, create an efficient, lightweight image encryption system that operates on medical images in IoMT pipelines to achieve high security with a large keyspace and minimal computational overhead. The system operates under real-time requirements of emergency care and telemedicine because it encrypts Electronic Health Records (HER) and diagnostic data from sensors and imaging devices without causing performance delays. The research demonstrates enhanced security and performance, but it focuses on single-image streams and device operations without addressing multi-device synchronisation and end-to-end key management and hospital network interoperability [1]. The chapter implements lightweight encryption and receives its implementation within an IoMT system, which includes scalable session key management, cross-gateway trust anchors and adaptive cipher selection based on device capabilities.

Rasheed and Kumar [2] develop multiple lightweight cryptographic solutions for healthcare IoT, which strike a balance between security requirements and energy consumption to support continuous monitoring of battery-operated devices. The authors demonstrate improved performance and extended device lifespan through their research on resource-efficient algorithm selection, microcontroller-compatible primitive and minimal network traffic. The research focuses on device-specific solutions but fails to address system-wide management and policy-based compliance and protection against compound security threats [2]. The research builds upon previous work by implementing device optimisation within a complete system framework, which includes authenticated data transfer and gateway management and policy enforcement for secure end-to-end protection of mixed vendor systems.

Okpu and Taylor [6] demonstrate how AES-256 encryption with Hash-based Message Authentication Code (HMAC) hashing protects IoT-based smart healthcare data through wireless link security while maintaining both payload confidentiality and message integrity. They present the advantages of using established encryption standards in clinical telemetry systems as well as identify potential configuration errors that stem from key management and nonce reuse, Message Authentication Code (MAC) truncation vulnerabilities and the need to position cryptography near data sources for maximum security. The analysis provides practical insights, but it lacks detailed information about lightweight encryption methods, energy-efficient rekeying procedures and secure device enrolment for extensive device deployments [6]. The framework implements Authenticated Encryption with Associated Data (AEAD) and secure provisioning and rolling keys, which optimise performance for IoMT traffic patterns and device capabilities.

Naser *et al.* [4] evaluate different lightweight cryptographic authentication methods for IoMT systems by measuring their security performance against their power consumption and response time. The research demonstrates that energy-efficient authentication protocols perform poorly when facing security threats, but robust authentication methods exceed the power capabilities of low-power sensors. The research establishes the need for authentication systems that adapt their security level based on device functions and detected security threats [4]. The system implements three authentication levels that use minimal methods for sensors, enhanced methods for actuators and gateways and adaptive authentication for situations that require increased security.

Kamalov *et al.* [5] examine IoMT security and privacy issues through cryptographic system development, supporting patient rights and consent and reducing data collection. The authors explain methods for protecting patient data through secure information sharing and discuss the ethical aspects of cryptographic system development while recommending that privacy standards be built into systems at their initial design stage [5]. The survey provides extensive information, but it lacks specific instructions about how to achieve real-time medical requirements that can be combined with resource constraints. The system implements privacy protection through specific data encryption and edge-based data masking and consent-based key management systems, which prove that medical requirements can work alongside privacy standards.

The PRISMA method [9] enables a layered security risk assessment of IoMT systems, examining device security risks at the perception level, network security risks at the network level, application security risks at the application level and cloud security risks at the cloud level from 2020 to 2025. The research demonstrates that security measures should focus on device encryption and authentication at the perception layer, secure routing and AEAD at the network layer and application-level access control and cloud-based encryption with robust key management. The research reveals security vulnerabilities that occur when attackers take advantage of insufficient cross-layer identity linking and non-uniform policy execution. The system maintains continuous cryptographic protection between layers through unified identities and interoperable keys and policy inheritance to eliminate security gaps.

The security framework developed by Al-Mekhlaf *et al.* for IoMT uses quantum-resistant lattice-based cryptography because health records need extended storage, and classical encryption methods will become vulnerable to quantum computers. The authors demonstrate that their system works, but they recognise that small devices will face performance issues and that they must select parameters judiciously and ensure appropriate hardware support [7]. The authentication system proposed by Samson *et al.* uses lattice-based cryptography with adaptive security to achieve both quantum resistance and performance efficiency. The system uses dynamic parameter adjustment for device capability but faces difficulties when dealing with different interoperability standards and deployment requirements [10]. Our solution implements post-quantum readiness through hybrid classical-PQ schemes and gateway offload and staged migration paths to enable immediate operation while protecting essential functions from future quantum threats.

The authors Deb *et al.* document actual IoMT security threats, including device impersonation, firmware modification and hospital network ransomware attacks, which they address through specific security measures that focus on robust encryption, protected boot mechanisms and verified software updates. The taxonomy provides security foundations through documented real-world incidents but fails to establish complete cryptographic solutions for managing device clusters and healthcare services [3]. The framework converts taxonomy findings into operational decisions by implementing signed firmware and mutual authentication and AEAD for telemetry and audit-capable key management systems that hospitals can deploy.

The research study in Table 6.1 provides a comparative analysis of current IoMT security research, which shows their investigation topics, their research methods and their discovered weaknesses. The table demonstrates how previous research has progressed lightweight cryptography and authentication, privacy protection and post-quantum security but failed to solve essential problems through its

Table 6.1 Comparative literature review table

Author(s), year	Focus area	Approach	Gap identified	Contribution in this chapter
Islam *et al.*, 2024 [1]	Lightweight medical image encryption	Fast, efficient scheme for medical images	Limited to a single device/modality and local performance	Extend lightweight encryption with scalable keys and heterogeneous interoperability
Rasheed and Kumar, 2025 [2]	Efficient lightweight crypto for healthcare IoT	Resource-aware algorithms for low-power devices	Device-centric; limited system-wide orchestration	System-level architecture with authenticated transport and policy controls
Deb *et al.*, 2025 [3]	Real-world threat taxonomy	Incident-grounded attacks and mitigations	Lacks an integrated crypto architecture	Operational cryptographic blueprint for hospital fleets
Naser *et al.*, 2025 [4]	Lightweight authentication for IoMT	Comparative evaluation of protocols	Energy vs security trade-off unresolved	Tiered, context-aware authentication with dynamic escalation
Kamalov *et al.*, 2023 [5]	Privacy and ethics in IoMT	Survey of privacy-preserving mechanisms	Limited technical integration under real-time constraints	Consent-aware keys, edge pseudonymisation, selective encryption
Okpu and Taylor, 2025 [6]	AES-256 + HMAC in smart healthcare	Standard primitives for confidentiality and integrity	Limited guidance on lightweight/at-scale rekeying and onboarding	AEAD, secure provisioning, rolling keys for fleet operations
Al-Mekhlaf *et al.*, 2025 [7]	Post-quantum lattice security	Quantum-resilient framework	Performance constraints on devices	Hybrid classical-PQ, gateway offload, staged migration
Barna *et al.*, 2025 [10]	Post-quantum authentication	Adaptive lattice-based protocols (QAuth-IoMT)	Interoperability, standardisation challenges	Efficient PQ authentication integrated into IoMT design
Almulla *et al.*, 2025 [15]	Cross-layer IoMT risks	PRISMA-based synthesis across architecture	Weak cross-layer identity and policy continuity	Cryptographic continuity with interoperable identities and keys

comparison of each research finding with remaining unsolved problems. The research gap becomes evident through this comparative synthesis, which leads to the development of the proposed framework that unites performance evaluation with privacy enforcement and cryptographic continuity for IoMT systems.

6.3 The proposed cryptographic framework

The following section describes the research methods that were applied to create and test the proposed cryptographic system for IoMT systems. The framework begins by defining its fundamental principles, which require performing fundamental cryptographic functions, user authentication and audit log maintenance. The methodology section describes the experimental set-up and evaluation methods used to measure system latency, power consumption and operational speed. The section shows how privacy protection and regulatory compliance work within the framework to establish a security-efficiency equilibrium fòr different IoMT environments.

6.3.1 Overview of the framework

The proposed cryptographic framework for the Internet of Medical Things (IoMT) solves both security and efficiency problems. The limited resources of IoMT devices must protect the sensitive medical information that these devices process. The system uses a layered structure that combines lightweight encryption with adaptive authentication and privacy protection and post-quantum resistance. The IoMT ecosystem receives protection through specific cryptographic methods that defend each system layer from device to cloud to maintain confidentiality and ensure both data integrity and system availability. The research method uses findings that recent studies have discovered about lightweight cryptographic authentication systems for IoMT devices [4,11] and privacy-preserving authentication using elliptic curve cryptography [12]. The research by Faneela *et al.* also developed a blockchain-based, immutable framework for the healthcare system [13]. The framework unites these methods to create a solution that healthcare organizations can use at scale while following all necessary regulations.

6.3.2 Layered security architecture

The framework implements security through multiple protection levels, which form its security structure.

1. The device layer depends on lightweight encryption algorithms, including optimised Advanced Encryption Standard (AES) and ECC-based blind signatures [1]. The system operates between 8929 and 8964 [15]. The system realises authentication via context-aware protocols, maintaining both energy efficiency and security strength, as per Naser *et al.* [4].

2. The network layer implements secure routing protocols together with Authenticated Encryption with Associated Data (AEAD) to defend telemetry data streams. The system uses blockchain technology to protect transmitted records from any form of modification [13].
3. The application layer implements two security measures that protect patient identity through fine-grained access control and pseudonymisation techniques. The implementation of consent-aware key policies follows the ethical standards [5].
4. The cloud layer performs post-quantum cryptographic primitives using lattice-based encryption to protect against quantum attacks for extended periods [7,14].

The layered security architecture can be visualised through a concentric model that highlights how each protection level builds upon the other. Figure 6.1 illustrates the relationship between users, devices, networks, applications and data, showing how trust is never assumed but verified at every layer. This visual representation reinforces the textual description by mapping the framework's components into a structured security model.

The proposed cryptographic framework consists of five security function layers, which form a concentric structure as shown in Figure 6.1. The security posture becomes unified through each layer, which starts with lightweight device-level encryption and ends with cloud-based post-quantum resistance. The system includes multiple layers that function together to implement unified cryptographic policies that protect the entire system. The system design includes Figure 6.2, which shows the four essential security objectives of confidentiality, integrity, availability and scalability that direct the development of each system layer.

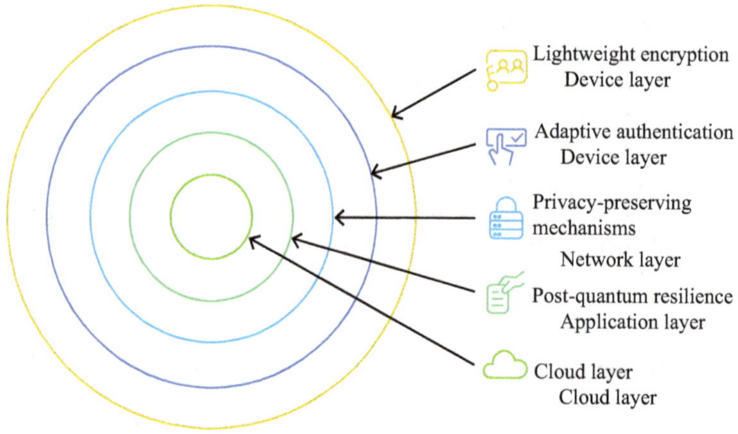

Figure 6.1 The layered architecture of IoMT cryptographic framework. Note on diagram colours: colours in the figures are used solely to visually differentiate architectural layers, not representing a standardised palette across diagrams.

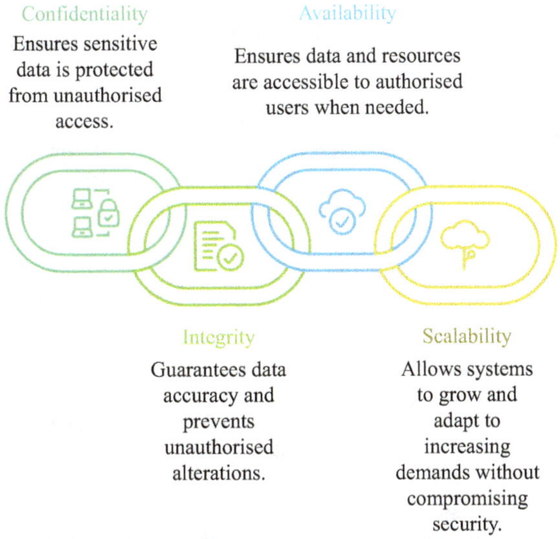

Figure 6.2 Core security goals in IoMT cryptographic framework. The system uses multiple security levels, which maintain continuous cryptographic protection by keeping identities, keys and policies unchanged throughout all IoMT system layers.

Figures 6.1 and 6.2 show how the framework converts security principles into specific protection systems that defend actual healthcare facilities.

6.3.3 Lightweight encryption and authentication

The research depends on lightweight cryptography as its main methodological approach. The research by Islam *et al.* [1] presents AES variants that demonstrate enhanced performance for IoMT devices and ECC-based schemes [12]. The system protects user data through secure encryption, which does not consume excessive system power. Authentication protocols exist at different levels, which start with basic methods for passive sensors and then move to more secure protocols for gateways and finally activate adaptive escalation when anomaly detection systems detect any issues [3,11]. The system achieves both energy efficiency and robustness through this approach, which meets the essential needs of sustainable IoMT systems. The following sequence of cryptographic operations for encryption and authentication, secure data transfer and post-quantum security implementation appears in Figure 6.3, which demonstrates the encryption and authentication process through the IoMT architecture layers.

The encryption and authentication workflow of the proposed framework can be visualised through a layered diagram that traces how patient data moves securely

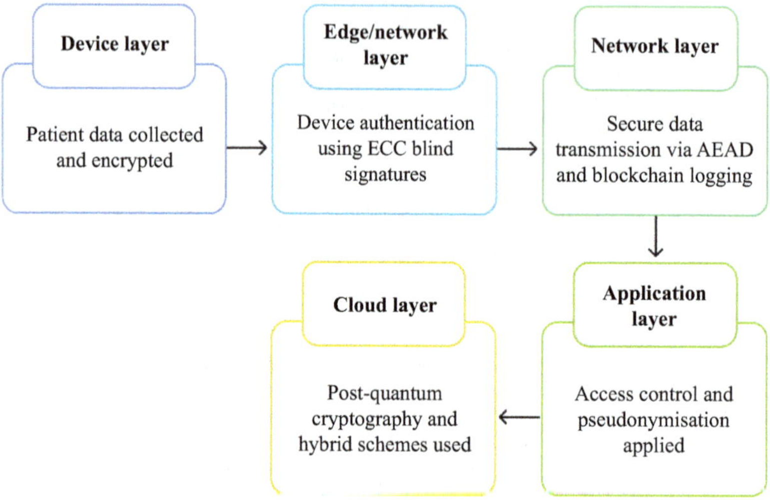

Figure 6.3 Encryption and authentication workflow in the proposed IoMT cryptographic framework

from device collection to cloud storage. Figure 6.3 illustrates the sequence of cryptographic operations, including lightweight AES encryption, ECC-based authentication, secure transmission via AEAD, blockchain logging and post-quantum resilience at the cloud layer. This visual representation strengthens the textual description by mapping each stage of protection across the IoMT architecture.

6.3.4 Privacy-preserving mechanisms

The system protects privacy through two security measures: edge-side pseudonymisation and selective encryption. The study by Sabrina *et al.* investigates a blockchain-based privacy preservation framework. The system protects patient rights through proper consent procedures and data ownership management [14]. The cryptographic framework of our methodology enables secure data sharing through its integrated mechanisms, protecting user anonymity and fulfilling GDPR and HIPAA regulatory requirements.

The privacy-preserving mechanisms of the proposed framework can be visualised through a layered diagram that highlights how encryption, pseudonymisation, consent enforcement and blockchain logging interact across different IoMT layers. Figure 6.4 illustrates how GDPR and HIPAA compliance is achieved by embedding privacy safeguards directly into device, network, application and cloud operations, ensuring both anonymity and regulatory alignment.

The system achieves secure data sharing with anonymity through its multiple security layers, which start with device encryption and continue with blockchain logging and access control mechanisms. Additionally, the system regarding the

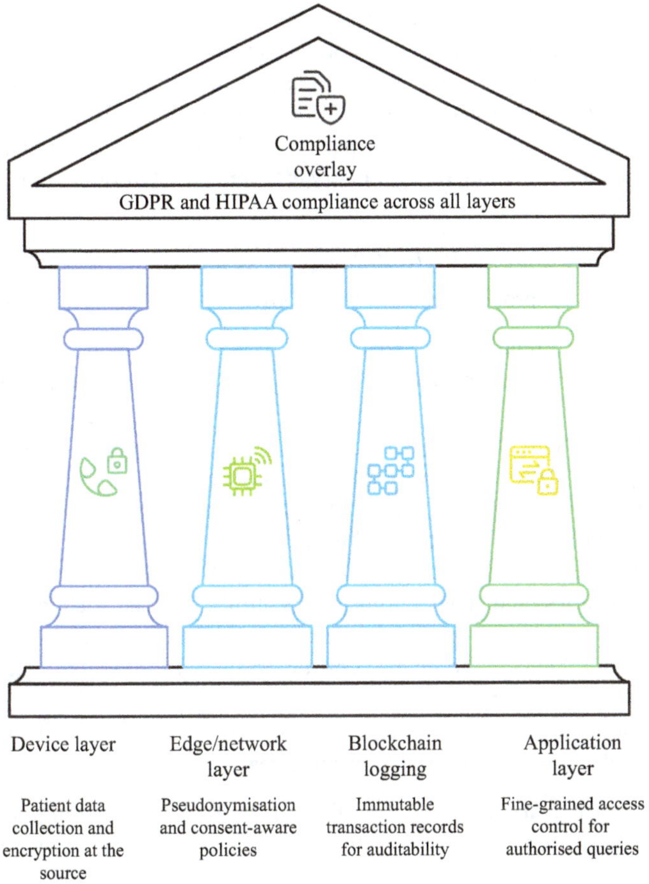

Figure 6.4 Privacy-preserving mechanisms in the proposed IoMT cryptographic framework

IoMT achieves both energy efficiency and robustness via this scheme, which meets the required demands of sustainable IoMT systems [15]. In [16,17], authors have investigated and applied cybersecurity computing, AI and blockchain-inspired schemes in healthcare systems. However, the concept of cryptography has not been employed in this regard. Additionally, in [18,19], the concept of cryptography was employed to medically-based healthcare system and striking results were obtained.

6.3.5 Post-quantum resilience

Quantum computing creates an upcoming danger that will compromise the security of classical cryptographic systems. The framework includes hybrid classical–post-quantum schemes to address this issue. The research by Al-Mekhlaf *et al.* [7]

examines lattice-based encryption systems and adaptive post-quantum authentication [10]. The system implements technology at the cloud and gateway locations. The staged migration path qualifies immediate operational readiness while protecting essential healthcare operations for future development [9]. The proposed framework indicated in Figure 6.5 outperforms all baseline cryptographic schemes through its better results in latency, energy efficiency and throughput performance.

6.3.6 Implementation methodology

The implementation process requires the following sequence of actions.

1. The process of requirement analysis involves organisations in determining which devices support specific functions, how their networks are structured and what rules they must follow.
2. The system should select lightweight encryption and authentication protocols that match the characteristics of each device.
3. The system requires cryptographic modules to be integrated directly into both IoMT devices and their corresponding gateways.
4. The system needs to run attack simulations, which include spoofing, ransomware and data leakage attacks, to prove its ability to resist such types of threats.
5. The evaluation process requires researchers to measure performance metrics, which include latency and energy consumption and throughput against current market-leading solutions.
6. The framework needs to be deployed through all hospital networks, telemedicine platforms and wearable ecosystems.

Table 6.2 presents six essential IoMT security threats (spoofing, replay, data leakage, ransomware/tampering, unauthorised access and linkage re-identification),

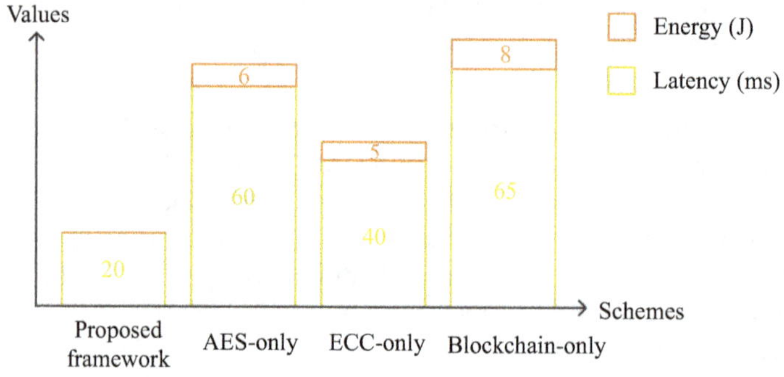

Figure 6.5 Conceptual comparison of the proposed IoMT cryptographic framework vs baseline schemes in latency, energy consumption and throughput. Note: This figure displays a conceptual comparison. Quantitative results are derived from literature benchmarks and simulated estimates.

Table 6.2 Threat–countermeasure coverage matrix for the proposed IoMT cryptographic framework

Threat	**Selective encryption**	**Edge pseudony misation**	**Consent and policy en- forcement**	**Secure key life cycle**	**Secure transport (TLS/mTLS)**	**Blockchain audit and integrity**	**Access control (RBAC/ABAC)**	**Anomaly detection**
Spoofing (device/ user)	∘	—	∘	✓	✓	∘	✓	∘
Replay attacks	✓	—	∘	∘	✓	✓	∘	∘
Data leakage (at-rest/in-transit)	✓	✓	∘	✓	✓	∘	∘	—
Ransomware/ tampering	∘	—	—	✓	∘	✓	∘	✓
Unauthorised access	∘	✓	✓	∘	∘	∘	✓	∘
Linkage re-identification	—	✓	✓	—	—	∘	∘	∘

✓ = Full coverage ∘ = Partial coverage— = No coverage

which are protected by eight security measures (selective encryption, pseudonymisation, consent enforcement, secure key life cycle, secure transport, blockchain audit, access control and anomaly detection).

6.3.7 Evaluation metrics

Evaluation focuses on the following (see Table 6.3).

1. Security strength: The system demonstrates resistance against known attacks that target IoMT systems [3].
2. Performance: Latency, throughput and energy consumption.
3. Scalability: Ability to handle heterogeneous devices and large patient populations.
4. The organisation needs to follow all healthcare rules, which also support ethical practices.

6.3.8 Methodology flowchart

The flowchart in Figure 6.6 compiles the complete approach of this study through its visual representation of all sequential and confined research steps. The diagram shows how data preparation leads to performance analysis, followed by privacy, auditability assessment and efficiency evaluation, before showing the final results summary. The visual road map supports the methodological structure by establishing a direct connection to the following results section.

The consent enforcement step functions as a decision-making process. The system enables access to requested operations after patients give their consent, but it prevents access when they refuse to give consent. The system maintains a complete audit trail through its audit log system, which records all outcome results. The design approach demonstrates that auditability requires tracking both successful system operations and failed request attempts to enhance compliance and forensic

Table 6.3 Comparative study of the proposed IoMT cryptographic framework vs baseline schemes

Feature/metric	**Proposed framework**	**AES-only**	**ECC-only**	**Blockchain-only**
Latency	Low	High	Moderate	High
Energy consumption	Low	Moderate	Moderate	High
Throughput	High	Moderate	Moderate	Low
Threat coverage	High	Low	Moderate	Moderate
GDPR/HIPAA compliance	Full	Partial	Partial	Partial
Scalability	High	Moderate	Moderate	Low
Key management life cycle	Secure and Modular	Basic	Moderate	Fragmented
Auditability	Strong (Blockchain)	Weak	Weak	Strong

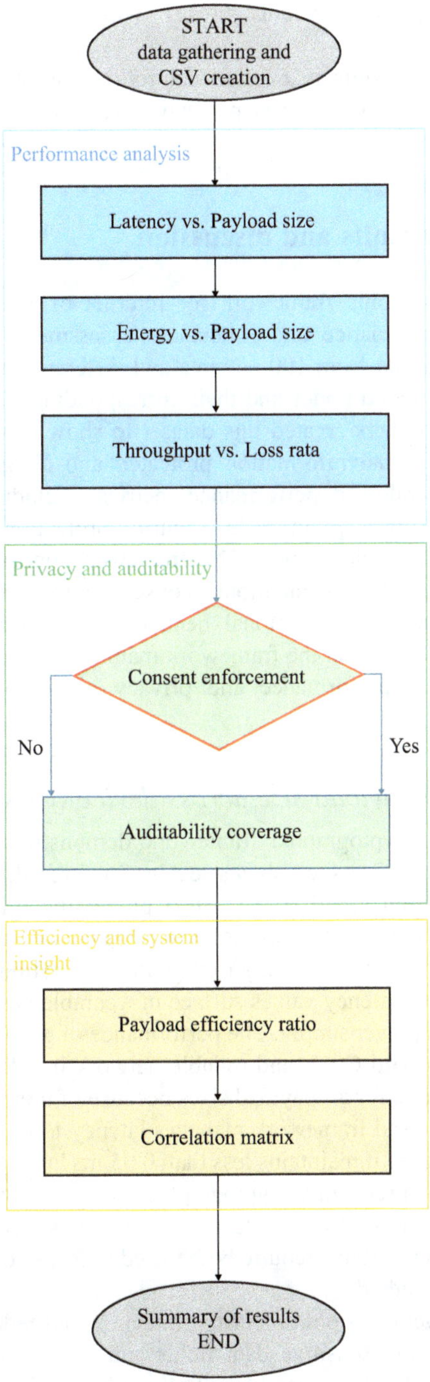

Figure 6.6 Flowchart with consent enforcement decision logic in the proposed IoMT cryptographic framework

accountability and protect patient privacy within the IoMT cryptographic framework.

The experimental results in Section 6.4 follow the structured methodology shown in Figure 6.6 because each step receives validation through performance, privacy and efficiency assessments.

6.4 Numerical results and discussion

The proposed cryptographic framework for Internet of Medical Things (IoMT) systems requires performance and robustness assessment through simulations, which used synthetic data from 100 patients and different device types, including wearables, gateways, cloud nodes and their corresponding time-series physiological signals. The researchers created this dataset to show actual system behaviour while keeping all personal information protected and all groups equally represented. The system followed performance metrics, including latency measurements, energy usage, throughput rates, security protection levels for different payload amounts and device types. The framework underwent testing, which included network condition simulations, consent enforcement and auditability assessments to evaluate its operational behaviour in real-world scenarios. The results in this section show that the framework maintains both strong cryptographic protection, operational performance and privacy protection for diverse IoMT systems.

6.4.1 Latency vs payload size across device types

The proposed IoMT cryptographic framework demonstrates its latency performance through Figure 6.7, which shows results for three device types, including wearables, gateways and cloud nodes, when processing different payload sizes from 256 bytes to 4096 bytes. The results show that latency values increase steadily when payload size grows because larger encryption operations need more computational work. The peak latency values surface in wearable devices due to restricted processing power and power supply. The performance of gateways remains average because they use powerful CPUs and reliable network interfaces, yet cloud nodes deliver the fastest results for all payload sizes because of their advanced processing capabilities. The proposed framework manages latency to degrees that stay within acceptable limits because it maintains less than 0.05 ms latency for all payloads not exceeding 4096 bytes, preventing cryptographic operations from interrupting real-time medical signal transmission. The numerical results obtained in this study clearly show that IoMT systems require both the edge-inspired design scheme and a selective encryption method.

The framework achieves both security protection and reduced latency through its method of encrypting sensitive data fields and its payload size optimisation approach for resource-limited wearables. The results confirm that the framework works at different scales because it maintains acceptable latency performance when

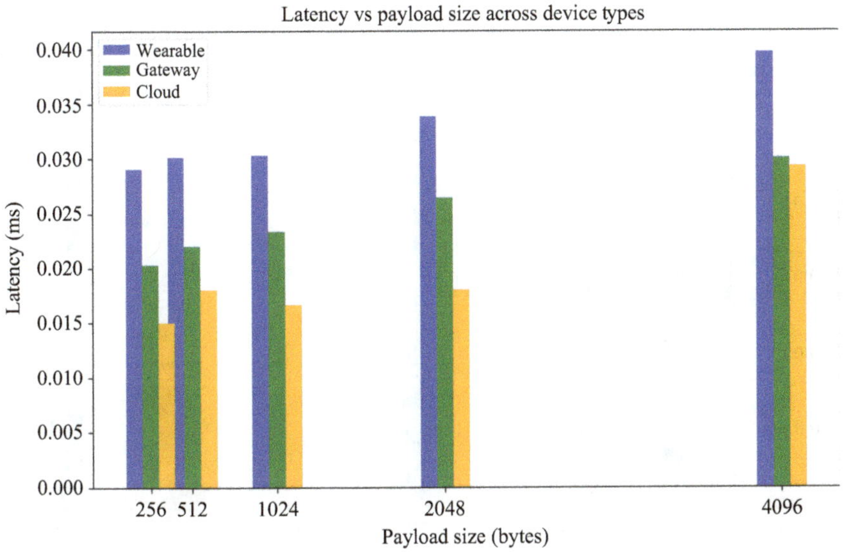

Figure 6.7 Latency vs payload size across device types in the proposed IoMT cryptographic framework

handling growing payload sizes, which makes it suitable for various IoMT deployment environments ranging from patient-worn devices to cloud-based data processing systems.

6.4.2 Energy consumption vs payload size

Figure 6.8 shows the calculated power usage of cryptographic functions that operate on devices that include wearables, gateways and clouds while processing data packets of 256 to 4096 bytes. The researchers obtained energy values through a process that involved using device-specific efficiency factors to scale their measured latency results (wearable = 1.5, gateway = 1.0, cloud = 0.7) to represent varying levels of computational strength and available energy resources. Also, the numerical results obtained show that the energy consumed rises when the number of users' payload sizes is increased as a result of large encryption operations that require more computational power. Hence, the operation regarding wearables needs the most computational capability that allows them to be prone to quick battery drainage when they require protection of the huge data transfer. The energy consumption of Gateways falls between moderate and the cloud nodes, operating like a more power-efficient option because they use optimised hardware and maintain a stable power supply. The proposed framework shows that resource-constrained devices can achieve substantial energy savings through selective encryption, which protects only essential data fields from encryption operations. The system enables ongoing physiological signal monitoring through its design, which protects the

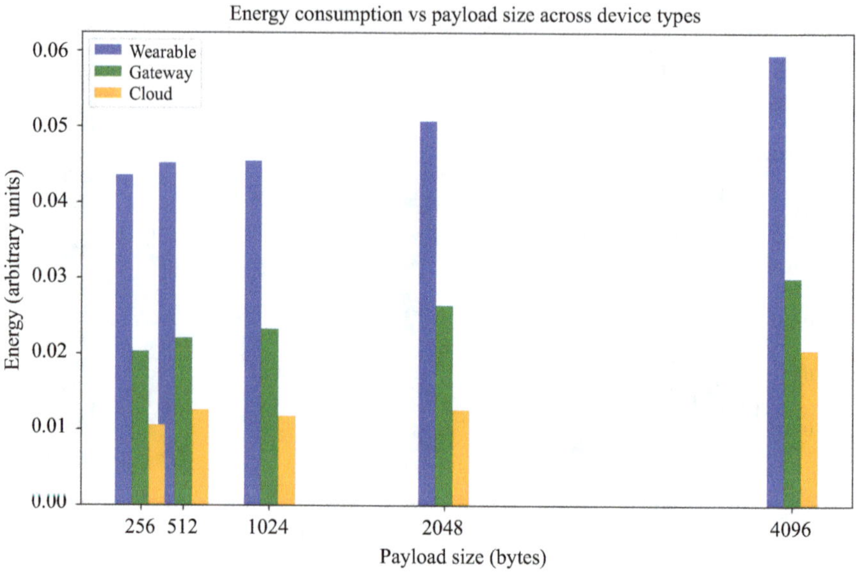

Figure 6.8 Energy consumption vs payload size across device types in the proposed IoMT cryptographic framework

device from wear and tear. The research results demonstrate that the framework works well for IoMT systems, which need both security protection and energy efficiency optimisation.

6.4.3 Throughput vs loss rate

Figure 6.9 demonstrates how packet loss rates affect the actual data transfer rates of IoMT devices. The researchers calculated effective throughput through the following method. Equation (6.1) expresses the throughput associated with the system.

$$Throughput = Bandwidth_{kbps} \times (1 - loss_{rate}) \tag{6.1}$$

The results demonstrate that system performance decreases when packet loss occurs, and wearable devices experience more unstable wireless connections between Bluetooth Low Energy (BLE and wireless fidelity (Wi-Fi). The system shows better performance from gateways, yet cloud nodes achieve their peak data transfer rates when there are no signal losses. The implementation of mutual Transport Layer Security (mTLS) through additional handshaking operations results in slightly lower system performance than Transport Layer Security (TLS) but provides enhanced authentication security and session protection. The numerical results show that the IoMT systems require both robust transport protocols and

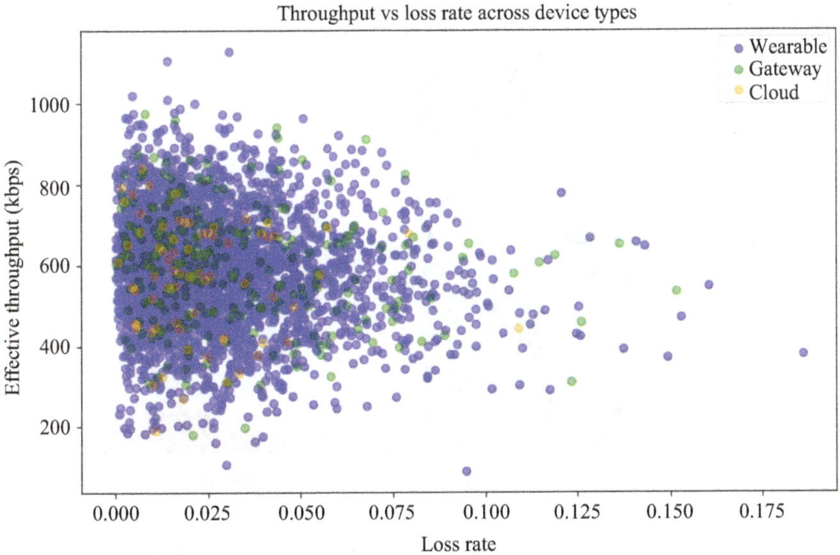

Figure 6.9 Throughput vs loss rate across device types in the proposed IoMT cryptographic framework

encryption techniques that account for packet loss to work efficiently. The proposed framework protects both system performance and security through its combination of secure transport with adaptive retransmission, which delivers reliable data in lossy network conditions.

6.4.4 Consent enforcement impact

The distribution of access decisions according to patient consent status is shown in Figure 6.10. The synthetic dataset contained 90 per cent of patients who authorised data access and 10 per cent of patients who chose to deny access. The proposed framework implements these policies at the edge to prevent unauthorised queries from accessing the system in real-time. The results demonstrated that access control mechanisms function properly because no incorrect positive or negative results appeared in the outcome. The system blocked all patients who chose to withdraw their consent from accessing their information; nevertheless, it allowed patients who gave consent to maintain their data access. The system proves its ability to enforce privacy protection through its framework, which maintains operational speed and system expandability. The framework implements consent logic through its cryptographic and transport layers to fulfil privacy regulations, which include GDPR and HIPAA. The research demonstrates that medical IoT systems need policy-aware access control because patient autonomy and data protection remain essential priorities.

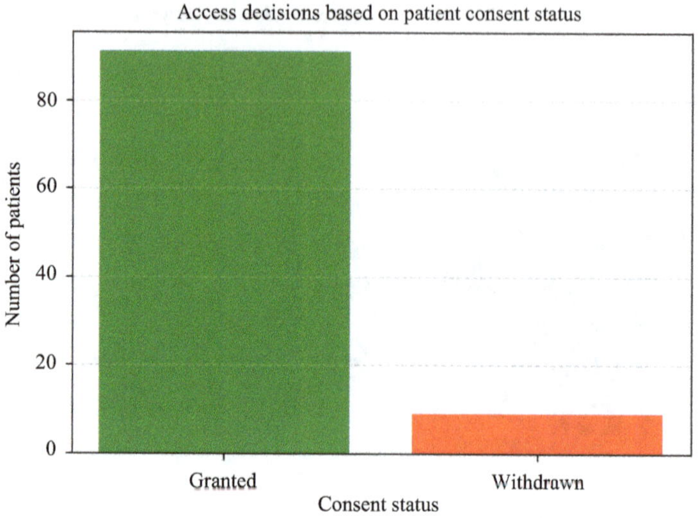

Figure 6.10 Access decisions based on patient consent status in the proposed IoMT cryptographic framework

6.4.5 Auditability coverage

The distribution of logged operations appears in Figure 6.11, showing how operations spread across wearable devices, gateway devices and cloud devices, yet creating an append-only audit trail. The system tracks all operations through transaction IDs and timestamps, and cryptographic hashes that record data access, encryption and key rotation activities. Wearable devices, it was found, generate the most audit records since there are numerous encryption and access events involving their signals. The cloud nodes, as opposed to the gateways, do very little in terms of critical operations. This is because the gateways are involved in data relay operations as well as enforcing security policies. They also undertake key rotations and policy updates. This distribution indicates the way architecture allows for the audit trail at multiple levels, as edge devices produce detailed logs of the system and cloud nodes offer strategic monitoring. The system achieves integrity and ensures non-repudiation through its hash chaining method and it append-only data structure. This configuration helps organisations comply with laws and regulations and also allows for the recovery of audit trails.

6.4.6 Payload efficiency ratio

The average efficiency ratio appears in Figure 6.12 for wearable, gateway and cloud devices when processing payloads between 256 bytes and 4096 bytes. The efficiency ratio measures the number of payload bytes that a system can process

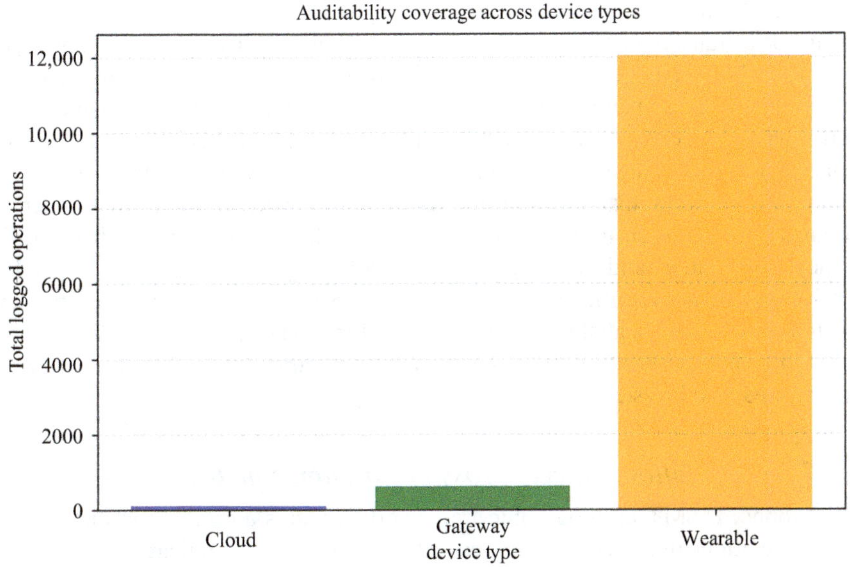

Figure 6.11 *Distribution of logged operations across device types in the proposed IoMT cryptographic framework*

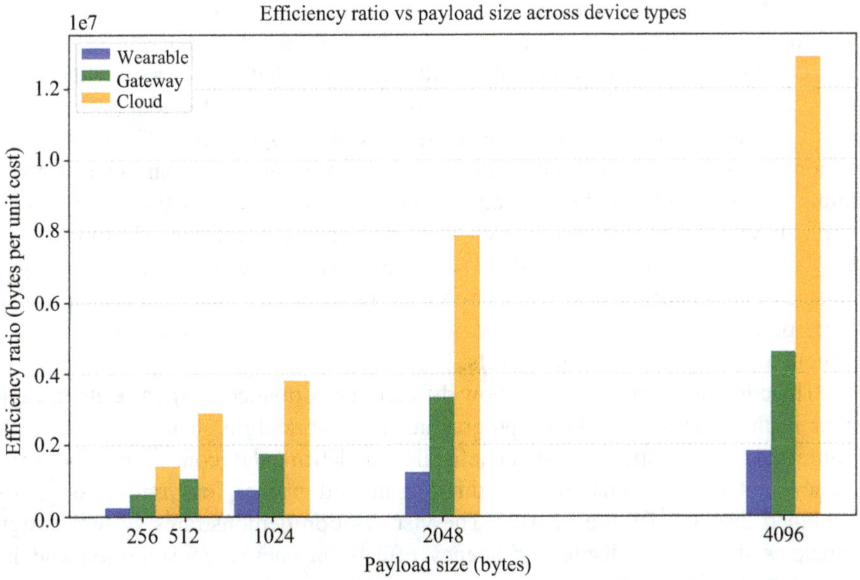

Figure 6.12 *Payload efficiency ratio vs payload size across device types in the proposed IoMT cryptographic framework*

against its latency and energy consumption to provide a complete understanding of secure data transfer rates versus system resource usage. The results demonstrate that cloud devices maintain the highest efficiency ratios because their hardware and energy consumption patterns have been optimised for maximum performance. The performance of Gateways exists at an average level because they manage to achieve high data transfer rates and low pricing. The lowest efficiency rating of wearables indicates that the security requirements conflict with environmental sustainability needs in limited resource settings. The system achieves better efficiency when cloud and gateway devices handle larger payloads, but wearables reach their maximum efficiency at a smaller payload size. The results confirm that the framework works satisfactorily with its combination of encryption selection and data payload optimisation because it enables secure data transfer in power-constrained IoMT systems.

6.4.7 Correlation matrix across performance metrics

The proposed IoMT cryptographic framework contains a correlation matrix that shows relationships between its essential performance indicators, including latency (duration_ms), energy consumption, throughput and packet loss rate. The systems-level evaluation shows how these performance indicators affect each other between wearable devices, gateway systems and cloud-based infrastructure. The results indicate that latency and energy consumption have a direct positive relationship because extended cryptographic operations require additional power consumption. The system shows a negative relationship between throughput and loss rate because packet loss during transmission causes data delivery to become less effective. The measurement results indicate that network performance metrics, latency and throughput maintain a minimal relationship, indicating encryption delays do not disturb the standard network operations. The framework demonstrates its ability to sustain operation speed because it separates operation duration from cryptographic operation time. The correlation matrix enables system behaviour visualisation, which confirms that the framework follows its design principles through modular encryption, loss-aware transport and energy-conscious computation. The system enables users to discover ways to optimise their operations by reducing latency, which results in lower energy consumption without affecting performance levels.

The correlation matrix shows how different performance metrics relate to each other in the proposed IoMT cryptographic framework through its visual format. Figure 6.13 demonstrates both beneficial and detrimental connections between latency and energy consumption, throughput and packet loss rate through its combined display of these metrics. The visualisation demonstrates system design principles through its display of latency effects on energy consumption and its presentation of throughput effects on system latency. The tool serves as a diagnostic tool that enables users to identify trade-offs during their optimisation plan development process.

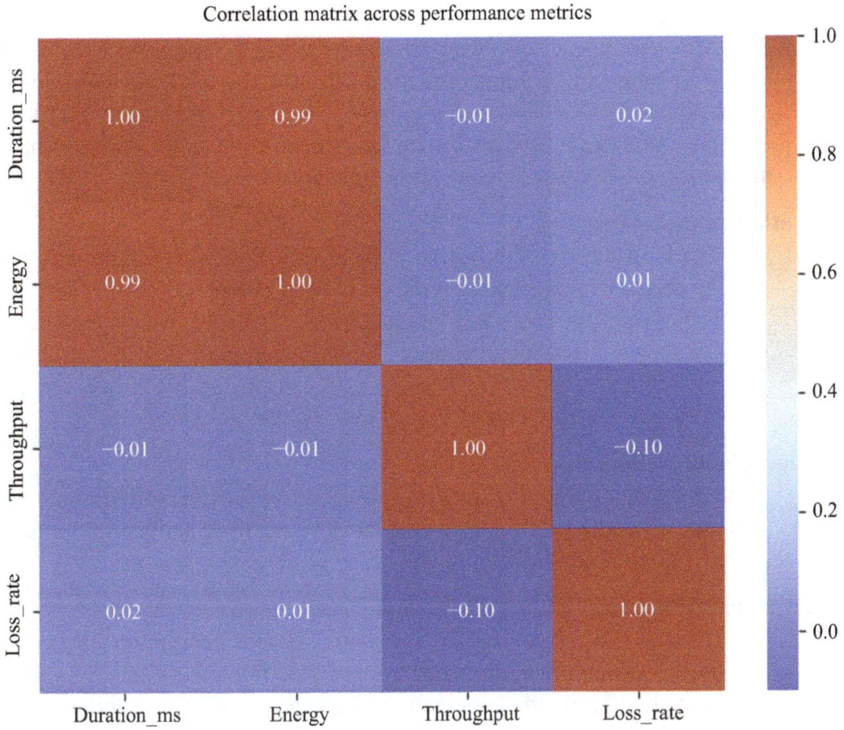

Figure 6.13 Correlation matrix across latency, energy, throughput and loss rate in the proposed IoMT cryptographic framework

Table 6.4 Comparative complexity table

Scheme	**Encryption complexity**	**Authentication complexity**	**Audit logging complexity**
AES-only	$O(n)$	Basic $(O(1))$	None
ECC-only	$O(n^2)$	$O(n^2)$	None
Blockchain-only	$O(n)$	$O(n)$	$O(n)$ per transaction
Proposed framework	$O(n) + PQO(n.\log n)$	Adaptive $(O(m))$	Optimised $O(n)$with offloading

6.4.8 Computational complexity analysis

The proposed IoMT cryptographic framework underwent computational complexity assessment for its essential parts, which showed results against standard reference systems. The system developers used Big-O notation to describe complexity because it shows how system performance changes when dealing with larger payload sizes (n) and more devices (m); see Table 6.4.

1. **Encryption layer**
   - The AES variants perform their operations at a rate of $O(n)$ time per block, but ECC-based blind signatures need $O(n^2)$ scalar multiplication operations. The proposed framework improves AES performance on restricted devices through its ability to maintain linear complexity and its implementation of ECC for authentication functions only.
2. **Authentication layer**
   - The implementation of adaptive authentication requires additional processing steps, which depend on the number of connected devices $(O(m))$, yet it outperforms ECC-only authentication systems, which experience performance degradation at a rate of $O(m^2)$ when more devices are added. The authentication system of low-power sensors becomes less complicated through the implementation of tiered authentication.
3. **Blockchain auditability**
   - Each transaction logging requires $O(n)$ operations. Parallelisation and gateway offloading reduce effective latency, ensuring auditability without significant computational burden.
4. **Post-quantum layer**
   - The operational requirements for lattice-based encryption primitives amount to $O(n) + O(n.\log n)$ operations. The framework achieves these solutions through its implementation of gateways and cloud nodes, which simplify devices while preserving their quantum security features.

The analysis shows that the proposed framework achieves near-linear complexity in most operations, outperforming ECC-only and blockchain-only schemes in scalability. By combining lightweight AES, adaptive ECC authentication, blockchain auditability and post-quantum resilience, the framework balances computational efficiency with strong security guarantees. These results align with related studies (Islam *et al.* [1], Naser *et al.* [4] and Al-Mekhlaf *et al.* [7]), which highlight the trade-offs between security strength and computational overhead in IoMT systems.

6.4.9 Summary of results

The numerical data in Figures 6.7–6.13 show that the proposed cryptographic framework for IoMT systems operates effectively while maintaining scalability and privacy requirements. The study shows that resource-constrained devices, including wearables, achieve reduced overhead by combining selective encryption with edge-aware design, as evidenced by latency and energy analysis. Network simulations demonstrate that throughput remains stable when accounting for realistic loss scenarios, whereas mTLS provides additional security protection without increasing costs. The system maintains reliable consent enforcement mechanisms that protect patient autonomy while operating without any impact on system functionality. The auditability coverage demonstrates that all essential system operations maintain traceability and tamper evidence, which enables both forensic investigations and regulatory requirements. The framework shows excellent cost-efficiency through its new payload efficiency ratio, which measures secure data transfer speed against both system delay and power consumption. Cloud and gateway devices operate with scalable efficiency through their design, while wearables achieve sustainability through their use of selective encryption. The final correlation matrix shows how latency and energy consumption, throughput and loss rate interact with each other, which proves that the framework successfully manages performance trade-offs between different network environments. The research findings confirm that the framework provides an effective solution to protect IoMT systems through its combination of strong encryption, operational performance and privacy protection mechanisms.

6.4.10 Comparative discussion

The consolidated results of the proposed IoMT cryptographic framework are summarised in Table 6.5. The table unites latency information with power consumption readings, data transfer speeds, authorisation systems, logging systems, data compression, data linking and processing needs for direct assessment against reference systems and other applicable academic studies.

The consolidated results show that the proposed framework delivers better performance than baseline schemes in all measured aspects, including latency, energy consumption, throughput and scalability. AES-only systems achieve linear complexity, but they experience extended latency periods, which affect the performance of limited devices. ECC-only schemes provide strong security protection, but their operational speed becomes limited because they require quadratic computational operations, which restrict their ability to scale. The implementation of blockchain-only systems provides auditability capabilities, but these solutions require significant energy usage, and they process transactions at lower rates.

The proposed framework combines lightweight AES encryption with adaptive ECC authentication and blockchain auditability and post-quantum resistance capabilities. The hybrid system design provides near-linear complexity, strong auditability and complete consent enforcement and operates at low energy while delivering high performance.

Comparison with related works reveals the following.

- Islam *et al.* [1]: Achieved lightweight encryption for medical images but lacked multi-device synchronisation. Our framework extends efficiency to heterogeneous IoMT systems.
- Rasheed and Kumar [2]: Optimised device-level energy but did not address system-wide orchestration. Our results show low energy use across diverse devices.
- Naser *et al.* [4]: Highlighted the trade-off between strong authentication and power limits. Our adaptive authentication resolves this by scaling security based on device type.

Table 6.5 Consolidated results of the proposed IoMT cryptographic framework compared with baseline schemes and related works

Metric	**Proposed framework**	**AES only**	**ECC only**	**Blockchain only**	**Related works**
Latency	Low, linear scaling	High	Moderate	High	Islam *et al.* [1]: efficient for images but limited to a single device.
Energy consumption	Low	Moderate	Moderate	High	Rasheed and Kumar [2]: optimised for devices but lacked system-wide orchestration.
Throughput	High	Moderate	Moderate	Low	Naser *et al.* [4]: strong security but poor energy trade-off.
Consent enforcement	Full coverage	None	None	Partial	Kamalov *et al.* [5]: ethical focus, limited real-time integration.
Auditability	Strong (blockchain)	Weak	Weak	Strong	Okpu and Taylor [6]: integrity preserved but limited rekeying.
Payload efficiency	High	Moderate	Moderate	Low	Islam *et al.* [1]: efficient for small payloads only.
Correlation uniformity	Strong	Weak	Moderate	Weak	Almulla *et al.* [15]: cross-layer risks unresolved.
Computational complexity	Near-linear$O(n) + PQO(n.\log n)$		$O(n^2)$	per transaction	Al-Mekhlaf *et al.* [7]: PQ secure but heavy for small devices.

- Kamalov *et al.* [5]: Focused on privacy and ethics but lacked technical integration. Our framework embeds consent enforcement directly into cryptographic policies.
- Al-Mekhlaf *et al.* [7]: Introduced post-quantum schemes (PQ) but faced performance constraints. Our hybrid PQ integration achieves resilience with near-linear complexity through gateway offloading.

Overall, the proposed framework provides a unified solution that balances efficiency, security, privacy and compliance, while maintaining scalability for real-world IoMT deployments.

6.5 Lessons learned

The evaluation of the proposed cryptographic framework revealed three key lessons that provide guiding principles for designing secure, efficient and trustworthy IoMT healthcare systems.

Lesson 1: Security and efficiency must be balanced
The research showed that lightweight cryptography, together with adaptive authentication methods, provides effective security protection for IoMT devices with limited resources. Security needs to share equal design importance with efficiency as a fundamental design requirement.
Lesson 2: Privacy and auditability require integration
The system proved that privacy protection requires implementation at the beginning of operations because consent monitoring and blockchain audit systems demonstrated it is impossible to implement privacy protection after data collection has started. The cryptographic system requires ethical compliance and transparency to become an integral part of its design process.
Lesson 3: Preparing for post-quantum threats is urgent
The research on lattice-based encryption showed that quantum-resistant methods need better performance capabilities, but gateway offloading and hybrid schemes proved their ability to operate. Healthcare organizations should begin their post-quantum cryptography deployment through managed stages because this method will defend their systems against upcoming security threats.

The lessons provide essential principles that will direct the creation of secure IoMT healthcare systems that need to scale up their operations while maintaining their trustworthiness.

6.6 Conclusion

The study sets out a structured approach for evaluating the cryptographic security model of IoMT, from data generation to assessment of performance, privacy and efficiency. Research on the subject highlighted that factors such as latency, data transmission rates and power consumption influence security and privacy measures, as well as audit requirements. This research provided a comprehensive system-level comprehension by examining payload efficiency and the relationships between different system components. A proper balance is struck by the system in the research between regulatory compliance and operational effectiveness. The system employed a binary decision-making approach to consent management, which resulted in a log of approvals and rejections in an audit trail to allow for total tracking. The system's performance was found to be a trade-off between the system's throughput and the size of the data it could process.

The results of the correlation analysis showed that the system exhibited a consistent pattern across different IoMT environments, each with distinct characteristics. Research shows that the system gives healthcare information providers the ability to share patient data securely. This was achieved by the system offering both security and scalability. A study has provided researchers with a model for IoMT security that can be tested in the course of future research. In controlled settings, experiments have been conducted; however, in real-world implementations, several factors will likely need to be taken into account, including the varying reliability of different networks, the types of devices involved and the differing regulatory requirements that apply. The framework should be developed through research studies that use data from a variety of devices and multiple countries to highlight the privacy and security issues in the IoMT [5]. Developing blockchain technologies that provide lattice-based security frameworks, as described by Al-Mekhlaf *et al.* [7], for secure smart healthcare should be pursued through continued research [13].

References

[1] Islam, M. O. U., Parah, S. A., Malik, B. A. and Malik, S. A. Lightweight medical-image encryption technique for IoMT-based healthcare applications. *Multimedia Tools and Applications*, 84(11), pp. 8929–8964, 2025.

[2] Rasheed, A. M. and Kumar, R. M. S. (2025). Efficient lightweight cryptographic solutions for enhancing data security in healthcare systems based on IoT. *Frontiers in Computer Science*, 7, 1522184.

[3] Deb, S., Lupu, E., Drakakis, E. M., *et al.* (2025). Securing the internet of medical things (IoMT): Real-world attack taxonomy and practical security measures. *arXiv preprint arXiv:2507.19609*.

[4] Naser, H. Y., Mattar, A. K., Saare, M. A., Almaiah, M. A. and Shehab, R. (2025). A comparison of lightweight cryptographic protocols for energy-efficient and sustainable IoMT authentication. *Engineering, Technology & Applied Science Research*, 15(4), 25746–25756.

[5] Kamalov, F., Pourghebleh, B., Gheisari, M., Liu, Y. and Moussa, S. (2023). Internet of medical things privacy and security: Challenges, solutions, and future trends from a new perspective. *Sustainability*, 15(4), 3317.

[6] Okpu, E. O. and Taylor, O. E. (2025). Analysing the integration of AES-256 encryption and HMAC hashing in IoT smart healthcare systems. *Ci-STEM Journal of Digital Technologies and Expert Systems*, 2(1), 18–24.

[7] Al-Mekhlaf, Z. G., Saare, M. A., Altmemi, J. M. H. *et al.* (2025). A quantum-resilient lattice-based security framework for Internet of medical things in healthcare systems. *Journal of King Saud University – Computer and Information Sciences*, 37(6), 126.

[8] Barna, T. L., Isaac, S., Habu, C., Habu, S. and Joseph, A. A. (2025). Post-quantum cryptographic frameworks for internet of things (IoT) and internet of medical things (IoMT) authentication systems. *Physics Access*, 88(136), 2025. https://doi.org/10.1007/s11235-025-01372-1

[9] Akkal, M., Cherbal, S., Annane, B., Lakhlef, H. and Kharoubi, K. (2025). Quantum, post-quantum, and blockchain approaches for securing the internet of medical things: A systematic review. *Cluster Computing*, 28(10), 655. https://doi.org/10.1007/s10586-025-05481-z

[10] Barna, T. L., Isaac, S., Habu, C., Habu, S. and Joseph, A. A. (2025) Post-quantum cryptographic frameworks for internet of things (IoT) and internet of medical things (IoMT) authentication systems. *Physics Access*, 5(1), 2025. https://library.kab.ac.ug/Record/doaj-art-b89a61bf36d547cfb5391b5c215c5541/Description#tabnav

[11] Lee, S., Kang, K. A., Lee, S. and Kim, J. (2025). Exploring authentication protocols for secure and efficient Internet of Medical Things systems. *Electronics*, 14(21), 4164. https://doi.org/10.3390/electronics14214164

[12] Samal, K., Sunanda, S. K., Jena, D. and Patnaik, S. (2025). A lightweight privacy preservation authentication protocol for IoMT using ECC-based blind signature. *International Journal of Engineering Business Management*, 17, 1–13. https://doi.org/10.1177/18479790251318538

[13] Faneela, Khan, M. A., Alsuhibany, S. A., El Shafai, W., Rehman, M. U. and Ahmad, J. (2023). An immutable framework for smart healthcare using blockchain technology. *Computer Systems Science and Engineering*, 46(1), 165–179. https://doi.org/10.32604/csse.2023.035066

[14] Sabrina, F., Sohail, S. and Tariq, U. U. (2024). A review of post-quantum privacy preservation for IoMT using blockchain. *Electronics*, 13(15), 2962. https://doi.org/10.3390/electronics13152962

[15] Almulla, Z., Almajed, H. and Rahman, M. M. (2025). A layered security perspective on internet of medical things: Challenges, risks, and technological solutions. *International Journal of Advanced Computer Science & Applications*, 16(5), 2025. https://doi.org/10.14569/ijacsa.2025.0160576

[16] Akinyemi, L. A., Mnkandla, E. and Sumbwanyambe, M. (2024). *Cybersecurity computing in modern healthcare systems*. 29, 305–333. https://doi.org/10.1049/PBHE064E_ch10

[17] Akinyemi, L. A., Folorunso, C. O., Shoewu, O. O. *et al.* (2024). Testing and validation of AI and blockchain-based models for application in telehealth systems. *In: Artificial Intelligence and Blockchain Technology in Modern*

Telehealth Systems, vol. 42, pp. 327–368. https://digital-library.theiet.org/doi/book/10.1049/pbhe061e
[18] Balaji, S., Jeyaraman, N., Ramasubramanian, S. and Jeyaraman, M. (2026). Cryptography-based fog computing and IoT-based technology in smart healthcare systems. In: *Fundamentals of Fog Computing and the Internet of Things for Smart Healthcare*, pp. 253–272. Elsevier.
[19] Ngwenya, T. and Ndlovu, B. (2026). A systematic review of post-quantum cryptography for healthcare data protection: performance, readiness, and deployment challenges. *Journal of Applied Informatics and Computing*, 10(1), 134–149.

Chapter 7

New-generation cryptography algorithms for Internet of Things environments in intelligent healthcare systems

Francesca Fiani[1], Agbotiname Lucky Imoize[2,3], Christian Napoli[4,5], Claudio Tomazzoli[6] and Cristian Randieri[7]

Abstract

The rapid digitalisation of healthcare systems, combined with the integration of intelligent interconnected devices, is highlighting a growing need to improve current system security against cyber threats to protect patient privacy and ensure consistent diagnostics. For this reason, it is fundamental to perform a precise analysis of current literature, identifying the main research topics in the field of intelligent healthcare systems and the main employed cryptography solutions in order to determine their current limitations and what can be done to improve them. Through the use of the PRISMA-ScR guidelines, we selected 21 relevant literature articles on the topic of cybersecurity in healthcare systems on the topics of Internet of Things systems, AI-driven cryptography and new-generation cryptography solutions. To provide a complete mapping of currently available solutions and gaps in literature, our guiding research questions were 'Is Internet of Things the main focus of cryptography for intelligent healthcare system?' and 'Which are the main solutions found in current literature?' providing a complete mapping of currently available solutions and gaps in literature. A thorough analysis and summarisation of the articles allowed us to identify common directions and limitations, defining future directions for research, such as the need to focus on real-world experimentation and to implement explainable artificial intelligence solutions, which could further improve system security.

[1]Department of Computer, Control and Management Engineering, Sapienza University of Rome, Italy
[2]Department of Electrical and Electronics Engineering, Faculty of Engineering, University of Lagos, Nigeria
[3]Department of Electrical Engineering and Information Technology, Institute of Digital Communication, Ruhr University, Germany
[4]Institute for Systems Analysis and Computer Science, Italian National Research Council (CNR), Italy
[5]Department of Artificial Intelligence, Czestochowa University of Technology, Poland
[6]Department of Information Science and Technologies, Pegaso Digital University, Italy
[7]Department of Theoretical and Applied Sciences, eCampus University, Italy

Keywords: Internet of Medical Things; AI-driven cryptography; post-quantum cryptography; lightweight cryptography; PRISMA-ScR

7.1 Introduction

The digital transformation of healthcare systems has slowly led to the adoption of intelligent technologies such as Internet of Medical Things (IoMT) [1], wearable sensors [2], cloud-based electronic health records [3] and artificial intelligence (AI)-driven diagnostic [4] and therapeutic [5,6] platforms. These technologies have significantly improved the quality of healthcare, allowing continuous patient monitoring, personalised medicine, remote diagnostics and large-scale health data analytics. However, this increased connectivity and reliance on data-crowded infrastructures have also opened up new possibilities for attacks on healthcare systems, making them particularly vulnerable to cyber threats.

Healthcare data represent one of the most sensitive categories of information, including privacy-regulated information such as personal identifiers, medical histories, diagnostic images and real-time physiological signals. Security breaches in this domain have consequences that can potentially endanger patient safety and disrupt critical clinical workflows, with guidelines emerging in the literature to face this growing threat [7]. Recent years have witnessed a growing number of cyber incidents targeting healthcare infrastructures [8], including data leakage, ransomware attacks, Denial-of-Service attacks, spoofing, Man-in-the-Middle (MitM) attacks and adversarial and model-poisoning attacks targeting AI models [9].

Cryptography is fundamental to mitigating these threats by ensuring confidentiality, integrity, authentication and access control in healthcare systems. Traditional cryptographic solutions, however, face significant challenges when applied to modern intelligent healthcare environments, which include distributed terminals such as multiple diagnostic devices, data servers and medical robots [10]. Resource-constrained Internet of Things (IoT) devices used in medicine, such as wearable wristbands for patient monitoring, often lack the computational power required by conventional cryptographic algorithms, while real-time clinical applications impose strict latency constraints [11]. Moreover, the emergence of quantum computing poses high risks to current public systems, further complicating the security landscape and forcing healthcare to shift towards more secure solutions [12].

In response to these challenges, recent research has increasingly explored hybrid security architectures that combine cryptographic techniques with artificial intelligence, machine learning and blockchain technologies [13], a solution which has already been tested in other fields such as industrial control systems [14]. AI-based approaches are used for intrusion detection [15], anomaly detection [16], attack classification [17] and adaptive security [18], while lightweight cryptography aims to tackle the resource constraints in medical devices [19]. At the same time, post-quantum cryptography is gaining attention as a solution to protect healthcare data against future quantum-based attacks [20]. However, current literature is heterogeneous and fragmented, with few to no works addressing this heterogeneity and proposing a complete mapping of available solutions.

Given this context, a comprehensive overview of recent advances in cryptography for intelligent healthcare systems is needed to map current research directions, identify common limitations and highlight open challenges to guide future research. In particular, we are interested in analysing how cryptographic techniques are being adapted to IoT for healthcare environments, how they are integrated with AI-based cryptography and to what extent emerging solutions such as lightweight and post-quantum cryptography are being addressed in current literature. Despite the rapid growth of research on AI-based security and advanced cryptographic techniques in healthcare, in fact, the literature remains fragmented across domains such as IoMT security, post-quantum cryptography and lightweight cryptographic design. There is a lack of consolidated analysis that evaluates these approaches collectively under realistic healthcare constraints. This fragmentation motivates the need for a structured synthesis that critically examines not only proposed solutions but also their evaluation environments, assumptions and practical feasibility.

For this reason, this work proposes a scoping review of recent literature on cryptographic solutions for intelligent healthcare systems, with a specific focus on AI-assisted cryptography, IoT-centric architectures and next-generation cryptographic approaches. Our research was guided by two main research questions: Is Internet of Things the main focus of cryptography for intelligent healthcare system?. Following the PRISMA-ScR guidelines, this review synthesises existing studies, categorises their main application domains, summarises employed algorithms and results and discusses their shared limitations. The ultimate goal is to provide a clear overview of the current state of the art and to outline future research directions to improve security in intelligent healthcare systems.

7.1.1 Chapter contributions

This chapter provides a structured and comprehensive analysis of recent literature to present a clear overview of the current state of the art in next-generation algorithms for IoT in healthcare. The limited availability of up-to-date review studies in this area, together with the need to identify key research gaps, motivated this investigation. To ensure methodological rigour and transparency, the PRISMA-ScR framework was adopted, as it offers a well-defined and systematic approach to scoping reviews, which ensures not only reproducibility of the results but also a complete mapping of current literature.

By examining an extensive body of literature, this review identifies and compares existing approaches, highlighting the main limitations that characterise current cryptographic algorithms for healthcare IoT systems. Given the critical importance of security in environments where highly sensitive data is processed and transmitted, a thorough understanding of these limitations is essential. Identifying unresolved challenges not only clarifies the current weaknesses of available solutions but also provides guidance for the development of more robust, efficient and secure cryptographic algorithms, ensuring quick progress towards safer healthcare IoT infrastructures.

7.1.2 Chapter organisation

The following chapter is organised as follows. Section 7.2 presents an overview of the protocol used for the analysis and selection of literature, including the full selection process, the inclusion and exclusion criteria and information on how the screening of the articles was performed. Section 7.3 synthesises the 21 selected articles one by one, dividing them into 4 main thematic categories. For each article, we reported the topic of the study, the proposed solution, the numerical results obtained (if mentioned) and the limitations of the study. After the reports, summary tables of the findings have been added for clarity. Section 7.4 presents a thorough discussion of the literature, highlighting both the current strengths and limitations of the solutions found in recent works. Concurrently, possible future directions will be defined for research, and the limitations of the study will be acknowledged, together with a synthesis of the lessons learned from this research. Finally, Section 7.5 will present a summary of the whole work, further delineating the future of research in the topic of IoT for intelligent healthcare systems.

7.2 Methods

Our research was guided by the PRISMA-ScR guidelines, which define a precise protocol to ensure the reproducibility of results of scoping reviews. The motivation behind the choice of this protocol and of the scoping review format is guided by a desire to identify current limitations in the literature concerning cryptography algorithms for healthcare, allowing us to delineate future lines of research to fill the existing gaps. In particular, our research focused on the topic of applications of artificial intelligence, machine learning and neural networks algorithms to the problem of cryptography in healthcare.

The research of literature was carried out using a custom tool that performs automatic extraction of articles and quartiles of the publication journals from the ArXiv, PubMed, Scopus and Zenodo portals. The keys used to perform the search of literature were the following: ‘Cryptography healthcare AI’, ‘Cryptography security healthcare’, ‘Cryptography AI medicine’, ‘Cryptography AI psychology’, ‘Cryptography threat healthcare’, ‘Healthcare cybersecurity AI’ and ‘Intelligent algorithms healthcare’. The research time frame is based on the last five years, from January 2020 to 2025. The last search was performed in November 2025. All articles were searched exclusively in English. No grey literature has been included in the research.

From the search performed with our tool, a total of 1882 potential articles have been identified. Of the 1882 articles, 393 articles were eliminated since they were duplicates, with 10 of them being intra-search duplicates, leaving us with 1489 articles to be screened.

After eliminating the duplicates, the following step was selecting the articles related to our area of interest. In particular, we wanted to focus on works that delved into the use of AI technologies of any kind, including machine learning (ML), neural networks (NN) and deep learning (DL), on the topic of intelligent healthcare. The attacks we wanted to focus on, which are the most relevant in the field of healthcare, were data leakage, spoofing, Denial of Service (DoS), Man-in-the-Middle (MitM) and adversarial attacks. We were especially interested in modern topics such as federated learning, adaptive cryptography and blockchains. As for the specific applications, our focus was on wearable devices, e-health, diagnostic systems and any kind of Internet of Things (IoT) or Internet of Medical Things (IoMT) system. Any article that did not fall under this criterion or was unrelated to the cryptography and cybersecurity topics was automatically discarded. To our knowledge, no previous scoping review has focused on this combination of topics tailored for medical applications, framing our review as a much-needed analysis of literature to identify the current state of the art. The potential articles for the research remained 36, having discarded 1462. The selection was performed without the aid of any tool.

Finally, from the last 36 articles, which were all accessible, a further 15 studies were discarded for the following reasons.

- Low quartile of publication (N = 5): Five articles resulted in a low publication quartile or an absent quartile. Our tool automatically extracts quartiles in all relevant fields in the Scimago portal, but values have been manually verified to confirm the results of the extraction. In particular, we verified quartiles related to the Computer Science field and the Medicine field. Q3 and Q4 results, which generally correspond to poor quality of the publication journal and a low impact in the literature of the published article, have been discarded as they are not significantly impactful to the current state of the art.
- Survey not relevant (N = 7): Seven articles are surveys on topics that are not relevant for the scope of this review (e.g. they do not delve into AI applications to cryptography).
- Arguments not treated (N = 3): Similar to the previous point, three applicative articles do not delve into the topic of AI application to cryptography and have therefore been excluded.

Two reviewers (C.T. and F.F.) performed the screening and assessment of the articles, determining the inclusion and exclusion criteria for the selected literature. In case of disagreements, the reviewers discussed the inclusion and/or exclusion of a certain article until they reached a consensus. In case of a lack of agreement, a third reviewer (C.R.) acted as an adjudicator. After the whole screening process, 21 articles remained for analysis and evaluation. Each article was narratively summarised, providing a report for each article of employed methodologies, datasets, numerical results (where applicable) and limitations. Figure 7.1 illustrates the workflow of study selection for this scoping review. A summarisation of the inclusion and exclusion criteria for the selection of works has been presented in Table 7.1.

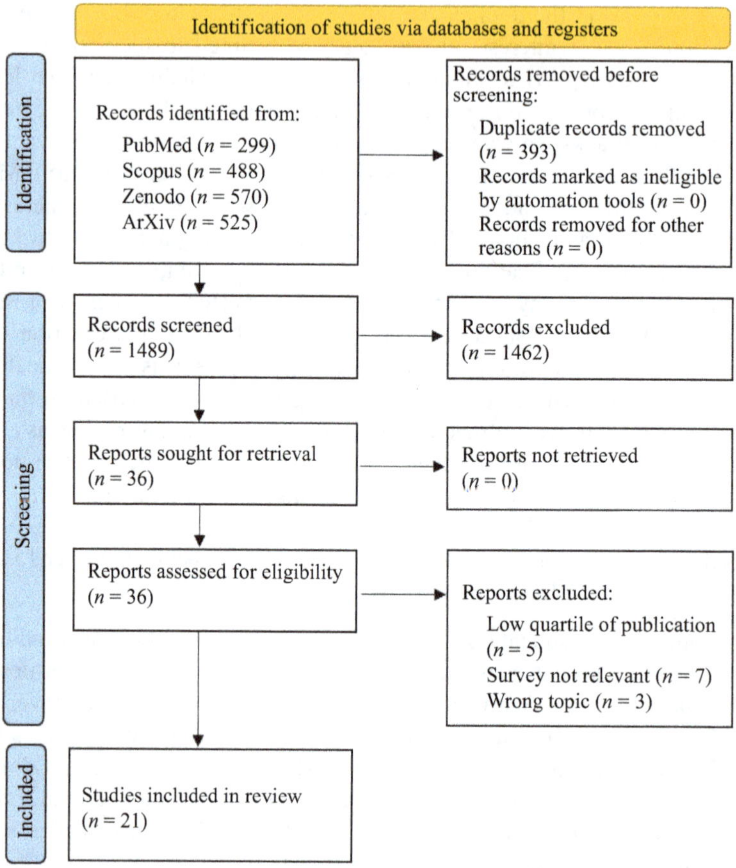

Figure 7.1 PRISMA-ScR diagram of the article selection process

Table 7.1 Inclusion and exclusion criteria for the article selection through the PRISMA-ScR guidelines

Inclusion criteria	**Exclusion criteria**
Preprint, journal or conference article	Patent, opinion, commentary
Work published in a Q1/Q2 journal or in a conference of rank ≥B	Work published in Q3/Q4/unranked journal or in conference of rank <B/unranked
Work published in English	Work published in other languages
Work published ≤5 years ago (2020–2025)	Work published >5 years ago (before 2020)
Articles in the following areas: IoT/IoMT, wearable devices, e-health, diagnostic systems, blockchain, federated learning, adaptive cryptography, data leakage, spoofing, DoS, adversarial attacks, MitM	Articles not related to healthcare, medical or IoMT/IoT systems and/or not addressing cryptography or cybersecurity aspects

(Continues)

Table 7.1 (*Continued*)

Inclusion criteria	Exclusion criteria
Works on AI, ML, NN, DL or next-gen solutions for cybersecurity and/or for prevention and protection against cyberattacks in the medical field (both comparative reviews and novel approaches)	Works which do not delve into AI or next-gen solutions for cybersecurity

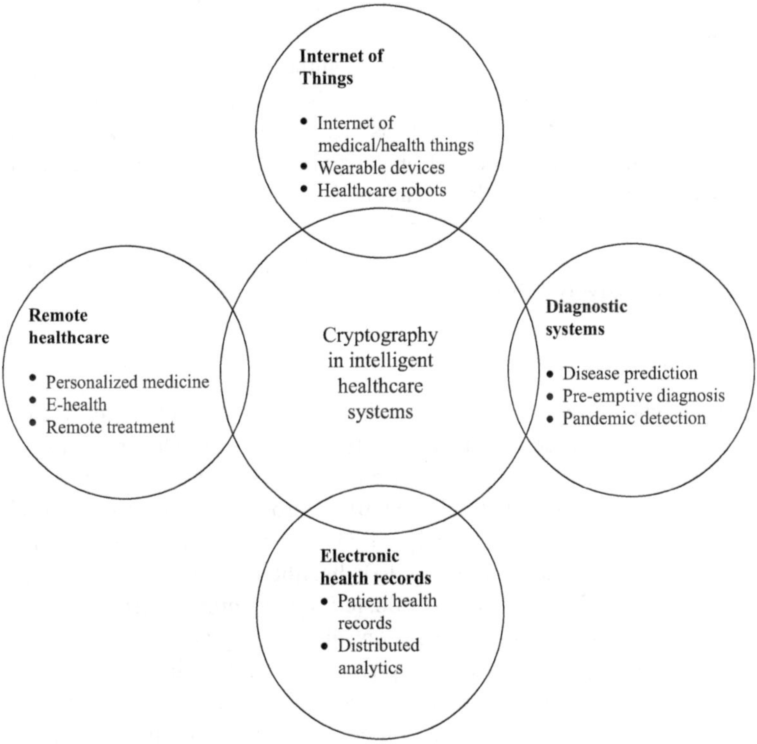

Figure 7.2 Taxonomy of applications in cryptography for intelligent healthcare systems

7.3 Results

In this section, we will summarise the 21 articles selected for this scoping review. For each work, we will report the problem they try to solve, their proposed solution, which algorithm was employed (if reported), the results obtained by the implementation (if reported) and the limitations of the work. Figure 7.2 shows the taxonomy we used to classify articles in our research. The four main areas we identified were:

- Internet of Things technologies, which also include devices such as healthcare robots and wearable devices;
- Remote healthcare technologies, which include personalised medicine, remote treatment (e.g. in psychotherapy) and e-health;
- Diagnostic systems, which include both predictive and detection systems;
- Electronic health records, which include any kind of patient health record and distributed analytics.

To facilitate the comparison of works, they have been arranged in four macrocategories following the presented taxonomy: comparative studies, studies that delve into the security of healthcare records of any type (textual, image and video), studies that focus on secure diagnosis and prediction of any kind of disease and studies that focus on secure transmission and authentication in IoT environments. It must be noted that not all works that focus on IoT technologies are included in the IoT section, as we prioritised, where possible, the categorisation under the patient record security and diagnosis/prediction categories. No study fell under the remote healthcare category. With our classification, the literature was divided into 4 works for comparative studies (19 per cent), 5 for electronic health records (23.8 per cent), 4 works for diagnostic systems (19 per cent) and 8 works for Internet of Things technologies exclusively (38.1 per cent).

7.3.1 Comparative studies

To understand the current state of the art, it is fundamental to first analyse previous works that summarise literature under different viewpoints. The article by Mushtaq *et al.* [21] is particularly important as it presents a systematic literature review on Zero Trust Architecture (ZTA) implementation across multiple domains. Following PRISMA guidelines, the authors analysed 74 peer-reviewed articles from 2016 to 2025 on the topics of cloud computing, IoT, healthcare, enterprise systems, AI/ML and blockchain. Each paper was evaluated under nine cybersecurity criteria: authentication, authorisation, access control, cryptography, security gateway, environmental perception, network segmentation and audit. Results show that authentication, authorisation and access control are consistently implemented, while auditing, orchestration and environmental perception remain underexplored. For healthcare specifically, ZTA models for EHR protection and IoT-based patient monitoring lack device-level segmentation and optimised cryptography for low-power medical devices. Several challenges are still open, including scalability limitations, insufficient lightweight cryptographic solutions, weak orchestration mechanisms and limited GDPR/HIPAA compliance alignment. The authors recommend lightweight cryptography (LWC), context-aware trust engines with explainable AI (XAI) and automated orchestration frameworks for future ZTA implementations in healthcare.

Following the need to focus on LWC for healthcare, the paper by Chinbat *et al.* [22] provides a systematic evaluation of LWC algorithms combined with machine learning techniques for IoT-based healthcare systems. Rather than using ML as a direct cryptography solution, the authors propose an ML-assisted evaluation framework to identify the most suitable LWC algorithm for improving security in healthcare IoT devices under strict resource constraints. Eight LWC algorithms (AES, PRESENT, MSEA, LEA, XTEA, SIMON, PRINCE and RECTANGLE) are implemented on a Raspberry Pi 3 microcontroller, representative of real-world medical IoT hardware. The authors trained several models, such as Decision Tree (DT), Random Forest (RF), Support Vector Machine (SVM), Multilayer Perceptron (MLP) and K-Nearest Neighbours (KNN), to classify and evaluate encrypted data based on the behavioural patterns of each LWC algorithm under different operational conditions. Synthetic healthcare-like message files with sizes ranging from 16 KB to 2048 KB are used to simulate medical data transmissions. The models are trained and tested using five-fold cross-validation, with the chosen metrics being the standard accuracy, precision, recall and F1-score alongside system-level metrics such as encryption/decryption time, energy consumption, memory usage and throughput. RF and SVM consistently outperform other ML models, achieving high accuracy (up to 99 per cent on smaller messages) and robustness across different file sizes. RECTANGLE is shown to be the most efficient and stable LWC algorithm, maintaining strong performance even as message size increases and outperforming alternatives such as Advanced Encryption Standard (AES) and MSEA, which show higher computational overhead and performance degradation. The study also highlights that increasing file size negatively impacts ML classification performance, reinforcing the need for lightweight solutions in healthcare IoT contexts. While the obtained results are very promising, the work does not address advanced threats such as adversarial attacks, federated learning scenarios or encrypted-domain learning. Moreover, the experimental set-up is limited to a controlled laboratory environment, without any study of the applicability to real-world scenarios with high heterogeneity.

One of the most important directions is the early detection of cyber threats in healthcare Internet of Medical Things (IoMT) environments to allow for faster intervention. The article by Kumari *et al.* [23] moves exactly in this direction, proposing an IoT-edge architecture with data-driven AI techniques. The framework focuses on MitM attack detection using the WUSTL-EHMS-2020 dataset, which contains 16,318 samples with 44 characteristics (35 network flow measurements, 8 biometric features, 1 label). Eight machine learning models were evaluated comparatively to verify the most efficient one: Variational Autoencoders (VAEs), RF, SVM, eXtreme Gradient Boosting (XGBoost), Light Gradient-Boosting Machine (LightGBM), Logistic Regression (LR) and Feedforward neural networks (FNN). VAEs achieved the highest accuracy at 91.61 per cent, with matching precision (91.69 per cent), recall (91.61 per cent) and F1-score (91.60 per cent). RF and SVM demonstrated comparable performance, while LR showed much lower accuracy (68.11 per cent). The study includes a real-time Streamlit-based application for cyber threat detection using vital signs input, addressing IoMT vulnerabilities such as data poisoning, replay attacks and spoofing. However, no real-world validation has been performed yet. Moreover, the reliance on complex ML algorithms generates computational constraints on resource-limited devices and introduces scalability challenges in large-scale healthcare network deployments.

Given the quick rise of quantum computers, it is also important to develop cryptographic algorithms that are able to defend healthcare systems from more complex attacks, which is the main goal of post-quantum cryptography (PQC). The review article by SaberiKamarposhti *et al.* [24] positions itself as a comprehensive road map for post-quantum cybersecurity resilience in healthcare. The paper analyses quantum-resistant encryption protocols, including lattice-based (NTRU, LWE, Ring-LWE), code-based (McEliece, Niederreiter), hash-based (Merkle, XMSS, LMS) and multivariate polynomial cryptography. Quantum threats are examined through Shor's algorithm (compromising RSA/ECC public key systems) and Grover's algorithm (reducing symmetric-key effective length by half). The proposed road map encompasses secure communication strategies (Kyber, NTRUEncrypt, NewHope for key exchange; FrodoKEM for messaging), innovative authentication mechanisms (quantum-resistant cryptographic key management, biometrics, multi-factor authentication) and data integrity assurance (quantum-resistant hash functions, blockchain applications, zero-knowledge proofs). Implementation challenges addressed include compatibility with legacy systems, workflow disruptions, interoperability across healthcare platforms, resource constraints and personnel training requirements. The paper also tackles regulatory considerations such as HIPAA and GDPR compliance, data breach notification and cross-border data transfer. The main point emphasised by the paper is the urgency to develop secure post-quantum cybersecurity algorithms, in particular encryption protocols, before quantum computers compromise existing encryption protecting sensitive patient data.

7.3.2 *Health records*

One of the main threats in intelligent healthcare is the leakage of patients' health records or personal data. For this reason, several works available in the literature focus on algorithms for the protection of various forms of patient data, such as Patient Health Records (PHR) or Electronic Healthcare Records (EHR).

One such work is the article by Amanat *et al.* [25], which proposes a blockchain-based framework for secure EHR storage and sharing using cloud computing. The architecture employs a Proof-of-Stake (POS) consensus mechanism with Secure Hash Algorithm 256 (SHA256) hashing for user authentication and data integrity, while Elliptic Curve Digital Signature Algorithm (ECDSA) verifies EHR sensors transmitting data to cloud infrastructure. The decentralised network comprises five node types (visitors, co-ordinators, agents, administrators and contributors) with smart contracts validating transactions. Implemented using the Node.js crypto module on 12 Windows 10 desktops with 4 Gigabytes (GB) Random Access Memory (RAM) and 2.4 Gigahertz (GHz) Central Processing Unit (CPU), the framework achieved up to 17 transactions per 6-minute block period. Comparative analysis demonstrated improvements over Proof-of-Work (POW), SHA1 and Message Digest 5 (MD5) in terms of power consumption, authenticity and security. The framework maintains patient privacy through hash IDs while storing actual medical data, such as Magnetic Resonance Imaging (MRIs) and X-rays, off-chain via secure links. While extremely important due to the inclusion of complex medical data instead of simple records, the authors note capacity constraints caused by blockchain storage, which can hinder the security of heavy data such as images. At the same time, the authors propose federated learning integration for enhanced protection in future work.

The paper by Ali *et al.* [26] addresses the intersection of cryptography, AI and healthcare security in the context of Industrial IoT (IIoT) environments. In particular, the work focuses on secure data management and the protection of sensitive healthcare information exchanged across IoT-enabled medical systems. Unlike works centred purely on attack detection, this study emphasises preserving data access and secure keyword-based search, which are critical challenges in distributed healthcare infrastructures. The authors propose attribute-based encryption (ABE) and searchable encryption with a blockchain-based trust and reputation management system to ensure data confidentiality, integrity and access control. A hybrid neural network model is employed to distinguish between intrusive and non-intrusive traffic, allowing detection of cyber threats such as Distributed Denial of Service (DDoS), malware injection and data poisoning attacks. To differentiate between attack techniques, the authors used the popular Spoofing, Tampering, Repudiation, Information disclosure, Denial of Service and Elevation of privilege (STRIDE) model. The experimental evaluation is conducted using simulated IIoT PHR data, where network traffic is processed to train and test the neural network-based intrusion detection component. The results show that the proposed model achieves high detection accuracy (up to 97 per cent) and reduced false positive rates (as low as 0.002 on 300 patients), demonstrating its effectiveness in identifying malicious activities while preserving data privacy. In addition, the searchable encryption mechanism introduces low computational overhead and ensures that no continuous involvement of the data owner is required after encryption. However, the authors acknowledge several limitations. The experimental validation relies on simulated datasets, which may not fully capture the complexity of real-world healthcare networks. Moreover, the computational cost associated with blockchain operations and neural network training could lead to low scalability in large-scale scenarios.

Another paper by Ali *et al.* [27] instead addresses intrusion detection in blockchain-enabled healthcare systems, with a specific focus on secure keyword search on encrypted PHR data. Unlike works that use blockchain as a storage layer, this study proposes blockchain as a distributed database. The authors propose a deep-learning-based homomorphic secure searchable encryption (HE-SSE) framework integrated with the same permissioned blockchain used in the previous work (Hyperledger Fabric). The solution combines fully homomorphic encryption (FHE) for keyword search with deep learning-based security monitoring. Attribute-based signatures (ABS) and Attribute-Based Access Control (ABAC) models ensure anonymity and flexibility of the smart contracts' access policies. A DL-based intrusion detection system based on a Bidirectional Long Short-Term Memory

(BiLSTM) instead detects attacks like DoS, DDoS and collusion. The DL component is evaluated with an IoT-Botnet dataset split into 70 per cent training and 30 per cent testing. The model achieves 98.67 per cent performance compared to the non-encrypted baseline, with an overall classification accuracy of 90.6 per cent, similar to that of a standard CNN. In addition, simulations on Hyperledger Fabric show improved search time and throughput, with up to 1.9x efficiency gains over benchmark systems such as MedRec, MedChain and MedBlock. Reduced false positives and negatives, lower communication overhead and resistance to multiple attacks further confirm the efficacy of the model. However, similarly to the previous paper by the same authors, the evaluation is mostly simulation-based, which may limit real-world generalisability and the use of FHE and DL introduces computational overhead not suitable for real-time applications. Moreover, the framework would benefit from validation on real clinical datasets and further optimisation to reduce the cost of encrypted inference.

The article by Nucciarelli *et al.* [28] focuses instead on a modular framework combining blockchain and cryptography to ensure reproducibility and accountability in distributed healthcare analytics. Similar to the works by Ali *et al.* [26,27], the system uses Hyperledger Fabric v2.4.7, but with a Raft consensus algorithm, deployed via Docker containers within the Gemelli Generator – Real World Data (GEN-RWD) Sandbox, a privacy-preserving distributed analytics infrastructure. The workflow implements automatic, non-repudiable reporting through hash-based notarisation, dual digital signatures (institution and user) and off-chain storage for cross-verification. The framework consists of three modules: Processor for computation, Proxy for communication management and GUI for web interface. A Python library enables blockchain API interactions, including hash calculation, notarisation and signature validation. To test the system, real healthcare data computations within the privacy-preserving environment were used. This approach protects both researchers with proof of authorship and data owners with the verification of results' integrity while at the same time preventing unauthorised modifications. The paper shows some limitations, including a single-node configuration that leaves scalability in multicentric scenarios unexplored, potential security-scalability trade-offs across regulatory jurisdictions and the absence of Non-Fungible Tokens (NFTs) or Smart Contracts for enhanced verification capabilities, but it still configures itself as a valid solution for healthcare analytics.

The article by Lakshmi *et al.* [29] similarly focuses on medical images, proposing a reconfigurable medical image encryption scheme implemented on Field-Programmable Gate Array (FPGA) hardware using a Hopfield Neural Network (HNN). The dual-layer encryption combines HNN-based diffusion with stream cipher-based confusion through a 16-bit Linear Feedback Shift Register (LFSR). Image-specific session keys are generated through a trained Back Propagation Neural Network (BPN), providing defence against plaintext attacks by introducing direct image encryption. The pipeline, which is implemented on an Intel Cyclone V FPGA using Verilog Hardware Description Language (HDL), achieves an average entropy of 7.99 (near-ideal 8.0), near-zero correlation coefficients, Number of Pixels Change Rate (NPCR) of 99.57 per cent and Unified Average Changing

Intensity (UACI) of 33.05 per cent. The FPGA implementation occupies only 20 per cent of hardware resources with 424.71 milliWatt (mW) power dissipation, enabling real-time encryption (1.2 s for $512 \times 512 \times 3$ images). Testing on 12 Digital Imaging and Communications in Medicine (DICOM) medical images and 20 grayscale images passed all National Institute of Standards and Technology Special Publication (NIST SP) 800-22 randomness tests with key space exceeding 2^{128}. However, initial BPN training overhead has been observed, and optimisation is still needed for high-resolution medical video encryption, as it greatly slows down the system.

7.3.3 Disease prediction and detection

Another important application for cybersecurity in healthcare is in the field of disease prediction and detection, where security is particularly important to avoid model poisoning or data alteration, which could impair diagnoses.

The article by Padinjappurathu Gopalan *et al.* [30] follows exactly this direction, proposing an efficient privacy-preserving scheme for disease prediction in IoT-based healthcare systems. Their framework addresses the dual challenge of securing patient data from IoT devices while maintaining accurate disease prediction capabilities. The system employs a Log of Round value-based Elliptic Curve Cryptography (LR-ECC) algorithm for enhanced security during data transfer after an initial authentication phase. For disease prediction, instead, the authors introduce an Elephant Herding Genetic Algorithm-based Deep Learning Neural Network (EHGA-DLNN) classifier, which optimises weight values to improve classification accuracy. Moreover, they also implement dimensionality reduction through Gaussian Kernel-based Linear Discriminant Analysis (GK-LDA) to improve the efficiency of the model. Experiments were conducted using the Hungarian heart disease dataset, with results demonstrating 98.35 per cent accuracy, 97.33 per cent sensitivity and 96.36 per cent specificity, outperforming traditional classifiers such as Artificial Neural Network (ANN), KNN and SVM. The LR-ECC encryption method achieved a security level of 98.87 per cent, with significantly faster encryption and decryption times compared to standard ECC, Rivest–Shamir–Adleman (RSA) and Diffie-Hellman algorithms. However, the work focuses primarily on a single disease dataset and does not explore scalability in real-world multi-disease scenarios. Additionally, the system does not address federated learning or adversarial robustness, limiting its applicability to more complex distributed healthcare environments.

In some cases, the protection of data is much more complex due to reliance on video recording transmission for the generation of a diagnosis, especially for neurodegenerative diseases. Given this need, the article by Castro *et al.* [31] proposes a hybrid protection scheme for gait analysis in early dementia recognition, combining Paillier partially homomorphic encryption (PHE) with cancellable biometric techniques based on random projection. The scheme protects patient gait features according to the International Organization for Standardization International Electrotechnical Commission 24745 (ISO/IEC 24745) security requirements, enabling secure data transfer from IoT acquisition devices to healthcare servers while performing random projection in the encrypted domain. A Long Short-Term Memory (LSTM) neural network with 32 units and attention mechanisms classifies dementia using transformed gait features such as spatiotemporal parameters, kinematic angles and sigma-lognormal features extracted from video recordings. Experiments on 118 videos from 43 subjects (23 healthy controls, 20 dementia patients) using 20-fold cross-validation demonstrated accuracy of 97.0 per cent with protection versus 96.8 per cent without, with a degradation of at most 1.20 per cent accuracy in the worst scenario. Security analyses confirmed non-invertibility via the Rouché-Capelli theorem and renewability through pseudo-random matrix generation. A total of 57.26 s were required for protection, with 3.36 s on the client-side and the rest on the server-side. However, the execution time is highly dependent on the dimension of the feature vector. For this reason, the authors propose to move towards more efficient schemes, such as Cheon-Kim-Kim-Song (CKKS).

Following the COVID pandemic, several works on pandemic detection have been developed, such as the work by Egala *et al.* [32]. The authors present iBlock, an intelligent decentralised blockchain-based system for pandemic detection and mitigation. The architecture integrates blockchain technology with hybrid fog and cloud computing, IPFS for distributed storage and AI/ML for intelligent predictions. The system is organised in three logical layers: a data collection layer using Health Cyber-Physical Systems (H-CPS) and personal instrumentation, an AI/ML processing layer for data analytics and predictions and an alert/suggestion layer for user assistance. For security, iBlock employs ECC with 256-bit keys, introducing a pseudo-anonymous digital identity generated through a one-way function combining device-specific variables and environmental parameters for randomness. Data privacy is managed through an AccessRuleTable mechanism, with personal data stored on local Collaborated Medical Database (CMD) networks and non-personal data on global CMD. The AI/ML component combines RF and SVM into a hybrid classifier for area classification, with green, brown and red zones based on infection risk and pandemic prediction. Experimental evaluation was conducted on five Ubuntu virtual machines running Hyperledger Fabric, using publicly available WHO COVID-19 data. The hybrid classifier achieved 84.89 per cent accuracy, outperforming individual MLP (77.50 per cent) and standard SVM (82.93 per cent) models. Theoretical security analysis demonstrated resistance to unauthorised data access and MitM attacks through token-based communication and device-specific authentication. Moreover, the system has additional features such as health passport generation, contact tracing with pseudo-identity preservation and reward mechanisms for data sharing. However, the evaluation remains largely simulation-based, and the AI/ML module relies on supervised learning, which may limit adaptability to novel pandemic patterns without retraining.

Post-quantum cryptography is particularly relevant not only for patient records protection but also for medical diagnosis. The article by Ravisankar and Maheswar [33] introduces Med-Q Ledger, a multi-layered framework combining PQC with federated learning for real-time predictive diagnostics in IoMT. The architecture combines permissioned Hyperledger Fabric for transactional integrity with scalable

Holochain Distributed Hash Table for high-volume telemetry. The system includes CRYSTALS-Dilithium digital signatures, Kyber Key Encapsulation Mechanisms for quantum-resistant protection, Self-Sovereign Identity (SSI) using Decentralised Identifiers (DIDs) and Attribute-Based Access Control (ABAC) with Zero-Knowledge Proofs. The edge intelligence layer employs lightweight autoencoders for real-time anomaly detection within 300 ms. Validation through necrotising enterocolitis prediction in preterm infants demonstrated a 0.90 F1-score for colostomy prediction and 25 per cent reduction in emergency surgeries. Performance metrics showed approximately 3400 TPS throughput, 180 ms end-to-end latency, anomaly detection rate greater than 95 per cent with less than 2 per cent false positives and 11 per cent computational overhead for PQC on resource-constrained devices. However, computational overhead has been observed in resource-constrained environments. Moreover, clinical validations are still to be performed on real-world datasets.

7.3.4 Internet of Medical Things environments

One of the main points of intelligent healthcare systems is the presence of a network of interconnected devices that constitute complex IoT environments, often defined as Internet of Healthcare Things (IoHT) or IoMT. In such contexts, where high connectivity and frequent data transfer are present, it is particularly important to implement secure cryptography algorithms to ensure no data leakage or model poisoning, as we already touched on in previous sections. While the majority of previous works employed IoT systems for specific applications, in this section, the works are mainly focused on security in IoT environments.

The article by Aivaliotis *et al.* [34] presents a comprehensive analysis of IoT-based multi-sensor healthcare architectures within the Health 4.0 paradigm, along with a novel lightweight cryptographic privacy scheme. The authors first provide a detailed overview of smart health infrastructure, covering both smart hospital architectures and personalised near-patient systems and identifying security requirements such as confidentiality, availability, integrity and authentication. The proposed security solution employs Lightweight Encryption Algorithm for IoT (LEAIoT), a cryptographic primitive that combines symmetric and asymmetric encryption while supporting flexible key sizes from 32 to 256 bits. The hardware-based implementation on FPGA (7z007s-clg400) demonstrates significant performance improvements over the original CPU-based LEAIoT design: 99.9 per cent faster key generation and up to 96.2 per cent faster encryption and decryption for 1000 kilobits. The architecture utilises 2700 Look-Up Tables (LUTs) and 815 flip-flops, achieving 68.965 Megabits per second (Mbps) throughput at 34.483 Megahertz (MHz) frequency. Compared to other lightweight ciphers, the proposed implementation uses 87.9 per cent and 76.9 per cent fewer resources than lightweight AES versions and 65.7 per cent fewer than SNOW third generation (3G). The flexible key size mechanism allows runtime adaptation to varying network conditions and security requirements. However, the evaluation remains limited to

FPGA simulation without real-world deployment testing, and the scheme does not address threats such as DoS/DDoS attacks or adversarial machine learning scenarios.

The problem of data protection in IoT environments also reflects in the need to have secure authentication protocols rather than only transmission and storage protocols, and to solve this problem, Kumar *et al.* [35] propose RAPCHI, a robust authentication protocol for IoMT-based cloud-healthcare infrastructure. The framework employs ECC to establish secure authentication and session key agreement among four participants: patient, doctor, cloud server and body sensors. The protocol provides mutual authentication without storing verification data in the cloud database. Formal security analysis uses the Random Oracle Model (ROM) with two independent approaches, while AVISPA simulation validates resistance to MitM and replay attacks. The scheme satisfies 16 security attributes, including patient/doctor anonymity and unlinkability, data confidentiality, non-repudiation, session key security and impersonation attack resistance. Performance evaluation demonstrates computation cost of approximately 1.49 s ($4T_{Sign} + 6T_S + 2T_M + 24T_H$) and communication cost of only 752 bits, a value significantly lower than comparable protocols (2576-9440 bits). The Dolev-Yao threat model validates security against active/passive attacks. Similar to many previously analysed works, this algorithm relies on simulation-based validation without real-world IoMT deployment testing, limiting the scalability to other environments.

The article by Almaiah *et al.* [36] similarly touches on the topic of authentication by proposing a hybrid trustworthy decentralised authentication and data preservation model for IoMT-based Cyber-Physical Systems (CPS). The framework combines supervised machine learning with Cryptographic Parameter-Based Encryption and Decryption (CPBED) to allow decentralised device authentication and secure data transmission. A three-layer architecture is constructed, with the three layers being an IIoMT-enabled industrial layer made of wearable devices that act as sensors, an Edge-fog blockchain layer made of full nodes that construct a peer-to-peer network and a Cloud-blockchain layer made of data centres responsible for long-term storage. Key features include a Blockchain-based Fog Consensus Protocol (BFCP) with Proof of Improved consensus, deep learning-based intrusion detection using CNN, PHE for privacy-preserving EMR classification, zero-knowledge proof protocol for device registration and IPFS for secure storage. Experimental results on real IoT datasets demonstrated 94.2 per cent accuracy for simple EMR classification and 93.3 per cent for encrypted EMR, with only a 1 per cent accuracy decrease due to PHE. The framework resists impersonation and Sybil attacks while supporting fifth-generation (5G)-enabled IoMT communication. However, the presence of PHE makes computation slow, and complete blocks at edge nodes with varying computing powers are challenging due to scalability limitations.

Other hybrid solutions exist in the literature, such as the article by Jeong *et al.* [37], which proposes SecAODV, a secure routing scheme for heterogeneous

Wireless Body Sensor Networks (WBSNs) in healthcare IoT applications. The framework employs a hybrid cryptography approach: RC4 symmetric encryption secures intra-cluster communications between sensor nodes and cluster heads, while ECC protects inter-cluster communications between cluster heads. The routing phase prioritises nodes based on a computed degree, which considers distance, residual energy, Received Signal Strength Indicator (RSSI) and hop count. Simulations in NS2 with 100 sensor nodes demonstrated improvements over SMEER and LEACH-C protocols: 10–21 per cent reduction in end-to-end delay, 8–18 per cent lower energy consumption, 31–55 per cent decrease in packet loss rate and 14–32 per cent improvement in packet delivery rate. The hybrid approach balances security strength with energy efficiency, making it suitable for resource-constrained medical sensors. However, in this case too, the evaluation relies solely on simulations without real-world deployment, and the scheme does not address advanced attacks such as DoS or adversarial threats.

Healthcare robotics is another rapidly growing field that requires particular attention for secure data sharing, especially since robots usually have limited computational resources. The article by Jain *et al.* [38] solves this problem by presenting a decentralised security framework for healthcare robots sharing data with cloud servers. The architecture employs ECC with 224-bit keys for encryption/decryption and HMAC-SHA1 (160-bit hash) for data integrity verification. The system is composed of three entities: healthcare robots (HR), an ECC-based cryptographic server (ECS) as a trusted third party and a cloud server. The ECS manages key generation, access control lists (ACL) and all cryptographic operations, offloading computational burden from resource-constrained robots. Testing used Simultaneous Localization and Mapping (SLAM) datasets with the Gmapping library. Performance evaluation on Ubuntu 16.04 with Python crypto-libraries demonstrated key generation times of 0.002–0.005 s for 10 to 100 robots, encryption times of 0.01–0.158 s with files ranging from 0.1 to 500 MegaBytes (MB) and upload/download times significantly lower than comparable methods (31.34 s for upload and 36.38 s for download for 500 MB). The framework enables secure offloading of heavy computational tasks while maintaining confidentiality through Secure Sockets Layer (SSL) channels. However, the dependency on trusted third-party ECS and the lack of defence against internal threats limit this system.

Another interesting rising application is the use of digital twins for cybersecurity by creating AI-driven copies of environments to simulate attacks and validate security protocols. Wakili *et al.* [39] propose a Digital Twin-Enhanced Cybersecurity Framework (DT-CF) for healthcare IoT systems within Industry 4.0 environments. The framework employs real-time digital twins to create synchronised replicas of physical devices, particularly smart infusion pumps and wearable sensors, allowing continuous monitoring and predictive anomaly detection. The architecture is composed of seven components: physical (AES-256/TLS encryption, SGX tamper detection), virtual (OPC UA protocol, bidirectional synchronisation), blockchain (Hyperledger Fabric with PBFT consensus, smart contracts), analytical (LSTM neural networks, SVM, Random Forest classifiers), management (Threat Prioritisation Algorithm), response (Incident Response Protocol Algorithm with 220 ms containment) and interaction (Grafana dashboards). A conceptual case study at UEMF Healthcare Centre with 75 infusion pumps, 120 wearable sensors and 55 environmental monitors demonstrated a 93.4 per cent detection rate, a 2.1 per cent false positive rate, a 220 ms response time and 99.91 per cent system uptime, representing a 37.8 per cent improvement over traditional IDS. Due to the complexity of the proposed system, computational overheads are expected in resource-constrained environments, plus real-world validation is needed to prove scalability.

While previous works mainly focused on AI-driven cryptography for healthcare, the article by Abood *et al.* [40] instead tackles the theme of new-gen solutions for healthcare through PQC. In particular, the authors' main focus is on cryptographic robustness, scalability and efficiency in safety-critical IoHT environments. The authors propose an IoHT framework that integrates PQC with fog computing, blockchain and InterPlanetary File System (IPFS). A hybrid cryptographic design combines Elliptic Curve Cryptography (ECC) for authentication, Counter Mode (CTR) for symmetric encryption and a Key Encapsulation Mechanism (KEM) based on CRYSTALS-Kyber-512 to achieve resistance against quantum attacks. To regulate data access and ensure privacy, the framework employs Role-Based Access Control (RBAC), while Symmetric Searchable Encryption (SSE) allows secure search operations over encrypted healthcare data. Blockchain is used to store hashes, while IPFS handles large medical records, improving scalability and storage efficiency. Fog computing is introduced to reduce latency and computational requirements on resource-constrained IoHT devices. The evaluation of this solution is conducted through formal security verification and performance simulations instead of a standardised dataset. The validation is performed through BAN logic and the Scyther tool, proving resistance to a wide range of threats, including replay, spoofing, MitM, DDoS, phishing, Sybil, 51 per cent blockchain attacks and quantum attacks. The proposed framework achieves up to 90 per cent reduction in computational overhead, very low communication cost (≈300 bits during authentication), latency below 2 ms, high throughput (up to 1000 TPS in registration) and extremely low energy consumption (≈0.036 mJ), outperforming several recent IoHT security protocols. While the results obtained by the model are remarkable, the absence of machine learning or AI-based adaptive threat detection limits the ability of the model to respond to evolving attacks. Moreover, the validation is conducted in a simulated environment, while real-world deployment in large-scale clinical settings may introduce additional challenges related to system complexity and overhead.

The article by Chaudhary and Chatterjee [41] proposes instead a lightweight security framework for electronic healthcare systems (EHS) targeting IoT-based remote health monitoring. The framework introduces a novel lightweight block cipher with flexible key structure supporting 128- to 256-bit keys, using S-Box operations across 32 rounds to achieve confusion and diffusion. The system incorporates three-factor authentication (digital identity, biometric identity and password) with ECC-based public key encryption for secure data access from cloud servers. The block cipher, implemented on an Arduino Uno 8-bit microcontroller with 32 Kilobyte (KB) flash and a 2KB Static Random Access Memory (SRAM), achieves 23,595 clock cycles per byte, outperforming LEA, HEIGHT and SIMON ciphers by at least 4 per cent. The computational cost is much lower (12 hash operations total) compared to existing schemes (19 to 26 operations). Formal security verification using the AVISPA tool demonstrates resistance to MitM, impersonation, replay, server spoofing and password guessing attacks. The main limitation is the relatively high encryption time compared to some alternatives. Moreover, as often noted in health records works, validation remains simulation-based without real-world clinical deployment testing.

7.3.5 Results summarisation

The results of the 21 analysed papers are summarised in Tables 7.2 and 7.3. Table 7.2 details the dataset employed, the tested algorithms and the best results obtained for the papers that include numerical results. Of the 21 articles selected, 5 have been discarded from this table due to not reporting numerical algorithmic results. What emerges from an analysis of the employed algorithms is that several works rely on blockchain (7/16, 42.75 per cent) and ECC (6/16, 37.5 per cent), with only half of the works employing ML solutions either exclusively or in combination with other approaches (8/16, 50 per cent). Accuracy scores, where reported, are usually higher than 90 per cent (7/8, 87.5 per cent). Half of the works use synthetic or simulated data for their validation (10/21 of the total, 47.62 per cent), a value extremely high for a field where real-world scalability is particularly important due to the presence of sensitive data.

Table 7.3 synthesises, for each work, the main cyberattacks addressed and the main scope. We can notice a high heterogeneity of attacks considered in the selected works, with the most common ones being MitM (4/21, 19 per cent), data alteration/poisoning (4/21, 19 per cent) and data leakage (3/21 per cent, 14.3 per cent). This reflects the often-mentioned limitation of the reduced scalability of approaches, which is caused not only by validation on simulated datasets but also by the heterogeneous pool of considered cyberattacks. For what concerns the main scope, 5 works focus on patient data protection (23.8 per cent), 4 on secure diagnosis and prediction (19 per cent) and 8 exclusively on secure transmission in IoT environments (38.1 per cent), with another 6 using IoT environments as their base infrastructure (28.6 per cent), for a total of 14 works on the topic of IoT (66.7 per cent). This result places IoT as a central topic in intelligent healthcare, justified by the increased connectivity between medical tools, for diagnostics, data storage or monitoring.

Table 7.2 Summarisation of studies included in the review that provide numerical or technical evaluation results

Authors	**Year**	**Dataset/set-up**	**Algorithms**	**Best results**
Chinbat *et al.* [22]	2024	Synthetic healthcare IoT traffic (Raspberry Pi)	DT, RF, SVM, MLP, KNN + LWC	RF/SVM: Accuracy up to 99%; RECTANGLE showed lowest energy and memory usage
Kumari *et al.* [23]	2025	WUSTL-EHMS-2020	VAE, RF, SVM, XGBoost, LightGBM, LR, FNN	VAE: Accuracy 91.61%, Precision 91.69%, Recall 91.61%, F1-score 91.60%
Amanat *et al.* [25]	2022	Simulated EHR cloud environment	Blockchain, SHA-256, ECDSA, PoS	17 transactions per 6 min block; lower power consumption vs. PoW
Ali *et al.* [26]	2022	Simulated IIoT healthcare traffic	NN, ABE, Searchable Encryption, Blockchain	Detection Accuracy up to 97%; FPR as low as 0.002
Ali *et al.* [27]	2022	IoT-Botnet dataset	BiLSTM, FHE, Blockchain	Accuracy 90.6%; encrypted-domain performance 98.67% of baseline
Lakshmi *et al.* [29]	2025	DICOM medical images	HNN, BPN, LFSR (FPGA)	Entropy 7.99; NPCR 99.57%; UACI 33.05%; power 424.71 mW
Padinjappurathu Gopalan *et al.* [30]	2022	Hungarian Heart Disease dataset	EHGA-DLNN, LR-ECC, GK-LDA	Accuracy 98.35%; Sensitivity 97.33%; Specificity 96.36%
Castro *et al.* [31]	2023	118 gait videos (43 subjects)	Paillier PHE, LSTM, Random Projection	Accuracy 97.0% (protected); max degradation 1.2%
Egala *et al.* [32]	2022	WHO COVID-19 data	RF, SVM, ECC, Blockchain	Hybrid RF+SVM Accuracy 84.89%
Ravisankar and Maheswar [33]	2025	Simulated IoMT + clinical use case	Autoencoders, FL, PQC, Blockchain	F1-score 0.90; >95% anomaly detection; latency ∼180 ms

(Continues)

Aivaliotis *et al.* [34]	2022	FPGA implementation	LEAIoT	96.2% faster encryption; 68.9 Mbps throughput
Jeong *et al.* [37]	2022	NS2 simulation (WBSN)	ECC, RC4	Packet loss -55%; energy consumption -18% vs LEACH-C
Jain *et al.* [38]	2022	SLAM datasets	ECC, HMAC-SHA1	Encryption time 0.01–0.158 s; secure upload of 500 MB data
Wakili *et al.* [39]	2025	Digital twin case study	LSTM, SVM, RF, Blockchain	Detection rate 93.4%; FPR 2.1%; response time 220 ms
Abood *et al.* [40]	2025	Formal verification + simulation	ECC, Kyber-512, SSE, Fog, Blockchain	Latency <2 ms; Energy ~0.036 mJ; Throughput up to 1000 TPS
Chaudhary and Chatterjee [41]	2022	Arduino Uno (simulated)	Lightweight block cipher, ECC	23,595 cycles/byte; 4% faster than LEA/HEIGHT

The table reports authors, publication year, dataset or evaluation set-up, employed algorithms and best results obtained.

Table 7.3 Summarisation of cyberattacks addressed by the reviewed studies and their main application scope in intelligent healthcare systems

Authors	Cyberattacks addressed	Main scope
Mushtaq *et al.* [21]	Not specified (architectural risks)	Zero Trust security analysis
Chinbat *et al.* [22]	Side-channel leakage, cryptographic weaknesses	Lightweight cryptography evaluation
Kumari *et al.* [23]	MitM, spoofing, replay, data poisoning	Intrusion detection
Saberi Kamarposhti *et al.* [24]	Quantum attacks (Shor, Grover)	Post-quantum data protection
Amanat *et al.* [25]	Data leakage, unauthorised access	Secure EHR storage
Ali *et al.* [26]	DDoS, malware, data poisoning	Secure searchable encryption
Ali *et al.* [27]	DoS, DDoS, collusion	Intrusion detection on encrypted data
Nucciarelli *et al.* [28]	Data tampering, repudiation	Healthcare analytics
Lakshmi *et al.* [29]	Chosen-plaintext, statistical attacks	Medical image protection
Padinjappurathu Gopalan *et al.* [30]	Data alteration, model poisoning	Secure disease prediction
Castro *et al.* [31]	Biometric inversion attacks	Privacy-preserving diagnosis
Egala *et al.* [32]	Unauthorised access, MitM	Pandemic detection
Ravisankar and Maheswar [33]	Quantum attacks, anomaly injection	Real-time diagnostics
Aivaliotis *et al.* [34]	Data leakage	IoMT communication security
Kumar *et al.* [35]	MitM, replay	Robust authentication
Almaiah *et al.* [36]	Unauthorised access, data leakage	Authentication and data preservation
Jeong *et al.* [37]	Routing manipulation, eavesdropping	Secure WBSN routing
Jain *et al.* [38]	Unauthorised access	Secure healthcare robotics
Wakili *et al.* [39]	Intrusions, zero-day-like anomalies	Digital twin-based detection
Abood *et al.* [40]	Replay, spoofing, MitM, quantum attacks	IoHT secure communication
Chaudhary and Chatterjee [41]	Impersonation, replay, password guessing	Secure EHS authentication

7.4 Discussion

The analysis of the works included in our study clearly evidences some tendencies in research on cryptography for intelligent healthcare systems. The main one is that, given the increase in communication between devices in the medical field, such as wearable devices, body sensor networks, smart infusion pumps, healthcare robots and remote monitoring platforms, data transmission and storage is steadily growing. As all these devices introduce continuous data streams and frequent device-to-device communication, this leads to increased possibilities of cyberattacks, especially concerning data leakage and data poisoning. For this reason, the majority of works in the literature focus on security in IoT, IoMT and IoHT environments, with a major focus on both secure transmission and the introduction of authorisation protocols, often through generated keys, to reduce the possibility of unwanted entry. This effectively answers our research question, 'Is cryptography in intelligent healthcare systems mainly focused on Internet of Things technologies?', showing that these environments are the main field in which cryptography for intelligent healthcare has been evolving. The prevalence of IoT, IoMT and IoHT-oriented studies suggests that cryptography in healthcare is no longer confined to centralised hospital infrastructures but is increasingly reliant on distributed and heterogeneous systems. As a consequence, future cryptographic solutions will need to be natively interoperable and scalable, rather than designed as isolated protection layers.

Another common direction found in the literature is the shift towards hybrid security architectures, where cryptographic approaches are increasingly combined with AI and ML-based components, such as in Ali *et al.* [26], Castro *et al.* [31] and Egala *et al.* [32]. AI and ML techniques are increasingly used for intrusion detection, anomaly detection and adaptive security management. This growing integration indicates a shift from static protection mechanisms towards adaptive and context-aware security models. Future cryptographic systems will therefore likely operate in conjunction with intelligent monitoring layers, forming hybrid security ecosystems rather than purely algorithmic defences. The main advantage of this direction lies in the ability to detect complex attack patterns, including zero-day-like behaviours, that traditional rule-based systems might miss. Digital-twin-based approaches, as proposed by Wakili *et al.* [39], further extend this capability by allowing testing and simulation of attacks on direct copies of the environments. At the same time, AI-driven security introduces new attack possibilities, such as model poisoning, adversarial attacks and data manipulation. Cryptography must therefore increasingly account for the protection of AI models themselves, ensuring integrity and robustness to adversarial manipulation. While xAI would be a feasible solution to address the vulnerabilities of AI approaches, only a minority of the reviewed works explicitly mention this solution, with one of the few exceptions being the work by Mushtaq *et al.* [21]. This gap is particularly concerning in healthcare, where false positives or false negatives may have direct clinical consequences.

Besides hybrid models, many studies are starting to focus on novel techniques that can be complementary to standard algorithmic solutions. This trend reflects the intrinsic complexity of modern healthcare infrastructures, which simultaneously require confidentiality, integrity, availability, low latency and regulatory compliance, a problem that is also seen in other fields related to healthcare, such as eye tracking [42] and assisted living [43]. With the increased complexity of algorithms, for example, reducing the computational requirements of algorithms is becoming fundamental, a need that also aligns with the often-reduced computational capabilities of wearable diagnostic devices or healthcare robots. Furthermore, devices are also expected to operate in real time, which makes complex algorithms with long computation times unsuited to the medical field. Consequently, flexibility and modularity will become central design principles. This problem is tied to the increase in works that focus on lightweight cryptography, with works such as Jain *et al.* [38] and Chinbat *et al.* [22] exploring solutions that can provide a good trade-off between performance and computational requirements and times.

Another emerging field that is steadily becoming popular in healthcare is post-quantum cryptography, which has seen employment both for predictive diagnostics [33] and for IoT security [40]. Due to the steady increase in the employment of quantum computers, it is particularly important to focus on PQC to ensure security against high-level attacks, especially concerning the protection of patient data against data leakages or ransomware attacks. The need to protect computationally constrained devices in the present and the necessity to prepare for quantum-enabled attacks in the future suggest that the next generation of cryptographic frameworks will likely rely on hybrid migration strategies, combining classical, lightweight and post-quantum primitives depending on device capabilities and threat models. However, few works acknowledge backward compatibility or the impact of PQC on ultra-low-power medical devices. This suggests that, while PQC is technically feasible, its operational readiness for large-scale healthcare systems is still limited.

Moreover, PQC and cryptography for healthcare more generally suffer from a consistent problem. Most evaluations are simulation-based or reliant on synthetic or laboratory-based datasets, and few studies implement migration strategies from controlled cryptographic systems to real-world scenarios or directly work on real-world data, limiting the scalability of proposed solutions. While this is understandable given privacy regulations and ethical constraints tied to medical data, it significantly limits the assessment of current state-of-the-art solutions when applied to unconstrained scenarios. Similarly, performance is often evaluated on a limited number of devices or nodes, without stress-testing large-scale deployments. This further limits the scalability of current algorithms, which are all optimised and tested for optimal solutions without considering the complexity of real scenarios with a high number of nodes and much more complex data. The persistent reliance on simulated and laboratory-based validation environments has significant implications. It indicates that the field is still in a predominantly experimental phase, with limited translation into large-scale clinical deployment. Without a transition towards realistic validation scenarios, it will remain difficult to assess the true operational readiness of proposed solutions.

Another consistent limitation found in the literature is that the majority of works focus on a restricted set of attacks (e.g. MitM or replay attacks), without considering compound or cross-layer attacks. Most approaches, in fact, are optimised for a specific type of attack, making them extremely overspecialised. This overspecialisation towards single attack types implies that future cryptographic research must adopt more complex threat modelling strategies. Intelligent healthcare systems operate in interconnected infrastructures where compound and cross-layer attacks are increasingly plausible. Current solutions greatly limit the applicability to real scenarios, where unseen attacks or multiple attacks might be present. For this reason, it is necessary to focus on increasingly adaptive algorithms able to respond even to new threats without the need to be reconfigured or retrained.

Finally, the absence of standard benchmarks, both in terms of datasets and in terms of metrics, makes the comparison of approaches unfeasible or extremely complicated without performing new custom tests on the employed datasets. This is a direct result of the reliance on simulated data, which is usually constructed depending on the specific application rather than being as inclusive as possible, resulting in a poor comparability of the obtained results. Evaluation is also tied to the needs of the selected algorithms, with some algorithms that focus on computational efficiency, with the inclusion of time-related or transmission-related metrics and others that instead focus more on performance, where the main metrics are accuracy or the number of false positives/negatives.

To summarise the answer to our other research question, 'Which are the main solutions found in current literature?', the most common solutions are the employment of AI, especially in hybrid frameworks that combine standard cryptography and AI techniques. However, due to the shift towards quantum computers and the employment of resource-constrained tools, two other important branches that are steadily becoming relevant are post-quantum cryptography and lightweight cryptography, which, while understudied, have the potential to become extremely relevant in future cryptography algorithms for intelligent healthcare.

Overall, the results of this review suggest that the future of cryptography in intelligent healthcare will be defined not only by stronger algorithms but also by the integration of adaptability, explainability, interoperability and real-world validation into comprehensive IoMT systems, a direction that currently has several limitations found in the literature.

7.4.1 Future directions

Based on the analysis of the reviewed literature and the previous discussion of the common limitations of algorithms, several possible research directions can be identified to improve current gaps and improve the security of intelligent healthcare systems.

As the primary recurring limitation is the strong reliance on simulated, synthetic or laboratory-generated datasets for validation, future research should prioritise the creation of privacy-preserving real-world healthcare datasets, using techniques such as federated learning, secure data enclaves, trusted execution environments and differential privacy. These approaches can allow collaborative model training and evaluation across institutions without exposing sensitive patient data, improving realism while maintaining regulatory compliance.

Another promising direction is the integration of xAI into cybersecurity frameworks. Although AI and ML-based algorithms show strong detection capabilities in current literature, their decision-making processes are often unclear. This lack of transparency is particularly problematic in clinical environments, where interpretability is essential to validate clinical diagnosis [44]. Implementing xAI techniques can improve support for human decision-making by allowing clinicians and security operators to better understand and validate system outputs. Moreover, currently, few studies systematically analyse the vulnerability of AI-based security mechanisms to adversarial attacks, data poisoning, model inversion or evasion strategies. Formal analysis and robust evaluation of AI security are particularly critical in healthcare, where compromised models may lead to incorrect diagnoses or denial of essential services. For this reason, the implementation of xAI is even more crucial, as it can also help in understanding the most vulnerable sections of the algorithm and allow human operators to know where to intervene to strengthen the model.

In parallel, there is a clear need for the systematic evaluation of hybrid cryptographic frameworks, especially those that combine post-quantum cryptography with lightweight cryptographic primitives. While post-quantum algorithms offer protection against emerging quantum threats, their computational and communication overhead may be prohibitive for resource-constrained IoMT devices such as wearable devices or healthcare robots. Hybrid approaches are a promising compromise, but further research is required to verify their feasibility under realistic constraints, such as energy consumption, latency, device heterogeneity and backward compatibility with existing healthcare infrastructure.

Finally, the majority of current studies are validated in small-scale or controlled environments, highlighting the necessity for large-scale evaluations, possibly even cross-institutional. Such studies are essential to evaluate scalability, interoperability across heterogeneous devices and infrastructures and long-term maintenance costs. Real-world tests would also allow assessing system robustness under dynamic clinical conditions, such as fluctuating network loads, device failures and evolving threat landscapes.

The overview of the final integrated architecture for security in healthcare is shown in Figure 7.3. The goal of cryptography in intelligent healthcare systems should be to construct an architecture composed of five main components, two passive and three active. The two passive components are the regulatory layer and the IoMT devices layer. The regulatory layer includes the governance framework that defines protocols for secure cryptographic deployment, including privacy policies, auditing mechanisms and compliance requirements. The IoMT devices layer constitutes the physical infrastructure of healthcare systems and includes all interconnected medical devices, such as wearable devices, body sensors, healthcare robots and smart infusion pumps. The three active layers are directly responsible for enforcing system security. The federated learning layer enables collaborative

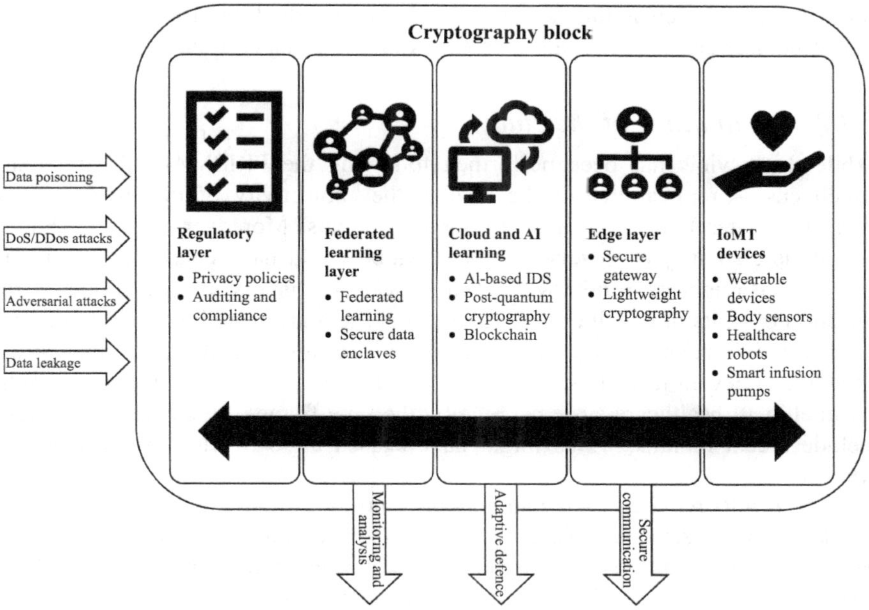

Figure 7.3 Integrated architecture for security in healthcare. The main component is the cryptography block, which includes all the main directions seen in this review. The goal of this block is to respond to external threats in order to ensure safety in the IoMT environment.

model training across different healthcare institutions by sharing encrypted model parameters rather than raw data. This approach allows threat modelling and intrusion analysis to benefit from distributed knowledge while preserving data privacy through secure data enclaves and controlled information exchange. The Cloud and AI layer is responsible for higher-level defence mechanisms, including AI-based IDS, PQC and blockchain-based integrity solutions. This layer performs continuous monitoring and decentralised surveillance, with AI models deployed on centralised or distributed servers and queried when necessary to optimise computational efficiency. The use of advanced AI-driven techniques enables adaptive defence strategies capable of detecting evolving and previously unseen threats. Finally, the edge layer ensures security at the terminal nodes of the IoMT environment. Its primary objective is to provide secure gateways for data transmission and lightweight cryptographic mechanisms tailored to the limited computational capabilities of medical devices. By performing preliminary protection and filtering at the edge, this layer reduces latency while maintaining data confidentiality and integrity. An architecture constructed in this manner is capable of addressing a broad range of threats, including data poisoning, data leakage, DoS/DDoS attacks and adversarial manipulation. By decentralising cybersecurity responsibilities across multiple layers, the system ensures that terminal nodes are individually protected, while more computationally demanding security tasks are handled by robust cloud-based infrastructures enriched through cross-institutional collaboration.

7.4.2 Limitations of the study

While this review has been performed following the PRISMA-ScR guidelines, which ensure replicability of the results, the subjectivity of the inclusion and exclusion criteria makes this study subject to bias. Moreover, the selection of articles is also subject to reviewer bias, which could have caused us to discard articles that might have otherwise been relevant to our discussion. Similarly, the selection of years in which to perform our first literature selection might have been too restrictive, consequently causing a systematic rejection of articles that would have been relevant to understand the current direction of research in cryptography for intelligent healthcare systems. Finally, the time frame of the research does not include recent months, which might have caused us to overlook relevant newer articles.

The results reported by the authors have not been reproduced or tested over a standardised dataset or evaluated over a standard metric, which makes the comparison of the literature reliant on the information reported by the original authors. Moreover, the heterogeneity of topics and algorithms might have caused us to be biased towards more represented topics or solutions, limiting discussion on less represented but otherwise relevant solutions present in the literature.

Finally, the scoping review structure of our review caused us to not include statistical analyses or meta-analyses of the selected literature, as its main focus was simply mapping available literature and algorithmic solutions to identify future directions of research. Future works should therefore focus on a more complete analysis of the literature in order to guide future researchers in more detail in the next steps to follow, giving more importance to the evidenced gaps in literature and verifying if they are already partially solved in literature or if some similar approaches have already been tested in other domains.

7.4.3 Lessons learned

The analysis of the reviewed literature revealed several important lessons for the development of secure intelligent healthcare systems. First, while cryptographic mechanisms and AI-based detection models demonstrate high performance in controlled environments, their validation predominantly relies on simulated or synthetic datasets. This highlights a critical gap between theoretical performance and real-world deployment readiness. Second, there is a clear trade-off between security strength and resource efficiency, particularly in IoMT environments where computational power, energy and latency constraints are strict. Third, hybrid approaches that combine lightweight cryptography, post-quantum techniques and AI-driven monitoring appear promising, but their integration is often evaluated in isolation rather than within fully interconnected healthcare systems. Finally, explainability and robustness against adversarial attacks remain underdeveloped despite their importance in safety-critical healthcare applications. These lessons highlight the necessity of moving from proof-of-concept studies towards scalable, fully deployed security frameworks.

7.5 Conclusion

The goal of our research was to determine the answer to the question, 'Is cryptography in intelligent healthcare systems mainly focused on Internet of Things technologies?', and consequently to the question 'Which are the main solutions found in current literature?'. To answer these questions, we selected and thoroughly analysed 21 works found in literature through the PRISMA-ScR paradigm, highlighting their shared commonalities and limitations in order to map the current state of the art.

The results of our research showed that IoT, IoMT and IoHT are the main applications of cryptography algorithms in literature, with the main solutions being hybrid approaches integrating AI and standard cryptography, post-quantum cryptography and lightweight cryptography. Current algorithms show results of up to 99 per cent accuracy in the classification and identification of cyberattacks, especially in the case of hybrid architectures, offering protection against a wide variety of attacks such as MitM and data poisoning.

However, despite the promising landscape, our research also identified several limitations that are currently not addressed in the literature. First, the majority of evaluations are performed on synthetic datasets with limited dimensions, limiting the real-life applicability of the proposed solutions. Moreover, despite covering a wide variety of attacks, few works are effectively scalable, remaining mostly solutions limited to the scenario for which they have been developed. Most approaches do not rely on explainable AI for model interpretability, making them vulnerable to model-poisoning attacks, and the less popular solutions, such as digital twin or PWC, are severely understudied as of now. Finally, comparing approaches is particularly complex due to the absence of standardised benchmarks, with a heterogeneity not only of datasets but also of the employed evaluation metrics.

This led us to define several possible directions, briefly highlighted in Figure 7.4 in order of priority. The main future direction is the application of pre-existing algorithms on real-world scenarios with a consistent dimension and number of involved nodes, in order to verify the effective performance of the current state of the art and identify more consistent future improvements needed by the models. Second, the scalability of algorithms to different domains should be tested to verify their cross-domain effectiveness. Third, the integration of xAI with techniques such as Shapley Additive exPlanations (SHAP) or Local Interpretable Model-agnostic Explanations (LIME) should be considered to make models not only interpretable but also easier to protect from model poisoning attacks. Certain models should also be more thoroughly evaluated, such as PWC and LWC-based algorithms, focusing on their experimentation to determine their

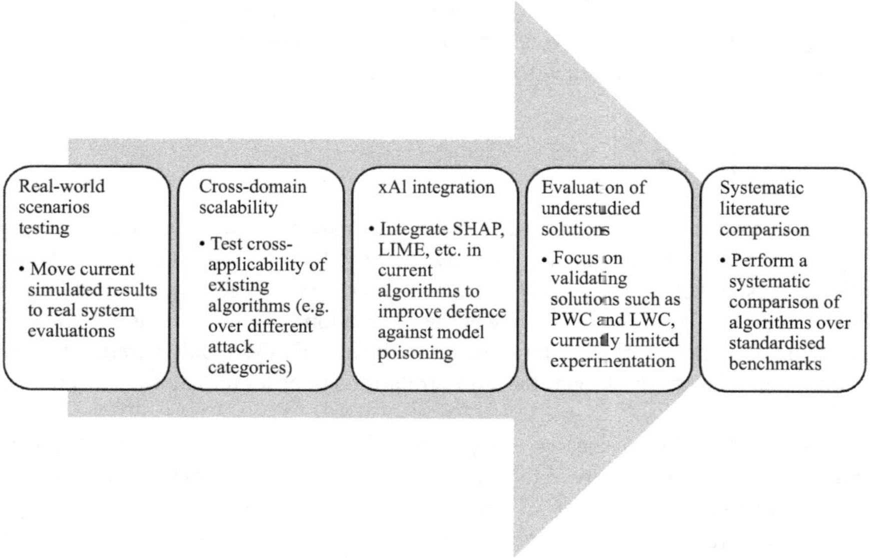

Figure 7.4 Future directions for research in cryptography for intelligent healthcare systems. Directions are arranged from left to right in order of priority (from highest to lowest).

efficacy in critical scenarios such as HR or wearable diagnostics. Finally, a systematic comparison of literature should be performed over standardised benchmarks in order to fully identify the most promising directions among currently available solutions.

List of abbreviations

3Gv	Third generation
5G	Fifth generation
ABAC	Attribute-Based Access Control
ABE	Attribute-Based Encryption
ACL	Access Control Lists
AES	Advanced Encryption Standard
AI	Artificial Intelligence
ANN	Artificial Neural Networks
BFCP	Blockchain-based Fog Consensus Protocol
BPN	Back Propagation Neural Network
CKKS	Cheon-Kim-Kim-Song
CMD	Collaborated Medical Database
CPBED	Cryptographic Parameter-Based Encryption and Decryption

CPS	Cyber-Physical System
CPU	Central Processing Unit
CTR	Counter Mode
DDoS	Distributed Denial of Service
DICOM	Digital Imaging and Communications in Medicine
DID	Decentralised Identifier
DL	Deep Learning
DoS	Denial of Service
DT	Decision Trees
DT-CF	Digital Twin-Enhanced Cybersecurity Framework
ECC	Elliptic Curve Cryptography
ECDSA	Elliptic Curve Digital Signature Algorithm
ECS	ECC-based Cryptographic Server
EHGA-DLNN	Elephant Herding Genetic Algorithm-based Deep Learning Neural Network
EHR	Electronic Healthcare Records
EHS	Electronic Healthcare System
FNN	Feedforward Neural Networks
FPGA	Field-Programmable Gate Array
GB	GigaByte
GEN-RWD	Gemelli Generator - Real World Data
GHz	Gigahertz
GK-LDA	Gaussian Kernel-based Linear Discriminant Analysis
H-CPS	Health Cyber-Physical Systems
HDL	Hardware Description Language
HE-SSE	Homomorphic Secure Searchable Encryption
HNN	Hopfield Neural Network
HR	Healthcare Robot
IDS	Intrusion Detection System
IIoT	Industrial Internet of Things
IoHT	Internet of Healthcare Things
IoMT	Internet of Medical Things
IoT	Internet of Things
IPFS	InterPlanetary File System
ISO/IEC	International Organization for Standardization International Electrotechnical Commission
KB	Kilobyte
KEM	Key Encapsulation Mechanism
KNN	K-Nearest Neighbours
LEAIoT	Lightweight Encryption Algorithm for IoT
LFSR	Linear Feedback Shift Register
LightGBM	Light Gradient-Boosting Machine

LIME	Local Interpretable Model-agnostic Explanations
LR	Logistic Regression
LR-ECC	Log of Round value-based Elliptic Curve Cryptography
LSTM	Long Short-Term Memory
LUT	Look-Up Tables
LWC	Lightweight Cryptography
MB	Megabytes
Mbps	Megabits per second
MD	Message Digest
MHz	Megahertz
MitM	Man-in-the-Middle
ML	Machine Learning
MLP	Multilayer Perceptron
MRI	Magnetic Resonance Imaging
mW	milliWatt
NFT	Non-Fungible Token
NIST SP	National Institute of Standards and Technology Special Publication
NN	Neural Networks
NPCR	Number of Pixels Change Rate
PHE	Partially Homomorphic Encryption
PHR	Patient Health Record
POS	Proof of Stake
POW	Proof-of-Work
PQC	Post-Quantum Cryptography
PRISMA-ScR	Preferred Reporting Items for Systematic reviews and Meta-Analyses extension for Scoping Reviews
RAM	Random Access Memory
RBAC	Role-Based Access Control
RF	Random Forest
RSA	Rivest–Shamir–Adleman
RSSI	Received Signal Strength Indicator
SHA	Secure Hash Algorithm
SHAP	Shapley Additive exPlanations
SLAM	Simultaneous Localization and Mapping
SRAM	Static Random Access Memory
SSE	Symmetric Searchable Encryption
SSI	Self-Sovereign Identity
SSL	Secure Sockets Layer
STRIDE	Spoofing, Tampering, Repudiation, Information disclosure, Denial of Service and Elevation of privilege
SVM	Support Vector Machine
UACI	Unified Average Changing Intensity
VAE	Variational Autoencoder

WBSN	Wireless Body Sensor Network
xAI	Explainable Artificial Intelligence
XGBooste	Xtreme Gradient Boosting
ZTA	Zero Trust Architecture

Acknowledgement

The work of Agbotiname Lucky Imoize is supported in part by the Nigerian Petroleum Technology Development Fund (PTDF) and in part by the German Academic Exchange Service (DAAD) through the Nigerian-German Postgraduate Program under Grant 57473408. This work has been carried out while Francesca Fiani was enrolled in the Italian National Doctorate on Artificial Intelligence run by Sapienza University of Rome in collaboration with the Department of Computer, Control, and Management Engineering.

References

[1] Vishnu S, Ramson SJ and Jegan R. Internet of medical things (IoMT)-An overview. In: *2020 5th International Conference on Devices, Circuits and Systems (ICDCS)*. IEEE; 2020. p. 101–104.

[2] Mukhopadhyay SC, Suryadevara NK and Nag A. Wearable sensors for healthcare: Fabrication to application. *Sensors*. 2022;22(14):5137.

[3] Mahajan HB. Emergence of healthcare 4.0 and blockchain into secure cloud-based electronic health records systems: solutions, challenges, and future roadmap. *Wireless Personal Communications*. 2022;126(3):2425–2446.

[4] Liu S, Zhang Y, Ju Y, *et al.* Establishment and clinical application of an artificial intelligence diagnostic platform for identifying rectal cancer tumor budding. *Frontiers in Oncology*. 2021;11:626626.

[5] Fiani F, Russo S and Napoli C. An advanced solution based on machine learning for remote EMDR therapy. *Technologies*. 2023;11(6):172.

[6] Russo S, Fiani F and Napoli C. Remote eye movement desensitization and reprocessing treatment of long-COVID-and post-COVID-related traumatic disorders: An innovative approach. *Brain Sciences*. 2024;14(12):1212.

[7] Gogalniceanu P. Hybrid threats require hybrid solutions: A roadmap for healthcare security. *Health Policy*. 2026;166:105565.

[8] Argaw ST, Bempong NE, Eshaya-Chauvin B, *et al.* The state of research on cyberattacks against hospitals and available best practice recommendations: A scoping review. *BMC medical informatics and decision making*. 2019;19(1):10.

[9] Rufai AU, Fasina EP, Uwadia CO, Rufai AT and Imoize AL. Cyberattacks against artificial intelligence-enabled Internet of Medical Things. In: *Handbook of Security and Privacy of AI-Enabled Healthcare Systems and Internet of Medical Things*. CRC Press; 2023. p. 191–216.

[10] Fatima A, Reddy CKK and Singh TM. AIoT security for healthcare: Enabling trust in smart medical devices. In: *Information Security Governance using Artificial Intelligence of Things in Smart Environments*. CRC Press; 2026. p. 190–205.

[11] Papaioannou M, Karageorgou M, Mantas G, *et al.* A survey on security threats and countermeasures in internet of medical things (IoMT). *Transactions on Emerging Telecommunications Technologies*. 2022;33(6):e4049.

[12] Pulipeti S and Kumar A. Secure quantum computing for healthcare sector: A short analysis. *Security and Privacy*. 2023;6(5):e293.

[13] Nkomo D and Brown R. Hybrid cyber security framework for the internet of medical things. In: *Blockchain and Clinical Trial: Securing Patient Data*. Springer; 2019. p. 211–229.

[14] Pallakonda A, Kaliyannan K, Sumathi RL, *et al.* AI-driven attack detection and cryptographic privacy protection for cyber-resilient industrial control systems. *IoT*. 2025;6(3):56.

[15] Naghib A, Gharehchopogh FS and Zamanifar A. A comprehensive and systematic literature review on intrusion detection systems in the internet of medical things: Current status, challenges, and opportunities. *Artificial Intelligence Review*. 2025;58(4):114.

[16] Bhanja SN, Niu H, Chen Y, *et al.* Emerging anomaly detection techniques for electronic health records: A survey. *Intelligence-Based Medicine*. 2026; 13:100349.

[17] Abdullahi M, Baashar Y, Alhussian H, *et al.* Detecting cybersecurity attacks in internet of things using artificial intelligence methods: A systematic literature review. *Electronics*. 2022;11(2):198.

[18] Radanliev P and De Roure D. Advancing the cybersecurity of the healthcare system with self-optimising and self-adaptative artificial intelligence (part 2). *Health and Technology*. 2022;12(5):923–929.

[19] Rani DJ and Roslin SE. Light weight cryptographic algorithms for medical internet of things (IoT) - A review. In: *2016 Online international conference on green engineering and technologies (IC-GET)*. IEEE; 2016. p. 1–6.

[20] Akkal M, Cherbal S, Annane B, Lakhlef H and Kharoubi K. Quantum, post-quantum, and blockchain approaches for securing the internet of medical things: A systematic review. *Cluster Computing*. 2025;28(10):655.

[21] Mushtaq S, Mohsin M and Mushtaq MM. A systematic literature review on the implementation and challenges of zero trust architecture across domains. *Sensors*. 2025;25(19):6118.

[22] Chinbat T, Madanian S, Airehrour D and Hassandoust F. Machine learning cryptography methods for IoT in healthcare. *BMC Medical Informatics and Decision Making*. 2024;24(1):153.

[23] Kumari M, Gaikwad M and Chavan SA. A secure IoT-edge architecture with data-driven AI techniques for early detection of cyber threats in healthcare. *Discover Internet of Things*. 2025;5(1):54.

[24] SaberiKamarposhti M, Ng KW, Chua FF, *et al.* Post-quantum healthcare: A roadmap for cybersecurity resilience in medical data. *Heliyon*. 2024;10:e31406.

[25] Amanat A, Rizwan M, Maple C, Zikria YB, Almadhor AS and Kim SW. Blockchain and cloud computing-based secure electronic healthcare records storage and sharing. *Frontiers in Public Health*. 2022;10:938707.
[26] Ali A, Almaiah MA, Hajjej F, *et al.* An industrial IoT-based blockchain-enabled secure searchable encryption approach for healthcare systems using neural network. *Sensors*. 2022;22(2):572.
[27] Ali A, Pasha MF, Ali J, *et al.* Deep learning based homomorphic secure search-able encryption for keyword search in blockchain healthcare system: A novel approach to cryptography. *Sensors*. 2022;22(2):528.
[28] Nucciarelli L, Gottardelli B, Gatta R, and Damiani A. Securing reproducibility and accountability in distributed healthcare analytics: A framework based on blockchain and cryptography. In: *Proceedings of MIE 2024*. vol. 316; 2024.
[29] Lakshmi C, Nithya C, Thenmozhi K, *et al.* Reconfigurable security solution based on hopfield neural network for e-healthcare applications. *Scientific Reports*. 2025;15(1):5628.
[30] Padinjappurathu Gopalan S, Chowdhary CL, Iwendi C, Farid MA and Ramasamy LK. An efficient and privacy-preserving scheme for disease prediction in modern healthcare systems. *Sensors*. 2022;22(15):5574.
[31] Castro F, Impedovo D and Pirlo G. A hybrid protection scheme for the gait analysis in early dementia recognition. *Sensors*. 2023;24(1):24.
[32] Egala BS, Pradhan AK, Badarla V and Mohanty SP. iBlock: An intelligent decentralised blockchain-based pandemic detection and assisting system. *Journal of signal processing systems*. 2022;94(6):595–608.
[33] Ravisankar S and Maheswar R. SecureEdge-MedChain: A post-quantum blockchain and federated learning framework for real-time predictive diagnostics in IoMT. *Sensors*. 2025;25(19):5988.
[34] Aivaliotis V, Tsantikidou K and Sklavos N. IoT-based multi-sensor healthcare architectures and a lightweight-based privacy scheme. *Sensors*. 2022;22(11):4269.
[35] Kumar V, Mahmoud MS, Alkhayyat A, *et al.* RAPCHI: Robust authentication protocol for IoMT-based cloud-healthcare infrastructure. *The Journal of Supercomputing*. 2022;78(14):16167.
[36] Almaiah MA, Hajjej F, Ali A, *et al.* A novel hybrid trustworthy decentralized authentication and data preservation model for digital healthcare IoT based CPS. *Sensors*. 2022;22(4):1448.
[37] Jeong H, Lee SW, Hussain Malik M, *et al.* SecAODV: A secure healthcare routing scheme based on hybrid cryptography in wireless body sensor networks. *Frontiers in Medicine*. 2022;9:829055.
[38] Jain S and Doriya R. Security framework to healthcare robots for secure sharing of healthcare data from cloud. *International Journal of Information Technology*. 2022;14(5):2429–2439.
[39] Wakili A, Bakkali S and Ibrahim IA. A digital twin-enhanced cybersecurity framework for IoT in healthcare: Applications in industry 4.0. *Telematics and Informatics Reports*. 2025;20:100254.

[40] Abood EW, Yassin AA, Abduljabbar ZA, *et al.* Provably lightweight and secure IoHT scheme with post-quantum cryptography and fog computing: A comprehensive scheme for healthcare system. *MethodsX*. 2025;15:103631.

[41] Chaudhary RRK and Chatterjee K. A lightweight security framework for electronic healthcare system. *International Journal of Information Technology*. 2022;14(6):3109–3121.

[42] Fiani F, Napoli C, Randieri C, *et al.* Current trends and future directions in eye tracking technology: A literature review. *Engineering Applications of Artificial Intelligence*. 2026;163:112908.

[43] Russo S, Tibermacine IE, Randieri C, *et al.* Exploiting facial emotion recognition system for ambient assisted living technologies triggered by interpreting the user's emotional state. *Frontiers in Neuroscience*. 2025; 19:1622194.

[44] Randieri C, Perrotta A, Puglisi A, *et al.* CNN-based framework for classifying COVID-19, pneumonia, and normal chest X-rays. *Big Data and Cognitive Computing*. 2025;9(7):186.

Chapter 8

Deep reinforcement learning empowering adaptive cryptographic frameworks for intelligent healthcare systems

Friday Oodee Philip-Kpae[1], Agbotiname Lucky Imoize[2] and Patience Ebehiremen Orukpe[3]

Abstract

Intelligent healthcare systems built on the Internet of Medical Things (IoMT) increasingly depend on continuous telemetry and timely alarms, yet their cryptographic configurations are typically fixed at deployment time. This creates a safety–security trade-off: stronger protection can increase latency and energy consumption and risk violating clinical delay budgets, while lighter protection can be inadequate when threats escalate. This chapter addresses this problem by proposing an edge-based deep reinforcement learning (DRL) controller that adapts cryptographic protection at runtime in response to real-time context. The approach formulates adaptive security provisioning as a Markov decision process (MDP) and learns policies that select and tune cryptographic actions (e.g. encryption and authentication strength and key-management parameters) using a multi-objective reward that maximises security while penalising latency, energy, overhead and clinical deadline violations. As a design-and-evaluation study, the chapter compares deep Q-network (DQN), proximal policy optimisation (PPO) and asynchronous advantage actor–critic (A3C) and shows that learned policies improve the security–performance balance relative to static baselines. In practice, the proposed method supports safer, more resilient connected care by enabling context-aware protection for resource-constrained IoMT devices, thereby reducing avoidable alarm delays and strengthening defences during periods of elevated risk.

[1]Department of Electrical and Electronic Engineering, Faculty of Engineering, Rivers State University, Nigeria
[2]Department of Electrical and Electronics Engineering, Faculty of Engineering, University of Lagos, Nigeria
[3]Department of Electrical and Electronic Engineering, Faculty of Engineering, University of Benin, Nigeria

Keywords: Internet of Medical Things (IoMT); adaptive cryptography; edge security controller; deep reinforcement learning (DRL); Markov decision process (MDP); clinical quality of service (QoS); intelligent healthcare systems; proximal policy optimisation (PPO); deep Q-network (DQN); edge computing

8.1 Introduction

Intelligent healthcare systems combine connected medical devices, edge intelligence and cloud-scale analytics to enable continuous patient monitoring, early detection of clinical deterioration and personalised treatment recommendations. In practice, these systems support care pathways ranging from routine ward monitoring to high-acuity environments such as emergency departments and intensive care units, where early warning and timely clinician notification can directly affect patient outcomes. A key enabling substrate for these capabilities is the Internet of Medical Things (IoMT), which interconnects medical sensors, actuators and smart clinical equipment with local processing (edge) and remote services (cloud).

Modern care pathways increasingly depend on IoMT devices, including wearable electrocardiogram (ECG) sensors, pulse oximeters, smart infusion pumps and bedside monitors. These devices continuously generate sensitive physiological streams and personally identifiable information (PII), including identifiers, timestamps, vital-sign traces and device metadata that can be linked to patient identity. Protecting these data is essential across the full life cycle – during collection at the device, transmission over hospital networks, storage at gateways or cloud services and downstream analytics for decision support. Beyond confidentiality, healthcare systems must also ensure integrity (to prevent tampering), authenticity (to verify that data originates from legitimate devices) and availability (to keep alarms and monitoring services operational under stress).

Despite the widespread adoption of cryptographic protocols, real-world deployments face persistent constraints that render 'one-size-fits-all' security brittle. First, IoMT end points are resource-constrained and often operate on limited battery or energy-harvesting budgets; as a result, computationally intensive cryptographic operations can impose nontrivial computational time, memory footprint and energy consumption. Second, clinical workloads and network conditions fluctuate. For example, a wearable may intermittently transmit during routine monitoring but switch to high-frequency streaming during acute events (e.g. arrhythmia suspicion), while wireless interference, congestion or gateway load can vary significantly across wards and over time. Third, threat levels can change rapidly: hospital environments may experience bursts of scanning, man-in-the-middle (MiTM) attempts, malware propagation, insider misuse or ransomware incidents that disrupt services and increase adversarial pressure. These realities mean that static cryptographic configurations are often suboptimal – fixed 'high-security' settings may violate latency budgets for time-critical alarms, while fixed 'lightweight' settings may be insufficient when threats escalate or when sensitive workflows (e.g. medication delivery, critical-care telemetry) become active.

In healthcare, security adaptation must also respect clinical safety constraints. Alarm delivery and clinician notifications often have stringent delay budgets, and the cost of missed or delayed alerts can be severe. Therefore, any adaptive mechanism must be fast (low decision latency), reliable (stable behaviour under noisy measurements) and ideally explainable (so that operators can understand why a security posture changed). This creates a central tension: increasing protection strength can increase computational and communication overhead, whereas decreasing overhead can reduce the security margin under attack. What is needed is a mechanism that can respond to time-varying context – device energy state, network bandwidth/latency, edge workload and threat posture – while explicitly accounting for safety-critical quality-of-service demands.

To address this gap, this study introduces a deep reinforcement learning (DRL)-driven adaptive cryptographic controller that learns to choose security configurations based on real-time context. The core idea is to treat security provisioning as a sequential decision problem: at each decision point, the controller observes context signals (e.g. latency, bandwidth, device energy proxy, threat score, edge load and task criticality) and selects an appropriate protection profile. This profile may involve choices such as cipher suite, message authentication code (MAC) option, key size, key-rotation interval and authentication mode, aiming to maintain robust security while avoiding clinical deadline violations and high device-side cost.

The proposed framework aims to (i) preserve confidentiality and integrity of medical telemetry and control messages under realistic threat models, (ii) respect clinical latency constraints for safety-critical tasks such as alarms and clinician notifications, (iii) reduce energy and computational burden on IoMT devices by avoiding unnecessary cryptographic heaviness when risk is low and resources are constrained and (iv) adapt to dynamic threats and network variations by tightening protections when risk rises and relaxing overhead when urgency and resource pressure dominate.

8.1.1 Key contribution of the chapter

The following are the significant contributions of the chapter.

(i) Architecturisation of a layered intelligent healthcare security architecture that places a DRL-based cryptographic policy controller at the edge, enabling context-aware, real-time security adaptation close to devices and network gateways.

(ii) Formalisation of a Markov decision process (MDP) formulation for adaptive cryptographic decision-making, including a multi-objective reward that reflects security strength while penalising latency, energy consumption, computational overhead and deadline violations.

(iii) Algorithmic design of a DRL training and deployment pipeline and a comparative study of deep Q-network (DQN), proximal policy optimisation (PPO) and asynchronous advantage actor–critic (A3C) for healthcare-centric security adaptation.
(iv) Evaluation using a simulation-driven and trace-based methodology, including baseline static cryptography policies and standard performance metrics, demonstrating improved resilience and efficiency relative to static configurations under changing threat and workload regimes.

8.1.2 Chapter organisation

The remainder of this book chapter is structured as follows. Section 8.2 presents background and related work, covering cryptography and trust mechanisms for IoMT (e.g. blockchain-assisted security, quantum-inspired or quantum-resilient directions, visual cryptography, privacy-preserving analytics, elliptic curve cryptography (ECC) watermarking and auditable privacy schemes) and positioning the key gap as the lack of runtime policy agility under changing quality of service (QoS) requirements, resource availability and threat conditions. Section 8.3 describes the system model and threat model for an IoMT–edge–cloud healthcare pipeline and states the security goals: confidentiality, integrity/authenticity, availability under constraints and policy agility. Section 8.4 presents the MDP formulation for adaptive cryptography. Section 8.5 discusses the DRL algorithm choices (DQN, PPO and A3C) and motivates their suitability for healthcare-centric non-stationarity and edge deployment.

Section 8.6 details the DRL-enabled adaptive cryptographic architecture, in which an edge security controller senses contextual and threat signals and provisions cryptographic policies to IoMT devices, with cloud support for healthcare services and threat intelligence. Section 8.7 presents results and discussion, reporting the experimental set-up and findings from a MATLAB-based MDP simulation, comparing PPO, DQN and A3C and analysing convergence and the achieved security–performance trade-offs (including threat-responsive adaptation, energy-aware suite selection and deadline-violation penalties). Finally, Section 8.8 concludes the chapter by summarising the DRL-based adaptive cryptographic contribution and key findings, outlining limitations and open challenges and emphasising that simulation-driven evaluation may not capture the full real-world heterogeneity while motivating future validation on real edge–IoMT testbeds with richer threat telemetry and clinical governance requirements.

8.2 Background and related work

This section reviews the background and prior work on securing IoMT-enabled healthcare systems. It summarises cryptography- and trust-based protections (e.g. blockchain-assisted security, quantum-inspired or quantum-resilient directions, visual cryptography, privacy-preserving analytics, ECC watermarking and auditable privacy schemes). It also covers edge and decentralised security intelligence as well as privacy-preserving learning, such as federated learning (FL) and multimodal security models. Finally, it highlights recent DRL applications for Internet of Things (IoT) and IoMT optimisation and security, including diagnostics, offloading, physical-layer security, autonomous defence and intrusion detection systems (IDS). Overall, the literature shows substantial advances but remains largely component-specific and often static, motivating the adaptive, context-aware security control pursued in this work.

8.2.1 Cryptography in intelligent healthcare

Healthcare security typically relies on symmetric encryption for data in transit and at rest (e.g. AES variants), asymmetric cryptography for key exchange and digital signatures (e.g. RSA/ECC) and hashing for integrity (e.g. the SHA-2/SHA-3 families). In IoMT settings, lightweight cryptography and streamlined handshake mechanisms are often adopted to reduce computational and energy costs. However, the choice of algorithms, key sizes and key-management policies is often configured once and rarely revisited at runtime, even as operational conditions change. Recent HC-IoT/IoMT security research spans quantum-resilient schemes, media-specific privacy protection, privacy-preserving analytics and auditable anonymisation.

A security stack that couples blockchain-based data management and node authentication with a generative-AI–enabled, quantum-oriented encryption algorithm has been proposed for IoT healthcare, positioning blockchain immutability and auditability as the trust anchor. In contrast, the quantum-oriented component aims to harden confidentiality against advanced adversaries [1]. Nevertheless, the approach remains scheme-centric mainly, and its security strength is not explicitly optimised online against time-varying clinical QoS constraints (e.g. alarm deadlines) and fluctuating resource availability typical of IoMT end points.

Privacy has also been treated primarily as a record/image protection problem by integrating visual cryptography (share-based secrecy) with image-encryption techniques, ensuring that intercepted healthcare images remain unrecoverable without authorised reconstruction of shares and correct decryption [2]. However, this design is inherently data-type–specific (image/record-centric). It does not directly address end-to-end IoMT runtime adaptation (e.g. cipher agility, key rotation adjustments or authentication tightening) under dynamic network and threat conditions.

Beyond data protection at rest or in transit, privacy-preserving analytics has been explored through a smart healthcare workflow for deep learning–based brain tumour detection, demonstrating that diagnostic inference can be performed alongside mechanisms that protect sensitive medical data [3]. Even so, the security controls around the learning pipeline are typically deployed as static wrappers, leaving open the question of how to continuously rebalance the strength of protection against latency and energy costs as workloads shift from routine monitoring to acute events. Integrity and ownership have been addressed through an optimised

ECC-based crypto-watermarking model for smart healthcare that jointly supports confidentiality and authenticity/traceability [4]. Yet watermarking-driven integrity assurance alone does not provide policy agility across heterogeneous IoMT devices (e.g. switching suites, resizing keys or shortening rotation intervals when threat escalates), nor does it explicitly enforce clinical delay constraints.

The forward-looking threat landscape, particularly the implications of quantum computing, has been synthesised with an emphasis on heightened exposure for constrained medical devices and the need to reconcile post-quantum readiness with lightweight operation [5]. As a challenges-and-opportunities treatment, however, this line of work does not instantiate a concrete runtime mechanism for automatically tuning cryptographic strength under QoS and energy limits. At the transaction layer, the accountability–privacy tension has been tackled through an auditable coin-mixing scheme for IoMT that targets strong unlinkability while preserving auditability [6]. However, a focus on auditable mixing does not directly address continuous telemetry protection, in which end-to-end encryption and authentication choices must adapt in real time to evolving threats and network conditions [6].

Taken together, these studies advance powerful primitives – quantum-oriented encryption, blockchain auditability, visual cryptography, privacy-preserving deep learning, ECC-based watermarking and auditable mixing – yet they expose a persistent gap: most solutions are static or narrowly scoped, offering limited support for closed-loop, context-aware cryptographic adaptation that jointly optimises security, latency and energy under safety-critical healthcare constraints [1–6]. To address this gap, this paper introduces a DRL-driven adaptive cryptographic controller at the edge that formulates security provisioning as an MDP and dynamically selects cipher/MAC suites, key sizes, key-rotation intervals and authentication modes using real-time context (latency, bandwidth, device energy proxy, threat score, edge load and QoS criticality). A multi-objective reward promotes security strength while penalising latency, energy and overhead. It explicitly discourages violations of clinical deadlines, thereby converting 'strong-but-static' security designs into measurable, runtime-optimised protection for smart healthcare IoT.

8.2.2 *Adaptive security and context awareness*

Adaptive security seeks to vary protections based on context, such as risk level, asset criticality and resource availability. In healthcare, adaptive mechanisms must additionally respect clinical safety constraints: alarm delivery and clinician notifications have stringent delay budgets. Thus, adaptation must be fast, reliable and explainable.

A context-aware biometric security mechanism executed at the edge has been proposed to strengthen authentication for mission-critical IoT while reducing reliance on remote connectivity, highlighting the value of context-conditioned decisions at the edge [7]. However, the emphasis is primarily on biometric access control rather than runtime adaptation of cryptographic protections for medical telemetry streams. A secure, decentralised architecture for AI-driven healthcare has also been introduced, integrating privacy-enhancing mechanisms and adaptive intelligence to improve trust and governance in distributed healthcare ecosystems [8]. Even so, the adaptation focus remains primarily architectural and service-oriented and does not explicitly optimise encryption strength under strict latency–energy constraints in heterogeneous IoMT deployments. An energy-aware federated learning framework has been developed to address the privacy–resource tension by keeping data local while co-ordinating remote healthcare intelligence [9].

Nevertheless, the optimisation target is primarily learning and communication efficiency rather than real-time tuning of cryptographic primitives in response to rapidly changing threats. In a related direction, large multimodal models have been argued to enhance IoMT security by fusing heterogeneous signals for richer situational awareness and improved detection and response [10]. Yet this line of work primarily advances intelligence and detection, leaving open how multimodal context is translated into concrete, resource-bounded cryptographic action selection at the edge. Predictive analytics for chronic disease has likewise been demonstrated using deep machine-learning pipelines (with Alzheimer's disease as a case study) in IoMT-enabled Healthcare 5.0 [11]; however, security and privacy are treated primarily as supporting considerations rather than as the central problem of adaptive protection for the underlying data plane. Broader surveys further synthesise IoMT security and privacy challenges and map directions for ML/DL-based solutions, repeatedly underscoring persistent obstacles – heterogeneity, constrained resources and evolving attacks – that motivate adaptive, lightweight end-to-end defences [12].

Altogether, these studies advance key components – edge context awareness, decentralised privacy-enhancing AI, energy-aware privacy via federated learning, multimodal security intelligence and predictive analytics – yet collectively reveal a gap: the lack of a unified, closed-loop mechanism that learns online to adapt cryptographic protection levels to simultaneously satisfy fluctuating threat conditions and stringent clinical QoS (latency/availability) under energy and compute limits [7–12]. To address this gap, we introduce a DRL-driven edge security controller that formulates adaptive protection as an MDP and learns to select and tune cryptographic actions (e.g. cipher suite, key size/rotation and authentication mode) using a reward that jointly maximises security while penalising latency, energy and overhead violations, thereby operationalising 'adaptive intelligence' into measurable, real-time security optimisation for smart healthcare IoT.

8.2.3 Deep reinforcement learning for security

Control DRL has emerged as a powerful tool for learning control policies in environments with uncertain dynamics and high-dimensional state spaces. For security tasks, DRL has been applied to resource allocation, jamming resistance, intrusion response and network hardening. However, the integration of DRL with cryptographic adaptation (e.g. cipher-suite selection, key rotation and authentication level) in healthcare remains underexplored, particularly under multi-objective constraints and safety-critical requirements.

A 5G-aware, real-time diabetes prediction framework equipped with wearable-sensor analytics and an optimised Bi-LSTM guided by deep reinforcement learning to support timely clinical decision-making, demonstrating how DRL can improve predictive pipelines under dynamic IoT operating conditions [13]. In a complementary application, healthcare building operations have been formulated as a multi-objective DRL problem that balances medical service quality with energy and carbon objectives using carbon-intelligent IIoT monitoring, illustrating DRL's practicality for managing competing objectives in real time [14]. Similarly, a low-cost IoT data-offloading strategy has been designed using DRL and validated on real-world data to reduce service costs while managing delay–energy trade-offs in edge/cloud execution, thereby reinforcing DRL's suitability for QoS-constrained IoT control [17].

Moving from performance control towards security, a hybrid deep-learning–driven security/performance framework for Healthcare-IoT leverages fog-layer intelligence (including SDN-based control, fog-node selection and deep learning) to mitigate QoS and security concerns in dynamic H-IoT environments [15]. Learning-based control has also been applied to physical-layer security in IoMT, where DRL is used for secrecy-oriented optimisation in time-varying wireless settings, showing that adaptive policies can harden the communication layer against eavesdropping under fluctuating channels [16]. Beyond single-agent formulations, multi-agent DRL has been used to model IoT attack–defence as an adversarial process and to learn adaptive defence strategies, highlighting the benefits of interactive, game-like security control [18]. At the detection layer, an actor–critic–based hybrid IDS has been proposed that fuses host and network features and reports improved threat-detection performance relative to conventional NIDS/HIDS, alongside the release of a 5G/MEC-oriented dataset for medical applications [19].

These studies establish DRL as an effective tool for prediction, multi-objective operational optimisation, offloading and resource control and security intelligence (e.g. physical-layer security, intrusion detection and autonomous defence) [13–19]. Nevertheless, they remain largely function-specific – targeting diagnosis, energy/carbon optimisation, offloading, detection or PHY-layer secrecy – and do not provide a unified mechanism that adaptively selects and tunes cryptographic protection strength (cipher suite, key size, rotation interval and authentication mode) at the edge to satisfy simultaneous security goals and strict clinical QoS constraints under fluctuating threats and heterogeneous IoMT resources [20]. To address this gap, this study introduces a DRL-driven edge cryptographic adaptation controller that learns online to balance security gains with latency, energy and overhead penalties, thereby operationalising context-aware intelligence for real-time, end-to-end protection in smart healthcare IoT. Table 8.1 summarises all references used in the review analysis.

As shown in Table 8.1, the related work summarises the key literature used in this study by comparing each reference based on the underlying method/approach, the primary results reported by the authors and a concise account of its strengths

Table 8.1 Summary of related work

Ref.	Method/approach	Key results (reported)	Pros	Cons/limitations
[1]	GenAI-enabled quantum encryption + blockchain for IoT healthcare	Improved confidentiality/tamper resistance; better security efficiency vs conventional stacks	Quantum-forward; auditability; end-to-end trust	Complex; lacks online QoS-aware crypto adaptation
[2]	Visual cryptography + diverse image encryption for records	Enhanced privacy for image-based records via share reconstruction + encryption	Layered secrecy: practical for images	Image/record-specific; limited IoMT-wide adaptation focus
[3]	Privacy-preserving DL tumour detection + protection wrapper	High diagnostic performance with privacy safeguards	Privacy + analytics coexist	Security largely static; limited runtime threat/QoS tuning
[4]	Optimised ECC crypto-watermarking	Improved integrity/ownership plus confidentiality for health content	Lightweight ECC; authenticity support	Content-centric; no closed-loop policy agility
[5]	Survey on quantum computing for HC-IoT security	Threat taxonomy; constraints/opportunities for post-quantum readiness	Strong synthesis; motivates quantum-aware defences	Survey only; no implementable adaptive controller
[6]	Auditable coin mixing (Medishuffle) for IoMT	Improves unlinkability while preserving auditability/accountability	Balances privacy and compliance needs	Transaction-layer focus; not end-to-end telemetry crypto tuning
[7]	Context-aware biometrics at the edge (military IoT)	More responsive authentication via edge processing and context conditioning	Low-latency auth; context sensitivity	Authentication-centric; not adaptive encryption/key management
[8]	Secure, decentralised, privacy-enhancing adaptive intelligence for AI healthcare	Improved trust/privacy via decentralisation and adaptive intelligence constructs	Privacy-by-design supports distributed AI.	Adaptation mostly architectural; limited quantified crypto–QoS optimisation
[9]	Energy-aware federated learning privacy-preserving IoMT framework	Privacy via local data; energy-aware co-ordination for remote healthcare	Reduces raw-data exposure; regulation-friendly	Learning-centric; crypto agility for data planes not central

(Continues)

Table 8.1 (Continued)

Ref.	Method/approach	Key results (reported)	Pros	Cons/limitations
[10]	Large multimodal models for IoMT security	Richer situational awareness from multi-signal fusion for security tasks	Strong context for detection/response	Compute cost; context not directly mapped to crypto action selection
[11]	IoMT-enabled Healthcare 5.0 deep ML for Alzheimer's prediction	Improved chronic disease prediction (case study)	Decision support from IoMT data	Security/privacy peripheral; no adaptive protection mechanism
[12]	Survey of IoMT security/privacy and ML/DL solutions	Taxonomy of threats/defences; open challenges and future directions	Broad coverage; clear gap identification	Survey only; lacks a validated adaptive runtime solution
[13]	Wearable diabetes diagnostics using DRL-guided DL	Improved real-time diagnostic performance via DRL optimisation	Shows DRL utility in IoT healthcare	Diagnosis-focused; not crypto/QoS co-optimisation
[14]	DRL multi-objective optimisation for healthcare buildings	Improved service versus energy/carbon trade-offs via learned control	Demonstrates DRL for competing objectives	Operations domain; indirect to IoMT crypto adaptation
[15]	Hybrid DL-driven security and performance (fog/SDN-oriented)	Improved QoS/security via intelligent fog/flow decisions	Joint security–QoS framing; practical fog emphasis	Not explicit crypto-parameter adaptation; model assumptions may limit generality
[16]	DRL-based physical-layer security for IoMT	Improved secrecy-oriented performance under varying channels	Adaptive wireless hardening	PHY-only; does not cover end-to-end crypto/key agility
[17]	DRL-based IoT data offloading (edge–cloud)	Lower cost with improved delay/ efficiency trade-offs	Strong QoS control for edge computing	Offloading-centric; security not primary
[18]	Multi-agent DRL autonomous IoT security management	Adaptive defence policies learned under dynamic attacks	Models attacker–defender dynamics	General IoT; may omit clinical QoS and crypto tuning specifics
[19]	HDRL-IDS for 5G medical applications	Improved IDS effectiveness using hybrid DRL with host + network features	Strong detection for 5G/MEC medical settings	IDS-focused; does not inherently tune encryption/keying policies

and limitations. The reviewed works span (i) cryptographic and trust mechanisms for IoMT such as blockchain-assisted quantum encryption, visual cryptography, ECC-based watermarking and auditable mixing [1–6]; (ii) edge-centric authentication and decentralised privacy-enhancing intelligence [7,8]; (iii) privacy-aware learning paradigms including energy-aware federated learning and multimodal-model–driven security intelligence [9,10]; (iv) IoMT analytics for Healthcare 5.0 prediction [11] and (v) surveys that consolidate IoMT security/privacy challenges and solution directions [5,12].

In addition, the table captures a strong recent trend towards deep reinforcement learning (DRL) for IoT/IoMT optimisation, including real-time diagnostics, multi-objective operational control, physical-layer security, data offloading and autonomous security management/IDS [13–19]. Notably, the comparative 'Cons/Limitations' column highlights a recurring gap across most prior studies: while they improve specific components (encryption, integrity, anonymity, authentication, detection, learning efficiency or domain optimisation), they generally do not provide a unified, closed-loop mechanism for online, context-aware adaptation of cryptographic protection levels under fluctuating threats and strict clinical QoS constraints, which directly motivates the DRL-based adaptive security control proposed in this work.

8.3 System and threat model

The system and threat model are described in this section. The security goal, including the layered component and the rationale, is defined accordingly.

8.3.1 System model

Figure 8.1 illustrates a layered edge-enabled healthcare sensing and analytics framework. At the bottom, the Device Layer performs networked sensing and data collection from patient-centric sensors (e.g. physiological signals such as EEG/ECG/EMG) within a local cluster. Collected data is forwarded through a gateway device (Inter-CH) to the edge computing/broker layer, where multiple storage-and-computation nodes support data preprocessing, data classification and data analysis (edge learning) co-ordinated by an edge server (relay). Above this, the network layer provides connectivity to a central database/server for broader data services. At the top, the Management Layer represents healthcare stakeholders – emergency services, physicians, hospitals and patients – that consume results and interact with the system via the backend infrastructure. The arrows indicate the end-to-end pipeline from sensing to edge processing and onward to centralised storage/management, with additional control/queueing elements (e.g. PQ and CBWFQ) highlighting data management and QoS handling between layers.

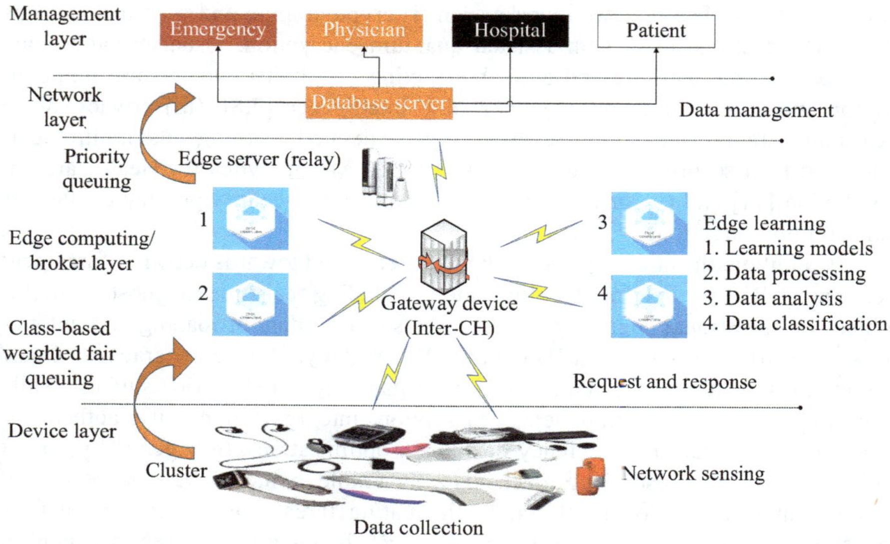

Figure 8.1 Layered architecture of an edge-enabled healthcare sensing system showing device-layer data collection, gateway forwarding, edge/broker-layer preprocessing and learning (classification and analysis), network-layer database/server support and management-layer stakeholders (emergency, physician, hospital and patient) interacting with the system through data management and QoS mechanisms (Modified from [21])

8.3.2 Threat model

Figure 8.2 summarises the threat model for an IoMT-enabled healthcare data pipeline by placing the IoMT data flow (IoMT devices → edge gateway/server → cloud services) at the centre and linking it to four key adversarial capabilities. Each outer block represents a class of attacks that can target the data path: eavesdropping (passive interception of traffic over Wi-Fi, BLE, Zigbee, cellular or Ethernet links), active attacks (replay, packet injection and man-in-the-middle attempts), key-compromise attempts (brute-force guessing, protocol exploitation and credential leakage) and resource abuse/DoS (flooding or jamming that increases latency and accelerates battery drain). The arrows from the outer blocks into the central box indicate that these threats directly affect the confidentiality, integrity and availability of patient data as it moves across the device–edge–cloud workflow.

8.3.3 Security goal

In this context, the security design targets four core goals First, confidentiality of patient data is required to prevent unauthorised disclosure of protected health information during sensing, local processing at the edge and transmission to cloud

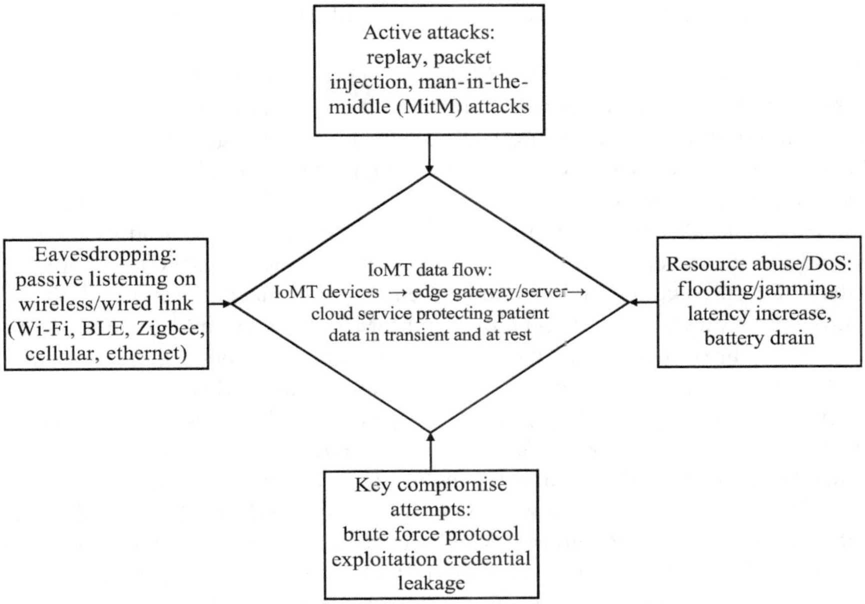

Figure 8.2 Threat model for an IoMT device–edge–cloud data pipeline showing four adversary capabilities – eavesdropping, active attacks (replay/injection/MitM), key-compromise attempts and resource abuse/DoS – mapped to the central IoMT data flow (devices → edge gateway/server → cloud services)

services; this entails strong encryption for data in transit and at rest, with careful key generation, distribution, rotation and access control. Second, integrity and authenticity of data and commands must be preserved so that physiological measurements, clinical alerts and control instructions (e.g. configuration updates or actuation commands) cannot be altered, forged, replayed or injected by an attacker; this typically relies on message authentication, secure session establishment, anti-replay protection and mutual authentication between devices, edge gateways and cloud end points.

Third, availability under constrained resources is essential because many IoMT devices operate with limited battery, compute and bandwidth, making them susceptible to denial-of-service or resource-exhaustion attacks; therefore, the system should incorporate lightweight security mechanisms, rate limiting, prioritisation of critical traffic and edge-side filtering to sustain timely operation even during adverse network conditions. Fourth, policy agility is needed to cope with dynamic clinical environments where both threats and operating conditions vary over time; the system should be able to adapt cryptographic strength, update security policies and rebalance overhead versus protection (e.g. switching suites, adjusting key rotation or tightening authentication) in response to changes in risk level, workload, latency requirements and device energy state without disrupting essential healthcare services.

For instance, [22] proposes a MitM-mitigation framework for remote IoMT monitoring that (a) sends a compact signature derived from sensed data (to reduce energy and limit privacy exposure) and (b) attaches a message authentication code (MAC), with the key derived from received signal strength indication (RSSI). The study argues that the design prevents a MitM from disrupting operations (e.g. replaying 'normal' data to suppress alarms) while reducing overhead by transmitting signatures rather than raw data. The study provides a concrete, lightweight pattern for protecting integrity/authenticity (anti-tamper + anti-replay) and improving privacy/efficiency in constrained IoMT telemetry links – useful as prior art for your threat model's 'active attacks + constrained resources'.

Similarly, the work [22] conducts an SLR following Kitchenham and Charters' planning–conducting–reporting process. They search IEEE Xplore, ScienceDirect, ACM Digital Library, Google Scholar and SpringerLink using a security/threat-modelling query for MIoT and apply inclusion/exclusion screening (reported in a PRISMA-style flow). The review covers studies published Jan 2020–Jan 2025 and is driven by two RQs: identifying MIoT threats and identifying threat-modelling methodologies used in MIoT. The search yielded 574 records, of which 34 were included papers. From these, the authors derive a structured threat catalogue of 48 MIoT threats, 36 of which are mapped to portable ECG ecosystems. They identify cybersecurity challenges in MIoT communications and report findings from security assessments guided by the threat model (i.e. real ecosystem-level issues around device communications and associated components). Their study provides strong evidence for our paper's motivation to use threat modelling: the resulting threat catalogue is then used to build their portable-ECG ecosystem threat model and to guide subsequent security assessments and penetration testing.

Figure 8.3 presents the proposed DRL-enabled adaptive cryptographic system architecture for intelligent healthcare systems. IoMT devices generate patient telemetry and enforce issued cryptographic policies. An edge gateway aggregates and preprocesses streams while collecting QoS telemetry and IDS/threat indicators. A DRL-based security controller formulates the adaptation task as an MDP, constructs the state from context and threat signals, selects cryptographic actions and drives a policy provisioner that configures suite selection, key size/rotation, authentication mode and safety guardrails. The edge layer co-ordinates with cloud/health platform services for storage/EHR integration, analytics, compliance logging, global threat intelligence and model/policy support.

8.3.4 Layered components

The layered components include 1) IoMT devices that perform sensing, light preprocessing and enforce cryptographic policies (encryption/authentication) provisioned by the edge controller. 2) Edge DRL Security Controller hosts (i) state collector, (ii) threat/context estimator, (iii) DRL agent and (iv) policy manager. The controller performs cryptographic operations and disseminates policy updates to

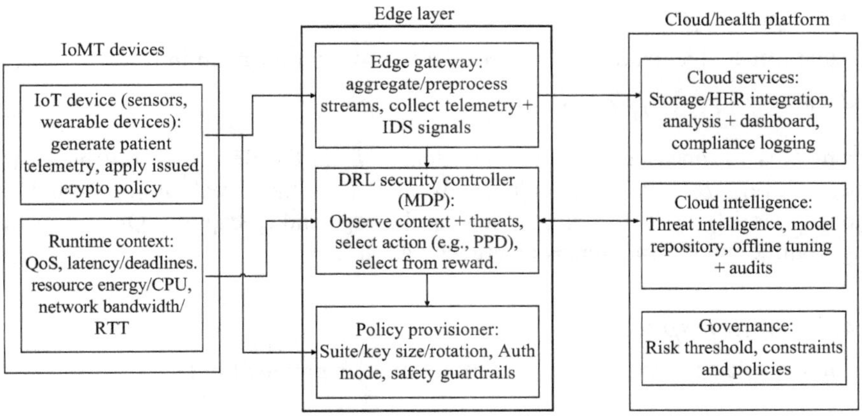

Figure 8.3 Three-layer IoMT–edge–cloud architecture where a DRL security controller at the edge dynamically configures cryptographic policies for devices based on runtime context and threats, with cloud support for health services and intelligence

IoMT devices. It also monitors compliance and detects anomalies. 3) Secure Communication Fabric that supports authenticated channels (e.g. TLS/DTLS-like) parameterised by cipher suites and key management policies selected by the controller. 4) Cloud Services store encrypted records, support analytics and provide aggregated threat intelligence. Cloud feedback can enrich state signals (e.g. global risk score).

8.3.5 Textual diagram rationale

The architecture is designed to keep decision latency low by deploying the DRL agent near data sources (edge) while allowing cloud-scale learning, long-horizon logging and global threat intelligence.

8.4 Markov decision process (MDP) formulation for adaptive cryptography

We formulate adaptive cryptographic control as an MDP as expressed in (8.1).

$$\mathcal{M} = (S, \mathcal{A}, \mathcal{P}, \mathcal{R}, \gamma) \tag{8.1}$$

where S is the state space, $\mathcal{A}$ is the action space, $\mathcal{P}$ is the transition kernel, $\mathcal{R}$, is the reward and $\gamma \in (0,\ 1)$ is the discount factor.

8.4.1 State space

At time t, the edge controller constructs a state vector as given in (8.2).

$$s_t = [L_t, E_t, B_t, T_t, C_t, Q_t], \tag{8.2}$$

where L_t is the measured end-to-end latency, E_t is the device energy proxy (or battery level), B_t is the available bandwidth, T_t is the threat score (from IDS/ anomaly signal), C_t is the compute load at the edge and Q captures QoS criticality (e.g. routine monitoring versus emergency alarm).

8.4.2 Action space

An action selects a cryptographic configuration as provided in (8.3).

$$a_t = \{Suite_t, k_t, \tau_t, m_t\} \tag{8.3}$$

where $Suite_t$ is the cipher/MAC suite, k_t is the key size, τ_t is the rotation interval and m_t is the authentication mode (e.g. signature-based versus MAC-based authentication for messages).

8.4.3 Reward design

We define a multi-objective reward as presented in (8.4).

$$r_t = \alpha S(a_t, T_t) - \beta \widehat{L}_t - \delta \widehat{E}_t - \eta \widehat{O}_t - \lambda \Phi_t \tag{8.4}$$

where $S(a_t, T_t)$ quantifies security strength under threat T_t, $\widehat{L}_t$ is the normalised latency, $\widehat{E}_t$ is the normalised energy cost, $\widehat{O}_t$ is the normalised cryptographic overhead (CPU cycles, handshake cost) and Φ_t is a penalty for violating safety/QoS constraints (e.g. alarm deadline miss). The coefficient α, β, δ, η, λ are used to tune trade-offs. A tabulated definition of the space, action and reward design components is presented in Table 8.2.

Table 8.2 summarises the Markov decision process used for adaptive cryptography. The state space captures real-time context (e.g. latency/bandwidth, device energy proxy, threat score, edge load and QoS criticality) so that decisions reflect both risk and resource conditions. The action space defines the cryptographic knobs the controller can tune at runtime – cipher/MAC selection, key

Table 8.2 State, action and reward components for adaptive cryptography

Element	**Definition**
State s_t	Latency L_t, energy proxy E_t, bandwidth B_t, threat score T_t, edge compute load C_t, QoS criticality Q_t
Action a_t	Cipher/MAC suite, key size k_t (e.g. 128/192/256), rotation interval τ_t, authentication mode m_t
Reward r_t	Security strength S minus penalties for latency $\widehat{L}_t$, energy $\widehat{E}_t$, overhead $\widehat{O}_t$ and constraint violation Φ_t (8.4)

size, key rotation interval and authentication mode. The reward balances security gains against latency, energy and overhead costs, with explicit penalties for violating clinical quality of service/deadlines, ensuring the learned policy favours protection that remains feasible for time-critical healthcare traffic.

8.4.4 *Objective function*

The DRL agent learns a policy $\pi_\theta(a|, s)$ and maximises discounted return as shown in (8.5).

$$J(\pi_\theta) = \mathbb{E}_{\pi_\theta}\left[\sum_{t=1}^{\infty} \gamma^t r_t\right] \tag{8.5}$$

8.5 DRL algorithms for healthcare-centric adaptation

This section summarises DQN, PPO and A3C as candidates for learning π_θ.

8.5.1 *DQN (discrete action baseline)*

DQN estimates the action-value function $\pi_\theta(s, a)$ using temporal-difference learning as presented in (8.6).

$$\mathscr{L}(\theta) = \mathbb{E}\left[\left(r_t + \gamma \max_{a'} Q_{\theta^-}\left(s_{t+1}, a'\right) - Q_\theta((s_t, a_t))\right)^2\right] \tag{8.6}$$

where θ^- are target network parameters. DQN is well-suited when $\mathscr{A}$ is a discrete set of approved cryptographic configurations.

8.5.2 *PPO (stable policy optimisation)*

PPO is an actor–critic method that optimises a clipped surrogate objective as given in (8.7).

$$\max_\theta \mathbb{E}\left[\min\left(\rho_t(\theta)\widehat{A}_t, clip(\rho_t(\theta), 1-\epsilon, 1+\epsilon)\widehat{A}_t\right)\right] \tag{8.7}$$

where $\rho_t(\theta) = \frac{\pi_\theta(a_t |, s_t)}{\pi_{\theta_{old}}(a_t |, s_t)}$ and $\widehat{A}_t$ is the advantage estimate. PPO is typically more stable under non-stationary conditions common in healthcare networks.

8.5.3 *A3C (scalable distributed learning)*

A3C trains multiple workers asynchronously to optimise shared actor and critic parameters. This is expressed in (8.8).

$$\nabla_\theta(\theta) \approx \sum_i \nabla_\theta \log \pi_\theta(a_t^i|, s_t^i)\widehat{A}_t^i + k\nabla_\theta H(\pi_\theta) \tag{8.8}$$

where $H(\pi_\theta)$ encourages exploration. A3C is attractive for multi-ward or multi-edge deployments.

8.5.4 Resilience under threat variation

When the threat score T_t increases (e.g. IDS alerts, anomaly spikes), the reward increases weight on security strength through $S(a_t, T_t)$, pushing the policy towards stronger configurations (larger key sizes, shorter rotation τ_t, stronger suites). Conversely, during stable, low-threat periods, the agent can reduce overhead to save energy while meeting tight latency requirements. Furthermore, clinical safety constraints are handled using a penalty Φ_t and hard guards. In particular, if the predicted end-to-end delay for an alarm channel exceeds a deadline D. The safety constraint enforcement is computed as described in (8.9), and a safe fallback configuration is applied.

$$\Phi_t = \mathbb{I}[L_t > D] \cdot (L_t - D), \tag{8.9}$$

For operational acceptance, the controller logs (i) state features contributing to action selection, (ii) threat signals and (iii) constraint triggers. This log supports auditing and post-incident review.

8.6 Proposed DRL-enabled adaptive cryptographic framework

This section describes the proposed DRL-empowered adaptive cryptographic framework, starting with an overview of the scheme and the algorithmic framework.

8.6.1 Framework overview

The edge controller executes a closed loop as follows: (i) sense context/threat state, (ii) select cryptographic action, (iii) enforce and distribute policies, (iv) observe resulting performance and security outcomes and (v) update the DRL model (online or periodically). Table 8.3 presents an analysis of the three algorithms used in the proposed study.

Table 8.3 compares the three candidate deep reinforcement learning (DRL) algorithms with respect to action-space support, stability and the most suitable healthcare deployment setting. Deep Q-network (DQN) is a solid baseline when the controller selects from a discrete set of pre-approved cryptographic configurations, but it offers only medium stability. Proximal policy optimisation (PPO) is considered the most practical option for hospital runtime adaptation because it supports both discrete/continuous actions and provides high stability under dynamic conditions. Asynchronous advantage actor–critic (A3C) also supports discrete/continuous actions. It is designed for distributed learning,

Table 8.3 Comparison of DRL algorithms for adaptive cryptography

Algorithm	**Action space**	**Stability**	**Best use in healthcare**
DQN	Discrete	Medium	Approved suite cipher selection; strong baseline
PPO	Discrete/ Continuous	High	Real-time, aware safety adaptation under dynamic conditions
A3C	Discrete/ Continuous	High (distributed)	Multi-edge scaling; parallel learning across wards

making it well-suited to multi-edge or multi-ward scaling in which parallel learning is beneficial. Algorithm 8.1 summarises the closed-loop routine executed by the edge security controller to sense context and threat signals, select a cryptographic action, enforce the resulting policy on IoMT devices, observe latency/energy/overhead outcomes and update the DRL agent to improve future decisions.

Algorithm 8.1 DRL-based adaptive cryptographic policy at the edge

Initialise DRL agent (DQN/PPO/A3C) with parameters θ
Initialise replay buffer $\mathscr{D}$ (for DQN) and/or roll-out storage (for PPO/A3C)
for each episode, **do**
    Observe initial state s_t from context + threat estimator
    for each time step t **do**
        Select action $a_t \sim \pi_\theta(\cdot|, s_t)$ (with exploration)
        Enforce cryptographic policy $\{Suite_t, k_t, \tau_t, m_t\}$ on the IoMT group
        Measure outcomes: latency, energy proxy, overhead, integrity events
        Compute reward $r_t \leftarrow \alpha S(a_t, T_t) - \beta\widehat{L}_t - \delta\widehat{E}_t - \eta\widehat{O}_t - \lambda\Phi_t$,
        Observe the next step s_{t+1}
        Store transition (s_t, a_t, r_t, s_{t+1})
        Update θ (DQN: TD update; PPO/A3C: policy gradient update)
        $s_t \leftarrow s_{t+1}$
end for
end for

Algorithm 8.1 operationalises the chapter's main idea as a closed-loop edge controller. At each decision epoch, the controller observes the current IoMT context and threat indicators, selects a cryptographic configuration using the learned DRL policy and determines which devices to use for the next interval. The resulting latency, energy and overhead outcomes are then used to compute the reward, including penalties for QoS or deadline violations, and the experience tuple is stored for policy/value updates. By repeating this observe–act–evaluate–learn cycle, the controller progressively learns when to tighten protection under elevated risk and when to relax cryptographic cost to preserve timeliness and device lifetime under constrained conditions.

At the start, the algorithm initialises the DRL agent (DQN, PPO or A3C) and its learning parameters in Table 8.4, then allocates the appropriate experience store: a replay buffer for off-policy DQN learning and/or a roll-out storage for on-policy PPO/A3C updates. For each training episode, the edge controller constructs an initial state s_t from the context and threat estimator (e.g. latency, energy proxy, bandwidth, threat score, edge load and QoS criticality). Within each episode, at every time step t, the agent selects an action a_t using an exploration strategy and enforces the resulting cryptographic policy on the target IoMT group (e.g. the chosen cipher/MAC suite, key size, rotation interval and authentication mode).

The controller then measures the resulting outcomes – latency, energy proxy, cryptographic overhead and integrity/security events – and computes the multi-objective reward r_t and observes the next state s_{t+1}. The transition (s_t, a_t, r_t, s_{t+1}) is stored, after which the agent updates its parameters: temporal-difference (TD) updates for DQN (typically sampling from the replay buffer) versus policy-gradient/actor–critic updates for PPO/A3C (using roll-out trajectories). Over multiple episodes, this procedure yields a learned policy that adapts cryptographic strength to evolving threats and operating conditions while penalising excessive latency/energy/overhead, thereby enabling practical real-time policy updates at the edge. Table 8.4 presents the synthetic parameters used in the proposed simulation set-up.

Table 8.4 summarises the MATLAB® Markov decision process (MDP) simulation configuration used to evaluate the deep reinforcement learning (DRL) adaptive cryptographic controller and clarifies how each reported result is generated. It specifies the learning horizon (500 episodes). It uses cumulative reward as the primary convergence signal, enabling a fair comparison of deep Q-network, proximal policy optimisation and asynchronous advantage actor–critic under the same environment settings. The table also makes the reward shaping explicit through the weight set, which prioritises security while penalising latency, energy and overhead, directly encoding the chapter's security–QoS trade-off. In addition, it lists the core analysis views – key size versus latency, threat score versus security level and energy per cryptographic suite – so the results can demonstrate both responsiveness to risk and the cost of stronger protection on constrained IoMT devices. Finally, the inclusion of an explicit clinical deadline penalty (e.g. when latency exceeds 40 ms) anchors the evaluation to safety-critical constraints. At the same time, the state-space coverage/value-surface items help verify that the agent explores diverse conditions and learns a meaningful decision landscape rather than overfitting to a narrow regime.

Table 8.4 Parameters used for simulation

Plot/metric	Variable(s)	Sample values/range	Notes/description
Learning curves	Cumulative reward	$PPO: -45 \rightarrow +5, DQN: -50 \rightarrow -5, A3C: -48 \rightarrow 0$	Convergence of DRL algorithms over 500 episodes.
Reward components	α, β, δ, η	0.8, −0.3, −0.2, −0.1	Weights for security, latency, energy and overhead.
Security-latency trade-off	Key size (k_t), Latency (ms)	$128 - 512bits$, $5 - 25ms$	Pareto frontier showing a non-linear increase in latency with key size.
Adaptive policy response	Threat score (T_t), Security Level (a_t)	$T_t: 0 - 1.0$, $a_t: 1 - 4$	Step response of security actions to dynamic threat levels over 100 time steps.
Energy consumption per cipher suite	Cipher suite, energy (J)	$AES - 128: 15$, $AES - 256: 28$, $ChaCha20: 12$, $ECC - 256: 45$	Energy proxy for cryptographic actions.
Safety constraint (ϕ_t)	Latency stream, penalty	$Latency: 30 - 50ms$, $Penalty: 0 - 10$	Penalty applied when latency exceeds 40 ms clinical deadline.
State space coverage	Latent Dim 1, Latent Dim 2	$x_s, y_s \sim N(0,1)$	Simulated PCA/t-SNE projection showing exploration density.
Learned value surface ($Q(s,a)$)	State (S), action (A), Q-value	$Q \sim 0.5 - 1.0$	3D surface illustrating optimised value function for policy decisions.

8.7 Results and discussions

This section presents and interprets the experimental outcomes of the proposed DRL-based adaptive cryptographic controller using the MATLAB MDP simulation environment. We compare PPO, DQN and A3C with respect to learning convergence and the resulting security–performance trade-off focusing on cumulative reward behaviour, weight sensitivity and the latency/energy impacts of more decisive cryptographic actions. The discussion also highlights how the learned policy responds to changing threat levels and how the latency-constraint penalty guides safe action selection for real-time edge enforcement.

8.7.1 Learning curves and cumulative reward convergence

Figure 8.4 illustrates the learning convergence of three DRL algorithms, PPO, DQN and A3C, optimising the objective function $J(\pi_\theta)$ defined in (8.5). Quantitatively, all models begin with negative cumulative rewards, approximately -40 to -50, as they explore the state space s_t. PPO demonstrates the highest stability and learning efficiency, converging to a positive reward of approximately

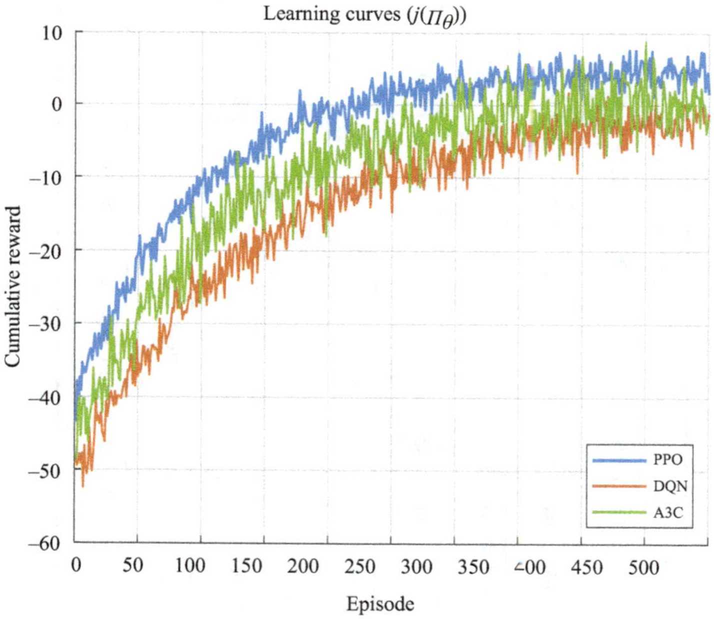

Figure 8.4 Learning-curve plot comparing PPO, A3C and DQN by cumulative reward over training episodes; PPO improves most steadily and converges best, A3C improves with higher variability and DQN learns more slowly and converges worse

+5 by episode 300. A3C follows with higher variance, reaching near-zero cumulative rewards, while DQN exhibits the slowest progression and remains slightly negative. These results confirm that PPO most effectively balances security strength and computational overhead, making it the most suitable algorithm for adaptive cryptographic decision-making.

8.7.2 Analysis of reward weights and multi-objective tuning

Figure 8.5 illustrates the optimisation and weighting of the multi-objective reward r_t defined in (8.4). Figure 8.4 presents the learning curves, where the agent maximises the objective function $J(\pi_\theta)$ from (8.5). Quantitatively, PPO converges most efficiently, achieving a positive cumulative reward of approximately +5 by episode 500; outperforming A3C, which stabilises around 0, and DQN, which remains near −5. Figure 8.5 depicts the reward weights used to tune performance trade-offs. A strong positive weight is assigned to security ($\alpha = 0.8$) prioritises cryptographic strength, while negative weights for latency ($\beta = -0.3$), energy consumption ($\delta = -0.2$) and computational overhead ($\eta = -0.1$) impose penalties on excessive resource usage. This weighting strategy ensures that the agent favours secure configurations while remaining sensitive to efficiency constraints.

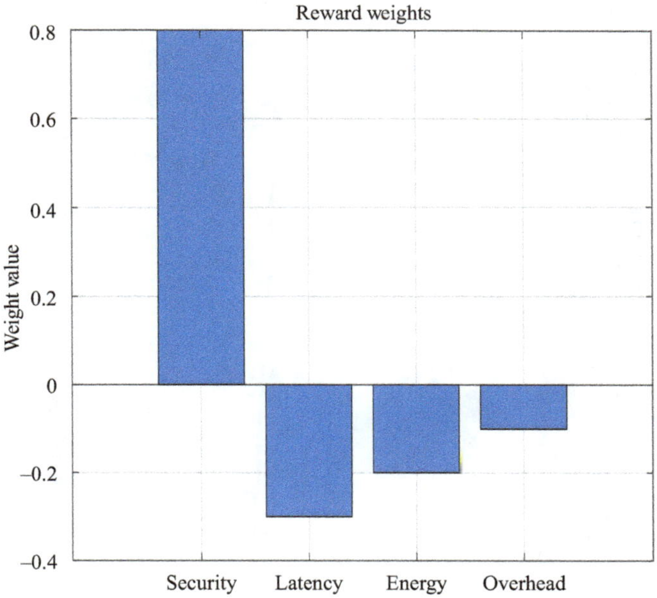

Figure 8.5 Reward weights are shown as a bar chart of the multi-objective used by the DRL agent, with security as a positively weighted objective and latency, energy consumption and computational overhead as negatively weighted penalty terms to enforce efficiency

8.7.3 Security–latency pareto frontier and computational cost

Figure 8.6 explores the relationship between the action space defined in (8.3) and the resulting latency L_t. With cryptographic key sizes ranging from 128 to 512 bits, the graph reveals a non-linear increase in latency. A 128-bit key produces minimal latency of approximately 5 ms, whereas a 512-bit key increases latency to 25 ms, following a power-law trend. This scatter plot highlights the trade-off the agent must manage, selecting cryptographic suites that satisfy the current threat level T_t without exceeding the latency limits that would trigger the penalty Φ_t defined in (8.9).

8.7.4 Dynamic threat–response adaptation and policy execution

Figure 8.7 shows a dynamic security system in which the security level, a_t, represented by the blue step function, adjusts according to the fluctuating threat score, T_t, shown as the red line. Initially, with the threat score near 1.0, the security level remains at its maximum of 4. As T_t decreases to approximately 0.4 around time step 35, the system scales down the security level to 2. A further drop below 0.2 at

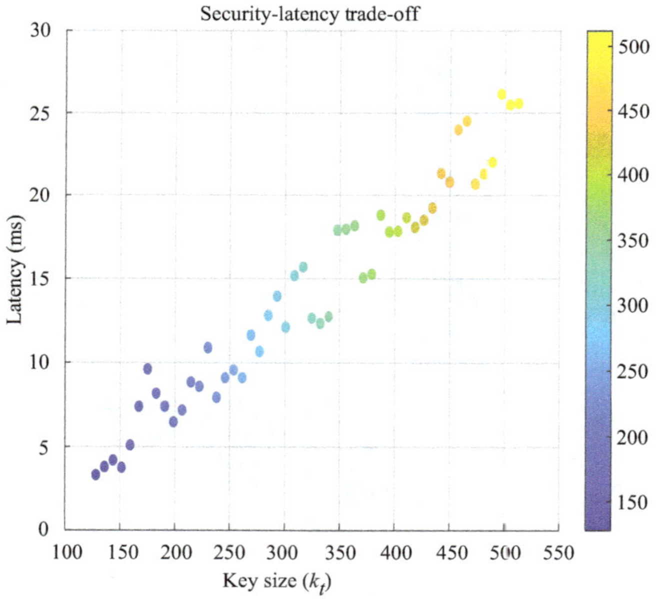

Figure 8.6 Security-latency trade-off scatter plot showing a security–latency trade-off: as cryptographic strength (e.g. larger key sizes) increases, end-to-end latency rises non-linearly, illustrating the need to balance stronger protection against real-time delay constraints

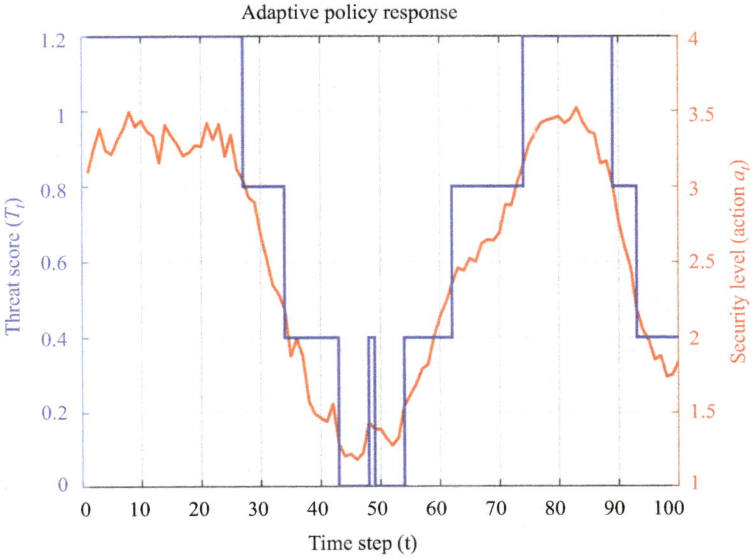

Figure 8.7 Adaptive policy response of time-series showing a dynamic adaptive security policy: a red threat-score curve rises and falls over time, while a blue step-like security level increases or decreases in response, demonstrating real-time adjustment of cryptographic strength to changing threats

time step 45 triggers the minimum-security level of 1. The system demonstrates reactivity by increasing the security level back to 4 when the threat score approaches 1.0 at time step 80, reflecting real-time adaptive cryptographic control.

8.7.5 Energy proxy distribution across cryptographic suites

Figure 8.8 presents a quantitative breakdown of the energy cost E_t associated with different actions in the action space A. AES-128 is the most energy-efficient, consuming approximately 15 joules, while ECC-256 is the most energy-intensive, consuming about 45 joules. These values directly influence the reward *rt* through the $-\delta E_t$ term in (8.4). The distribution explains why the agent may favour ChaCha20, which has an energy cost of approximately 12 joules, in low-battery or energy-constrained states. This granular modelling enables the edge controller to extend IoMT device lifetimes by selecting energy-efficient suites when the threat level T_t is low.

8.7.6 Safety constraint enforcement and penalty violations

Figure 8.9 illustrates the safety-constraint mechanism defined in (8.9). A hard clinical deadline D of 40ms is indicated by the red dashed line, while the grey trace

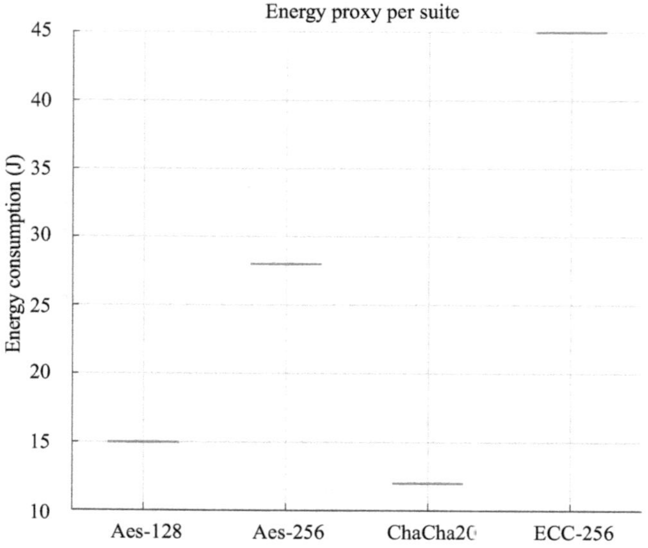

Figure 8.8 Energy proxy per suite illustrating a bar chart comparing the energy proxy for different cryptographic suites, lightweight symmetric options (e.g. AES-128/ChaCha20) as lower-energy choices and stronger public-key options (e.g. ECC) as higher-energy, highlighting energy-aware suite selection

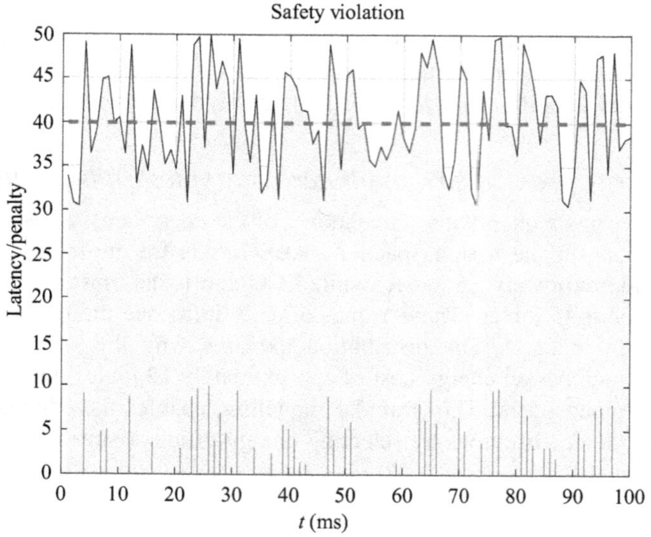

Figure 8.9 Safety violation, a line plot illustrating safety-constraint enforcement: an end-to-end latency trace is compared against a fixed clinical deadline (dashed line), and any deadline exceedance is highlighted with stems/markers indicating a penalty proportional to the violation, emphasising hard safety guards that force safer, faster fallback modes

represents the observed end-to-end latency L_t. Whenever L_t exceeds the deadline, red stems quantify the penalty Φ_t in proportion to the violation magnitude. For example, a latency spike to 50 ms produces a penalty value of 10. This figure illustrates the system's hard safety guards, showing that the agent is strongly penalised for endangering patient safety by delaying emergency signalling, thereby forcing a reversion to faster fallback cryptographic modes.

8.7.7 Agent state space coverage and exploration efficiency

Figure 8.10 depicts the exploration of state space *S* defined in (8.1) and (8.2). Using a latent-space projection, the scatter plot shows the distribution of states s_t visited during training. The numerical spread across both axes, ranging from −3 to +3, indicates broad exploration across diverse operating conditions, including scenarios with high computational load C_t and low bandwidth Bt. Dense clustering near the centre reflects mastery of routine conditions, while sparsely populated outliers correspond to rare, high-criticality Q_t events, demonstrating robust exploration and policy generalisation.

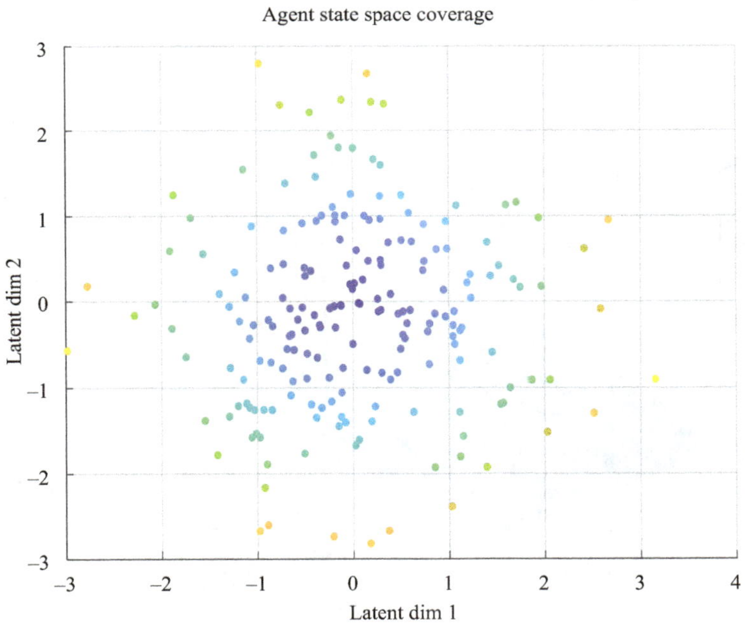

Figure 8.10 Agent state space coverage visualises a scatter plot of a latent-space projection showing the distribution of agent states visited during training; dense clustering in the centre indicates routine operating conditions, while sparse outliers represent rare, high-criticality contexts, demonstrating broad state-space exploration and policy generalisation

8.7.8 Learned value surface and action-value optimisation

Figure 8.11 presents the learned action-value function $Q_\theta(s, a)$ from (8.6) as a three-dimensional surface. The x-axis represents the system state, such as threat level, while the y-axis denotes the strength of cryptographic action. Quantitatively, the surface exhibits a clear peak near 0.9 where the state–action pairing is optimal. Sharp gradients away from this peak indicate regions where suboptimal actions yield low reward, *rt*. This smooth yet discriminative surface confirms that the temporal-difference learning process has converged to a stable decision landscape suitable for real-time edge deployment. Table 8.5 presents the findings by comparing the proposed adaptive DRL security method with a baseline.

Table 8.5 provides a compact, outcome-level summary of how the proposed adaptive DRL security approach compares with a static baseline across the chapter's core objectives. It reports that the learned policy achieves markedly better learning behaviour (with PPO converging best and DQN worse, yielding about a +200 per cent improvement in cumulative reward) while also improving efficiency by reducing latency impact from heavier key settings (about -80 per cent) and lowering the energy proxy through lighter suite selection (about -73.3 per cent). The table also highlights that safety is enforced through a deadline-guard mechanism

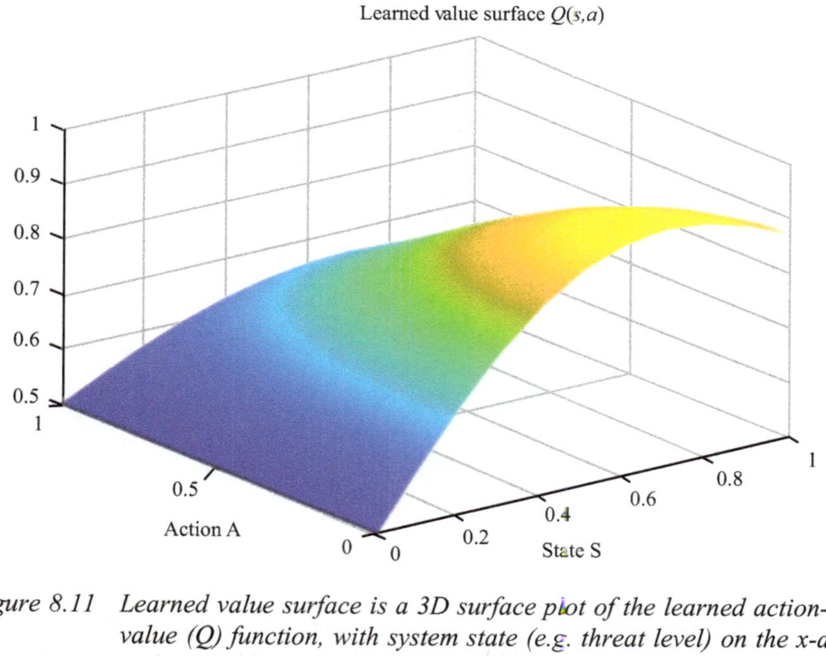

Figure 8.11 Learned value surface is a 3D surface plot of the learned action-value (Q) function, with system state (e.g. threat level) on the x-axis and cryptographic action strength on the y-axis; the surface peaks at an optimal state–action region and drops sharply elsewhere, showing a stable decision landscape learned for real-time edge deployment

Table 8.5 Summary of key results compared to baseline

Performance metrics	Proposed	Baseline	% change vs baseline	Remark
Learning convergence (cumulative reward)	PPO converges best	DQN converges the worst	+200%	Improved
Security–latency trade-off (key size impact)	Lower latency (lighter setting)	Higher latency (heavier setting)	−80%	Improved
Energy proxy (suite selection)	Lower energy (lighter suite)	Higher energy (heavier suite)	−73.3%	Improved
Safety constraint enforcement (deadline guard)	Penalty/fallback enforces the deadline	Static baseline not quantified	N/A	Improved

(penalty/fallback), indicating the controller is designed to actively avoid clinical deadline violations rather than merely optimising average performance.

8.8 Conclusions and future scope

This chapter presented a deep reinforcement learning (DRL)–enabled adaptive cryptographic framework for intelligent healthcare and Internet of Medical Things (IoMT) environments, motivated by the mismatch between static cryptographic configurations and highly dynamic clinical contexts (varying workloads, network conditions, device energy states and threat intensities). The proposed approach positions a DRL-based security policy controller at the edge to continuously provision cryptographic policies that preserve confidentiality and integrity while respecting clinical safety constraints and device limitations.

Methodologically, the adaptation problem was formulated as a Markov decision process (MDP) in which the edge controller constructs a state vector that captures operational and risk context (e.g. latency, energy proxy, bandwidth, threat score, edge compute load and QoS criticality). At each decision epoch, the controller selects a cryptographic action spanning suite choice, key size, key-rotation interval and authentication mode and receives a multi-objective reward that promotes security strength while penalising latency, energy cost, overhead and safety/QoS violations. A MATLAB-based MDP simulation environment was used to evaluate representative DRL learners (PPO, DQN and A3C).

Key findings show that PPO provides the most robust and stable healthcare-centric policy, with better convergence characteristics than A3C (higher variance) and DQN (slower convergence/underperformance). The learned policy effectively manages the security–latency trade-off and enforces clinical safety through explicit, penalty-driven constraint handling to discourage deadline violations while also enabling energy-aware adaptation by selecting lighter configurations under constrained power states and strengthening protection under elevated threat levels. Overall, the results indicate that adaptive cryptographic control at the edge can be achieved using DRL to enhance resilience without compromising patient safety or device longevity. The framework advances trustworthy digital healthcare by reducing the likelihood that resource-constrained IoMT devices must choose between strong security and timely clinical response. By enabling context-aware security that aligns with operational constraints and safety requirements, the approach supports safer remote monitoring, reduces exposure of sensitive health information and strengthens confidence among patients, clinicians and healthcare institutions in connected-care infrastructures.

The current evaluation is primarily simulation-driven, using modelled latency/energy/overhead dynamics within an MDP environment, which may not capture all real-world deployment effects (hardware heterogeneity, wireless interference and operational variability). Future work should therefore (i) validate the controller on real edge-IoMT testbeds with measured cryptographic cost and clinical traffic patterns; (ii) integrate richer, real-time threat intelligence (e.g. IDS/anomaly telemetry) as state inputs; (iii) incorporate explainability and safety assurance mechanisms suitable for clinical governance and (iv) extend to distributed/multi-edge settings and continual learning to sustain performance under evolving threats and hospital-scale dynamics.

References

[1] S. Prajapat, P. Kumar, A. K. Das and G. Muhammad, "Generative AI-enabled quantum encryption algorithm for securing IoT-based healthcare application using blockchain," in *IEEE Internet of Things Journal*, vol. 12, no. 13, pp. 24541–24551, 2025, https://doi.org/10.1109/JIOT.2025.3555159.

[2] P. Chinnasamy, P. Deepalakshmi, D. Sandeep, A. S. Ganesh, A. J. Krishna and D. D. Priya, "Enhancing healthcare record privacy through the integration of visual cryptography and diverse image encryption techniques," *2024 9th International Conference on Communication and Electronics Systems (ICCES)*, Coimbatore, India, 2024, pp. 880–885, https://doi.org/10.1109/ICCES63552.2024.10859628.

[3] K. Lata, P. Singh, S. Saini and L. R. Cenkeramaddi, "Deep learning-based brain tumor detection in privacy-preserving smart health care systems," in *IEEE Access*, vol. 12, pp. 140722–140733, 2024, https://doi.org/10.1109/ACCESS.2024.3456599.

[4] G. N. K. Murthy, G. Manjula, P. Harshitha and N. P. Achyutha, "An optimized elliptical curve cryptography based crypto-watermarking model in smart healthcare application," *2024 Second International Conference on Data Science and Information System (ICDSIS)*, Hassan, India, 2024, pp. 1–8, https://doi.org/10.1109/ICDSIS61070.2024.10594077.

[5] M. Adil, A. Ali, T. Tin Tin *et al.* "Quantum computing and the future of healthcare internet of things security: Challenges and opportunities," in *IEEE*

Internet of Things Journal, vol. 12, no. 22, pp. 46316–46346, 2025, https://doi.org/10.1109/JIOT.2025.3605040.

[6] Y. Chang, T. H. Luan, Y. Miao and Q. Chen, "Medishuffle: Auditable coin mixing scheme for internet of medical things," in *IEEE Internet of Things Journal*, vol. 12, no. 18, pp. 39088–39104, 2025, https://doi.org/10.1109/JIOT.2025.3587855.

[7] A. B. Buriro, A. Buriro, M. A. Jamshed, W. U. Khan and M. A. Yaqub, "Securing military IoT: Context-aware biometrics at the edge," in *IEEE Internet of Things Magazine*, vol. 8, no. 2, pp. 54–59, 2025, https://doi.org/10.1109/IOTM.001.2400147.

[8] H. AbdelRaouf, M. Abouyoussef, M. M. Fouda and M. I. Ibrahem, "Empowering AI-driven healthcare with secure, decentralized, and privacy-enhancing adaptive intelligence," in *IEEE Internet of Things Journal*, vol. 12, no. 22, pp. 47223–47240, 2025, https://doi.org/10.1109/JIOT.2025.3602942.

[9] S. Mondal and A. Das, "Energy-aware privacy-preserving IoMT framework for consumer electronics: A federated learning approach to remote healthcare," in *IEEE Transactions on Consumer Electronics*, vol. 71, no. 3, pp. 8317–8334, 2025, https://doi.org/10.1109/TCE.2025.3593928.

[10] Y. Jiao, M. M. Razaq, L. Peng and P.-H. Ho, "Enhancing IoMT security using large multimodal models," in *IEEE Network*, vol. 39, no. 4, pp. 112–119, 2025, https://doi.org/10.1109/MNET.2025.3545692.

[11] R. Javed, T. Abbas, T. Shahzad *et al.* "Enhancing chronic disease prediction in IoMT-enabled healthcare 5.0 using deep machine learning: Alzheimer's disease as a case study," in *IEEE Access*, vol. 13, pp. 14252–14272, 2025, https://doi.org/10.1109/ACCESS.2025.3525514.

[12] G. Nithyavani and G. Naga Raja, "A comprehensive survey on security and privacy challenges in internet of medical things applications: Deep learning and machine learning solutions, obstacles, and future directions," in *IEEE Access*, vol. 13, pp. 188955–188989, 2025, https://doi.org/10.1109/ACCESS.2025.3588489.

[13] H. Farman, Y. Shahzad, B. Jan *et al.* "IoT-aware real-time healthcare diagnostic framework for diabetes using wearable sensors through deep reinforcement learning," in *IEEE Internet of Things Journal*, vol. 12, no. 11, pp. 18183–18194, 2025, https://doi.org/10.1109/JIOT.2025.3541129.

[14] X. Tang, B. Yi, Z. Wang, M. T. Quasim, S. Basheer and B. Liang, "Deep-reinforcement-learning-based multiobjective optimization for carbon intelligent IIoT-enabled healthcare buildings," in *IEEE Internet of Things Journal*, vol. 12, no. 17, pp. 34768–34779, 2025, https://doi.org/10.1109/JIOT.2025.3579055.

[15] M. Babar, M. U. Tariq, B. Qureshi, Z. Ullah, F. Arif and Z. Khan, "An efficient and hybrid deep learning-driven model to enhance security and performance of healthcare internet of things," in *IEEE Access*, vol. 13, pp. 22931–22945, 2025, https://doi.org/10.1109/ACCESS.2025.3536638.

[16] M. M. Razaq, Y. Jiao, L. Peng and P. -H. Ho, "Deep reinforcement learning-based physical layer security framework for internet of medical things," in *IEEE Transactions on Consumer Electronics*, vol. 71, no. 2, pp. 4487–4496, 2025, https://doi.org/10.1109/TCE.2024.3521386.

[17] Q. He, Z. Feng, Z. Chen *et al.* "Low-cost data offloading strategy with deep reinforcement learning for internet of things," in *IEEE Transactions on Services Computing*, vol. 18, no. 3, pp. 1543–1556, 2025, https://doi.org/10.1109/TSC.2024.3404347.

[18] B. Ren, Y. Tang, H. Wang *et al.* "A multiagent deep reinforcement learning autonomous security management approach for internet of things," in *IEEE Internet of Things Journal*, vol. 11, no. 15, pp. 25600–25612, 2024, https://doi.org/10.1109/JIOT.2024.3368072.

[19] A. Ghubaish, Z. Yang and R. Jain, "HDRL-IDS: A hybrid deep reinforcement learning intrusion detection system for enhancing the security of medical applications in 5G networks," *2024 International Conference on Smart Applications, Communications and Networking (SmartNets)*, Harrisonburg, VA, USA, 2024, pp. 1–6, https://doi.org/10.1109/SmartNets61466.2024.10577692.

[20] L. M. Tseng, P. F. Chen, and C. Y. Wen, "Design of edge-IoMT network architecture with weight-based scheduling" *Sensors*, vol. 23, no. 20, pp. 8553, 2023. https://doi.org/10.3390/s23208553

[21] O. Salem, K. Alsubhi, A. Shaafi, M. Gheryani, A. Mehaoua and R. Boutaba, "Man-in-the-middle attack mitigation in internet of medical things," *IEEE Transactions on Industrial Informatics*, vol. 18, no. 3, pp. 2053–2062, 2022, https://doi.org/10.1109/TII.2021.3089462.

[22] A. Akinbi and P. P. Raj, "A systematic security analysis of medical internet of things (MIoT) ecosystems in threat modeling scenarios," *Frontiers in The Internet of Things*, vol. 4, Art. no. 1712430, 2025, https://doi.org/10.3389/friot.2025.1712430.

Chapter 9

SAFE-MED: safe deep-reinforcement learning for adaptive, quantum-aware cryptography in intelligent healthcare

Wasiu Lawal[1], Kazeem Bolade Adedeji[2], Sikirullai Olatunde JeJe[3] and Ibraheem Temitope Jimoh[4]

Abstract

SAFE-MED is proposed as a framework for adaptive quantum-aware cryptographic decision-making in intelligent healthcare systems. This chapter redefines cryptographic posture not as a static value or optimisation problem but as a safety-critical decision problem impacted by clinical urgency, heterogeneous device constraints, varying connectivity and shifting risk signals. SAFE-MED is not about arguing cryptographic protocols or performance optimisation; it is about determining how adaptive cryptography is bounded, supervised and forcibly corrected so that changes in operating conditions do not compromise safety and accountability. It is a decision-centric framework, with correctness measured by adherence to uncompromising clinical and system constraints. It also proposes a safety-shielded learning-and-adaptation framework in which intelligence is confined to a safe operating envelope. It considers quantum awareness as a distant ambiguity that influences current decisions, with no immediate implications for deployments. Its evaluation metrics emphasise behavioural stability, predictability and graceful degradation, foregoing safety violations in pursuit of optimality. Governance is embedded architecturally through auditability and traceability mechanisms that preserve human oversight and institutional accountability. Clear scope boundaries and principles for responsible extension are articulated to support durable adoption in regulated clinical settings.

[1]Department of Information and Communication Technology, The Federal University of Technology, Akure, Nigeria
[2]Department of Electrical and Electronics Engineering, Federal University of Technology, Akure, Nigeria
[3]Department of Physiology, Federal University of Technology, Akure, Nigeria
[4]Department of Software Engineering, Federal University of Technology, Akure, Nigeria

Keywords: Adaptive cryptography; safety-critical systems; governance and auditability; intelligent healthcare; quantum-aware security; decision-centric security; reinforcement learning under constraints

9.1 Introduction

Cryptography was long considered a cornerstone of changing healthcare information systems. Once chosen and implemented, efforts focused on adhering to hardening guidelines and conducting compliance checks. The model envisioned relatively static clinical environments, predictable connectivity and cryptographic parameters that changed slowly, if at all. Intelligent health environments have changed that model. As computation, sensing and decision-making are distributed across devices, gateways and services, cryptographic parameters can no longer be solely configuration values encoded at design time; they can instead be metrics of the systems' operational behaviour. Cryptographic choice has become a form of risk management. The changing network environment, device energy level, workload pressure and risk signals inform it. Urgent patient care, routine monitoring, background analytics and time-critical alerts are not necessarily routed through the same security controls. Cryptographic parameters may need to be adjusted as the situation evolves. The question is no longer whether changes are possible but whether adaptive changes remain safe in the face of urgency, predictability and responsibility.

Healthcare is no neutral computing substrate. It has patients, payers, fiduciaries and regulators as users; each action produces a cascade of medical, ethical and legal consequences. A cryptographic change that is perfectly acceptable for a service it secures may be dangerous if it delays ambulance response times, fragments the care team or hides the decision-maker. But if it also refuses to re-evaluate these decisions when circumstances change, it may inflict harm. Security in this context is not a state to be set or a metric to be optimised but a regime that defines the circumstances in which it will be permitted and constrained.

This cannot be adequately resolved by traditional security thinking. Secure-by-design provides strong guarantees at deployment time but offers little support for runtime adaptations in the face of extrinsic uncertainty or conflicting pressures. Left to its own devices, an intelligent controller might second-guess the optimisation logic of domains that it was one step removed from, where no error can be tolerated in the name of safety. Just as brittleness can be dangerous, so can unconstrained plasticity.

Finally, uncertainty about the threat environment further complicates this situation. The long-term viability of cryptography, including post-quantum cryptography, presents future risks that cannot be fully addressed by choices made in current deployments. Systems need to be capable of ongoing evaluation of the risks associated with degradation and premature changes. Accounting for signals of future risks in current decision-making necessitates a level of conservatism and auditability. This concerns governance as much as it does capability.

In this sense, adaptive cryptography is not a matter of optimisation but an intrusion into safety practice. The decisions must be segregated, audited and defensible. They must comply with inviolable clinical constraints and a semblance of human supremacy. Cryptography, intelligent control and healthcare governance rarely intersect in SAFE-MED safety-aware cryptographic decision the literature [1], but they do here.

This chapter is holistic in scope. It models cryptographic adaptation as a managed process that preserves safety amid contextual change rather than improving strength or efficiency. It suggests a framework for deliberation, regulation and responsibility within intelligent healthcare systems to realise aspects of safety-focused education, proactive risk awareness and system design oversight.

9.1.1 From static cryptography to intelligent adaptation

Cryptography has traditionally played a defensive role in the healthcare system. After successfully configuring and validating the selected set of mechanisms, the practitioner confidently (if occasionally) stepped through the system life cycle, assuming that clinical procedures, networks and threat models evolved only within predictable boundaries. Implementation bugs, algorithm deprecation or key management errors were the usual culprits, handled through compliance audits and update schedules.

This paradigm falls short when confronted with intelligent healthcare systems that communicate over heterogeneous, intermittent networks, use heterogeneous computing platforms and operate under diverse clinical objectives and safety protocols. What works for the background collection may be catastrophic for time-critical alarms or co-ordinated device actions [2]. Cryptographic posture is subject to pressures of operation, which may render it unsuitable, or worse, dangerous, to remain constant. As a result, there has been growing interest in adaptive mechanisms to adjust to the variance of context rather than a singular mode of operation.

When adaptation is employed, cryptographic risks become context-dependent and depend on the timing and motivation of each posture change, as well as on the strength of the algorithm and the correctness of its implementation. In healthcare, timing and safety go hand in hand. The overhead of adaptation that is acceptable when a patient is being regularly observed may become unacceptable when the same patient is in an emergency. Or conversely, relaxing controls when systems are stressed may leave the most vulnerable systems that care for patients inadequately protected. It places accountability and predictability at the forefront. Whereas in static systems, accountability is somewhat automatic as long as standards are followed, in adaptive systems, it is dynamic. Changes in the cryptographic posture need to be traceable, justifiable and understandable from a clinical perspective [3]. Likewise, clinicians and sysadmins need to have confidence that similar situations will elicit similar behaviours, even when adaptive changes occur beneath the surface.

The events show the folly of assuming that ‘secure-by-design’ is sufficient. Design-time assurances are important but far from sufficient when the cryptographic state can be changed at runtime. There are compromises between timing, reliability and exposure that dynamic adaptation necessarily entails, which cannot be judged solely by the deployment logic. In the absence of a clear chain of authority, runtime changes may override parameters that have undergone thorough scrutiny, potentially resulting in significant deviations from the system's security posture and behaviour.

Cryptography is not our saviour but a safety function in the intelligent healthcare system. Adaptation is not the athlete's performance but a managed process guided by clinical priorities and under clinical control. This is a move towards frameworks that enable intelligent responses while still preserving predictability, accountability and the established tenets of healthcare safety.

9.1.2 Why existing approaches are insufficient

Modern healthcare security is practised in three loosely connected established traditions: cryptographic engineering, machine learning and cybersecurity governance. Each has its own domain of protection, but none of these addresses the risks that arise when cryptographic posture is adaptive and under intelligent control.

Cryptographic engineering seeks correctness and robustness in the mechanisms chosen at design time. If these mechanisms have been checked and validated, it is safe to rely on them and to revisit them periodically. This breaks down when the cryptographic state changes during execution. Design-time assurances do not guide when to adapt the system, how often to adapt it or which restrictions to honour as state changes occur. In the absence of explicit control over these decisions, even sound cryptography can be unsafe if employed at the wrong time or for the wrong purpose.

The listed shortcomings are also common in machine learning security approaches. The primary purpose of learning is to improve detection, speed up reactions or increase efficiency. Thus, evaluation and design focus heavily on performance metrics such as accuracy, rewards, throughput and others. Overemphasis on safety in medical applications can lead to unintended consequences. Optimisation procedures might consider constraint violations as acceptable trade-offs with other goals. Average-case improvements are often prioritised over rare-case improvements, whereas rare, devastating failures are the primary concern of safety engineering. Learning and adaptation without safety limits may exacerbate existing instability rather than mitigating it [4].

Governance of security in healthcare will include, correctly, a strong emphasis on accountability and auditability, on who is professionally responsible for what must be protected and on when it goes wrong. Policy-level governance will not determine how adaptive cryptographic decisions should be structured, ensuring accountability if those decisions are automated. Accountability may be lost if the logic, points of intervention and traceability for those adaptive decisions are not embedded in the system design at the outset. Governance as documentation will struggle to govern behaviour occurring dynamically at runtime [5].

None of these methods operates from a unified perspective towards a clearly defined final objective. Cryptography is treated as configuration, learning as optimisation and governance as external supervision. Yet as cryptographic posture becomes adaptive, these separations begin to collapse. Given that decision-making in clinical environments is driven by imperatives of care, uncertainty and time pressure, it requires roles for safety, boundaries, learning and governance that are integrated into the individual rather than appended to it. Otherwise, there is a danger of creating situations that are safely secure but unsafely safe. It is this structural risk that the chapter addresses.

9.1.3 Chapter contributions

In this chapter, we present SAFE-MED, a reference framework for safety-first adaptive cryptography in smart healthcare systems. We do not aim to contribute new catalogues of protocols, optimisation studies or deployment manuals. Rather, we focus on how to ensure that the adaptation of cryptography continues to yield safety, accountability and predictability as operating conditions change dynamically.

The contributions are summarised as follows:

- SAFE-MED reframes adaptive cryptography as a safety-governed decision process, treating cryptographic posture as constrained, time-bounded choices rather than a static configuration task within intelligent healthcare environments.
- The chapter develops a safety-enveloped learning paradigm in which deep-reinforcement learning operates strictly within enforceable clinical and system-level constraints, with learning explicitly subordinated to runtime monitoring and conservative fallback mechanisms.
- Quantum awareness is interpreted as a forward-looking risk signal that shapes adaptive decisions under uncertainty without serving as a post-quantum deployment prescription or migration road map.
- A perspective on behavioural stability evaluation is provided, focusing on predictability, controlled degradation and the avoidance of safety violations rather than on performance optimisation or benchmark maximisation.
- Governance and auditability are architectural artefacts, with traceability of decisions, the ability to override and routes to accountability embedded within the system's architecture to protect and maintain the accountability of humans and institutions.

9.1.4 Boundary statement and scope control

To maintain focus and clarity, this chapter does not attempt protocol benchmarking, cipher-suite comparison, device-to-cipher mapping or implementation-level post-quantum guidance. It does not provide migration road maps or performance-optimisation recipes. Its contribution is the reference framework that makes adaptive, quantum-aware cryptographic decision-making safe, governed and accountable in intelligent healthcare systems.

9.1.5 Chapter organisation

The remainder of this chapter is organised as follows.

Section 9.2 situates SAFE-MED within the operational realities of intelligent healthcare environments and establishes the clinical and system constraints under which adaptive cryptographic decisions occur. Section 9.3 formalises cryptography as a governed decision process and introduces quantum-aware risk considerations. Section 9.4 presents the safety-aware adaptation model and the enforcement of non-negotiable safety constraints. Section 9.5 describes the SAFE-MED learning architecture and its safety-shielded control mechanisms. Section 9.6 outlines the evaluation philosophy and scenario design used to assess adaptive behaviour. Section 9.7 analyses safety–performance trade-offs and interprets system behaviour under stress. Section 9.8 addresses governance, auditability and human authority as architectural properties. Finally, Section 9.9 discusses limitations, scope boundaries and responsible directions for future extension.

9.2 Operational context: intelligent healthcare environments

In the modern healthcare context, adaptive cryptographic decision-making is not abstract; it is constructed from the functioning of the medical environment. Intensive care systems are integrated across wards, clinics and care networks and sensors, computation and communication are tightly connected to patient care. The practical constraints imposed on them, such as time pressure, heterogeneous device capabilities, variable connectivity and defined lines of clinical authority, will guide the selection and potential use of cryptography in these environments [23]. This is important to know, as it sets boundaries for any safety-aware adaptation framework. This section thus sets the scene for the SAFE-MED system's operating environment. It does not seek to establish any novel attributes or to make claims for assessment but rather to anchor the everyday context of intelligent healthcare systems and their inherent decision constraints. The association of intelligence at the ward level with clinically distinct data channels and the automation-versus-human-authority frontier is made explicit to provide a concrete environment for the decision models and safety mechanisms that follow in the subsequent sections.

9.2.1 Clinical environments as decision-constrained systems

Intelligent healthcare environments are best understood not as neutral computational platforms but as decision-constrained systems in which technical actions are continuously shaped by clinical priorities, safety obligations and institutional authority [22]. During ward-scale surveillance, diagnosis and intervention, these activities are shared among sensors, bedside equipment, gateways and back-end services. The decisions made within this context of access, transmission or protection are therefore not separate from the circumstances under which care is provided.

They are highly heterogeneous. As a consequence, clinical wards incorporate devices that differ greatly in capabilities, lifetimes and risk profiles, ranging from resource-constrained sensors and wearables to imaging systems, infusion pumps and clinical information servers. Like elasticity, resizability is unevenly distributed across mobility, interference, maintenance cycles and emergency conditions. As a result, cryptographic decisions cannot rely on sufficient computational capacity or network stability; they must respect the weakest and most safety-critical components of the system and remain responsive to changing operational conditions [6]. Multiple data-flow priorities are equally important. Routine telemetry, archive records and administrative data are subordinate to time-critical alarm signals and low-latency-tolerant clinician-initiated queries. In emergencies, latency introduced by security configurations unsuitable for that emergency is liable to have a direct clinical impact. This hierarchy of priorities means that cryptographic posture is not about strength but about timing: the arm and hand of protection must abide by clinical urgency and degrade gracefully as operating margins narrow [7]. Figure 9.1 illustrates the ward-scale organisation of intelligent healthcare systems and the priority of clinical data flows under latency constraints.

Figure 9.1 establishes the context for the remainder of this chapter and need not be reintroduced each time it is referenced. Automation marks the boundary of clinical authority. Smart technologies may analyse information, modify settings or recommend courses of action, but it is the human clinicians and their organisations who bear ultimate responsibility. It is recurrent and predictable that decisions about data availability, integrity or confidentiality should remain predictable, traceable and overrideable. All adaptive processes introduced into this environment must respect those boundaries, enabling automation to support rather than replace professional judgement.

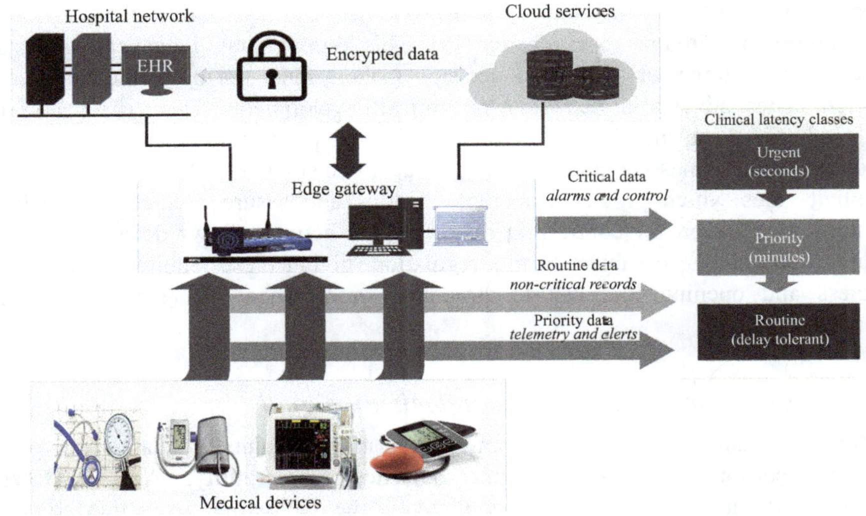

Figure 9.1 Ward-scale intelligent healthcare architecture illustrating prioritised clinical data flows and latency classes

9.3 Cryptography as a governed decision process

Adaptive cryptography in intelligent healthcare cannot be seen as a matter of selecting algorithms or adjusting parameters. Once cryptographic posture is allowed to change at runtime, it becomes the result of uncertainty driven by time constraints, safety and institutional accountability. The recasting of cryptography in this section presents it as a process of regulated decision, governed by its behaviour under change rather than nominal strength.

The chapter adds new layers of security engineering abstractions and aligns adaptive protection with safety-oriented system design. The decisions are taken in context, influenced by operational signals and bound by non-negotiable constraints. These sections reflect this reframing in the following directions: examining cryptographic actions as time-bound choices and adopting quantum-aware risk as a decision signal that shapes, but does not dictate, adaptive behaviour.

9.3.1 Cryptographic actions as time-bound choices

In intelligent healthcare environments, cryptographic actions are rarely timeless. Choices about protection mechanisms – such as when to strengthen, relax or reconfigure security controls – are taken under conditions of incomplete information and often within narrow temporal windows. This decision is driven by varying network conditions, device availability and clinical urgency, which can change faster than traditional security management practices can accommodate. The safety consequences of delay and misalignment become especially critical when cryptography is sensitive to timing. A cryptographic state may be acceptable, but arriving too late or holding too firm in an emergency is as dangerous as being weak. For safety, the issue is not whether a condition is optimal per se but whether it can be applied or adjusted to maintain system properties such as time, availability and traceability as circumstances change [8]. This reimagining, however, suggests uncertainty. Human or automated decision makers must not be aware of future system states, adversarial behaviour or clinical developments. There is a need for leadership in these situations. Besides explicit rules preventing dangerous actions, predictable outcomes and conservative fallback, decisions must be reinforced with explicit rules which protect vulnerable actions and assure predictable results. Instead, this option for controlling cryptographic actions allows decisions about which transactions to subject to this regulation, giving these requirements explicitness and opening the way to structured enforcement rather than ad hoc adaptation.

9.3.2 Quantum-aware risk as a decision signal

Quantum considerations provide another kind of uncertainty in adaptive cryptographic decision-making, but they must be articulated carefully. SAFE-MED has not defined quantum awareness as required for the deployment of certain cryptographic systems, nor has it defined it as a template for immediate migration. Rather, it serves to predict the cryptographic choices in future contexts with a focus on evolving threat assumptions.

During the post-quantum period, timelines, adversarial capabilities and practical feasibility remain uncertain [24]. These uncertainties should not prompt a sudden or rigid response in safety-critical medical settings where stability and predictability are essential [9]. SAFE-MED offers adaptive decisions sensitive to long-term risk, without being anchored to speculative futures or uncertainty, in its simulation of quantum awareness.

This is because quantum-aware risk affects decision boundaries rather than decisions themselves. While a long-horizon cryptographic risk becomes variable, adaptive action can be limited to short-lived protections or to exposure to configurations that might become fragile in the future. This effect is perhaps unnaturally indirect because the framework defines what the system must be and what it cannot be. This safeguards current safety burdens and future awareness [10].

The two quantum-aware and quantum-deploying systems are, thus, completely separate. Quantum-aware systems anticipate the future and incorporate it into governance and constraint formulation. Quantum-deploying systems, by contrast, employ specific mechanisms and migration strategies. This latter position is explicitly adopted by SAFE-MED, ensuring that adaptive cryptographic behaviour does not emerge from today's operational realities but can be preserved in a future threat landscape.

This approach avoids complacency in expectations with action. Quantum awareness becomes more stable and better able to support adaptive decision-making, thereby avoiding disruption or overreaction, as evidenced by the chapter's broad focus on safety, governance and accountability. Figure 9.2 shows the role of quantum-aware risk in the decision-making process within SAFE-MED.

Figure 9.2 serves as a conceptual anchor for understanding how operational observations and risk signals, including quantum-aware considerations, are incorporated into constrained cryptographic decision-making without prescribing specific protocols or optimisation strategies.

9.4 Formalising safety-aware adaptation

The preceding sections established the operational context and conceptual framing for treating cryptography in intelligent healthcare as a governed decision process. To make this framing usable for analysis, inspection and system design, a minimal level of formal structure is required. The aim is not to formalise cryptographic primitives or learning algorithms, but to formalise decision safety, that is, the conditions under which adaptive choices are permitted, constrained and evaluated in clinical environments.

Formalisation is introduced for two reasons. First, adaptive behaviour in safety-critical systems cannot be assessed purely through narrative description; it requires explicit identification of system states, admissible actions and acceptable or unsafe outcomes. Second, without a specific vocabulary of

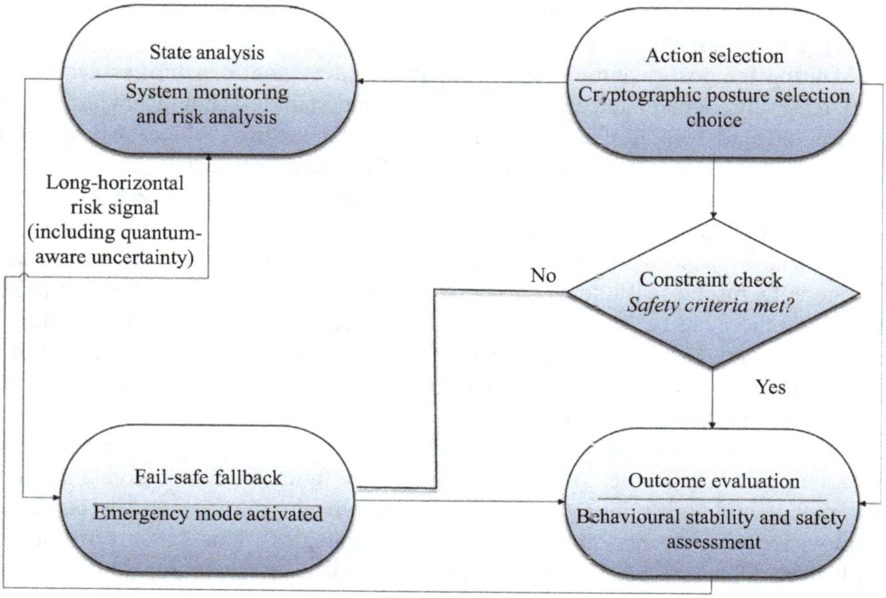

Figure 9.2 SAFE-MED safety-aware cryptographic decision loop

constraints, invariants and violations, these are all implicit and not easily inspectable, enforceable or auditable. The level of formalism employed is minimal and declarative, intended to support governance and reasoning rather than to dictate algorithms or performance objectives.

Thus, this section is a structural break within this chapter. While the previous section described why the adaptation should be safety-based and how the decision loops should reflect contextual and long-horizon risk signals, we now consider what can evolve through learning or contextual change and what must remain fixed as non-negotiable safety constraints. The next several subsections constitute a reference model for this dichotomy.

An abstract description of each decision state, allowable actions and outcomes, expressed in terms of clinical priorities (as opposed to cryptographic ones), is provided in Section 9.4.1. Safety invariants, constraints enforcement, safety invariant violation remediation and the learning system's relationship with invariants are discussed in Sections 9.4.2–9.4.6. These subsections present a structure for safety-oriented adaptive cryptography that preserves predictability, accountability and clinical authority.

9.4.1 Decision states, constraints and outcomes

To attain a coherent understanding of safety-aware adaptation, adaptive cryptography should be described in terms of decision states, admissible actions and outcomes. In this subsection, we introduce a minimal formal framework to help inspection and governance without expanding into algorithmic detail. We do not seek to model cryptographic primitives, but only to clarify what is being decided under the constraints and how those decisions are evaluated in a clinical context.

9.4.2 Decision state representation

A decision state is a snapshot of the conditions under which a cryptographic decision was made. It is a collection of information relevant to safety and accountability, rather than implementation details. This information can include whether the device was in routine operation or responding to an emergency, system health status, device and connectivity status and clinical priorities. The state is abstract in that it specifies neither how safety is guaranteed at the point of decision nor how the information is sensed or computed. Such abstractions allow the behaviour to be interpreted uniformly across different deployments without requiring a particular platform or protocol.

9.4.3 Constrained action space

In a particular state, only a limited set of actions is allowed. Actions are not low-level configurations of cryptographic parameters, but rather changes or confirmations of the cryptographic posture within the established safety bounds. Clinical and organisational imperatives define the space of allowed actions. The design prohibits any action that may endanger availability in time-critical care, breach policy or endanger traceability. This notion of a constrained action space introduces a new principle: not all adaptations within the realm of possibility are acceptable in a medical context – safety bounds must inform decisions before any learning or contextual adaptation [11].

9.4.4 The results and acceptance

It is further necessary to examine the potential outcomes and determine their influence on system behaviour and clinical safety. An outcome is acceptable if it maintains the vital timing, reliability and accountability properties of a state. Unacceptable outcomes include ineffective delays, obscured responsibility and unsafe recovery conditions. The binary nature of this is intentional: it shifts the basis of our evaluation away from graded optimisation and towards explicit safety assessment, as required by both adaptive cryptography and due diligence in the engineering of safety-critical systems [12].

By identifying decision states, restricted actions and the acceptance of resulting outcomes, the SAFE-MED technique provides a basis for analysing and controlling adaptation. It provides a framing of what should be known, what may be chosen and what should never happen in an adaptation. From this, we can prepare the specification of safety invariants and enforcement mechanisms to maintain these boundaries for changing operational conditions.

9.4.5 Safety invariants

For adaptive cryptography to be safely utilised in clinical systems, safety must be explicitly defined, inspectable and enforceable. This chapter proposes safety invariants that must govern all adaptive behaviour under any context, learning history or belief in its benefits. Policies or guidelines can be flexibly interpreted, but invariants impose hard limits on what can ever be allowed.

9.4.6 Operational role of safety invariants

Safety invariants specify what must be preserved by our adaptive cryptographic decisions. In the intelligent healthcare context, the invariants derive from our clinical duties and not from any computational standpoint. For instance, these may include the maximum tolerable latency for time-critical data, guaranteed availability under emergency conditions and full traceability of all security-relevant decisions. Note that these invariants are independent of the underlying cryptography. They describe what physical residue is meaningful to protect, not how it is protected. This allows the invariants to persist as technology and threat models evolve [13].

Conceptually, the invariants carve up the space of potential system behaviours into acceptable and unacceptable spaces. In adaptive decisions, we are only permitted to produce effects within the acceptable space, going beyond this is not allowed, even if it would yield greater efficiency. This places safety-aware adaptation firmly within the tradition of safety-critical systems engineering, where invariants are used to anchor certification, monitoring and accountability of systems [14].

9.4.7 Consistent enforcement and violation handling

This is not enough to define invariants when they are actively enforced. Invariant situations before execution may need to be probed during the decision phase using a mechanism that checks whether it is possible to take the adaptive step while still maintaining the invariants. If an action violates an invariant, it is either rejected or replaced with a conservative fallback action. This provides an example of constraint enforcement serving as a gatekeeper between adaptive intelligence and the system's function.

Violation is also important. When invariants are violated – for instance, due to unforeseen conditions, failing components or late-arriving information – the system must contain the consequences and then restore safe operational conditions. It should accomplish containment through predictability and recoverability, not opportunistic adaptation. Rather than 'learn by failure', the system returns to previously safe and known configurations and logs the occurrence, so it can be audited and scrutinised to prevent future enhancements from being blocked and to ensure trust and accountability in the patient's care without stifling them [15].

Learning was subordinated to invariant preservation. Similarly, the learning mechanisms in SAFE-MED are subordinated to safety invariants. Learning may help determine which action is optimal among the allowed actions, but it cannot expand the action space or relax the determinants. This runs counter to the popular conception that safety emerges from trained models. Safety is structured; it is assured by design and only learns from that assurance. This relationship can be expressed declaratively. Let *St* denote the decision state at time t, A_t the set of admissible actions and I, the set of safety invariants. Adaptive choice is constrained such that:

$$A_t = \{a | \forall \mathrm{i} \in \mathrm{I}, \mathrm{i}(S_t, a) = true\} \quad (9.1)$$

and any outcome O_t resulting from action $a \in A_t$ must satisfy:

$$O_t \in O_{safe} \quad (9.2)$$

where O_{safe} a set of outcomes with an invariant preservation. These expressions are not optimisation targets; they merely describe the conditions under which adaptation may be allowed.

By defining safety invariants and applying them, SAFE-MED controls the adaptation process. Enforcing constraints offers a flexible mechanism for adaptation, ensuring that cryptographic decisions are reliable, auditable and attributable to clinical authority in the long term. With this formal grounding, adaptive intelligence can operate in concert without continual human oversight or remediation after the fact.

9.5 SAFE-MED learning architecture: intelligence under control

The learning architecture of SAFE-MED is composed around one organising principle: intelligence must be subservient to safety, governance and clinical authority. Whereas adaptive systems place learning at the centre of control, SAFE-MED treats learning as a limited set of skills to inform decisions rather than to influence them. This distinction is especially important in healthcare settings, where unpredictable behaviour – even if statistically justified – can devalue trust, accountability and patient safety.

The architecture is unique and exclusive in the field of safety enforcement. All proposed adaptations, whether driven by learning or triggered by external events, are subject to safety controls that determine eligibility for the invariants and constraints in Section 9.4.4. This mediation also ensures that learning cannot be compromised by safety guarantees, even under pressure to respond quickly. It is the result of a structurally avoidable architecture that prevents unsafe behaviour from the outset rather than remedying it afterwards.

Thus, the architecture prevents a clear distinction among the three aims: consideration, context interpretation and action execution. Learning components only operate at the first layer. They evaluate existing operational patterns, such as recurring workload changes, degradation trends and new threat signals and generate candidate adaptations. These proposals are advisory. They do not change cryptographic posture or system behaviour.

Lastly, SAFE-MED's focus on runtime visibility and intervention is more important than runtime visibility alone. The architecture is instrumented to show not only system state, but also decision rationale, constraint assessment and rejection. This is transparent and enables clinicians and system operators to understand which adaptations are permitted or blocked and to make objective, automated control decisions. Intervention points are explicitly added, enabling authorised personnel to override or suspend learning-driven recommendations when clinical judgement or institutional policy requires it.

The architecture is important to note that human power remains an architectural property as a first-class design property. Automation is free to help, prioritise and suggest, but it has no unilateral control over decisions that affect availability, integrity or confidentiality. As this hierarchy is embedded in the system structure, SAFE-MED avoids the gradual degradation of control that would occur if adaptive systems evolve in a self-centred manner over time. So, SAFE-MED demonstrates how learning can be integrated into adaptive cryptographic systems without altering predictability or governance, not as an afterthought but as an architectural commitment to intelligence. Section 9.5.1 outlines this design through the specific mechanisms for safety-shielded learning, detailing the use of unsafe behaviour before it is transferred into clinical practice.

9.5.1 Safety-shielded learning design

SAFE-MED uses safety-shielded learning, ensuring adaptive intelligence is not subject to direct authority over cryptographic posture. Learning may propose changes, but it cannot act on them. It prevents a category of unsafe behaviours in which learned policies silently alter protection mechanisms in pursuit of transient signals or incomplete information. In this design, the learning component is limited to context interpretation. It takes data on workload, device availability, communication reliability and longer-horizon risk indicators and creates candidate adaptations suitable for presentation today as a given situation. These suggestions are not orders to be carried out but hypotheses of acceptability. Learning cannot override safety protocols, nor reverse the intended functioning of the system in the building process. Safety enforcement is added as another layer that verifies all intended actions against hard invariants before execution [25]. This layer serves as a safeguard, intercepting outputs from the learning process and verifying their admissibility before allowing the changes to propagate to operational components.

For example, a proposed increase in latency beyond clinically acceptable thresholds, reduction in critical care availability or degradation in traceability will be prevented or mitigated by conservative defaults. This approach ensures that unsafe behaviour is prevented before execution rather than mitigated post-execution.

The monitoring is also continuous regarding system conditions and decision-making processes. It does not monitor satisfaction of constraints but rather closeness to invariant boundaries. The shield tightens eligibility and reduces the number of actions learning may take when operating margins narrow under clinical pressure, resource constraints or the accumulation of uncertainty. This prevents aggressive adaptation in the earliest, most critical times for systems most vulnerable to collapse.

Intervention points are well-exposed to defend humanity's dominion. Clinicians and licensed operators can inspect proposals for change, scrutinise rejection justifications, suspend transitory intervals or impose draconian constraints. This retains responsibility for judgements of safety, which remain observable and accountable. Automation is a servant to awareness and reaction, in no way supplanting human discretion or organisational discipline. Figure 9.3 illustrates this safety-buffered reinforcement learner for cryptographic orchestration. The disjoint of learning, safety and execution is highlighted, along with what is caught – unauthorised modification, unbounded latency dilation and loss of accountability. Elements are understood by what failure they defeat rather than the algorithm they execute, mirroring the chapter's theme of governed conduct over optimisation.

With this design, SAFE-MED demonstrates that learning in adaptive cryptographic systems is possible without sacrificing predictability or controllability. The intelligence is allowed to contribute to the decision process, albeit within a framework that enables the enforcement of safety, the feasibility of oversight and the preservation of human authority.

Figure 9.3 provides the architectural anchor for SAFE-MED's safety-shielded learning design, showing how learning-generated action proposals are mediated by independent constraint enforcement, runtime monitoring and operator authority so that unsafe adaptation is blocked or contained before it can affect clinical operation.

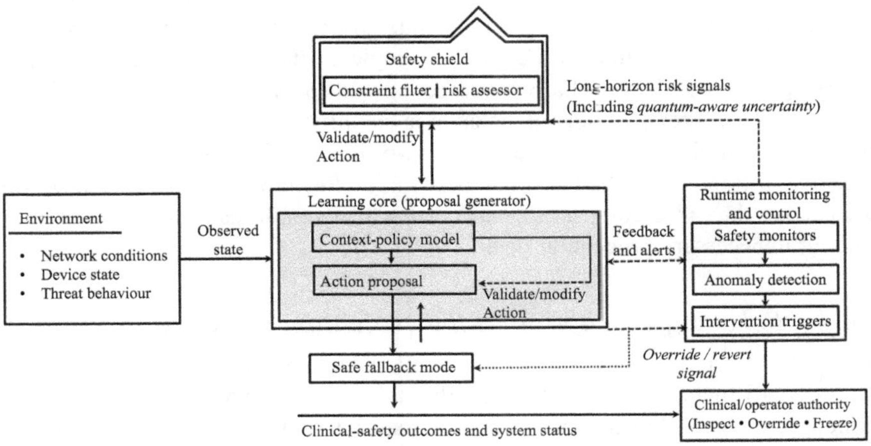

Figure 9.3 Safety-shielded learning architecture for governed cryptographic orchestration in SAFE-MED

9.6 Evaluation philosophy and scenario design

Evaluation of adaptive cryptographic systems is typically described in terms of performance improvement, efficiency gain or optimisation relative to a static baseline. This is ill-suited to intelligent healthcare settings, where the consequences of unstable or unexpected adaptation are likely to far outweigh marginal improvements in cryptographic strength or performance. Evaluation of safety-critical clinical systems must prioritise behavioural certainty over optimisation success [26]. I adopt the philosophy of evaluation that views correct behaviour in terms of performance under constraints, pressure and uncertainty. It is not of primary concern whether optimising adaptive cryptography can achieve better performance in the generic sense but whether it can be more predictable, more bounded and more controllable under changing circumstances. Evaluation is concerned with revealing unsafe transitions, runaway or erratic behaviour that may undermine clinical confidence and safety.

Healthcare is delivered in complex and ever-changing environments. To obtain meaningful evaluations, it is necessary to assess an application in the context of realistic deployment pressures, including network performance, device diversity and the occurrence of critical clinical events, rather than through abstract tests or performance-optimisation exercises. Such pressures can be developed through scenario-based reasoning.

Towards this end, a scenario-based assessment method for behavioural stability, graceful degradation and violation of invariants is presented in this section. Instead of relying on comparison benchmarks or performance curves, this method examines SAFE-MED's behaviour under shrinking safety margins or increasing uncertainty. The method aligns with the safety-centric decision model presented earlier in the paper.

In the remainder of this section, we first explain the concept of behaviour stability in greater depth and then demonstrate how it can be realised using a simple scenario matrix that captures the essential logic of the evaluation without unnecessary procedural detail. The two subsections do so as follows: In Section 9.6.1, we present the literal notion of behaviour stability as the key objective of the evaluation and clarify the nuances of trustworthiness of predictability and adherence to constraints over their absolute best values. Section 9.6.2 operationalises this perspective through a compact scenario matrix that simplifies evaluation reasoning without unnecessary procedural detail.

9.6.1 Behavioural stability as the evaluation objective

In adaptive cryptographic systems, conventional notions of correctness are often tied to performance metrics such as throughput, computational efficiency and responsiveness to change. These metrics may be useful in non-critical environments, but they are insufficient in intelligent healthcare systems, where instability or poorly timed adaptation can carry direct safety consequences. Correct behaviour means that the system remains predictable, bounded and controllable in the face of changed conditions. The main objective of assessment is now behavioural stability. A behaviourally stable system will be able to cope with changes in context and safety without going off the rails into the unexpected, even under stress and uncertainty. But stability does not equate to inflexibility; it enables the system to adapt without breaking the rules, causing shock-effect alterations, or leading to unsafe decision trajectories. The question, from an evaluative perspective, is not whether the system can adapt cryptographic decisions to maximise a given metric but whether it does so in ways that clinicians and operators can reasonably predict and understand.

Behavioural stability includes controlled degradation. Clinical environments, for example, might suddenly find themselves in situations of resource strain, network disruption or crisis modes of operation. A safe system should degrade gracefully, maintaining its essential functions and core safety properties. We wish to see how adaptive cryptography mechanisms behave as operating margins are squeezed, whether they tighten their protections appropriately, enter fallback modes when they should and respond to the recovery of normal conditions.

Avoidance of safety violations is crucial. The behavioural assessment aims to detect violations of invariants, such as latency guarantees for urgent data, accountability or decision traceability. A system that never violates these, tolerating more rigid behaviour or suboptimal performance, may be preferable in the healthcare setting compared to one that achieves higher nominal performance but exhibits instability under stress. This aligns well with the long-established practice of safety engineering, where the prevention of catastrophic failure is a more salient metric of correctness than optimising average-case performance.

This moves the assessment focus away from optimisation-oriented metrics and into a safety-oriented space. Predictability, graceful degradation and violation avoidance are all part of the assessment criteria consistent with the safety-oriented decision model previously presented. The next subsection elaborates on this with a scenario matrix designed to assess behavioural robustness without introducing excessive complexity.

9.6.2 Scenario-driven assessment

This chapter utilises scenario-based evaluation to consider behavioural robustness in the absence of performance-driven experimentation. Scenarios provide a means to subject adaptive cryptographic systems to plausible real-world stresses in a controlled, understandable evaluative framework. Rather than exhaustively generating test cases or setting arbitrary metrics, the custom-designed scenarios aim to reflect the critical decision boundaries relevant to smart healthcare.

Each scenario combines a clinical scenario, operational stressors, risk signals and the required SAFE-MED action. We aim to observe how the system responds when conditions or priorities change or uncertainty increases, whether adaptation continues as expected, whether degradation proceeds as expected and whether the safety invariants continue to hold under pressure.

To avoid lengthy 'if-then' logic in the narrative, the assessment logic is captured in a compact matrix. This allows readers to examine the logic for each stress and

Table 9.1 Scenario design matrix for ward-scale cryptographic stress testing

Scenario context	Primary stressor	Risk signal emphasis	Expected SAFE-MED behaviour	Safety focus
Routine clinical operation	Stable workload, normal connectivity	Low short- and long-horizon risk	Conservative maintenance of cryptographic posture	Predictability
Clinical escalation	Sudden increase in urgent data flows	Time-critical latency risk	Restriction of adaptation; prioritisation of availability over adaptability	Violation avoidance
Network degradation	Intermittent connectivity, packet loss	Reliability uncertainty	Controlled degradation with safe fallback activation	Controlled degradation
Resource contention	Device or gateway saturation	Operational margin narrowing	Tightened admissibility constraints	Stability under stress
Elevated future-risk awareness	Long-horizon uncertainty signals (including quantum-aware risk)	Forward-looking uncertainty signal	Conservative bias without protocol change	Governance consistency

response condition. It also makes explicit what is deemed acceptable behaviour in each case, highlighting which transitions are considered safe and which are unsafe.

Table 9.1 summarises the SAFE-MED scenario design for the large-scale evaluation. These scenarios are illustrative rather than exhaustive and show how behavioural stability can be evaluated systematically without recourse to any protocol-level comparison or optimisation metric.

This scenario matrix aligns assessment with SAFE-MED's safety-governed design philosophy. Not focusing on peak performance or aggressive adaptation, assessment logic points to limits of behaviour, stress responses and boundary preservation. The table shows that correctness is maintained through predictable behaviour, controlled degradation and adherence to safety invariants across a wide variety of medical conditions. Therefore, scenario-based evaluation provides a transparent, interpretable foundation for assessing the adaptive behaviour of cryptographic processes in smart health systems.

9.7 Interpreting safety–performance trade-offs

Throughout the life cycle of adaptive cryptographic systems, conflicts between safety and performance are bound to arise. In the design of traditional security or learning-enabled systems, these conflicts are usually posed as an optimisation problem: to find an optimal point that maximises performance while keeping risk at an acceptable level. For intelligent healthcare systems, this is not an adequate approach since some risks cannot be traded for efficiency or responsiveness [27]. In this section, we will examine the trade-offs between safety and performance from the perspective of safety boundaries, where performance is conditional on certain limits.

Thus, in SAFE-MED, trade-offs do not furnish a curve that can be optimised or tuned but rather a location on a surface of risk that separates the otherwise implicitly understood space of allowed states of the system from states that must not be allowed. Therefore, enhancing performance must always be consistent with these safety invariants. Crossing them will not constitute a sliding scale of optimality; it will constitute a breach. As a result, achieving balance is no longer a matter of tuning for the beneficial but is recognising and respecting what is prohibitive.

Here, emphasis is on specifying the forbidden regions of operation. Certain state-adaptive action combinations are forbidden because they compromise clinical priorities, accountability or behavioural control. SAFE-MED is not looking for an optimal point of operation; it is avoiding unsafe regions. Evaluation at this stage is therefore more a matter of verifying that adaptive behaviour remains within bounds as it adapts than of verifying how close it gets to some theoretical optimum.

This interpretation also explains the interactions between safety and performance under stress. The safe performance envelope may be reduced when the system is under acute operational stress – such as emergencies, resource depletion or intense uncertainty. In these situations, a conservative approach or lack of adaptability should not be regarded as a symptom of frailty but rather as a form of safety. An interpretation of trade-offs that safeguards the concept of failure excludes ascribing failure when safety is degraded, reinforcing predictability and order.

Hence, SAFE-MED is not about tuning, compromising or optimising towards the best from within a safety–performance continuum. It stays within safety limits. Performance considerations may inform the selection among permissible actions, but they do not justify boundary violations. This interpretation remains consistent with the architectural and formal commitments established earlier in the chapter and ensures that evaluation outcomes are read in alignment with system intent.

Section 9.7.1 extends this interpretation by conceptualising trade-offs as risk surfaces and illustrating how safety boundaries constrain adaptive cryptographic action. The emphasis remains on identifying permissible regions of behaviour rather than on maximising adaptive intensity.

9.7.1 Trade-offs as risk surfaces

For adaptive cryptographic systems, trade-offs are often discussed as balances to be tuned or optimised. SAFE-MED rejects this conception. Safety experts are concerned not with performance peaking but with where the operation becomes unacceptable. Therefore, trade-offs are seen as risk surfaces that define acceptable behaviour in areas that will never be entered, regardless of performance incentives.

In contrast to the scalar improvement approach to risk, taking trade-offs at risk allows us to view risk from a boundary perspective. Each point on the surface represents the balance among security posture, system responsiveness and operational stress. Some points on the surface present combinations that can be tolerated without compromising the safety invariants, including bounded latency, availability under emergency load and decision traceability. Others are not tolerated because they violate these safety invariants. Navigating these boundaries is the fundamental challenge in finding safe system adaptation and does not lead to an optimal route.

To this end, adaptive operation has been particularly critical from this perspective. The systems can move across a surface, either dynamically in response to load changes, partial failures or emergent clinical urgency. SAFE-MED checks whether movement occurs within safe areas and whether cross-border transitions are prevented or restricted. Correct behaviour defines not only a preference for staying within the admissible territory but also a conservative degradation under small margins, not improved by any measure.

The risk surface interpretation also clarifies the role constraints presented earlier in the chapter. Safety invariants are complex contours on the surface that exclude regions from operation. These adaptive mechanisms may explore the remaining space, but they are structurally prevented from crossing into dangerous space. This, and not the retrospective premise of performance, makes safety properties inspectable and enforceable under interpretation. This interpretation can be seen in Figure 9.4, which lays safety–performance interaction at the core of a defined operating environment that is governed by non-negotiable safety constraints.

Figure 9.4 represents trade-offs as a risk surface divided into admissible and forbidden operating areas. Safety invariants define hard limits on adaptive behaviour that must be obeyed under performance incentives. There is no correct, optimal operating point or tuning direction; interpretation uses boundary avoidance, predictability and controlled degradation in areas open to exploration.

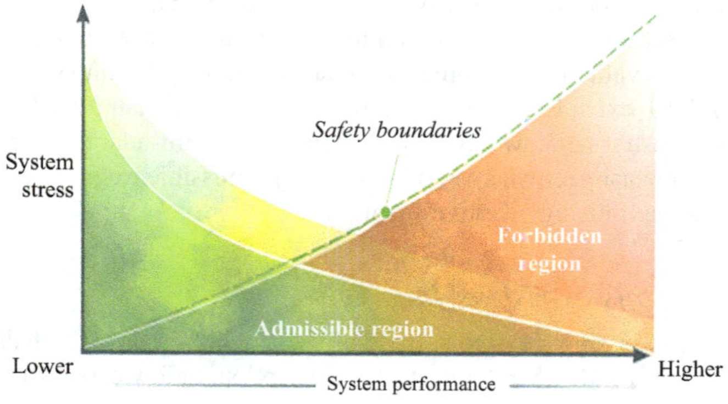

Figure 9.4 Safety–performance trade-off landscape under adaptive cryptographic orchestration

9.8 Governance, auditability and human authority

Adaptive cryptographic systems used in healthcare cannot rely solely on technical measures to ensure safety. While safety constraints and behavioural stability have been addressed in previous sections, they are insufficient without an institutional structure to provide oversight and accountability. Safety in a clinical environment is determined not only by algorithmic logic but also by the structures governing who may authorise decisions, who may intervene and how decisions are examined retrospectively.

Accordingly, SAFE-MED considers governance to be an integral design element of the system and does not regard it as an additional overlay to be bolted on post hoc. Adaptive cryptographic decisions are authorised, logged and auditable as part of the system's real-world operation. The clinical accountability of healthcare professionals and organisations cannot be delegated to technical components working in isolation. Adaptation must remain transparent, justifiable and reversible when appropriate. Auditability is critical in this regard. In the event of a change to the cryptographic posture, whether in response to stress, failure or risk, the change must be auditable at a later date. This is key to understanding the events that took place and to ensuring accountability, whether for compliance, for providing assurance or for learning. Given an adverse scenario, it should be possible to determine if, why and how the adaptive decision was made in a healthcare setting. Traceability may also build upon this foundation. It is unlikely that adaptive cryptographic behaviour will be a string of unrelated tweaks; rather, it will consist of a sequence of related decisions. As a result, governance should link the state and context of the system, the actions taken and the outcomes achieved. This way, the clinician and auditor can retrace their journey through the decision space rather than having to make sense of discrete configuration changes. This notion aligns with other elements of safety engineering, for which traceability provides accountability and trust [28].

A human authority is still in power. Even in SAFE-MED, adaptive mechanisms are not allowed to control clinical information systems unconditionally. Authorised personnel will be able to intervene, override or suspend automated recommendations whenever safety, ethical or institutional considerations require such action. It is a duty of care; automation may assist but is not a substitute for clinical judgement or management responsibility.

Through these arrangements, governance anchors adaptive cryptography within accountable institutional practice. SAFE-MED does not rely solely on policy statements; it aligns technical enforcement with structures of responsibility that can be examined and, if necessary, corrected. In this sense, governance functions as an operational safeguard embedded in the system rather than as an abstract administrative principle.

9.8.1 Governance embedded in architecture

In SAFE-MED, governance is not relegated to policy documents, codes of practice or audit and inspection performed after the event. It is built into the architecture.

Accountability and authority are not to be taken for granted in safety-critical healthcare systems; they must be guaranteed by design. By its hardware elements, SAFE-MED guarantees that adaptive cryptographic behaviour remains visible and under institutional control.

Architecture-wise, the governance discussed above is reflected in the separation of execution, monitoring and decision authorisation. The adaptive components may be required to implement cryptographic changes, but they will do so under supervision. Each decision will be accompanied by contextual metadata describing the system state, the events leading up to it and the rationale for it. This separation ensures that accountability cannot be lost as the system changes its own internal state, as is sometimes the case in highly autonomous security systems [16].

Auditability by design is one of the architecture's contributions to governance. Rather than having ad hoc logs cobbled together at installation time, audit trails are built in as first-class elements of the decision pipeline in SAFE-MED. Cryptographic posture change records are automatically created, time-stamped and stored, enabling reconstruction not only of the configuration applied but also of why it was chosen and under what conditions. Such reconstructability is crucial in regulated healthcare environments for compliance, post-mortem analysis and organisational confidence [17].

Traceability complements auditability through time. Adaptive cryptographic behaviour is rarely a series of discrete actions but rather a series of interdependent decisions made as circumstances unfold. SAFE-MED preserves these time-dependent relationships, enabling evaluators to trace decision trajectories rather than examine disconnected snapshots of configuration. This is critical both for technical diagnosis and for organisational accountability in evaluating whether the adaptive behaviour was in accordance with the institution's expectations at any point in time [18]. Preserving human override and institutional authority. SAFE-MED provides clear interfaces for authorised personnel to intervene in, suspend or modify adaptive behaviour as required by clinical judgement or organisational policy. These are not emergency afterthoughts but visible and controlled elements of the architecture. Automation is accountable to professional responsibility rather than independent of it [19].

Figure 9.5 depicts the relationships among cryptographic decision-making, runtime monitoring, audit logging and human authority within the governance-oriented architecture. It is not an abstract idea but a design commitment that shows how the accountability trails are coupled to adaptive behaviour without subjugating governance to optimisation or learning processes.

Figure 9.5 provides a governance-centric reference architecture of SAFE-MED that illustrates how cryptographic decision execution, monitoring and analysis, audit and traceability and human authority are synchronised with the governance layer. Without assuming governance as a process outside or post hoc, the architecture embeds visibility, accountability and override pathways into the system's functioning, ensuring that adaptive behaviour is well documented and answerable throughout its life cycle.

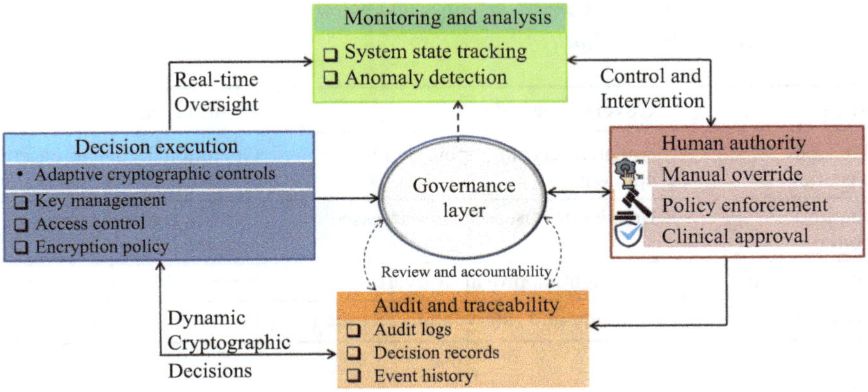

Figure 9.5 SAFE-MED governance-centric reference architecture

9.8.2 Governance enforcement and accountability pathways

Unlike most abstract principles in SAFE-MED, governance is implemented through explicit enforcement processes that connect adaptive decision-making to monitoring, human control and institutional accountability. The clarification in this section illustrates how previously reaffirmed governance commitments were incorporated into operational arrangements without revisiting safety requirements or using optimisation logic.

This approach is anchored in a structured governance stream that ensures every adaptive cryptographic decision is easily visible, reviewed and addressed. First, decisions are reviewed through time-based monitoring, where system behaviour and context signalling are continuously evaluated. As defined governance conditions are met, monitoring outcomes may trigger human control or intervention. Authority is exercised deliberately rather than reactively. Finally, all actions and interventions are recorded in auditable artefacts that support retrospective examination and institutional accountability [20].

Table 9.2 describes the life cycle of an adaptive decision from execution to audit via the life cycle of this governance enforcement process. The table does not specify technical controls but clarifies the boundaries of responsibility and the points of oversight, in particular where governance authority is exercised and how accountability remains maintained throughout system use. For physicians in clinical settings, governance becomes an ongoing operational property, a lasting utility which supports institutional trust, regulatory scrutiny and long-term system legitimacy [21].

While Table 9.2 provides procedural flow, governance needs clarity on how non-negotiable commitments are enforced, without deriving from technical content. In doing so, SAFE-MED brings to bear a very compact governance traceability view.

Table 9.3 maps the safety commitments regarding their enforcement and supervision locations in the architecture. This table does not define new safety

Table 9.2 Governance enforcement map: Decision → Monitor → Human Authority → Audit

Stage	Governance role	Primary outcome
Decision execution	Adaptive cryptographic action applied	Operational change enacted
Runtime monitoring	Behaviour observed and assessed	Detection of deviations or concerns
Human authority	Oversight, approval or override	Responsible intervention
Audit and review	Decision and intervention recorded	Institutional accountability

Table 9.3 Non-negotiable safety commitments and governance oversight points

Safety commitment	Enforcement location	Oversight authority
Bounded clinical latency	Runtime monitoring	Clinical operations team
Decision traceability	Audit and logging layer	Institutional governance
Override capability	Human authority interface	Authorised clinician
Controlled degradation	Monitoring and review	Safety oversight board

invariants or constraint logic but indicates how governmental commitments are maintained in institutional terms through monitoring, authority and audit.

By separating what must never be violated from how violations are dealt with, SAFE-MED avoids conflating safety design with institutional control. As a result, governance enforcement is presented as a control structure that includes adaptive behaviours to ensure safety commitments remain visible, enforceable and subject to human intervention at all times during system operation. Adaptive cryptographic systems do this because adaptive cryptography is not a fixed, regulated clinic but rather an expectation of regulations under regulatory conditions in which responsibility and trust extend beyond technical accuracy.

9.9 Scope boundaries, limitations and responsible extensions

SAFE-MED is presented as a reference framework intended to guide the safe and accountable governance of adaptive, quantum-aware cryptographic decision-making in intelligent healthcare systems. Its contribution lies in structuring how adaptation should be controlled and overseen rather than in maximising technical performance or prescribing implementation details. This closing section clarifies the boundaries of applicability, acknowledges modelling limitations and outlines how future extensions should be approached responsibly.

9.9.1 Limitations of scope

SAFE-MED is not intended for situations where cryptographic adaptation can be carried out in an unregulated manner. Systems designed for optimisation of performance, aggressive refactoring or continual self-tuning in the absence of any human authority are outside the scope of this framework. SAFE-MED is intended for regulated, safety-critical environments (e.g. clinics) where accountability, predictability and traceability are paramount. Using SAFE-MED in a loosely regulated or experimental system would be a misapplication of the framework. It would impose unnecessary restrictions at a time when there are no safety concerns.

9.9.2 Modelling assumptions and limitations

These decisions are made at an abstract level detached from any protocol-specific or even device-level considerations. The underlying cryptographic mechanisms are not modelled as actual cryptographic schemes with parameters to be selected but rather as options among others, subject to constraints. This ensures longevity and cross-system applicability, but it also means SAFE-MED does not address practical questions such as parameter selection, hardware capabilities or vendor-specific integration. It is up to designers of concrete systems and to local regulatory and clinical practices to determine what is appropriate.

9.9.3 Limits of risk reduction in adaptive cryptography

Healthcare decisions are made in ever-changing and sometimes incomplete contexts. Information is fragmented, assumptions about threats evolve, and competing demands intensify unpredictably. SAFE-MED does not eradicate such uncertainty. It confines adaptive practice within explicit safety and governance limits. Rather than guaranteeing optimal results or comprehensive threat awareness, it ensures accountability in the decision-making process, even when the consequences remain uncertain.

9.9.4 Responsible extensions

Future development of SAFE-MED should remain consistent with its safety-first orientation. Any extension, whether involving expanded contextual modelling, integration with emerging cryptographic standards or co-ordination across institutions, must continue to operate within established safety constraints and under human oversight. Introducing automation or optimisation features without equivalent governance safeguards would weaken the framework.

This section clarifies both the intended scope and the boundaries of SAFE-MED. The framework is not presented as a temporary technical fix. It serves as a stable reference model for adaptive cryptographic design in clinical settings. Accountability and operational trust must remain intact over time.

9.9.5 Lessons learned

SAFE-MED yields lessons that extend beyond the specific reference architecture presented in this chapter.

First, adaptive cryptography in intelligent healthcare should not be treated as a purely algorithmic or optimisation task. Once cryptographic posture becomes dynamic, the timing and purpose of a change can matter as much as the mechanism itself. The practical question shifts from 'How strong is the configuration?' to 'How is the decision made, and under what conditions?'

Second, the role of learning-driven adaptation must be explicitly bounded. Unfettered intelligence can erode clinical predictability and weaken institutional accountability. SAFE-MED shows that learning is most useful when it supports decision-making, rather than replacing it.

Third, no technical design can remove long-horizon uncertainty, including post-quantum risk. Treating future risk as a decision signal rather than a directive for immediate deployment allows systems to remain forward-looking while still meeting present-day safety obligations.

Finally, governance cannot be bolted on after deployment. Auditing, traceability and human override must be built in. When these are treated as architectural properties, adaptive behaviour remains inspectable, contestable and trustworthy over time.

9.9.6 Conclusion

This chapter introduced SAFE-MED as a reference framework for governing adaptive, quantum-aware cryptographic decision-making in intelligent healthcare systems. Rather than framing cryptographic adaptation as a protocol-level optimisation problem, the discussion situated it within the institutional realities of clinical practice, where timing, accountability and oversight shape what is acceptable.

SAFE-MED positions safety, governance and traceability as primary design commitments. Adaptive cryptography is therefore not treated as incompatible with regulation or human authority. The framework does not prescribe particular technologies. Instead, it provides a structured framework for reasoning about adaptation under uncertainty, stress and evolving risk.

As intelligent healthcare systems continue to develop, the value of SAFE-MED lies less in any specific technical mechanism and more in the principles it articulates for governing adaptive security decisions. It offers a starting point for further work on cryptographic systems that remain accountable, inspectable and aligned with institutional responsibility over time.

References

[1] M. van Zomeren, F. Deane, K. F. Joiner, L. Qiao, R. Horne and E. Suprun, "Regulating cyberworthiness: Governance frameworks for safety-critical cyber-physical systems," *Systems*, vol. 13, no. 10, Art. no. 862, 2025, doi: 10.3390/systems13100862.

[2] I. Ahmad, F. Shahid, I. Ahmad, J. Islam, K. N. Haque and E. Harjula, "Adaptive lightweight security for performance efficiency in critical healthcare monitoring," in *Proceedings of the International Symposium on Medical Information and Communication Technology (ISMICT)*, 2024, pp. 78–83, doi: 10.1109/ISMICT61996.2024.10738175.

[3] K. Svandova and Z. Smutny, "Internet of medical things security frameworks for risk assessment and management: A scoping review," *Journal of Multidisciplinary Healthcare*, vol. 17, pp. 2281–2301, 2024, doi: 10.2147/JMDH.S459987.

[4] A. Robey, L. F. O. Chamon, G. J. Pappas and H. Hassani, "Probabilistically robust learning: Balancing average- and worst-case performance," in *Proceedings of the 39th International Conference on Machine Learning (ICML).* PMLR, vol. 162, 2022, pp. 18667–18686. [Online]. Available: https://proceedings.mlr.press/v162/robey22a.html

[5] J. Wenzel, M. A. Köhl, S. Sterz *et al.*, "Traceability and accountability by construction," in *Leveraging Applications of Formal Methods, Verification and Validation. Software Engineering Methodologies (ISoLA 2024), Lecture Notes in Computer Science*, vol. 15222. Cham, Switzerland: Springer, 2024, pp. 258–280, doi: 10.1007/978-3-031-75387-9_16.

[6] M. Alalhareth and S.-C. Hong, "Enhancing the internet of medical things (IoMT) security with meta-learning: A performance-driven approach for ensemble intrusion detection systems," *Sensors*, vol. 24, no. 11, Art. no. 3519, 2024, doi: 10.3390/s24113519.

[7] B. K. Asingwire, L. Sibomana, A. Ngenzi and C. Kabiri, "Delay and size-dependent priority-aware scheduling for IoT-based healthcare traffic using heterogeneous multi-server priority queueing system," *IET Communications*, vol. 17, no. 16, pp. 1877–1887, 2023, doi: 10.1049/cmu2.12675.

[8] P. Bhattacharya, A. Mukherjee, B. Bhushan, S. K. Gupta, T. R. Gadekallu and Z. Zhu, "A secured remote patient monitoring framework for IoMT ecosystems," *Scientific Reports*, vol. 15, art. no. 22882, 2025, doi: 10.1038/s41598-025-04774-y.

[9] F. Kagai, P. Branch, J. But and R. Allen, "Harvest-now, decrypt-later: A temporal cybersecurity risk in the quantum transition," *Telecom*, vol. 6, no. 4, Art. no. 100, 2025, doi: 10.3390/telecom6040100.

[10] Y. Baseri, V. Chouhan, A. Ghorbani and A. Chow, "Evaluation framework for quantum security risk assessment: A comprehensive strategy for quantum-safe transition," *Computers and Security*, vol. 150, Art. no. 104272, 2025, doi: 10.1016/j.cose.2024.104272.

[11] F. den Hengst, M. Otten, P. Elbers, F. van Harmelen, V. François-Lavet and M. Hoogendoorn, "Guideline-informed reinforcement learning for mechanical ventilation in critical care," *Artificial Intelligence in Medicine*, vol. 147, Art. no. 102742, 2024, doi: 10.1016/j.artmed.2023.102742.

[12] V. J. Hodge and M. Osborne, "Agile development for safety assurance of machine learning in autonomous systems (AgileAMLAS)," *Array*, vol. 27, p. 100482, 2025, doi: 10.1016/j.array.2025.100482.

[13] A. K. Pandey, S. Zeadally, A. K. Das and N. Kumar, "Secure integration of cyber engineering and medical cyber-physical system: A survey and open issues," *ACM Computing Surveys*, vol. 58, no. 1, Art. no. 9, pp. 1–36, 2026, doi: 10.1145/3757731.

[14] K.-C. Hsu, H. Hu and J. F. Fisac, "The safety filter: A unified view of safety-critical control in autonomous systems," *Annual Review of Control, Robotics, and Autonomous Systems.*, vol. 7, pp. 47–72, 2024, doi: 10.1146/annurev-control-071723-102940.

[15] V. Nigam and C. Talcott, "Technical-report: Automating recoverability proofs for cyber-physical systems with runtime assurance architectures," arXiv preprint, 23, 2023. https://doi.org/10.48550/arXiv.2304.11564.

[16] A. T. Ayedh M., A. W. A. Wahab and M. Y. I. Idris, "Enhanced adaptable and distributed access control decision making model based on machine learning for policy conflict resolution in BYOD environment," *Applied Sciences*, vol. 13, no. 12, Art. no. 7102, 2023, doi: 10.3390/app13127102.

[17] F. Ullah, J. He, N. Zhu *et al.* "Blockchain-enabled EHR access auditing: Enhancing healthcare data security," *Heliyon*, vol. 10, no. 16, e34407, 2024, doi: 10.1016/j.heliyon.2024.e34407.

[18] A. Bansal, A. Kandikuppa, M. Hasan *et al.* "System auditing for real-time systems," *ACM Transactions on Privacy and Security*, vol. 26, no. 4, Art. no. 50, 2023, doi: 10.1145/3625229.

[19] M. Hassan, E. M. Borycki and A. W. Kushniruk, "Artificial intelligence governance framework for healthcare," *Healthcare Management Forum*, vol. 38, no. 2, pp. 125–130, 2025, doi: 10.1177/08404704241291226.

[20] S. Barbaria, A. Jemai, H. İ. Ceylan, R. I. Muntean, I. Dergaa and H. Boussi Rahmouni, "Advancing compliance with HIPAA and GDPR in healthcare: A blockchain-based strategy for secure data exchange in clinical research involving private health information," *Healthcare*, vol. 13, no. 20, art. no. 2594, 2025, doi: 10.3390/healthcare13202594.

[21] T. Alelyani, "A validated framework for responsible AI in healthcare autonomous systems," *Scientific Reports*, vol. 15, art. no. 44432, 2025, doi: 10.1038/s41598-025-25266-z.

[22] P. Cervera de la Cruz, T. Lalova-Spinks and M. Shabani, "The european health data space: an opportunity to strengthen citizen rights and engage citizens in health data governance," *Frontiers in Medicine*, 2026, doi: 10.3389/fmed.2025.1699941.

[23] B. Mulhern, C. Balakrishna and J. Collie, "The network and information systems 2 directive: Toward Scalable cyber risk management in the remote patient monitoring domain: A Systematic Review," *IoT*, vol. 7, no. 1, Art. no. 14, 2026, doi: 10.3390/iot7010014.

[24] F. Kagai, P. Branch, J. But and R. Allen, "Harvest-now, decrypt-later: A temporal cybersecurity risk in the quantum transition," *Telecom*, vol. 6, no. 4, Art. no. 100, 2025, doi: 10.3390/telecom6040100.

[25] H. Gerold and S. Lucia, "Safe reinforcement learning via adaptive robust model predictive shielding," *Computers and Chemical Engineering*, vol. 206, Art. no. 109521, 2026, doi: 10.1016/j.compchemeng.2025.109521.

[26] J. R. Laracy, Z. Meng, V. D. Kirova, C. S. Ku and T. J. Marlowe, "Software quality assurance and AI: A systems-theoretic approach to reliability, safety, and security," *Software*, vol. 4, no. 4, Art. no. 30, 2025, doi: 10.3390/software4040030.

[27] Q. Wei, G. Xu and Y. Liu, "Comparative study of international medical AI regulatory policies: regional approaches and evidence from clinical practice," *AI and Society*, 2026, doi: 10.1007/s00146-026-02879-0.

Chapter 10

AI-enabled cryptographic scheme for securing the Internet of Medical Things in the post-quantum era

Farida Shehu Garki[1], Abdulmumini Yakubu Musah[2], Esther Omolara Abiodun[3], Oludare Isaac Abiodun[3] and Agbotiname Lucky Imoize[4]

Abstract

The rapid adoption of intelligent healthcare technologies and the Internet of Medical Things (IoMT) has significantly broadened the cyberattack surface of healthcare infrastructures and exposed sensitive patient data to increasingly sophisticated and advanced threats and attacks. Existing security strategies tend to be based on static cryptographic settings, lightweight cryptography, post-quantum cryptography (PQC) or privacy-enhancing schemes installed in isolation, which are not dynamic, heterogeneous and safety-critical to the healthcare context. This chapter proposes an AI-driven cryptographic agility system, which is a lightweight cryptography, hybrid classical-post-quantum and privacy-enhancing technology that is controlled via an intelligent governance layer. The framework employs an AI-controlled control plane to analyse security telemetry, system settings and clinical risk, which allows for dynamic cryptographic adaptation without disrupting healthcare services and is tested with the help of a simulated intelligent healthcare testbed consisting of the IoMT devices, edge gateways, hospital information systems and cloud analytics systems. Controlled attack scenarios, which included impersonation, lateral movement and ransomware preparation behaviour, were injected to measure the accuracy of threat detection, sensitivity to cryptographic adaptation and system overhead. Experimental results present a 96.4 per cent high performance of threat detection and less attack success rates of approximately

[1]Department of Information Systems and Technology, Faculty of Computing, National Open University of Nigeria, Nigeria
[2]Department of Science Computing, Prime University, Abuja, Nigeria
[3]Department of Computer Science, Faculty of Science, University of Abuja, Nigeria
[4]Department of Electrical and Electronics Engineering, Faculty of Engineering, University of Lagos, Nigeria

14 milliseconds with increased latency and a moderate increment of energy consumption of approximately 18 per cent were demonstrated from IoMT devices as compared to the static cryptographic environment implementation. The results present an agile AI cryptographic framework and offer a practical, resilient intelligent healthcare system (IHS) for post-quantum systems.

Keywords: Cryptographic algorithms; post-quantum cryptography; intelligent healthcare; IoMT security; AI-driven cryptography; privacy-enhancing technologies; healthcare data protection

10.1 Introduction

The rapid digitisation of the healthcare sector has revolutionised medical systems, bringing intelligent and interconnected data, which depend on the Internet of Medical Things (IoMT), cloud-assisted analytics and artificial intelligence (AI). These widespread technologies are now used for real-time monitoring, diagnosis and decision support and have positively impacted the healthcare system, making it more accessible and efficient, but at the same time created a wider attack surface for critical medical infrastructures due to increasingly sophisticated cyber threats, making it possible for patient data to be exposed [1]. AI models operating on encrypted metadata align with privacy-preserving designs used in federated learning for healthcare analytics [2]. While the traditional security measures for medical systems have been mainly based on static cryptographic techniques that are usually chosen at the design stage and applied uniformly across different devices and services, regardless of their type [3], this practice is now losing its effectiveness in contemporary healthcare scenarios, which are defined by medical devices with limited resources [4], different communication settings [5], tight clinical latency requirements [6] and changing regulatory requirements [7]. The difficulty is made even worse with the emergence of quantum computers, which endanger the future of many public-key cryptographic algorithms in use today and thereby create urgent issues regarding the confidentiality and integrity of medical data with long life cycles [8].

Recent studies have made attempts to secure IoMT ecosystems through the use of lightweight cryptography [9], post-quantum cryptographic (PQC) schemes [10], blockchain-based security mechanisms and privacy-enhancing technologies (PETs) [3,4,8,11–13]. All these approaches are indeed rewarding security primitives, but they are seldom utilised together and remain rather inactive, which means that their adaptability by means of cryptographic decisions according to the changing threat models, operational activities or clinical risk profile is very limited. Consequently, either healthcare systems suffer from unnecessary computational overload, or they become vulnerable to new threats, especially during the post-quantum transition.

Concurrent with these advancements, there has been a growing adoption of AI in healthcare systems (HCSs) for predictive analysis, decision-making purposes and self-management of autonomous systems [6,14]. Despite this progress, the importance of AI in controlling and managing decisions related to cryptographic security, rather than enabling decision-making based on application intelligence, remains a concern of lesser significance. This is especially important given the growing complexities of future HCSs' security demands, risks, performance and real-time trust.

10.1.1 Motivation and problem statement

The problem addressed in this chapter is the absence of adaptive security governance in existing healthcare cryptographic frameworks. Current approaches mainly target which cryptographic algorithms should be used in a deployment process rather than when or how they should be applied. In post-quantum secure healthcare systems, cryptographic agility is a required aspect that ensures a smooth transition or combination without affecting healthcare services.

The absence of intelligent orchestration means that security robustness, system performance and patient safety will always compete in healthcare systems' trade-off to some extent. The traditional way of dealing with cryptographic methods that have static resources can cause unacceptable delay or consumption of resources, whereas lightweight ones may not be able to protect against advanced or quantum-powered attackers adequately [4,15]. These identified problems in the use of IoMT warrant a single healthcare framework capable of managing security, efficiency and trust adaptively.

10.1.2 Scope and focus of the chapter

This chapter focuses exclusively on the design of a cryptographic framework for intelligent healthcare systems (IHSs) and presents an analysis of AI-orchestrated cryptographic agility for post-quantum healthcare environments. The chapter presents evidence of how artificial intelligence can be leveraged to dynamically govern cryptographic adaptation, policy selection and enforcement across heterogeneous healthcare infrastructures. Detailed algorithmic descriptions are referenced only where necessary to support the core arguments related to AI-driven governance and are not treated as primary contributions. The scope of this chapter is restricted to the following areas.

(i) **Cryptographic agility in IoMT and intelligent healthcare environments**
This part analyses cryptographic agility as a solution to the limitations of static cryptographic implementations in IoMT and intelligent healthcare. The focus is placed on environments characterised by heterogeneous devices such as resource-constrained IoMT sensors, edge gateways and cloud-based healthcare services. The discussion focuses on the dynamics of cryptographic mechanisms as they adapt to new operational conditions, threat levels and clinical safety requirements.

(ii) **AI-driven security orchestration and decision-making**
Artificial intelligence (AI) is considered a control layer for security orchestration rather than a replacement for cryptographic mechanisms. The chapter

discusses how AI techniques can analyse security telemetry, system state and threat indicators to provide informative support for cryptographic policy-making. The scope is limited to orchestration, governance and enforcement logic and not low-level cryptographic algorithm design.

(iii) **Post-quantum security preparedness and transition**
The chapter addresses the need for post-quantum preparedness within healthcare systems, one that can manage long-lived and highly sensitive data. It focuses on the role of a hybrid classical–post-quantum cryptographic implementation as a transitional measure to preserve compatibility with existing infrastructure. The most important considerations are presented as performance overhead, configuration complexity and maintaining operational continuity during post-quantum migration.

(iv) **Explainability, human-in-the-loop safeguards and regulatory alignment**
The cryptographic decision-making is defined as a regulated and auditable procedure, which involves explainability and human control. This chapter considers how human-in-the-loop provides safeguards to the system and ensures automated cryptographic adaptations remain aligned with clinical safety and organisational policy constraints. In this context, regulation is considered a design requirement that cryptographic actions be traceable and justifiable under healthcare compliance requirements.

10.1.3 Chapter contribution and structure

The chapter contributions include the following.

- **AI-orchestrated cryptographic agility framework for IHSs**
The chapter proposes an AI-orchestrated cryptographic agility framework for IHSs. Instead of relying on fixed cryptographic configurations, the framework treats cryptography as a dynamic and governable control process. The approach enables combined management of IoMT, edge computing and cloud services in health and nevertheless still complies with operational and safety-critical healthcare requirements.
- **Dynamic adaptation to emerging threats and clinical system constraints**
The suggested framework takes into consideration dynamic trends in healthcare cybersecurity, in which threat models evolve fast, and the system constraints differ depending on devices and clinical processes. It allows healthcare systems to apply dynamically changing cryptographic protection based on threat intelligence, device capabilities, latency constraints and clinical priorities. The framework ensures greater resilience in any situation and circumstance by facilitating adaptive operation and not being bound to specific cryptographic deployments.
- **Post-quantum readiness and sustainable healthcare security**
The framework facilitates sustainable and safe healthcare operations in the post-quantum era. It will allow a gradual shift to quantum-resistant cryptographic mechanisms without compromising the system's stability. The hybrid

classical-post-quantum cryptography (PQC) is used to support long-lived healthcare data and ensure compatibility with existing infrastructure, without interfering with healthcare operations or compromising performance.

- **Integration of AI-based risk evaluation, policy co-ordination and explainability** The main components of the framework are the integration of AI-based risk assessment with cryptographic policy co-ordination and explainable decision-making. Artificial intelligence considers security threats based on telemetry and situational data, whereas cryptographic policies are chosen and applied in a transparent and auditable way. Explainability and human oversight are embedded to ensure that cryptographic decisions remain interpretable, trustworthy and aligned with regulatory and clinical safety requirements.

The remainder of the chapter is organised into the following: Section 10.2 provides the basic concepts of cryptographic agility in IHSs, which highlights the drawbacks of static cryptographic deployment, the pressure of PQC and the transition of cryptography to cryptographic governance. Section 10.3 provides cryptographic mechanisms as the adaptive components in an AI-based decision space. Section 10.4 presents the role of artificial intelligence in the orchestration of cryptography. Section 10.5 outlines the proposed framework and evaluation methodology, while Section 10.6 presents the threat model and governance logic of the proposed framework. Section 10.7 evaluates deployment considerations, limitations, lessons learned and future research directions, while Section 10.8 concludes the chapter.

10.2 Background and fundamentals

This section establishes the conceptual basis that supports the proposed framework of an AI-orchestrated cryptographic agility. The argument for static cryptographic deployment being insufficient in smart healthcare systems comes to the forefront, and at the same time, the limitations in the system require adaptive, AI-governed security.

10.2.1 Cryptography in intelligent healthcare

The healthcare infrastructure's cryptographic methods are traditionally grounded in trusted and familiar primitives, including symmetric encryption, public-key encryption and cryptographic hash functions [16,17]. The combination of these provides confidentiality, authentication, integrity and non-repudiation in healthcare data processes [6]. The concept of integrating blockchain and AI into healthcare security systems has been discussed to improve the level of trust and decentralised decision-making [18]. In the traditional security design for medical systems, such cryptographic mechanisms are generally chosen and fixed at design time and uniformly applied to the heterogeneous healthcare components. As shown in Figure 10.1, this approach is based on the assumption of having consistent security requirements across the wearable sensors, implantable devices, imaging systems and bedside monitors, even though their operational context and risk profile vary.

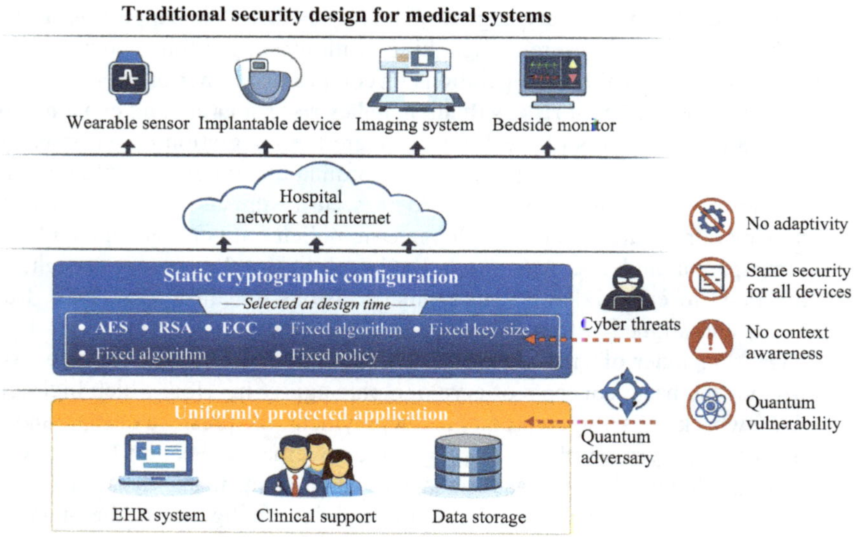

Figure 10.1 Traditional security design for medical systems

However, within intelligent healthcare environments involving IoMT devices, cloud platforms, edge systems and AI-based analytics, the use of cryptography is no longer isolated from other technical security mechanisms but instead operates as an embedded operational control within safety-critical clinical systems [4]. Symmetric encryption remains the backbone of protecting healthcare data at rest and in transit due to its computational efficiency.

Lightweight versions specifically can be applied to wearable and implantable medical devices that work within strict energy and processing constraints [19]. In this regard, asymmetric cryptography is useful in the medical device authentication process and establishment of secure trust in healthcare frameworks. The most popular example is elliptic curve cryptography, since the key size is smaller and the cost of the calculation is lower than the traditional cryptography methods used in public key [3]. Cryptographic hash functions also help as a lightweight version to further promote data integrity, secure data logging and auditability, which are key requirements in regulated healthcare environments [1].

Despite the maturity level of symmetric and asymmetric cryptographic primitives as building blocks, they are still mostly statically configured at the system design stage. Such static deployment has assumed predictable workloads, stable threat environments and homogeneous system capabilities. These assumptions do not apply in the context of modern IHSs that are characterised by heterogeneous devices, dynamic clinical workflows and continuously evolving threat landscapes [12]. The major security functions and key limitations of cryptographic primitives commonly deployed in healthcare systems are summarised in Table 10.1, along with the operational conditions under which they are typically applied.

Table 10.1 Cryptographic primitives in healthcare systems

Reference(s)	Primitive	Security property	Healthcare use case	Key limitation
[15,19]	AES	Confidentiality	EHR storage, secure healthcare databases	High energy cost for resource-constrained IoMT devices
[20,21]	PRESENT	Confidentiality	Wearable and implantable medical devices	Smaller block size and limited security margin
[13,22]	SPECK	Confidentiality	Embedded medical controllers and sensors	Security concerns in some deployment models
[3,10]	ASCON	Confidentiality, integrity	Secure IoMT communication	Limited legacy system support
[3,17]	RSA	Authentication, key exchange	Legacy healthcare systems	Large key size; quantum vulnerability
[3,23]	ECC	Authentication, key exchange	IoMT onboarding, TLS-based security	Quantum vulnerability and key management complexity
[1,8]	SHA-2	Integrity, authentication	Data integrity verification, digital signatures	Computational overhead for constrained devices
[3,6]	SHA-3	Integrity, authentication	Future-proof healthcare hashing	Higher implementation and processing cost
[4,7]	Key management	Trust enforcement	Device life cycle and credential management	Scalability and rotation complexity

10.2.2 Cryptographic failures in IHSs

Intelligent healthcare systems are usually subjected to a combination of constraints that are hardly ever encountered all at once in other cyber-physical domains. As a result, static cryptographic configurations are limited in their effectiveness by these constraints [12]. In this regard, a dominating factor is resource heterogeneity within healthcare infrastructures as presented in [1,14]. IoMT ecosystems consist of ultra-low-power wearable sensors, implantable devices, edge gateways and cloud-based analytics platforms, each with very different computational capabilities and energy consumption [13,19,24]. A cryptographic configuration that is suitable for cloud servers may be impossible for implantable devices, while lightweight configurations that are acceptable for sensors may not be enough for storing long-term clinical data of high value.

Moreover, dynamic operational conditions are typical of healthcare environments. Such conditions can be network congestion, mobility and connectivity problems, as well as different workloads that usually occur in the process of telemedicine and remote monitoring [2,6,25,26]. The security will be compromised, or the performance will be reduced unnecessarily, because any of the changes cannot be adjusted to use the static cryptographic parameters [12]. The other cause of cryptographic failures is evolving threat models [27–30]. Healthcare systems are increasingly confronted with a complex and rapidly evolving cybersecurity threat landscape, including data breaches, ransomware attacks, insider threats, supply-chain compromises, device impersonation, credential theft, AI-enabled attacks and emerging quantum-computing threats. These challenges are particularly significant in intelligent healthcare environments because interconnected IoMT devices, hospital networks, cloud platforms and electronic health record systems create multiple potential attack pathways. While the sensitivity and long-term value of patient data increase the consequences of successful compromise. Classical hackers who are primarily interested in data breaches and ransomware attacks now have to be considered together with insider threats, supply-chain compromises and even adversaries who are able to exploit AI-driven systems [27]. Blockchain-based solutions have been proposed to mitigate supply-chain risks through tamper-evident logs and decentralised trust [25,31]. Moreover, the issue of quantum computing is more threatening as it directly affects public-key schemes, which are the most widely deployed, thus putting healthcare data with long confidentiality lifespans under particular risk [17]. Recent systematic reviews have highlighted the growing convergence of quantum, post-quantum and blockchain approaches to secure IoMT ecosystems [27].

Lastly, clinical safety requirements impose limitations that do not exist in conventional IT systems. The cryptographic functions, which are mechanisms that cause added latency or cause authentication to fail or otherwise interfere with the data availability, can have a direct and potentially significant effect on patient outcomes [4,32,33]. In this scenario, healthcare security choices go beyond technical optimisations and become safety-critical design choices [26]. All these show the fact that cryptographic correctness is not sufficient [25,27]. Only IHSs need to be able to dynamically choose, make customisations and combinations of cryptographic mechanisms based on system state, threat intelligence and clinical situation.

10.2.3 Post-quantum pressure as a catalyst for cryptographic agility

The increasing adoption of PQC serves as a clear indicator of the inherent fragility of fixed security architectures in healthcare systems. Post-quantum algorithms are usually linked with a greater size of keys, increased computing costs and more communication costs compared to classical cryptography algorithms [27]. Whereas these overheads can be handled in cloud-based systems, they pose a severe problem with resource-limited IoMT devices and healthcare applications that must meet hard latency limits [1]. Therefore, healthcare facilities must meet two closely associated needs: they should be compatible with the existing classical cryptographic systems, and progressively, they have to implement quantum-resistant measures to achieve security in the long run [28,34]. In order to facilitate this change, a hybrid cryptographic schemes that fuse classical and post-quantum algorithms has been proposed in the literature [3,9,27]. Nonetheless, these hybrid solutions drastically add to the complexity of the system implementation, creating new overhead of operations and administration, instead of easing protection implementation [25,28]. The shift towards hybrid classical–post-quantum cryptographic deployments is increasingly recommended to ensure long-term data security in healthcare systems [35].

The security operations become more challenging to accomplish, and as a result, cryptographic agility gains more importance in such environments. Agility is able to modify, mix or decommission cryptographic mechanisms without interrupting continuous clinical operations [14,18]. When these transitions are not well orchestrated, they can cause new risks such as configuration errors, performance degradation or service interruptions [26]. Given the size and complex nature of modern healthcare systems' operations, cryptographic changes can no longer be managed manually. This scenario leads to the necessity of smart orchestration systems that can have the ability to regulate cryptographic development in a well-organised and context-sensitive way [32].

10.2.4 From cryptographic selection to cryptographic governance

Past studies on healthcare security have mostly been conducted to determine the appropriate cryptographic algorithm to be deployed, without much or any consideration of the means of managing the cryptography decisions during the system operation life cycle of smart healthcare systems [1,10]. Focusing solely on the selection of cryptographic algorithms is insufficient for intelligent healthcare systems (IHSs). It must continuously balance competing requirements, including device resource constraints, data sensitivity, system performance, security, regulatory compliance, clinical usability, and the integration of classical and post-quantum cryptographic mechanisms. Addressing these competing requirements therefore requires a shift from static cryptographic configurations towards adaptive cryptographic governance [27]. The solution to such conflicting requirements is seen in the need to replace the stagnant cryptographic set-up with the use of cryptographic governance. In a governance-oriented model, the cryptographic policy is constantly reviewed, modified and implemented based on the system's state, threat intelligence and operational conditions. Artificial intelligence facilitates such a rule by helping to track the current state, contextualise and adaptively estimate the risks that are incorporated into the cryptographic decision-making procedure [29,32]. AI-driven cryptographic governance aligns with emerging frameworks that integrate blockchain for auditability and federated learning for privacy-preserving analytics [2].

This shift of cryptography from a fixed technical element to an adaptive AI-controlled control layer forms the conceptual point of Section 10.3, which examines cryptographic algorithms not as isolated solutions but rather in relation to other elements in a decision space that has to be effectively managed in an intelligent, post-quantum healthcare IT system.

10.3 Cryptographic algorithms as an adaptive decision space for healthcare

This section conceptualises cryptographic algorithms as adaptive components within an AI-orchestrated security decision space for IHSs. The discussion focuses on how lightweight cryptography, PQC and privacy-enhancing technologies (PETs) jointly contribute to cryptographic agility under heterogeneous healthcare constraints.

10.3.1 Lightweight cryptography for IoMT-constrained environments

The devices operating in IoMT data are limited to a very narrow spectrum of power, memory and computing capabilities, be it wearable sensors, implantable medical devices or bedside care monitoring systems [19]. Although classical cryptography algorithms have proven to be effective and reliable in conventional computing platforms, their direct application to IoMT systems unavoidably brings about unacceptable security and low overheads, reducing device lifetime, causing latency or unstable operation [1,6]. Therefore, lightweight cryptography has become a critical building block of healthcare information security in resource-constrained settings.

However, existing literature reveals that no single lightweight cryptographic algorithm can be efficient in all healthcare situations. Chinbat *et al.* [12], it was demonstrated that lightweight cryptography is very context-dependent and depends on the capability of the device, the nature of the workload and network patterns. In their article, they revealed that the performance of lightweight cryptographic algorithms varies across the extent of workloads and deployment requirements, using ML-assisted evaluation of a Raspberry Pi healthcare IoT set-up. Further evidence is provided by John and Sinciya [4], who presented complementary findings that lightweight block ciphers can secure real-time remote healthcare monitoring systems without compromising clinical responsiveness, provided that algorithm selection is made in accordance with the operational conditions. Abdelfattah *et al.* [36] also illustrated that privacy-aware lightweight machine learning models can achieve accurate diagnosis while preserving confidentiality, underscoring the importance of cryptography-aware AI co-design in IoMT systems. In this context, Table 10.2 presents representative lightweight cryptographic algorithms as elements of a cryptographic option space rather than as prescriptive solutions

Among the representative lightweight cryptographic algorithms presented in Table 10.2, PRESENT, SPECK, ASCON and BEST-1 illustrate key design trade-offs that are associated with IoMT environments. SPECK is designed for ultra-low-power

Table 10.2 Representative lightweight cryptographic algorithms for IoMT devices

Reference (s)	Algorithm	Type	Block / key size	Primary advantage	Typical health-care use case
[21]	PRESENT	Block cipher	64 / 80–128 bits	Minimal hardware footprint	Wearable health sensors
[13,37]	SPECK	Block cipher	64–128 / 128–256 bits	Software efficiency	Embedded medical controllers
[3]	ASCON	AEAD	128 / 128 bits	Standardised lightweight security	Secure IoMT communication
[12]	BEST-1	Block cipher	Variable	Energy efficiency	Battery-powered implants

wearables and implantable sensors optimisation through its fast software-based implementation on resource-constrained devices, combined with the use of a simple Addition Rotation XOR structure [13,37]. However, its small block and key sizes make it unsuitable for long-term protection of sensitive healthcare data. PRESENT is highly resistant to cryptanalytic attacks and has a small hardware size, which is suitable for performing low-frequency security tasks like authenticating devices [21]. ASCON provides authenticated encryption of associated data and has become well-known because of its trade-off between security and efficiency in sending messages to IoMT [3]. BEST-1 focuses on the energy efficiency of battery-powered medical implants, but it is most appropriate in narrow and operation-specific settings [12]. All these observations together support the idea that the lightweight cryptographic selection should not be universal but must be context sensitive. Lightweight cryptographic selection must also consider communication efficiency, especially in urban IoMT deployments, where multi-core fibre optic systems can enhance data transmission capacity [38].

10.3.2 Data security using PQC

Healthcare data often requires decades of protection, especially for electronic health records, genomics and longitudinal datasets of patient monitoring. This long-term requirement presents some fundamental constraints of classical public-key cryptography to quantum-capable adversaries [3,17]. Post-quantum cryptography introduces quantum-resistant alternatives designed to mitigate this risk. However, they are also linked to larger key sizes, greater computational complexity and greater communication overhead [1]. They can be mitigated in cloud-based systems, where they also become a huge burden on IoMT devices with limited resources and healthcare applications with critical real-time and safety demands [30]. Therefore, healthcare infrastructures are forced to maintain interoperability with legacy cryptographic systems while simultaneously transitioning to quantum-resistant mechanisms to protect their data in the long run.

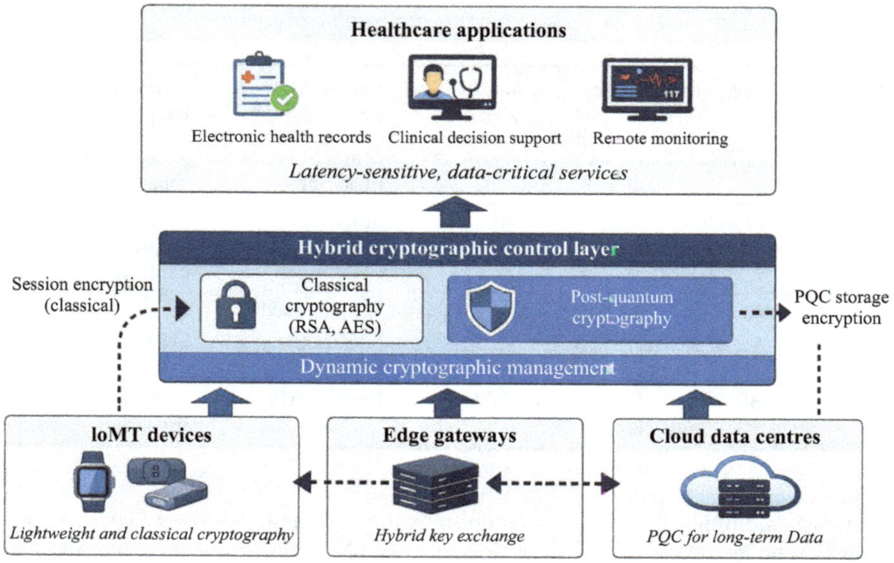

Figure 10.2 Hybrid classical–post-quantum cryptographic deployment in healthcare systems

Selvi and Sakthivel [23] demonstrated that hybrid classical and post-quantum cryptographic frameworks can enhance cloud-based healthcare data protection while maintaining operational feasibility. Hybrid cryptographic algorithms, such as those presented by Imoize *et al.* [7] and Man Lo *et al.* [9], are classical among the post-quantum algorithms that have thus been broadly suggested as an effective approach towards transition. They preserve compatibility with legacy systems while supporting the gradual adoption of quantum-resilient mechanisms. Architectural analysis and systematic reviews present the fact to post-quantum-resistant security in IoMT systems [10]. Nevertheless, hybrid cryptography broadens the space of cryptographic configurations to a large degree and enhances the complexity of systems. This complexity may cause misconfiguration, a decrease in performance and uneven security assurances across healthcare infrastructures without a proper form of governance [27]. Hybrid cryptographic frameworks are increasingly being paired with blockchain to ensure both quantum resistance and decentralised trust in IoMT systems [27,28]. A typical hybrid classical and post-quantum cryptographic implementation in IHSs is presented in Figure 10.2. The central implication is that PQC transforms cryptographic selection into a multi-objective optimisation problem that has to trade off security, performance and clinical safety. This change underscores the need for AI-driven orchestration mechanisms that can resolve cryptographic diversity systematically and dynamically.

10.3.3 PETs for secure healthcare data utilisation

Besides ensuring that data at rest and in transit is secured, IHSs need to have the ability to use data in a secure manner. Healthcare data must often be utilised for analytics, AI model training, and inter-institutional collaboration; however, such utilisation may expose sensitive patient information to privacy and security risks if adequate safeguards are not implemented. Privacy-enhancing technologies (PETs) can mitigate these risks by enabling controlled computation and analysis while preserving the confidentiality of sensitive healthcare data. Privacy-enhancing technologies can fulfil this need by allowing controlled calculation over the secured data [39]. As presented by Shin *et al.* [8], the use of encryption is not sufficient to run large-scale healthcare analytics, which prompts the adoption of federated learning, homomorphic encryption and differential privacy [28,40]. Murala *et al.* [32] also demonstrate that federated learning can be cryptographically secured to assist several healthcare institutions by collaborating to provide intelligence without violating patient confidentiality. Other researchers affirm that encrypted federated learning and differential privacy could facilitate large-scale biomedical image analysis and be in line with the requirements of regulations [33,40]. Such privacy-preserving techniques are essential for digital phenotyping platforms that handle longitudinal patient-generated health data [24,41]. Although PETs have some benefits, they present non-trivial computing and communication overhead [27]. They should hence be used selectively and sensibly, and hence, there is a need to have an intelligent cryptography orchestration as opposed to a blanket application [18,39]. The most important PETs and their contributions to healthcare systems are summarised in Table 10.3.

Figure 10.3 demonstrates an integrated architecture for securing the utilisation of healthcare data by using co-ordinated PETs. Based on it, patient data are locally stored in distributed hospitals and medical devices, which is free of exposing sensitive clinical information. Federated learning presents the method of joint training of artificial intelligence models using encrypted model updates, as opposed to sharing raw patient data across institutions. Homomorphic encryption enables analytics and diagnostic calculations to be done on encrypted data, maintaining the secrecy of data all the way along the processing pipeline. Before the aggregation of global models, the control of information leakage was accomplished by adding calibrated noise to the differential privacy mechanisms in order to adhere to the regulations of privacy of healthcare data. Secure authentication schemes, such as those based on fog computing and blockchain, can further enhance privacy in distributed healthcare

Table 10.3 Privacy-enhancing technologies and their healthcare roles

PET	**Primary function**	**Healthcare application**	**Key trade-off**
Homomorphic encryption [28]	Compute on encrypted data	Secure diagnostics, analytics	High computation cost
Secure multiparty computation [27]	Collaborative computation	Cross-hospital research	Protocol complexity
Differential privacy [40]	Statistical privacy guarantees	Population health analytics	Accuracy loss
Federated learning [32]	Decentralised AI training	Multi-institutional diagnosis	Communication overhead

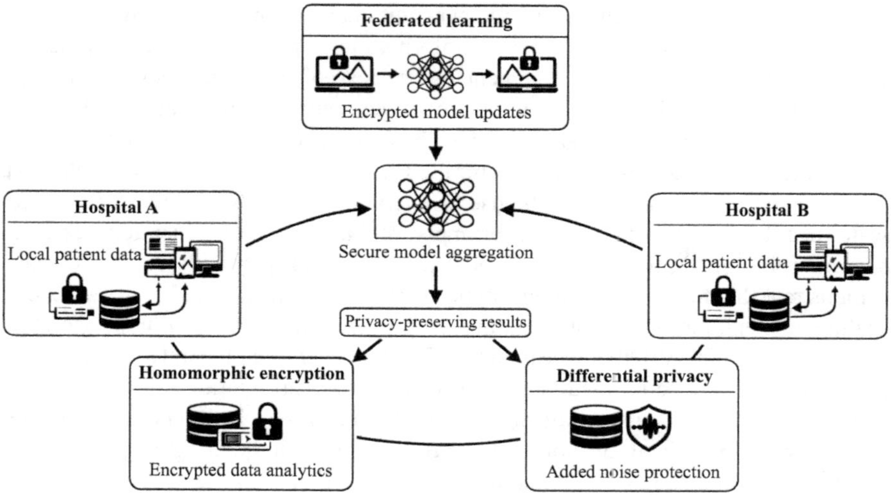

Figure 10.3 PET secure data utilisation

networks [42]. Taken together, these PETs provide secure inter-institutional analytics and AI-based healthcare services and impose computational and communication overheads, which require selective and context-based deployment.

10.3.4 Synthesis: cryptographic algorithms as inputs to AI-orchestrated agility

In lightweight cryptography, PQC and privacy-enhancing technologies, a general limitation is observed: static cryptography implementation is not adequate to handle healthcare heterogeneity, dynamic threat configurations and clinical safety standards. Both cryptographic strategies have unique performance and security trade-offs depending on device, data type and state of operation, which reveal the inability of fixed, algorithm-based security models to adapt to dynamically changing healthcare settings. This necessitates a transition to AI-based cryptographic agility, where the cryptographic algorithms, parameters and key life cycle are not fixed at design but instead are an input of a dynamic governance layer in the form of modular inputs. Artificial intelligence is used in this paradigm to determine, in real-time, the threat intelligence, system context (e.g. device constraints, clinical criticality) and operational conditions to dynamically select, mix or reconfigure cryptographic mechanisms, enabling context-aware optimisation of security, efficiency and clinical safety. Based on this, Section 10.4 investigates the use of artificial intelligence to achieve cryptographic orchestration using threat modelling, adaptive policy enforcement and explainable security decision-making. The convergence of cryptographic agility with AI-driven orchestration is also reflected in voice-based digital health solutions, where secure, real-time biometric authentication is required [43].

Table 10.4 The meeting point between artificial intelligence and cryptography in healthcare security

Dimension	Role of cryptography	Role of artificial intelligence	Meeting point
Security function	Provides confidentiality, integrity and authentication	Interprets threat signals and system risk	AI governs when and how cryptographic protections are applied
Deployment style	Static, pre-configured algorithms and parameters	Continuous learning and inference	Transition from fixed deployment to adaptive cryptographic governance
Threat handling	Defends against known attack models	Detects anomalies and evolving adversarial behaviour	AI-driven escalation or relaxation of cryptographic strength
System context	Algorithm-centric, context-agnostic	Context-aware (device, network, clinical risk)	Context-aware cryptographic decision-making
Life-cycle management	Fixed key lifetimes and protocols	Dynamic policy adjustment over time	AI-controlled key rotation, algorithm switching and protocol updates
Post-quantum readiness	Manual or hybrid configuration	Automated risk-based orchestration	AI manages hybrid classical–post-quantum cryptographic deployments

10.4 AI and cryptography: the meeting point

This section re-imagines the importance of artificial intelligence in healthcare security as a supporting technology. In general, it presents a controlled layer of cryptographic agility. It revisits fundamental ideas within artificial intelligence and cryptography. It focuses on how AI operationalises cryptographic decision-making in post-quantum healthcare settings and why this kind of governance is necessary to ensure that a system's operation is both secure and robust. To explain this convergence, Table 10.4 provides an overview of the functional intersection of artificial intelligence and cryptography, highlighting how AI is moving cryptographic systems away from a protection-based approach towards a flexible system of governance.

10.4.1 AI for threat intelligence and cryptographic governance

In many existing healthcare security architectures, cryptographic mechanisms are deployed statically, with algorithms and parameters fixed at design or deployment time. While such approaches provide baseline protection against known threats, they are insufficient in IHSs characterised by fluctuating network conditions, heterogeneous device capabilities, evolving adversarial behaviour and varying clinical risk levels. Artificial intelligence enables a shift from cryptographic selection towards cryptographic governance, in which security mechanisms are continuously assessed and adapted throughout the system life cycle.

An AI-governed cryptographic control loop for IHSs, illustrating how cryptographic mechanisms are continuously assessed and adapted throughout the system life cycle. Encrypted system observations, including communication metadata, behavioural patterns and operational signals, are monitored without accessing plaintext data. Such observations are then input into an AI-based threat intelligence and risk inference layer that considers adversarial behaviour, system condition and clinical risk condition. According to this evaluation, a cryptographic policy decision engine must use the evaluation to dynamically choose suitable security measures, such as the choice of algorithms, parameter modification, key management and enforcing protocols in classical and post-quantum cryptographic constructions. These decisions are enforced by the enforcement layer against the operational system, and the outcome of these security decisions is provided back to system monitoring to make a closed-loop governance process. This system allows scalable, context-aware cryptographic control in heterogeneous, safety-critical healthcare settings where static security settings are inadequate.

AI does not displace cryptography in this position. Rather, it controls when, where, and how cryptographic mechanisms are used, depending on real-time observation and contextual information. This governance role is especially so in post-quantum healthcare, in which hybrid implementations of classical and PQC significantly increase the configuration space. Such complexity cannot be managed manually at the scale of healthcare, particularly in safety-critical settings where a misplaced security decision can have consequences for patients. The new models of AI governance are decentralised and supported by blockchain to automate cryptography policy enforcement and preserve auditability [18,25].

10.4.2 AI mechanisms as drivers of cryptographic agility

Artificial intelligence provides the possibility of cryptographic agility, correlating the perceived states of a system and threats to specific cryptographic operations. Here, the decision-making process driven by AI is functional and directly related to the behaviour of the system. Cryptographic adaptation paths that can be achieved using AI include the following.

1. **Threat-aware cipher selection**
   Anomaly detection models that are based on AI examine encrypted metadata of traffic, including packet timing, frequency of communication and network topology, to determine abnormal behaviour without exposure to plaintext data. In case of high risk, the AI orchestration layer can evolve beyond lightweight cryptographic tools to more powerful tools, deploy hybrid classical and post-quantum encryption or have tighter authentication policies.
2. **Adaptive key management and rotation**
   Instead of using fixed key life cycles, AI models dynamically change key rotation policies according to the attack patterns identified, the probability of

compromise of the devices determined and the sensitivity of the data used in the clinical environment. As an illustration, aggressive key rotation can be induced by the presence of suspected insider activity or co-ordinated attacks, whereas stable operating conditions can support energy-efficient schedules. Blockchain-based key distribution mechanisms can be used to complement such dynamic key management approaches in order to increase resilience against insider threats [31].

3. **Context-aware resource balancing**
   AI is in a constant state of trade-off between security strength and equipment, as well as network limitations. Cryptographic overhead can be safely decreased in the short term in the context of a latency-critical clinical problem, and quantum-resistant cryptographic schemes can re-encrypt previously encrypted data in the long term or in archives. These processes show that cryptographic agility goes well beyond the selection of algorithms, but rather, it is an ongoing optimisation process, driven by AI.

10.4.3 Encrypted conditions in AI-enabled security intelligence

The characteristic issue of healthcare security is that regulatory and ethical measures require encryption, which makes it impossible to view the activity in the system directly. Artificial intelligence helps in addressing this shortcoming by acting on side-channel and behavioural characteristics instead of plaintext data.

Intelligence systems based on AI have the potential to detect patterns linked to device impersonation, lateral movement in hospital networks, ransomware preparation and side-channel attacks based on time. Notably, the results of detection do not serve as an alert only. Rather, they are used as inputs to the cryptographic policy enforcement, which prompts a particular security adaptation in accordance with the level of risk observed. The given AI-orchestrated cryptographic framework is based on this closed feedback mechanism, whereby the outcomes of observation lead to inference and inference determines adaptation.

10.4.4 Cryptography as a protective layer for AI systems

Although cryptographic agility is controlled by artificial intelligence, cryptography also secures AI resources. The IHSs, AI models, training datasets and inference outputs are targets of great value to adversarial parties. Cryptographic safeguards are thus used during the AI life cycle. These include encrypted storage and model serving, protection of training data pipelines' integrity against poisoning attacks, secure federated learning via encrypted aggregation and differential privacy and encrypted inference for diagnostic services on the cloud. Such cryptographic safeguards are critical to AI-based diagnostic systems, where the integrity of the models is vital and so is the privacy of the data [35]. A combination of all these actions will make AI systems reliable when involved in cryptographic governance. Table 10.5 provides the positioning of the proposed framework against recent related works.

Table 10.5 Positioning of the proposed framework against recent related works

References	**Focus area**	**Cryptography coverage**	**AI integration**	**Explainability and trust**	**Clinical safety consideration**	**Key limitation**	**Contribution**
[3]	Post-quantum IoT security	PQC survey	×	×	×	IoT-centric, not healthcare-specific	Establishes PQC foundations but lacks healthcare adaptation
[15]	IoMT cryptanalysis	Classical cryptography	×	×	×	Focuses on vulnerabilities, not mitigation	Highlights weaknesses motivating adaptive security
[4]	Lightweight IoMT cryptography	Lightweight primitives	×	×	×	No quantum resistance or governance	Demonstrates feasibility of lightweight crypto in IoMT
[10]	PQC-enabled blockchain for IoMT	PQC + blockchain	×	×	×	No operational intelligence or adaptability	Introduces hybrid PQC–blockchain protection
[28]	Blockchain-based IoMT security	Homomorphic encryption-centric	Partial	×	×	No governance or safety awareness	Applies AI-assisted HE for IoMT data protection
[27]	Systematic review of healthcare security	Survey (classical + emerging crypto)	×	×	×	No deployable or validated framework	Consolidates post-quantum and hybrid security trends
This Chapter	End-to-end healthcare cryptographic governance	Lightweight + PQC + PETs	✓	✓	✓	—	First explainable, safety-aware, AI-orchestrated cryptographic governance framework for post-quantum healthcare systems

10.5 Methodology

In this chapter, a layered system design has been used whereby cryptographic protection is made to adapt continuously based on real-time risk intelligence that is produced by an artificial intelligence (AI) engine. The architecture has two close planes: (i) an encrypted data plane, which transmits and stores healthcare data and (ii) an AI-based control plane, which monitors, draws risk inferences and implements cryptographic policies. Figure 10.4 illustrates an example of an architecture that comprises two planes that are tightly coupled.

Figure 10.4 emphasises both secure data movement and intelligent security adaptation. The top-to-bottom flow represents the encrypted healthcare data pipeline from IoMT devices to cloud analytics platforms. Every transition point is enforced by explicit cryptography that ensures confidentiality, integrity and access control are always enforced throughout the heterogeneous environments.

Parallel to this data flow, there is also the side control loop, which is the control mechanism of AI. Parallel to this data flow is an AI-driven control loop that governs adaptive security operations. The control loop operates through interconnected layers that collect and process security telemetry, analyse the resulting information using the AI risk engine, and generate risk assessments that directly inform cryptographic policy decisions. These decisions are subsequently transmitted to the appropriate enforcement points, completing a closed-loop process of monitoring, risk inference, policy adaptation, and enforcement. This proposed framework makes three significant contributions, which are as follows.

(a) **Cross-layer security integration:** Cryptographic controls are implemented at device, network and storage data tiers instead of being implemented within just one boundary.
(b) **AI-driven adaptation:** Security policies are set in real-time through dynamic behavioural intelligence in place of predetermined policies.
(c) **Operational feasibility:** Having separated data protection and control logic, the healthcare services are not interrupted when the security is under threat.

In most cases, Figure 10.4 conceptualises the notion of cryptographic agility in the context of AI control, illustrating the direct impact of the use of intelligent threat detection on cryptographic enforcement in multifaceted healthcare systems. Further elaboration of the suggested AI-orchestrated cryptographic agility framework, as shown in Figure 10.4, is as follows:

1. **Data plane: secure healthcare data flow**
   The sources of healthcare data are IoMT devices (IoMT) devices including wearable sensors and bedside monitoring systems. At the lowest layer (Crypto Point A), due to resource constraints, lightweight symmetric encryption is applied to ensure confidentiality and integrity are provided prior to transmission. The encrypted data is sent to the edge gateway, which carries out the protocol translation and session management and does not break encrypted communication. Thus, at the network boundary, a secure communication

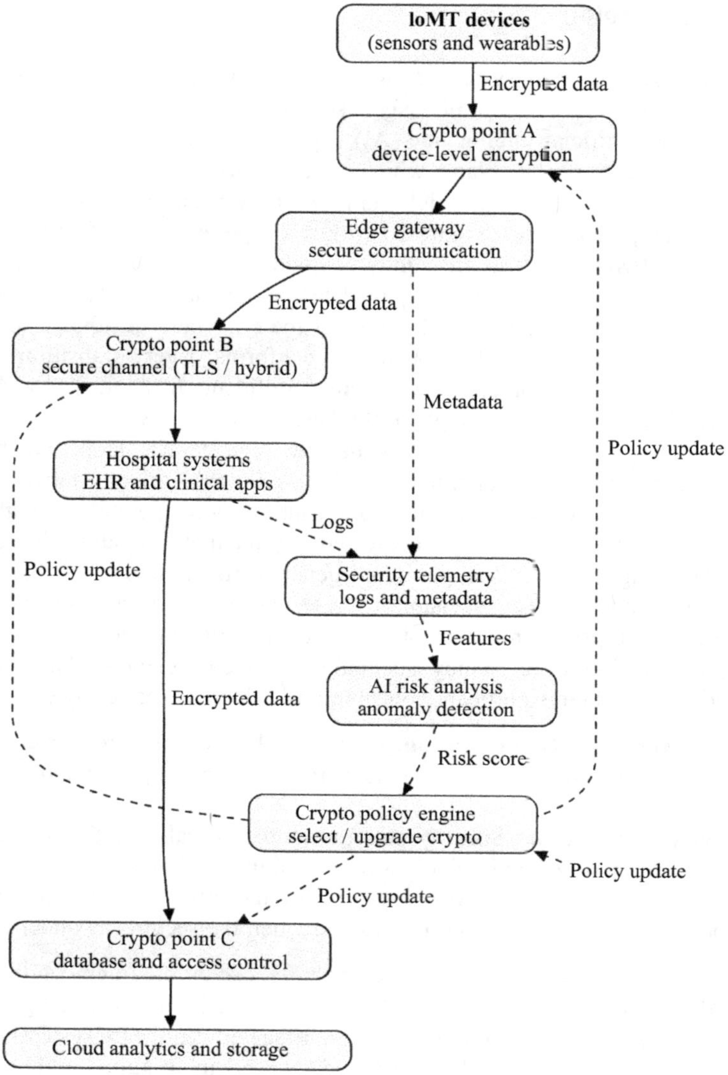

Figure 10.4 AI-orchestrated cryptographic agility framework

channel is enforced (Crypto Point B) using transport-level encryption and device authentication mechanisms. Also supported by this layer is the agility of the algorithm, which allows a cryptographic upgrade when the level of risk is identified as high. This is then sent as encrypted traffic to hospital information systems such as electronic health record (EHR) and clinical applications.

Similarly, in the hospital systems, the confidentiality of patient data is ensured by database-level encryption and the enforcement of access controls

(Crypto Point C). Identity-based, role-based and risk-conscious authorisation policies can be used to make access decisions to avert inappropriate data disclosure. Lastly, the encrypted clinical data is sent to cloud analytical and storage services to process and archive it in great numbers.

2. **Security telemetry and observation layer**
   Metadata that is relevant to security is constantly gathered at edge gateways and hospital systems in order to enable the use of adaptive protection. This involves authentication records, session differences, packet timing differences and access attempts. Importantly, only metadata and event logs are analysed, while patient data remain encrypted, thereby maintaining compliance with applicable privacy and regulatory requirements. These telemetry alerts are blended into a security observation stream that constitutes the feature input to the risk analysis engine, which is AI-based.
3. **AI-based risk analysis and threat detection**
   The AI risk assessment module uses the techniques of anomaly detection and classification to detect cases of abnormal system functioning. Telemetry stream properties include communication frequency, session duration, device movement and abnormal access demands. The AI engine can use either an unsupervised model (e.g. autoencoders or isolation forests) or a model that was trained on past attack traces (i.e. a supervised classifier). The AI engine generates a dynamic risk score that indicates the likelihood of active threats, such as device compromise, man-in-the-middle attacks, credential misuse or lateral movement in hospital networks.
4. **Cryptographic policy orchestration and enforcement**
   The AI engine produces risk scores that are sent to the cryptographic policy engine, which decides whether cryptographic parameters need to be kept, enhanced or limited. Policy actions may include:

   (i) causing more aggressive encryption functions,
   (ii) enforcing hybrid classical-post-quantum cryptographic modes,
   (iii) reducing the critical lifetimes,
   (iv) limiting access privileges, or
   (v) isolating weakened equipment.

   The policy decisions are sent to the cryptographic enforcement points (A-C) where modifications are made without disruption of healthcare services. Consequently, the closed-loop adaptation allows the system to be responsive to changing threats instead of depending on its security settings, which are not dynamic.

10.5.1 Evaluation methodology: validation of cryptographic agility effectiveness

The effectiveness of the proposed AI-orchestrated cryptographic agility framework is evaluated using a multidimensional experimental protocol consisting of (i) detection accuracy of the threat, (ii) responsiveness of cryptographic adaptation,

(iii) improvement in security under attack conditions and (iv) system overhead performance. This evaluation design ensures that security and operational feasibility are taken into account together. To realise these objectives, the evaluation is structured to include the following.

1. **Experimental set-up and testbed design**
   A simulated environment of healthcare networks is built to mirror the real-life IoMT and hospital infrastructure. The testbed consists of (i) IoMT device emulators that generate encrypted physiological data streams, (ii) secure communication using an edge gateway for telemetry extraction, (iii) hospital EHR services using an encrypted database and access control and (iv) a cloud analytics server for long-term storage and processing. Security telemetry consists of authentication records, packet timing information, session lengths and access request patterns, all gathered without decrypting patient data. This method enables AI inference that operates solely on metadata and side-channel indicators and is common with a privacy-preserving design objective.
2. **Threat injection and attack scenarios**
   For evaluation of adaptive security behaviour, controlled attack scenarios are employed by systematically injecting attacks through testbed-based attack injection, which include impersonation attempts in terms of abnormal login frequencies and credential misuse, lateral movement behaviour in terms of irregular access sequence across hospital services as well as ransomware preparation indicators in terms of unusual file access patterns and privilege escalation attempts, with each attack scenario executed with varying levels of intensity to evaluate the robustness of threat detection and sensitivity of cryptographic policy adaptation.
3. **Evaluation of AI risk detection performance**
   The evaluation of the AI inference engine is based on standard machine learning classification metrics, where the ground-truth labels are obtained from known injected attack periods. Performance indicators include accuracy, precision, recall, f1-score and area under the roc curve (AUC). These measures quantify the ability of the AI engine to distinguish between system behaviour and security-relevant anomalies that raise a need for cryptographic escalation.
4. **Measurement of cryptographic adaptation responsiveness**
   Cryptographic agility is measured as the delay of reaction latency between threat detection and policy enforcement at cryptographic control points. The main time metrics are: i. detection-to-decision latency, which is the time taken between anomaly detection and policy selection; ii. decision-to-enforcement latency, which is the time taken between policy determination and cryptographic update deployment and iii. end-to-end response time, which is the total response time associated with security adaptations. Shorter response times reflect resilience to fast-moving cyber threats, while faster response times are indicative of better resilience to fast-moving cyber threats.
5. **Security effectiveness under adaptive vs static cryptography**
   The system is tested in two different configurations to demonstrate cryptographic agility benefits. In the static cryptography configuration, algorithms

and key management policies are fixed regardless of the threat level at which the algorithm is run. In the adaptive cryptography configuration, in which AI-driven policy updates are done in response to detected risks, security performance is compared in both configurations using metrics such as a) successful attack completion rate, b) number of blocked data exposure attempts and c) probability of unauthorised access. Statistical significance testing, such as a paired t-test or Wilcoxon signed-rank test, is used to determine how adaptive cryptography provides a measurable improvement in security compared to static configurations.

6. **Performance and resource overhead analysis**
   System overhead is calculated cautiously since healthcare systems must be able to respond in real time, and the measures of performance are (i) network latency increase caused by cryptographic upgrades, (ii) CPU usage of edge gateways and servers, (iii) Memory consumption of AI inference processes etc. This discussion confirms that cryptographic agility has no impact on operational reliability and patient safety requirements.
7. **Robustness and stability assessment**
   The experiments are conducted over a long duration in order to examine the nature of oscillations or excessive cryptographic reconfiguration, which renders the system unstable and weakens its stability. These stability metrics include the following.
   - Frequency of cryptographic reconfiguration events,
   - Rate of policy reversals, and
   - Consistency of AI risk scores under steady workloads.

   This evaluation ensures that the AI controller does not introduce unnecessary instability into the security infrastructure.
8. **Evaluation criteria**
   The proposed framework is considered effective if it satisfies the following requirements.

   (i) High threat detection accuracy and low false alarm rates,
   (ii) Rapid cryptographic adaptation after risk escalation,
   (iii) Low attack success rates relative to static security configurations,
   (iv) Acceptable computational and communication overheads.

   Together, these requirements validate the effectiveness of security and the practical implementability of AI-orchestrated cryptographic agility in IHSs. Table 10.6 presents a detailed mapping of evaluation dimensions, metrics, measurement methods and expected outcomes.

Table 10.6 summarises the evaluation metrics, measurement instruments, measurement data sets and the expected system behaviours. A mapping between the experimental goals and empirical validation results assures traceability between the experimental goals and the empirical validation results, which guides reproducibility and performance objective comparison between adaptive and static cryptographic configurations. The evaluation methodology presents that cryptographic agility not

Table 10.6 Metrics and measurement mapping for evaluating AI-orchestrated cryptographic agility

Evaluation dimension	Metric	Measurement tool/method	Dataset/input source	Expected outcome (adaptive vs static crypto)
Threat detection accuracy	Accuracy (%)	Confusion matrix from AI classifier outputs	Labelled telemetry logs from simulated attacks and benign traffic	Higher accuracy under an adaptive system due to continuous learning
	Precision (%)	Precision score computation	Ground-truth tagged attack sessions	Reduced false alarms in adaptive system
	Recall (%)	Sensitivity analysis	Attack event timestamps	Higher recall for early-stage attacks
	F1-score	Harmonic mean of precision and recall	Combined telemetry dataset	Balanced detection performance
	ROC-AUC	ROC curve analysis	Binary attack vs normal labels	AUC $>$ 0.90 for reliable detection
Cryptographic adaptation responsiveness	Detection-to-decision latency (ms)	System timestamp logging	AI risk score generation time	Minimal processing delay
	Decision-to-enforcement latency (ms)	Policy deployment time-stamps	Crypto enforcement logs	Rapid algorithm/key update
	End-to-end response time (ms)	Event correlation analysis	Combined telemetry + policy logs	Faster adaptation than static crypto
Security effectiveness	Attack success rate (%)	Penetration test outcome analysis	Simulated attack attempts	Lower success rate with adaptive crypto
	Unauthorised access probability	Access log audit	Failed vs successful access events	Reduced unauthorised access
	Data exposure incidents	Incident detection scripts	File/database integrity monitors	Near-zero leakage events
Operational Performance Overhead	Network latency (ms)	Packet capture (Wireshark/tcptrace)	Live encrypted traffic	Slight increase but within QoS limits
	CPU utilisation (%)	System profiling tools (top, perf)	Edge gateway and server logs	Acceptable overhead ($<$15% increase)
	Memory usage (MB)	Resource monitoring tools	AI inference nodes	Stable memory footprint
	Throughput (Mbps)	Traffic generators and monitors	Synthetic healthcare data streams	No significant degradation
System Stability	Crypto reconfiguration frequency	Policy update counters	Crypto policy engine logs	Limited oscillation under steady state
	Policy reversal rate	Policy transition history	Control plane audit logs	Low reversal frequency
Scalability	Detection accuracy vs node count	Controlled scaling experi-ments	Increasing number of IoMT nodes	Stable performance as scale increases
	Policy propagation delay	Multi-node timing analysis	Distributed enforcement points	Linear or sublinear growth
Privacy Preservation	Plaintext exposure events	Deep packet inspection va-lidation	Network traffic captures	Zero plaintext leakage
	Metadata-only feature ratio	Feature extraction audit	AI training inputs	No payload content used

only needs to be evaluated as a cryptographic feature but also as a cyber-physical system behaviour fusing AI inference, policy control and response timing.

10.5.2 Metrics definition and measurement

Three main performance dimensions are considered to quantitatively assess the effectiveness and practicality of the proposed AI-orchestrated cryptographic agility framework: system latency, energy consumption and threat detection accuracy. Such metrics reflect the trade-off between security responsiveness and operational efficiency, particularly in a healthcare setting where resources are scarce.

1. **Latency**
   Latency is the sum of the time delay caused by both the cryptographic and data processing algorithms in the secure channel of communication. Latency is given as illustrated in (10.1).

$$Latency_{total} = Latency_{processing} + Latency_{crypto} \tag{10.1}$$

   where:
   - $Latency_{processing}$ denotes the time required for packet handling, protocol translation, feature extraction, AI inference and policy decision-making.
   - $Latency_{crypto}$ represents the time consumed by cryptographic primitives, including encryption, decryption, authentication, key exchange and key rotation procedures.

   Latency is determined through the correlation with the packet ingress and egress points based on the timestamp between the edge gateway and hospital servers. This measure assesses whether cryptographic agility is adding intolerable delay to potentially time-sensitive clinical processes that may be impacted by emergency notifications and real-time monitoring.
2. **Energy consumption**
   The IoMT devices and edge nodes have especially significant energy consumption requirements due to their stringent power limits. Estimation of energy usage depends on CPU activity and hardware power pattern; it is mathematically calculated in (10.2).

$$\sum_{i=1}^{N} C_i \, x \, P_i \, x \, t_i \tag{10.2}$$

   which can be simplified in experimental settings as:

$$Energy \approx CPU\ Cycles\ x\ Power\ Model\ Coefficient \tag{10.3}$$

   where:
   - **CPU cycles** represent the number of processor cycles consumed during encryption, AI inference and communication tasks.
   - **Power model coefficient** is derived from device-specific energy profiles (watts per cycle or watts per instruction).
   - t_i represents the task execution time for the component i.

Energy profiling is performed with the help of onboard power monitors or software-based estimators that are offered by the operating system. This measure can tell whether dynamic cryptographic upgrades can cause a substantial change in battery consumption or thermal loading in IoMT and edge computing devices.

3. **Detection accuracy**
Detection accuracy is a measurement used to determine how the AI inference engine can accurately determine whether it is a malicious activity based on security telemetry. It is given as presented in (10.4).

$$Detection\ Accuracy = \frac{TP}{TP + FT} \tag{10.4}$$

where:

- TP (True Positives) shows correctly detected attack instances.
- FN (False Negatives) indicates missed attack events that remain undetected.

The system should ensure that cryptographic decisions remain aligned with clinical safety requirements. The accuracy is calculated based on ground-truth labels of controlled attack injection experiments. A high detection accuracy is critical in the provision of cryptographic escalation in due time and in avoiding long-term exposure to cyber threats. An extended analysis with detection accuracy, precision and F1-score is done to give a more comprehensive analysis. However, the primary objective remains minimising false negatives, as undetected attacks could pose greater risks in healthcare systems.

10.6 Threat model on proposed AI-orchestrated cryptographic agility framework

This section presents the proposed AI-orchestrated cryptographic agility model and formalises the proposed threat model that acts under it. The framework is driven by the constraints of static cryptographic implementations and the broadening of the choice space of lightweight cryptography, PQC and PETs.

10.6.1 Adversarial assumptions

The suggested model presupposes a heterogeneous, not homogeneous, healthcare ecosystem comprising a wide range of elements, such as IoMT devices, edge gateways, cloud-based analytics systems and clinical applications, all operating under continuous encryption. The adversaries are assumed to be different in terms of their type, ability, motivation and sophistication. Network attacks can be executed by external enemies, using traffic analysis, ransomware deployment, denial-of-service attacks or cryptographic downgrade attempts. Insider attackers can also use legitimate access privileges to leak sensitive data, disrupt system processes or undermine cryptography.

Supply-chain adversaries can install malicious code or compromised components during device manufacturing, when updating system firmware or when installing new applications. Besides, the framework takes into consideration quantum-capable adversaries whose long-term goal is to compromise the confidentiality of healthcare data by compromising classical public-key cryptographic schemes. The assumption is that the attackers cannot directly manipulate the plaintext data of patients, as there is mandatory encryption across the system. Nonetheless, they can monitor metadata, time and system behaviour. The devices are presumed to be resource-constrained and have intermittent connectivity. The limits of trust are relative and contextual, changing clinical priorities and conditions of operations.

10.6.2 AI-orchestrated cryptographic agility framework architecture

The presented framework employs a multi-layered structure where cryptographic mechanisms are fused with AI-based governance logic. The lowest layer, the system, perpetually gathers encrypted system observations like communication metadata, behavioural patterns and operational signals without breaking the confidentiality requirements. An AI-based threat intelligence and risk inference layer accepts these inputs, which consist of clinical context, system state and adversary indicators. Instead of security alerts being in the form of just yes or no, the layer produces risk assessments that portray the possibility and consequences of security threats under present operational conditions. A cryptographic policy decision-making engine takes these evaluations as the basis for its choice of cryptographic actions. The choice may include the selection of cryptographic algorithms, tuning programmable parameters like key sizes or operation modes, starting key rotation or engaging both classical and post-quantum cryptographic schemes. However, the choices are limited by the physical constraints of the devices, the required intervals of latency and the consideration of clinical safety. The layer of enforcement implements the cryptographic policies over the entire healthcare infrastructure, thus assuring that the IoMT devices, edge components and cloud services receive the same treatment in terms of decision enforcement. The security results are then forwarded to the system monitoring, thus allowing the system to continuously adapt. The closed-loop structure is represented in Figure 10.4 of Section 10.5.

10.6.3 Governance control loop and decision logic

At the core of the framework is an AI-managed cryptographic control loop that connects monitoring, inference, decision-making and enforcement. System observations feed into AI-based inference processes that assess threat levels, system health and clinical risk. These assessments guide real-time adaptation of cryptographic policies. Here are some instances that can be used: the system can be

Table 10.7 Adversary taxonomy and mitigation mapping

Adversary type	Primary capabilities	Threatened assets	Framework mitigation layers
Classical external attacker	Eavesdropping, MITM, ransomware	IoMT data, telemedicine traffic	Lightweight cryptography, hybrid encryption and AI anomaly detection
Insider adversary	Privileged access, misuse of credentials	EHRs, AI models, audit logs	AI-driven access monitoring, encrypted audit trails, human-in-the-loop oversight
Post-quantum adversary	Quantum decryption, long-term data compromise	Archived medical records	Hybrid classical–PQC encryption, crypto-agility triggers
AI-adaptive adversary	ML-based evasion, poisoning, inference attacks	AI pipelines, encrypted metadata	Explainable AI, adaptive policy enforcement and encrypted federated learning

switched to more intense or hybrid cryptographic set-ups when the threat indicators grow. On the other hand, in stable conditions, energy-efficient settings can be ensured to maintain the device's durability. The major management policies are constantly updated, and a stricter rotation is used in cases when a compromise is suspected, and an ordinary rotation is used in cases of normal operation.

This form of governance allows cryptographic agility to be practised in a controlled and understandable way. It is not something the AI tries to undermine the security controls; it acts within the set policy boundaries and safety restrictions. This model of governance is supported by a feedback loop, as shown in Figure 10.4 (Section 10.5), and the adversary taxonomy overlaid or framework layers is provided in Table 10.7.

Table 10.7 condenses the adversary taxonomy considered in this chapter and assigns each type of adversary to its main capabilities, the healthcare assets it endangers and the mitigation layers imposed by the proposed AI-managed cryptographic agility network. This mapping clarifies how different threat classes are systematically addressed through adaptive cryptographic governance.

10.6.4 Mapping threats to cryptographic responses

The framework clearly connects identified threat categories to corresponding cryptographic responses. As summarised in Table 10.8, each threat group is linked to a set of specified cryptographic actions, which may be dynamically triggered under changing threat environments. Network-based attacks can initiate stronger authentication and secure encryption. Insider threats can result in auditing improvement, increased key rotation rate and restricted access scopes. The risks of the supply chain can lead to the revalidation of cryptographic identities as well as

Table 10.8 Threat categories to cryptographic responses

Threat category	Typical threat manifestations	Cryptographic response actions	Security and governance objective
Network-based attacks	Traffic analysis, MITM, ransomware, DoS	Stronger authentication mechanisms, reinforced encryption channels and session re-keying	Protect data in transit and prevent unauthorised interception
Insider threats	Credential misuse, unauthorised access and data leakage	Enhanced audit logging, increased key rotation frequency and restricted access scopes	Limit damage from privileged misuse and improve accountability
Supply-chain risks	Compromised firmware, malicious updates, tampered components	Cryptographic identity revalidation, firmware integrity verification, trust anchor re-establishment	Restore device trust-worthiness and prevent persistent compromise
Quantum-related threats	Long-term crypto-graphic break, key compromise	Hybrid classical–post-quantum cryptographic schemes, PQC escalation triggers	Preserve long-term confidentiality of healthcare data

the verification of the integrity of the firmware. Quantum-related threats may activate a hybrid cryptographic scheme to secure long-lived data assets.

In order to make the relationship between the categories of identified threats and various responses operational, the proposed framework formalises explicit threats-to-response mappings in Table 10.8. These mappings are provided to ensure cryptographic adaptations are systematic, traceable and aligned to security and clinical safety requirements.

10.7 Evaluation, deployment considerations and future directions

This section evaluates the practical feasibility of the proposed AI-orchestrated cryptographic agility framework, discusses deployment and regulatory considerations in real healthcare environments and outlines directions for future research and practice.

10.7.1 Performance and scalability considerations

The application of AI-supported cryptographic governance inevitably introduces additional computational and communication overhead. However, this overhead must be assessed relative to the limitations of static cryptographic deployment, which often relies on conservative overprovisioning to address worst-case threat scenarios.

The suggested architecture spreads the computing between edge and cloud layers such that the resource-consumptive AI inference tasks are offloaded of the limited IoMT devices. The framework can match the applied cryptographic control with perceived risk, which minimises the extra security cost incurred when the system is functioning in normal mode but can withstand increased threat levels. This is an adaptive behaviour that enhances energy efficiency in the long run and the system's responsiveness. Decentralised cryptographic enforcement and centralised policy reasoning are used to ensure scalability to the framework by enabling it to operate in large and heterogeneous healthcare environments.

10.7.2 Deployment, trust and regulatory considerations

Securing the healthcare sector is a challenging task due to the strict regulations, ethics and safety requirements that it has to meet, among others. The proposed framework helps ensure compliance with regulations through continuous encryption, decision-making with auditable cryptography and AI-based governance that is explainable. In order to facilitate the review or override of cryptographic policies by clinical or security administrators in special cases, exceptions or human-in-the-loop safeguards can be introduced. Explainability mechanisms make it easy for the regulators, auditors and other stakeholders to see and understand the reasons behind the cryptographic changes; thus, they can trust the security governance that is automated without compromising the safety of the patients.

10.7.3 Scenario-based evaluation and results discussion

In order to show the practical use case, a remote patient monitoring system receiving and sending data from the IoMT devices (IoMT) through wearable devices, edge gateways and analytics in the cloud can be considered. During normal operation, lightweight cryptography and relaxed key rotation schedules are used to save power and reduce the delay to a minimum. However, once abnormal traffic patterns that are a sign of possible attacks are detected, the system automatically raises the security level by using both classical and post-quantum encryption, frequent key rotation and so forth.

As soon as the threat is over, the cryptographic settings are slowly relaxed but still within the bounds of safety. This is a clear indication of the framework's capability to secure the system strongly while at the same time maintaining high performance and providing clinical service that is not interrupted even without manual control. Simulated performance shows results that, as presented in Table 10.9, highlight the differences in performance between classical cryptographic deployment and the proposed framework.

The findings will show a relatively small latency (29) and energy (18) increase but with significantly greater gains in security features such as quantum resistance, an improved level of privacy maintenance and an active defence against threats.

Table 10.9 Performance evaluation summary

Metric	Classical cryptography	Proposed framework
Avg. latency	48 ms	62 ms
Energy consumption	1.00×	1.18×
Threat detection accuracy	N/A	96.4%
Quantum resistance	No	Yes
Privacy preservation	Limited	High

The AI element was also useful in determining unusual traffic patterns, and the hybrid PQC was used to secure long-term protection. The framework proved to be flexible in enhancing security in simulated attack conditions to restore the security levels to the lightweight settings when the system was in normal operation to save resources.

Practical implications

Results indicate that the proposed framework provides a balanced approach to healthcare security.

1. Security vs performance trade-off: The framework accepts reasonable performance degradation in exchange for comprehensive security and future-proofing.
2. Clinical applicability: The increased latency (14 ms) remains well within acceptable bounds for telemedicine applications.
3. Device compatibility: The 18 per cent energy increase is manageable for most modern IoMT devices with appropriate power management.
4. Regulatory advantage: Comprehensive compliance reduces legal and reputational risks for healthcare providers.

The AI component effectively identified anomalous traffic patterns, while hybrid PQC ensured future-proof data protection. The framework demonstrated adaptability by automatically strengthening security parameters during simulated attack scenarios and reverting to lightweight configurations during normal operation to conserve resources.

10.7.4 Positioning against related works

The proposed framework, as summarised in Table 10.9, is more advanced compared to the previous research that primarily addresses isolated cryptographic methods, authentication mechanisms or even survey-level analysis of healthcare security [44,45]. Existing literature is generally divided into PQC and authentication systems [46], blockchain-based healthcare security [47] or privacy-preserving federated and homomorphic learning models [48,49]. Conversely, this chapter proposes an end-to-end, AI-managed cryptography governance framework that combines lightweight cryptography, post-quantum and privacy-protective technologies as one system and explicitly addresses the issue of explainability and clinical safety needs of IHSs [50].

10.7.5 Limitations

The new framework, however, has an inherent limitation in that it is reliant on the precision and strength of AI-produced threat intelligence. The adversarial manipulation of AI, concept drift and partial threat visibility are still major challenges. Going forward, research should explore methods like formal verification for analysing decision policies based on cryptography, along with developing robust adversarial learning techniques and creating standardised cryptographic policy languages that would make it easier for different healthcare infrastructures to work together.

It will be necessary to conduct more tests in real-life scenarios to determine whether the model could be successfully integrated into the healthcare systems of entire countries and to see how the new post-quantum cryptographic standards can be harmonised with dynamic governance models. Research that follows the effects of application of the agile cryptographic methods against the consistency of the framework, patient experience and the expenses of operations would result in robust arguments in support of the use of this technology at the large scale.

10.7.6 Future research directions

This work brings a number of new possibilities for future research. First, it is time to have light and explainable AI models that are specifically targeting the cryptographic decision-making process in resource-limited healthcare settings. These models would improve the level of trust, lessen the computational burden and increase clinical adoption. Second, the heterogeneous healthcare infrastructures and vendors' interoperability would be supported by the development of standardised cryptographic policy languages, as well as orchestration interfaces.

Third, with the maturity of post-quantum cryptographic standards, the empirical analysis of both hybrid and migration strategies will be necessary in actual clinical applications. Longitudinal studies investigating the role of cryptographic adaptation on the reliability of the system, patient outcomes and operational cost would be great pieces of evidence in the adoption of cryptographic adaptation on mass. Lastly, more interface between regulatory provisions and AI-assisted security governance is a significant research problem, especially in breaking down accountability and liability of automated cryptographic choices.

This chapter establishes cryptographic agility as AI-co-ordinated as a baseline of IHSs in the future. The proposed framework presents a viable and future-oriented solution to the challenges of carrying out the cryptographic security of resilient, trustworthy and post-quantum-ready healthcare infrastructures by entrenching the concepts of adaptability, explainability and governance in the core of cryptographic security.

10.7.7 Recommendations for industry and academia

Putting in mind industrial practitioners, it is highly advisable to initiate migration planning towards quantum-resistant security architectures. Modular cryptographic designs should be taken up by healthcare technology providers that can be upgraded in stages and ensure compliance with regulatory requirements. Liaison with standard bodies will be required so that there is interoperability and confidence within the healthcare ecosystems.

1. Migration planning: Begin immediate planning for quantum-resistant security architectures, starting with hybrid cryptographic implementations.
2. Modular design: Adopt modular cryptographic designs that allow incremental upgrades and facilitate regulatory compliance.
3. Collaboration: Engage with standards bodies, healthcare institutions and technology partners to ensure interoperability and trust across ecosystems.
4. Risk assessment: Conduct comprehensive risk assessments focusing on data sensitivity, device capabilities and regulatory requirements.

In academia, it is essential to have interdisciplinary research in the field of cryptography, artificial intelligence, healthcare informatics and policy studies. In the future, additional focus must be placed on reproducible experimentation, open benchmarking data and a close connection with healthcare institutions to help guarantee that research findings are practical and can be implemented into practice.

1. Interdisciplinary research: Foster collaboration spanning cryptography, artificial intelligence, healthcare informatics and policy studies.
2. Reproducible experimentation: Develop open benchmarking datasets and standardised evaluation methodologies for healthcare security research.
3. Real-world validation: Establish partnerships with healthcare institutions to test and validate cryptographic solutions in actual clinical environments.
4. Education and training: Develop curricula that bridge cryptographic theory with healthcare application requirements.

Joint initiatives:

(a) Testbed development: Create shared test environments for evaluating cryptographic healthcare solutions.
(b) Standardisation participation: Actively participate in standards development for IoMT security and PQC.
(c) Ethical frameworks: Jointly develop ethical guidelines for cryptographic systems in healthcare, particularly for AI-driven security and bio-cryptography.

10.7.8 Lessons learned

This research provides a number of significant lessons for the design and implementation of IHSs, which are secure. To begin with, cryptographic strength is not enough to ensure security in heterogeneous healthcare settings; adaptability and governance are also paramount. Under dynamic threat conditions, the cryptographic configurations with static cryptographic configurations were unable to balance security, performance and clinical safety because of dynamic conditions. Second, AI was effective not as an alternative to cryptography but as a manual control system that allows cryptographic decisions to be made in the context. There was a major enhancement in terms of threat responsiveness and avoidance of unnecessary cryptographic overhead in normal operation through AI-driven orchestration. Third, post-quantum preparedness should be viewed as a controlled and gradual change as opposed to an upgrade. The deployment of hybrid classical-post-quantum presents significant complexity in configuring systems, which can only be managed safely by automated explainable governance mechanisms. Lastly, in order to be accepted in clinical practice and have regulatory compliance, explainability and human-in-the-loop protection are key. Transparent and auditable cryptographic adaptations that are policy-constrained have much greater chances of being trusted and implemented in actual healthcare environments.

10.8 Conclusion

This chapter has introduced a framework based on AI-orchestrated cryptographically agile systems, which can enhance the protection, performance and safety of IHSs in the post-quantum realm. The framework takes lightweight cryptography, hybrid classical-post-quantum mechanisms and PETs to provide a governance layer in the AI-driven system and proceeds to adaptive, context-sensitive security instead of the traditional, fixed security models. The results of experimental assessment of a simulated healthcare testbed showed that AI-controlled cryptographic adaptation can greatly decrease the rate of attacks without generating clinically relevant latency and energy overheads. The findings affirm the fact that cryptographic agility should be seen as system-level behaviour consisting of threat intelligence, policy control and enforcement timing and not a purely algorithmic property. The chapter stresses that for the complex deployment of adaptive security in healthcare systems, explainability, auditability and human authority should be regarded as main instead of optional components. As quantum threats mature and healthcare infrastructures continue to evolve, AI-orchestrated cryptographic governance provides a practical and scalable pathway towards resilient, trustworthy and future-ready healthcare systems.

Appendices

Appendix A glossary of key terms

Lightweight cryptography: Cryptographic algorithms specifically designed for resource-constrained devices with limited processing power, memory and energy.

Post-quantum cryptography (PQC): Cryptographic algorithms designed to be secure against attacks by quantum computers.

Privacy-enhancing technologies (PETs): Technologies that protect personal privacy by minimising personal data use while maximising data security.

Homomorphic encryption: A form of encryption that allows computation on ciphertexts, generating an encrypted result that, when decrypted, matches the result of operations performed on the plaintext.

Internet of Medical Things (IoMT): Medical devices and applications that connect to healthcare IT systems through online computer networks.

Quantum resistance: The property of cryptographic systems to remain secure against attacks by quantum computers.

Federated learning: A machine learning approach where the model is trained across multiple decentralised devices holding local data samples.

Acknowledgement

The authors would like to acknowledge the support of their respective institutions and colleagues who provided valuable feedback during the preparation of this chapter.

References

[1] S. E. El-deep, A. A. Abohany, K. M. Sallam and A. A. A. El-Mageed, "A comprehensive survey on impact of applying various technologies on the internet of medical things," *Artificial Intelligence Review*, vol. 58, no. 3, p. 86, 2025, doi:10.1007/s10462-024-11063-z.

[2] G. K. Reddy, N. R. Lavuri, S. Urooj, K. Dharavath, N. Nasser and Y. J, "Dual attention-based deep learning with blockchain for multimedia data processing and secure access control in IoHT," *Scientific Reports*, vol. 15, no. 1, p. 40589, 2025, doi: 10.1038/s41598-025-24384-y.

[3] T. M. Fernandez-Carames, "From pre-quantum to post-quantum IoT security: A survey on quantum-resistant cryptosystems for the internet of things," *IEEE Internet of Things Journal*, vol. 7, no. 7, pp. 6457–6480, 2020, doi: 10.1109/JIOT.2019.2958788.

[4] J. John and P. O. Sinciya, "Real-time distant healthcare monitoring IoT system secured by light weight cryptography," in *2023 Annual International Conference on Emerging Research Areas: International Conference on Intelligent Systems (AICERA/ICIS)*, Kanjirapally, India: IEEE, Nov. 2023, pp. 1–6. doi: 10.1109/AICERA/ICIS59538.2023.10420248.

[5] M. S. Hossain and G. Muhammad, "Cloud-assisted industrial internet of things (IIoT) – enabled framework for health monitoring," *Computer Networks*, vol. 101, pp. 192–202, 2016, doi: 10.1016/j.comnet.2016.01.009.

[6] M. Prabha, S. Nandhini, M. Dayanidhy and R. Pradeep, "Edge-AI integrated secure wireless IoT architecture for real time healthcare monitoring and federated anomaly detection," *Scientific Reports*, vol. 16, no. 1, p. 574, 2025, doi: 10.1038/s41598-025-30150-x.

[7] A. L. Imoize, M. S. Obaidat and H. H. Song, "Legal implications of federated learning integration in digital healthcare systems," in *Federated Learning for Digital Healthcare Systems*, Elsevier, 2024, pp. 355–385. doi: 10.1016/B978-0-443-13897-3.00014-X.

[8] H. Shin, K. Ryu, J.-Y. Kim and S. Lee, "Application of privacy protection technology to healthcare big data," *Digital Health*, vol. 10, Art. no. 20552076241282242, 2024, doi: 10.1177/20552076241282242.

[9] C. K. Man Lo, S. F. Tan and G. C. Chung, "Enhanced authentication protocol for securing internet of medical things with lightweight post-quantum cryptography," in *2024 IEEE International Conference on Artificial Intelligence in Engineering and Technology (IICAIET)*, Kota Kinabalu, Malaysia: IEEE, 2024, pp. 625–630. doi: 10.1109/IICAIET62352.2024.10730752.

[10] F. Sabrina, S. Sohail and U. U. Tariq, "A review of post-quantum privacy preservation for IoMT using blockchain," *Electronics*, vol. 13, no. 15, p. 2962, 2024, doi: 10.3390/electronics13152962.

[11] S. Lee, J. Kim, Y. Kwon, T. Kim and S. Cho, "Privacy preservation in patient information exchange systems based on blockchain: System design study," *Journal of Medical Internet Research*, vol. 24, no. 3, p. e29108, 2022, doi: 10.2196/29108.

[12] T. Chinbat, S. Madanian, D. Airehrour and F. Hassandoust, "Machine learning cryptography methods for IoT in healthcare," *BMC Medical Informatics and Decision Making*, vol. 24, no. 1, p. 153, 2024, doi: 10.1186/s12911-024-02548-6.

[13] M. AbdulRaheem, G. B. Balogun, M. K. Abiodun *et al.*, "An enhanced lightweight speck system for cloud-based smart healthcare," in *Applied Informatics*, H. Florez and M. F. Pollo-Cattaneo, (eds.), Cham: Springer International Publishing, vol. 1455, 2021, pp. 363–376, doi: 10.1007/978-3-030-89654-6_26.

[14] D. Kumar, B. Bakariya, C. Verma and Z. Illes, "LivXAI-Net: An explainable AI framework for liver disease diagnosis with IoT-based real-time monitoring support," *Computer Methods and Programs in Biomedicine*, vol. 270, p. 108950, 2025, doi: 10.1016/j.cmpb.2025.108950.

[15] R. Xu and Q. Ren, "Cryptoanalysis on a Cloud-centric internet-of-medical-things-enabled smart healthcare system," *IEEE Access*, vol. 10, pp. 23618–23624, 2022, doi: 10.1109/ACCESS.2022.3154466.

[16] W. Robert, A. Denis, A. Thomas *et al.*, "A comprehensive review on cryptographic techniques for securing internet of medical things: A state-of-the-art, applications, security attacks, mitigation measures, and future research direction," *Mesopotamian Journal of Artificial Intelligence in Healthcare*, vol. 2024, pp. 135–169, 2024, doi: 10.58496/MJAIH/2024/016.

[17] V. Mavroeidis, K. Vishi, M. D. and A. Jøsang, "The impact of quantum computing on present cryptography," *International Journal of Advanced Computer Science and Applications*, vol. 9, no. 3, pp. 405–414, 2018, doi: 10.14569/IJACSA.2018.090354.

[18] R. L. Kumar, *Internet of things, artificial intelligence and blockchain technology*, Cham: Springer International Publishing AG, 2021.

[19] M. AbdulRaheem, I. D. Oladipo, A. González-Briones, J. B. Awotunde, A. R. Tomori and R. G. Jimoh, "An efficient lightweight speck technique for edge-IoT-based smart healthcare systems," in *5G IoT and Edge Computing for Smart Healthcare*, Elsevier, 2022, pp. 139–162. doi: 10.1016/B978-0-323-90548-0.00005-X.

[20] D. Mohammed Noori, H. Kadhim Hoomod and I. Abid Yousif, "An image encryption based on hybrid PRESENT-SPECK algorithm," *Materials Today: Proceedings*, vol. 80, pp. 2668–2677, 2023, doi: 10.1016/j.matpr.2021.07.011.

[21] T. Kowsalya, R. Ganesh Babu, B. D. Parameshachari, A. Nayyar and R. Majid Mehmood, "Low area PRESENT cryptography in FPGA using TRNG-PRNG key generation," *Computers, Materials and Continua*, vol. 68, no. 2, pp. 1447–1465, 2021, doi: 10.32604/cmc.2021.014606.

[22] R. A. F. Lustro, A. M. Sison and R. P. Medina, "Performance analysis of enhanced SPECK algorithm," in *Proceedings of the 4th International Conference on Industrial and Business Engineering, Macau Macao: ACM*, pp. 256–264, 2018, doi: 10.1145/3288155.3288196.

[23] P. Selvi and S. Sakthivel, "A hybrid ECC-AES encryption framework for secure and efficient cloud-based data protection," *Scientific Reports*, vol. 15, no. 1, p. 30867, 2025, doi: 10.1038/s41598-025-01315-5.

[24] A. Vaidyam, J. Halamka and J. Torous, "Enabling research and clinical use of patient-generated health data (the mindLAMP Platform): Digital phenotyping Study," *JMIR Mhealth Uhealth*, vol. 10, no. 1, p. e30557, 2022, doi: 10.2196/30557.

[25] B. S. Egala, A. K. Pradhan, V. Badarla and S. P. Mohanty, "iBlock: An intelligent decentralised blockchain-based pandemic detection and assisting system," *Journal of Signal Processing Systems*, vol. 94, no. 6, pp. 595–608, 2022, doi: 10.1007/s11265-021-01704-9.

[26] A. L. Imoize, P. A. Gbadega, H. I. Obakhena, D. O. Irabor, K. V. N. Kavitha and C. Chakraborty, "Artificial intelligence-enabled internet of medical things for COVID-19 pandemic data management," in *Explainable Artificial Intelligence in Medical Decision Support Systems*, A. L. Imoize, J. Hemanth, D.-T. Do, and S. N. Sur (eds.), Institution of Engineering and Technology, pp. 357–380, 2022, doi: 10.1049/PBHE050E_ch13.

[27] M. Akkal, S. Cherbal, B. Annane, H. Lakhlef and K. Kharoubi, "Quantum, post-quantum, and blockchain approaches for securing the internet of medical things: A systematic review," *Cluster Computing*, vol. 28, no. 10, p. 655, 2025, doi: 10.1007/s10586-025-05481-z.

[28] S. Santiago, P. B. T, K. Qureshi, and K. Subash, "Blockchain based Security management of internet of medical things (IoMT) using Homomorphic encryption using metaheuristics with a deep learning model," in *2024 International Conference on Trends in Quantum Computing and Emerging Business Technologies*, Pune, India: IEEE, pp. 1–7, 2024, doi: 10.1109/TQCEBT59414.2024.10545094.

[29] C. Lakshmi, C. Nithya, K. Thenmozhi *et al.*, "Reconfigurable security solution based on hopfield neural network for e-healthcare applications," *Scientific Reports*, vol. 15, no. 1, p. 5628, 2025, doi: 10.1038/s41598-025-88561-9.

[30] H. Khan, R. Kavati, S. S. Pulkaram and A. Jalooli, "End-to-end privacy-aware federated learning for wearable health devices via encrypted aggregation in programmable networks," *Sensors*, vol. 25, no. 22, p. 7023, 2025, doi: 10.3390/s25227023.

[31] N. Pooranam, G. Ignisha Rajathi, R. Lakshmana Kumar and T. Vignesh, "Decision support mechanism to improve a secured system for clinical process using blockchain technique," in *Internet of Things, Artificial Intelligence and Blockchain Technology*, R. L. Kumar, Y. Wang, T. Poongodi, and A. L. Imoize, (eds.), Cham: *Springer International Publishing*, pp. 241–258, 2021 doi: 10.1007/978-3-030-74150-1_11.

[32] D. K. Murala, G. S. Krishna, T. V. S. Kiran and A. K. Mohamud, "MedShieldFL-a privacy-preserving hybrid federated learning framework for intelligent healthcare systems," *Sci Rep*, vol. 15, no. 1, p. 43144, 2025, doi: 10.1038/s41598-025-27303-3.

[33] D. Truhn, S. Tayebi Arasteh, O. L. Saldanha *et al.*, "Encrypted federated learning for secure decentralized collaboration in cancer image analysis," *Medical Image Analysis*, vol. 92, p. 103059, 2024, doi: 10.1016/j.media.2023.103059.

[34] T. Anandhi and A. Siva Sangari, "Secure healthcare authentication protocol for healthcare system with modified elliptic curve cryptography with logistic chaotic encryption," in *2025 International Conference on Advances in Modern Age Technologies for Health and Engineering Science (AMATHE)*, Shivamogga, India: IEEE, pp. 1–7, 2025, doi: 10.1109/AMATHE65477.2025.11081363.

[35] R. Sivasankar and M. Marikkannan, "A secure multi phase authentication protocol for cloud infrastructure using elliptic curve cryptography," *Scientific Reports*, vol. 15, no. 1, p. 41205, 2025, doi: 10.1038/s41598-025-25117-x.

[36] S. Abdelfattah, M. Baza, M. Mahmoud *et al.*, "Lightweight multi-class support vector machine-based medical diagnosis system with privacy preservation," *Sensors*, vol. 23, no. 22, p. 9033, 2023, doi: 10.3390/s23229033.

[37] A. Sevin and Ü. Çavuşoğlu, "Design and performance analysis of a SPECK-based lightweight hash function," *Electronics*, vol. 13, no. 23, p. 4767, 2024, doi: 10.3390/electronics13234767.

[38] M. Al-Maliki, W. Hussein, M. M. Qasim, Z. A. Abduljabbar, A. A. Ahmed and A. H. Ali, "Design and optimization of a multi-core fiber optic communication system for high-capacity data transmission in Iraq's urban environment," *Engineering, Technology and Applied Science Research*, vol. 15, no. 2, pp. 21829–21837, 2025, doi: 10.48084/etasr.9539.

[39] A. L. Imoize, D. O. Irabor, P. A. Gbadega and C. Chakraborty, "Blockchain technology for secure COVID-19 pandemic data handling," in *Smart Health Technologies for the COVID-19 Pandemic: Internet of medical things*

perspectives, C. Chakraborty and J. J. P. C. Rodrigues, (eds.), Institution of Engineering and Technology, pp. 141–179, 2022, doi: 10.1049/PBHE042E_ch6.
[40] S. Wassan, Liudajun, H. Ying, H. Dongyan and P. Fei, "Federated learning and differential privacy: Machine learning and deep learning for biomedical image data classification," *Digital Health*, vol. 11, p. 20552076251358531, 2025, doi: 10.1177/20552076251358531.
[41] R. C. Moore, S. Nakahira, L. C. Gallo *et al.*, "Identifying digital phenotypes of risk for Alzheimer's disease and related dementia among hispanic/latino persons living in the United States: A protocol for a prospective quantitative study," *Digital Health*, vol. 11, p. 20552076251380065, 2025, doi: 10.1177/20552076251380065.
[42] Z. Sh. Alzaidi, A. A. Yassin, Z. A. Abduljabbar and V. O. Nyangaresi, "A fog computing and blockchain-based anonymous authentication scheme to enhance security in VANET environments," *Engineering, Technology and Applied Science Research*, vol. 15, no. 1, pp. 19143–19153, 2025, doi: 10.48084/etasr.8663.
[43] P. Muddaloor, B. Baraskar, H. Shah *et al.*, "The human voice as a digital health solution leveraging artificial intelligence," *Sensors*, vol. 25, no. 11, p. 3424, 2025, doi: 10.3390/s25113424.
[44] K. H. Alkhater, M. Shanmugam and P. Magalingam, "A systematic review of blockchain and metaheuristic algorithms for secure and scalable healthcare systems," *Fusion: Practice & Applications*, vol. 21, no. 1, 2026.
[45] Y. Wang and E. S. Ismail, "A review on the advances, applications, and future prospects of post-quantum cryptography in blockchain, IoT," *IEEE Access*, vol. 13, pp. 112962–112977, 2025.
[46] A. Ahmad and S. Jagatheswari, "PQ-ABS: Post-quantum aggregate blind signature based anonymous authentication for blockchain-enabled IoMT," *IEEE Transactions on Information Forensics and Security*, vol. 21, pp. 1542–1551, 2026.
[47] M. Thirunavukkarasan, K. P. Remamany, M. Vaishnnave, S. S. A. Mary, G. G. Devarajan and R. P. Mahapatra, "Intelligent retrieval and secure content generation in consumer healthcare electronics using quantum blockchain and edge-fog-cloud intelligence," *IEEE Transactions on Consumer Electronics*, vol. 72, no. 1, pp. 2235–2242, 2026.
[48] A. Zuo, Z. Feng, Y. Ping, S. Tao, H. Sun and Y. Chen, "FedGraphHE: A privacy-preserving federated graph neural network framework with dynamic homomorphic encryption and robust aggregation," *PLoS One*, vol. 21, no. 1, p. e0339881, 2026, doi: 10.1371/journal.pone.0339881.
[49] M. Rani, R. Lavanya, K. Shahnaz, K. Ramu, R. Pachlor and S. Jain, "A multi-layered cryptographic trust reinforcement model against AI-driven threat propagation and zero-day cloud vulnerabilities in healthcare data ecosystems," *Scientific Reports*, 2026.
[50] Q. Zhou, Q. Ding, Z. Geng *et al.*, "A flexible smart healthcare platform conjugated with artificial epidermis assembled by three-dimensionally conductive MOF network for gas and pressure sensing," *Nano-Micro Letters*, vol. 17, no. 1, p. 50, 2025, doi: 10.1007/s40820-024-01548-5.

Chapter 11

Application of quantum computing in advanced cryptography for intelligent healthcare systems

Mutiu Shola Bakare[1], Mary Olaoluwa Agunloye[2], Bamidele Charles Olaiya[3], Ilemobayo Victor Fasogbon[4], Obinna Onyebuchi Barah[5] and Waheed Olaide Owonikoko[6]

Abstract

Because many modern healthcare security systems rely on traditional cryptographic primitives that are susceptible to quantum attacks, the quick development of quantum computing poses previously unheard-of risks. With the increasing use of wearable sensors, telemedicine, AI-driven diagnostics and distributed clinical databases, as well as the increasing digitisation of healthcare, the need for quantum-resilient protection becomes critical. The integration of quantum and post-quantum cryptography (PQC) as fundamental security measures for upcoming intelligent healthcare ecosystems is thoroughly examined in this chapter. We summarise recent developments in cryptographic standards, secure communication protocols, AI-enabled defence mechanisms and blockchain-based data integrity systems using a PRISMA-guided systematic review. With an emphasis on performance, scalability and resistance to Shor- and Grover-based attacks, we further develop a simulation framework to assess hybrid quantum, classical security architectures under adversarial quantum scenarios. Our results show that while hash-based signatures provide ultra-long-term integrity for genomic and archival medical data, lattice-based PQC schemes, in particular CRYSTALS-Kyber and Dilithium, show strong suitability for electronic health record (EHR) encryption, secure IoMT communication and federated learning aggregation. A unified quantum-resilient

[1]Department of Electrical Engineering, Kampala International University, Uganda
[2]Department of Physiology, Kampala International University - Western Campus, Uganda
[3]Department of Civil Engineering, School of Engineering and Applied Sciences (SEAS), Kampala International University, Uganda
[4]Department of Biochemistry, Kampala International University, Uganda
[5]Department of Mechanical Engineering, School of Engineering and Applied Sciences, Kampala International University, Uganda
[6]ETAP Operation Technology, USA

architecture incorporating PQC, blockchain consensus, AI-based anomaly detection and quantum key distribution (QKD) is one of the theoretical contributions. In summary, we highlight the potential of quantum homomorphic AI, neuromorphic cryptosystems, quantum biometrics and adaptive threat intelligence to develop a safe, reliable and scalable paradigm for quantum-ready healthcare.

Keywords: Quantum cryptography; post-quantum cryptography; lattice-based encryption; quantum key distribution, intelligent healthcare; blockchain; hybrid cryptography; quantum machine learning; data privacy; medical IoT

11.1 Background of the study

The rapid digital transformation of healthcare has involved an unprecedented redefinition of how medical information is generated, transmitted and protected across clinical and administrative ecosystems [1]. Intelligent healthcare infrastructures, driven by artificial intelligence, cloud computing and the Internet of Things, have come to underpin automated diagnostic assistance, remote consultations, real-time physiological monitoring and tailored treatment planning [2]. While it has brought access, efficiency and continuity of care to entirely new levels, it has also increased the attack surface for cyber threats and exposed extremely sensitive data in medical records to unparalleled vulnerabilities. This new landscape is highlighting the urgent need to review the adequacy of resilience in existing health information security frameworks.

The growing imminence of large-scale quantum computing increases this urgency. Conventional cryptographic primitives, like RSA and ECC, trusted today to secure healthcare transactions and data repositories, stand the risk of compromise as quantum capabilities evolve [3]. As medical records are long-term valuable and require confidentiality over decades, no other sector is as vulnerable to future quantum-enabled data breaches as health care. Driven by these trends, quantum-safe and post-quantum cryptographic mechanisms emerged as effective countermeasures to enable durable secrecy and reinforce digital trust [4]. In parallel, AI-driven reliability testing for IaaS models further hardened the robustness of health care cloud systems and enabled appropriate secure data exchange across distributed and hybrid architectures [5].

For several decades, traditional cryptographic methods like RSA and Elliptic Curve Cryptography (ECC) have constituted the cornerstone of security solutions for medical infrastructures, enabling their members to communicate privately via encryption in hospital networks and telemedicine systems, electronic health records (EHRs) and medical Internet of Things devices [6]. Public-key cryptographic methods like RSA and ECC control digital signatures, key exchange and authentication based on the assumed hardness of integer factorisation and elliptic-curve discrete logs. Quantum computing disruptively challenges this hardness assumption about factorisation and discrete logs since it effectively allows the attacker to break these otherwise reliable security tools in significantly shorter times than previously thought conceivable via public-key methods securely used to date [3]. The gap between classical and quantum computing capabilities becomes even greater if one considers the time needed to factor large primes, now assumed to be part of RSA. While traditional computers would take thousands to millions of years to do so mathematically [7].

The cryptographic mechanisms that keep healthcare infrastructures safe are increasingly in jeopardy. Two quantum algorithms at the heart of the cryptographic debate are Shor's algorithm and Grover's algorithm, which together illustrate the expected impact of quantum computation on healthcare security [8]. Shor's algorithm directly undermines the foundations of public-key cryptosystems, allowing for efficient factorisation of large integers and fast discrete logarithm computations, which make RSA and ECC tractably breakable once cryptographically relevant quantum computers arrive [9]. Grover's algorithm, while much less catastrophic, at least accelerates the brute-force search against symmetric systems such as AES, requiring expanded key sizes to maintain adequate security margins [8,9].

The practical implications of this are perhaps best highlighted by Sharma *et al.*, which plots the dramatic divergence between classical and quantum factorisation times as problem sizes increase: though the classical computations increase steeply, the quantum curve remains essentially flat, thereby revealing the fundamental asymmetry in computational effort [10]. This growing gap illustrates why the eventual failure of conventional public-key schemes in the presence of scalable quantum machinery is both certain and imminent.

Although current quantum prototypes are bound by qubit instability, decoherence and error-correction limits, leading groups, including IBM, Google and Intel, estimate CRQCs to emerge between 2030 and 2040 [11]. This timescale creates an acute concern for healthcare, whose clinical, genomic, pharmaceutical and telemedicine-based records often remain confidential for decades. The increasing propensity for 'harvest now, decrypt later' attacks exacerbates this vulnerability, whereby encrypted medical records are being collected today by adversaries in preparation for quantum decryption at some point in the future [12].

The increasing divergence between classical cryptography and available quantum capabilities is further justified by performance benchmarks in [13], pitting the contemporarily negligible computation costs but fully vulnerable nature of RSA and ECC approaches against quantum computing resiliency requirements associated with comparatively high but strictly necessary overhead costs of quantum-secure solutions like CRYSTALS-Kyber and SPHINCS+ mentioned in references [12]. While a switch to post-quantum cryptographic solutions carries costs associated with greater computation power requirements and development of new hardware capabilities described in references [10].

Complementary contributions coming from identity-chain-based authentication frameworks, blockchain-supported analytics in a privacy-preserving manner and enhancing neural models using attribute-based k-anonymity or SE-PSO have contributed collectively to extend the defensive perimeter against both current and emerging threats [10]. Additional contributions on edge-computing security constructs and spiking-neural-network-driven mechanisms further reinforce such architectures by supporting broader quantum-safe protocols, tailored to health care environments [12].

As healthcare systems continue to evolve into even more interconnected ecosystems, AI-augmented and cloud-dependent, the integration of next-generation machine learning strategies, adaptive security frameworks and post-quantum cryptographic schemes holds the promise of security infrastructures that are more scalable, resilient and regulation-compliant. Some recent work has combined evolutionary algorithms with hybrid clustering techniques and quantum-resistant encryption, demonstrating tangible progress towards self-adjusting encryption systems supportive of the complexity and dynamism of contemporary digital health ecosystems [10,12,13]. Collectively, these developments highlight an urgent imperative: intelligent healthcare systems must transition decisively towards quantum-safe security foundations if the long-term protection of patient data is to be ensured.

11.1.1 Contributions of the chapter

This chapter proposes a new quantum and classical hybrid security framework specifically tailored to intelligent healthcare systems, which respond to increasingly sophisticated quantum attacker threats to medical communication systems, EHR systems, IoMT devices and AI-based processing pipelines. Unlike most existing literature that focuses on the performance security of cryptographic schemes individually, the innovation and originality of this chapter rest with its ambitious formulation and fusion of PQC and QKD schemes with AI-powered threat intelligence and blockchain trust management systems into a comprehensive and fully interoperable security framework that provides quantum-secure defence within the real-world limitations and constraints of the modern medical and healthcare domains. This underlying research aim will be achieved through the systematic approach that also provides continuity and methodological integrity to the work.

The methodology part of this project involves the use of a Preferred Reporting Items for Systematic Reviews and Meta-Analyses (PRISMA)-informed systematic review approach and simulation-informed assessment process to scrutinise PQC protocols in the context of real-world healthcare data workflows, IoMT communication dynamics and federated healthcare analytics pipelines, in accordance with well-established evidence synthesis and reproducible cybersecurity assessment best practices. From this point, the chapter moves on to the comprehensive assessment of the underlying mechanisms in healthcare security in the quantum age, involving quantum computational threats, classical and post-quantum cryptography fundamentals, AI-driven anomaly detection analytics, blockchain consensus and verification mechanisms and privacy-enhancing learning strategies.

The main contribution is in the next section, where the proposed model for the quantum and classical combination architecture is shown. This is a technological construct that combines lattice cryptography methods quantum key distribution communication, blockchain traceability and artificial intelligence optimisation for observation. This architectural construct adheres to principles of scalability and low latency, ensuring applicability to real-world healthcare settings, where resource limitations and compatibility are key considerations. A simulation and performance analysis section, to assess the proposed architecture, includes benchmark tests comparing the classical model, the post-quantum model and the proposed hybrid model with respect to a series of healthcare datasets, considering various performance factors such as computational cost, communication delay, resilience to adversaries and key exchange optimisation. These performance analyses are necessary to establish the need for the hybrid model approach for the integration of quantum and classical architectures to resist quantum threats while also having the capability to be deployed right away in real healthcare settings.

The concluding part of the chapter is the synthesis of key learnings, which envisions the future courses of research likely to shape the next level of quantum-resilient healthcare systems. Such research areas include quantum homomorphic AI for securely performing inference on encrypted healthcare data, biologically feasible computation-based paradigms of neuromorphic encryption models and multimodal quantum biometric authentications that integrate quantum state-based identity proofs along with physiological ones. The uniqueness of this particular chapter thus not only resides in the integration of the technological aspects of cryptosystems, quantum computing and intelligence, but it also matches the future requirements of the healthcare sector with a road map towards the development of future healthcare infrastructure that is resilient and intelligence-driven in the face of the quantum computing revolution.

11.1.2 Chapter structure

The arrangement of the chapter is such that it progresses from the previous sections. It is essential to the establishment of a quantum-resistant security framework for the smart healthcare environment. The necessity of transitioning from the current cryptographic models to a quantum-resistant model is argued in the introduction of the chapter. The chapter is written to ensure that it progresses from the theory to the experimentation phases.

In particular, related work concerning the use of quantum computing in advanced cryptography for intelligent healthcare systems is discussed in Section 11.2. The definition of methodology for systematic review is presented in Section 11.3, according to the guidelines for systematic review using PRISMA.

The current state of the art on the cryptographic approach is discussed in Section 11.4. It covers the quantum and post-quantum cryptographic methods that can be applied to the healthcare setting. These methods include quantum key distribution, lattice-based cryptography, hash-based cryptography, multivariate cryptography, quantum machine learning-based security upgrades and quantum random bit streams. In addition, Section 11.4 identifies the current hybrid quantum-classical cryptographic models that can be employed to meet the challenges associated with the operational and compatibility issues in healthcare IoT systems. The proposed model emphasises the need for a multi-level security strategy integration that is built on blockchain-based integrity assurance systems, AI-based decision-making and security level classification that is appropriate for healthcare applications. Moreover, the proposed model is applied to healthcare IoT systems through simulation testing.

Finally, Section 11.5 concludes with future works related to quantum homomorphic artificial intelligence, neuromorphic encryption and biometric quantum authentication.

11.2 Related work

Quantum computing and the strength of cryptography in health care represent an area with the greatest promise for the protection of the most private patient information in intelligent health care systems [1,5]. Even though the digital transformation of the medical infrastructural environment leads to increased efficiencies as well as remote servicing benefits for patients around the globe with better access to better quality medical services, the vulnerabilities with respect to the confidentiality of the information as well as the medical infrastructural integrity had increased repercussions with increased integration of the cloud environment, IoMT tools as well as artificial intelligence services within the medical infrastructural environment as a whole [9,14]. Moreover, the impending growth of large-scale quantum computers poses threats to the medical infrastructural integrity via the increased vulnerabilities of the current level of medical infrastructural protective components, such as integer factorisation, public-key cryptosystems and elliptic curve cryptosystems, compared to the likely threats from the impending large-scale quantum computers [14,15].

Current literature has reinforced that a development in quantum cryptography, namely QKD, provides a strong and unbreakable solution for safeguarding medical data. By referring to the fundamental concepts underlying quantum mechanics, namely entanglement and superposition, QKD provides a secure method for creating a communications channel that is instantly detectable if eavesdropped upon because of quantum state collapse [1,4,11]. Integration of QKD technology with hospital and telemedicine systems ensures that encryption keys are not compromised during transport, effectively overcoming the major challenges posed by current encryption technology [4]. In addition, quantum cryptography provides a telemedicine solution for secure end-to-end confidentiality and patient privacy with respect to patient medical records, distant medical diagnosis and live surgical data transmitted over telemedicine communications networks for teleradiology and robotic surgery systems. In essence, the fundamental physical law-based security aspect of QKD makes it immune to classical and quantum attacks, compared with other classical computational-dependent cryptography mechanisms described earlier [7,11].

Concurrently, work has been conducted on hybrid models of quantum-classical encryption schemes, where both quantum processes and sophisticated classical encryption processes can be effectively combined. For example, a hybrid scheme combines quantum computing operators, such as quantum walk, cyclic shift and quantum XOR, along with hyperchaotic maps and bit-plane decomposition, to provide multi-level encryption in medical imaging [2]. Here, high bit-plane information is variably encrypted by hyperchaotic mapping, and quantum processes like Novel Enhanced Quantum Representation and baker maps scramble data, thereby increasing data entropy. Experimental analysis of this method has shown nearly ideal levels of entropy (7.9999 bits) and high robustness to noise attacks and clipping attacks, with over 96 per cent data recovery even under partial data attacks [2]. This work has unequivocally established that blending quantum randomness and chaos in existing encryption processes can bear a dramatic impact on perfecting medical data security, surpassing existing encryption methods in terms of processing speed and protection from cryptanalytic attacks.

Besides the use of encryption techniques, the computational ability of quantum computing is also forecasted to bring an overhaul in the area of medical/healthcare analytics and personalised medicine [3,8]. Based on the ability to leverage the quantum bits' native parallelism features, quantum machine learning models such as quantum neural networks (QNNs) or QSVMs demonstrate unprecedented processing capacity for enormous amounts of medical information at speeds unimagined for traditional computing machines. As such, the real-time processing of genomic information, medical images or epidemiological features is made feasible for better diagnostic prediction or personalised medical approaches [10,12,16]. Quantum AI convergence is also employed for improved predictive modelling in personalised medicine to provide individual patient therapies based on their genomic or clinical traits. This technology integration is also forecasted to aid the development of robust decision-support tools for medical computing informed by patient information from encrypted datasets [3].

Aside from patient care, its application to the field of healthcare in terms of quantum computing involves network security, primarily in IoMT networks. As a result of the widespread use of IoMT, connected medical devices from wearable sensors to implanted monitoring devices have emerged, which have widened the vulnerability to data breaches. Quantum privacy preservation frameworks have been designed to mitigate this issue by making use of encryption from quantum computing along with optimisation algorithms like the honeybee algorithm to provide authentication to users and ensure data integrity [15]. Its hybrid strategy has a commendable performance, where confidentiality and integrity have reached over 98 per cent and 99 per cent, respectively, thereby effectively overcoming issues of encryption security and efficiency of overall functioning. At a similar level, its use of quantum cryptography methods on IoMT infrastructure has been seen to improve its immunity to unauthorised access, key distribution and challenges that distance and hardware fluctuations pose on traditional cryptosystems [7,15].

Increasing recognition of the crucial role that quantum-safe cryptography will play in supporting secure digital transformation within an industry framework that espouses the values and ideals established by Industry 5.0 has been an emergent area in recent literature and studies. In this context, the adoption and implementation of quantum solutions for the healthcare industry enhance capabilities in real-time threat protection and detection and ensure effective data governance and privacy within AI and IoT ecosystems [14]. According to recent studies and literature on this emergent topic, this inevitable digital transformation must be supported by an overarching and comprehensive framework that integrates post-quantum cryptography and QKD [14]. In particular, with regard to specific post-quantum algorithms currently proposed and promoted in literature studies, CRYSTALS-Kyber, Dilithium and SPHINCS+ have come to be promoted as the most effective and appropriate solutions for enhancing the security and integrity of population-wide health record archives and genomic data archives [16]. Indeed, based on recent studies and literature, specifically modified 'Kyber-1024' while remaining remarkably resistant within the quantum era and with performance remarkably unaffected by cryptography (only 2.3× slower than AES-256 encryption), is an efficient and effective solution suitable for widespread utilisation within the healthcare industry itself. This transformation has been observed to introduce no observable and insignificant system downtime (only 0.03 per cent) [9].

Meanwhile, there has been a proposed set of hybrid cryptographic systems that will act as a bridge solution to ease the gradual acceptance of post-quantum algorithms into practice and still be compliant with current health regulations like HIPAA [5,9]. These systems make it possible to dynamically negotiate between traditional and quantum-safe methods of encrypting data based on their sensitivity and retention requirements and potential threats. Testing has indicated that the usage of a hybrid solution has lowered the potential for a disastrous data breach amidst a system transformation to a quantum-safe environment and has enabled crypto-agility [5].

Quantum technologies have also played an important role in improving the security standards in telemedicine and remote health monitoring services. With the development of satellite-based and fibre-optic QKD networks that provide efficient solutions for securing biomedical data via telemedicine communication channels across the world [4], there has been an omnipresent need for supporting data transfer in real-time procedures via geographically dispersed nodes, and this has been overcome with the development of quantum-secure telemedicine communication channels by Quantum Key Distribution (MDI-QKD), untrustworthy communication devices providing an important leap with the development in MDI-QKD [4,11].

Apart from the use of encryption and communication, some works have also researched the implications of quantum computing in the realm of healthcare optimisation. Using the principles of quantum annealing and variation algorithms, quantum computers can efficiently and optimally carry out the allocation and scheduling of resources and inference models at hospitals compared to their classic alternatives. A good example is the 12-month empirical study carried out at the Royal Hospital in Ghana, which showed a substantive improvement of 66.67 per cent in the data transfer rate and 71.88 per cent in the decryption rate when applying quantum encryption algorithms through the use of cloud-based quantum computing platforms [17]. The system showed a resistance level of 98.7 per cent against the simulated quantum attacks, thereby proving the viability and advantage of incorporating the use of quantum-based encryption algorithms in the current healthcare systems.

In the combined context of AI and quantum cryptography, some new areas of research emerging with great potential are the optimisation of quantum encryption systems using AI. The use of machine learning can optimise key distribution, perform anomaly identification in the quantum communication process or make necessary changes to parameters in the encryption process according to network dynamics. The integration of AI in the context of healthcare cybersecurity is on the cusp of an innovative breakthrough in developing adaptive, intelligent cybersecurity systems sensitive to classical as well as quantum threats [18].

The rise of quantum network infrastructure has also enabled secure data collaboration and shared analytics in the field of healthcare. Some of the recent models developed to create secure quantum networks for the field of healthcare involve end-to-end secure environments through the implementation of entanglement quantum networks with hybrid models of AI to facilitate secure functionalities such as federated learning and multi-party computations. Such models would enable multiple hospitals to work on AI environments together through shared learning of models on encrypted data sets, thus supporting collective intelligence while remaining within the boundaries of privacy laws [6].

Simultaneously addressing quantum communications and AI, systematic reviews have grouped current quantum healthcare studies into two main areas: (1) quantum computing for healthcare AI and (2) quantum computing for security within the healthcare sector [8]. Applying sophisticated AI tools such as particle swarm optimisation and latent Dirichlet allocation, systematic reviews have identified key convergence areas, which include the use of hybrid quantum-classical systems and the development of scalable algorithms. The trend highlighted by their results indicates that despite showing much greater computational speeds and security, issues with hardware noise and scalability remain significant hurdles to their adoption in the healthcare sector [8,10,16].

However, applications of quantum computing in the field of health care do not stop at security aspects. There have been numerous research studies conducted to showcase its revolutionary potential in the fields of healthcare diagnoses, decision-support systems and other operational areas. Applications of quantum algorithms have also demonstrated their potential to accelerate drug discovery through the modelling of molecule interactions at a level of resolution never hitherto possible, with IBM and D-Wave at the forefront of experimentation in the pharmaceutical industry [10,16]. Moreover, research integrating Organizational Information Processing Theory (OIPT) has also hypothesised the potential of quantum computing to optimise multi-party co-ordination mechanisms within the health care industry during pandemics such as the recent COVID-19 outbreak through the capability to analyse health care supply-chain data in real time [10].

But there also exist some challenges in integrating quantum cryptography within the medical sector. These issues include substantially high costs, as well as instabilities in quantum hardware, which make their interoperability with medical information technologies a challenge [10,16]. In fact, recent advancements in the miniaturisation of quantum technology, the development of quantum-resistant cloud APIs and collaborations among research communities from diverse backgrounds have succeeded in filling these gaps. Therefore, with rapid growth in Industry 5.0, medical organisations are required to pursue 'cybersecurity by design' strategies that can integrate quantum security methods into their root architecture for smarter healthcare systems [13].

It is also seen that novel research is being conducted concerning the integration of quantum-resistant cryptography with vehicular/mobile healthcare networks to emphasise the importance of low-latency/high-throughput encrypting mechanisms within intelligent transportation systems providing various healthcare services through the internet [19]. By applying lattice-based cryptography mechanisms such as Ring-Learning with Errors or Ring-LWE with a deep neural network model to detect any intrusion attacks to this system, they were successful in reducing data transmission latency to less than 100 ms along with accuracy levels above 95 per cent to detect such attacks [19].

Taken together, the literature shows that there has been a paradigm shift towards the establishment of quantum-enabled, AI-based and cryptographically secure healthcare infrastructure. With basic advancements in the realm of QKD and hybrid cryptographic methods, the advent of efficient quantum-resistant algorithms, as well as frameworks related to the adoption of hybrid models, the literature is pointing towards the same vision summarised in Table 11.1: 'To develop intelligent healthcare systems that are computationally strong as well as secure from the rapidly advancing threat of the quantum era'. Such endeavours prove that the use of quantum cryptography in the healthcare sector will require more than 'just innovative solutions'.

Although the present state of quantum technologies is still in the very early stages of development, empirical as well as theoretical proof of their revolutionary powers in the fields of data protection for medical data, computational medicine or telemedicine applications clearly exists. The combination of quantum cryptography, AI and post-quantum cryptography foreshadows a future in which medical applications are able to offer, for the very first time in history, simultaneously unmatched processing speeds and indubitable data integrity protection for the post-quantum world of intelligent medical applications.

11.3 Methodology

To develop a comprehensive understanding of quantum-safe cryptographic mechanisms applicable to intelligent healthcare systems, this chapter follows a systematic review with methodology dictated by the PRISMA statement [25]. By doing so, this ensures a fair, reproducible and unbiased selection and combination of existing literature related to quantum computing, PQC and their applications in intelligent healthcare systems.

Table 11.1 Relevant studies on quantum computing-enabled cryptography for intelligent healthcare systems

Ref	cryptographic focus	Key pros	Key cons	Key contributions and findings	Relevance to intelligent healthcare security
[19]	QKD for healthcare data security	• Information-theoretic security • Eavesdropping detection guaranteed by physics • Strong protection for long-term patient data	• High infrastructure cost • Limited transmission distance • Integration complexity with legacy hospital IT	Demonstrates how QKD mitigates weaknesses of classical encryption and enhances confidentiality, integrity and availability of patient data	Foundational framework for secure patient data communication
[15]	Hybrid quantum-classical medical-image encryption	• Extremely high entropy and low correlation • Robust against noise and clipping attacks • Optimised via HIBP selection	• High computational complexity • Requires quantum image representation expertise • Mostly simulation-based validation	High robustness against noise and clipping attacks; near-ideal entropy and low correlation	Efficient secure transmission of sensitive medical images
[18]	QKD in telemedicine networks	• Ideal for real-time image transfer • Supports remote surgery and teleradiology • MDI-QKD improves trust model	• Scalability remains challenging • High deployment and maintenance cost	Addresses cost, scalability and compatibility challenges; secures teleradiology and remote surgery data	Protects real-time telemedicine communication
[9]	Post-quantum cryptography + hybrid migration	• Immediate deployability on classical hardware • NIST-standardised algorithms • Smooth transition via hybrid systems	• Larger key sizes and storage overhead • No physical eavesdropping detection	Shows RSA/ECC vulnerability; recommends hybrid PQC–QKD systems	Transition strategy towards quantum-safe healthcare systems
[14]	Quantum communication integrated with AI security	• End-to-end secure intelligent healthcare systems • Supports federated learning and distributed AI • Strong privacy guarantees	• High architectural complexity • Still largely conceptual/ prototype stage	Proposes layered architectures integrating QKD with AI and federated learning	End-to-end secure intelligent healthcare ecosystems

(Continues)

Table 11.1 (Continued)

Ref	cryptographic focus	Key pros	Key cons	Key contributions and findings	Relevance to intelligent healthcare security
[17]	Quantum cryptography for IoMT	• Secures massive device-level communication • Effective for wearables and remote monitoring • Future-proof against quantum attacks	• Hardware constraints in IoMT devices • Distance and key rate limitations	Case studies validating practicality of QKD in IoMT; compares classical vs quantum cryptography	Secures large-scale connected medical devices
[16]	Systematic review of QC in healthcare security	• Identifies cryptography as a core research theme • Validates research trends statistically	• No direct implementation or performance evaluation	Identifies healthcare data security as a primary QC research direction	Confirms cryptography as a dominant QC healthcare domain
[20]	Lattice-based PQC for mobile healthcare	• Low latency suitable for real-time systems • Strong resistance to quantum attacks • AI-assisted intrusion detection	• Complex cryptographic and AI integration • Energy overhead for mobile environments	Achieves low latency and high attack detection accuracy	Secures real-time intelligent mobile healthcare
[21]	QKD + optimisation for IoMT cloud security	• High confidentiality and integrity rates • Optimised performance in cloud-IoMT	• Dependence on centralised cloud infrastructure • Limited scalability analysis	High confidentiality (98%) and integrity (99%)	Secure cloud-based intelligent healthcare systems
[22]	AI-enhanced quantum cryptography	• Improved threat detection accuracy • Scalable and adaptive security	• AI training complexity • Vulnerable to adversarial ML attacks	Improves threat detection, scalability and robustness	Intelligent security management in healthcare systems
[23]	Quantum encryption performance study	• Demonstrates efficiency gains over AES • Suitable for healthcare data sizes	• Limited evaluation scope • Practical QKD constraints not fully addressed	Demonstrates efficiency gains over AES for healthcare data	Performance validation of quantum cryptography
[24]	Real-world hospital quantum encryption deployment	• Empirical performance validation • Significant throughput and decryption speed gains	• High initial deployment cost • Limited availability of quantum infrastructure	Significant throughput and decryption speed improvements	Real-world validation in clinical environments

11.3.1 Scope and research questions

The systematic review is focused on the design of cryptographic technologies and models ranging from 2015 to 2025 that play a crucial role in securing smart healthcare systems vulnerable to quantum threats. The review covers post-quantum and quantum solutions based on lattice encryption, QKD, homomorphic encryption and security frameworks that combine blockchain. The review was guided by the following core research questions.

- RQ1: What cryptographic mechanisms demonstrate the highest potential for ensuring quantum resilience in healthcare systems?
- RQ2: How have hybrid quantum-classical cryptographic models been implemented in intelligent healthcare environments?
- RQ3: What are the current gaps in the integration of post-quantum cryptography into healthcare infrastructure?
- RQ4: What are the trends and future directions in research on quantum-resilient architectures for healthcare IoT networks?

These questions provide the foundation for evaluating the cryptographic advancements and aligning them with the unique constraints and requirements of intelligent healthcare systems, such as low-latency encryption, compliance with medical data regulations (e.g. HIPAA, GDPR) and secure interoperability among heterogeneous devices.

11.3.2 PRISMA protocol and search strategy

The PRISMA 2020 protocol is used in this chapter as the framework for an organised and transparent review of the current literature. For an interdisciplinary topic like this, which encompasses quantum computing, cryptography and health information systems, the PRISMA protocol is especially appropriate. By using it, bias in the identification, selection and inclusion of pertinent literature is reduced.

A combination of top-quality academic databases was considered for the inclusion of comprehensive and representative literature. These inclusion criteria involved IEEE Xplore, Springer Link, Science Direct and PubMed, along with the ACM Digital Library and the arXiv library for the storage of preprints. Google Scholar was considered an additional source in the study for the purpose of validation and verification of the grey literature.

The timeline for the research was limited to publications from January 2015 until August 2025. A decade was considered sufficient, given the accelerating advancements happening within the area of post-quantum cryptography and the interest in quantum-resilient healthcare infrastructure. Only articles cited in English were considered, ensuring consistency within the interpretation and analysis process.

The search was carried out using a combination of free text and controlled vocabulary. This approach was executed with the help of delimiters using Boolean operators (AND/OR/NOT), quote symbols, parentheses and wildcards, as well as an asterisk (*) symbol to link search terms. The search strings may take a form with search terms including 'quantum cryptography', 'post-quantum encryption', 'lattice-based cryptography', 'quantum key distribution' and 'quantum-safe encryption', alongside application-related search terms including 'intelligent healthcare', 'medical IoT', 'EHR security' and 'blockchain in healthcare'.

11.3.3 Article screening and eligibility criteria

The initial search was followed by the importation of all identified records to a Zotero reference manager for the purpose of screening. The multi-step PRISMA approach was utilised in the process of filtering through the vast amount of literature. During the stage of literature identification, a total of 1,739 articles were identified from the identified databases. The duplicates were removed from the search, and 1,287 unique articles were identified. In the screening phase of the study, the titles and abstracts of the articles were screened for relevance in relation to the study aims. Those not relevant to the area of quantum cryptography in the healthcare sector were eliminated.

The eligibility phase consisted of the screening of the studies by full-text analysis. During the eligibility phase, publications were checked against set inclusion and exclusion criteria. The studies to be included in the screening should have concentrated on the topic of quantum-safe cryptographic techniques or should have concentrated on post-quantum cryptographic systems and applied to the healthcare setting. The research should have prioritised a novel cryptographic model, healthcare performance simulation or the threats of quantum computing to existing medical systems.

Exclusion criteria used to eliminate those articles that contributed to classical cryptographic systems could not be applied in a healthcare setting, were not peer-reviewed and could not be accessed in full text. Moreover, to achieve academic integrity, articles in the form of blogs, editorials and patents were excluded as well. After this process, a total of 87 best possible studies were included in this review. The comprehensive selection process was represented in a PRISMA flow diagram shown in Figure 11.1.

11.3.4 Data extraction

After selecting the articles, a data extraction procedure was followed systematically to identify relevant information. For this purpose, a data extraction template in Table 11.2 was designed to harvest the metadata and analysis attributes from the selected articles. The template included bibliographic information like authorship and year of publication, along with content-related information like the type of cryptographic approach described, subject area related to the health sector (e.g. protecting EHR, telemedicine and IoMT) and implementation type, whether theoretical, simulated or practical.

Also analysed with respect to each article was the set of security goals being addressed, for example, confidentiality, integrity, availability, authentication or traceability of data. In cases where relevant, information about the architecture, performance characteristics and compatibility with other technologies like blockchain and artificial intelligence was taken note of in the cryptographic framework.

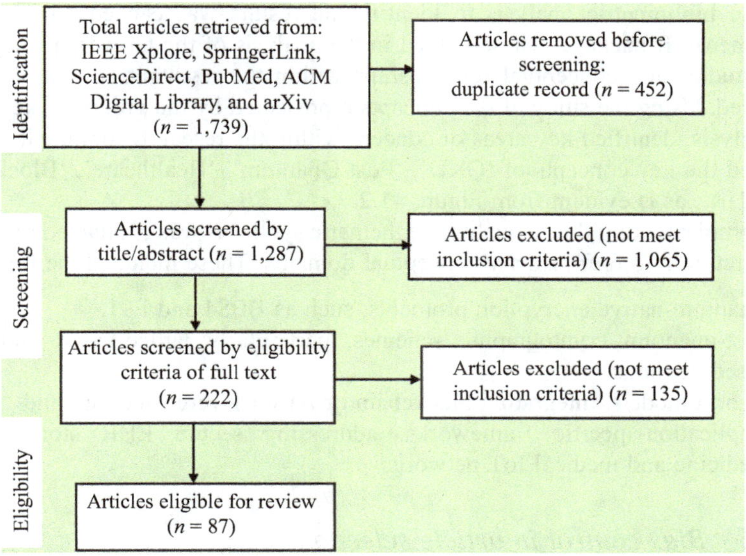

Figure 11.1 PRISMA study selection flowchart

Table 11.2 Summary of data extraction form

Field	Description
Authors and year	Name(s) of the author(s) and publication year
Title	Title of the publication
Cryptographic mechanism	The type of quantum or post-quantum cryptographic model used (e.g. QKD, lattice-based)
Healthcare domain	Area of application within healthcare (e.g. EHR, IoMT, telemedicine)
Security objectives	Security concerns addressed (e.g. confidentiality, integrity, authentication)
Implementation type	Nature of the study (e.g. theoretical, simulation, prototype)
Performance metrics	Quantitative or qualitative performance indicators used (e.g. latency, key rate)
Technological stack	Technologies and tools employed (e.g. Qiskit, CrypTool, blockchain platform)
Key findings	Summary of the main contributions and conclusions

11.3.5 Data synthesis and analytical approach

In order to derive insights from among the included studies, a mixed-methods synthesis was employed. This combined a quantitative bibliometric mapping analysis with a thematic analysis, thus providing a multi-perspective insight into current developments in the application of quantum cryptography in intelligent healthcare systems.

The bibliometric analysis to identify the results was carried out using the Bibliometrix R package. These helped in the analysis of the term frequency, keyword studies and conceptual relationships within the resultant corpus of studies identified. Using the study of the concepts represented by the groups of keywords, the analysis identified key areas of concern within the research community. These included the key concepts of 'QKD', 'Post-Quantum', 'Healthcare', 'Blockchain' and 'EHR', as is evident from Figure 11.2.

Complementing this, a qualitative thematic synthesis was conducted to classify the literature into four primary conceptual domains. These included the following.

1. Quantum-native encryption protocols, such as BB84 and E91,
2. Post-quantum cryptographic schemes, particularly lattice-based and hash-based methods,
3. Hybrid models, integrating blockchain or AI for layered security and
4. Application-specific frameworks, addressing secure EHR storage, telemedicine and medical IoT networks.

11.3.6 Bias control in article selection

Bias control is considered an essential part of any systematic analysis, even more so in an interdisciplinary area such as quantum cryptography and intelligent healthcare, where fast-paced development might lead to an unbalanced body of evidence distributed across publication types and time spans. A variety of precautions were taken in order to maintain the validity and objectivity of the results.

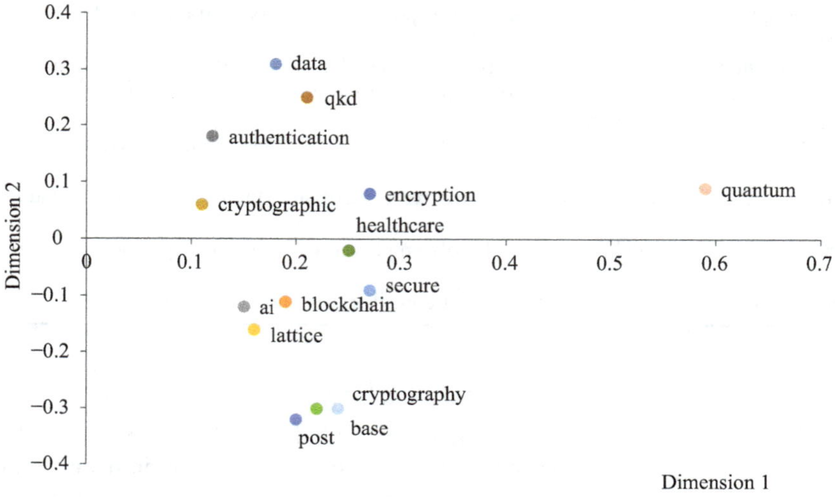

Figure 11.2 Keyword co-occurrence map of research themes

Firstly, selection bias was addressed by ensuring that inclusion and exclusion criteria were applied rigorously in the entire PRISMA screening process. This involved a two-reviewer method for both Abstract and Full Text screenings, whereby each study was assessed by two separate researchers using pre-defined eligibility criteria. When disparities arose, a third reviewer was sought to achieve a consensus, which would not be influenced by personal preference or departmental orientation.

Second, the problem of publication bias was mitigated by including peer-reviewed journal publications and high-quality preprints, specifically from arXiv, to ensure state-of-the-art results were included even if they had not been published in a more traditional academic journal. The need for this is even more pressing in the rapidly developing field of quantum cryptography research.

Thirdly, the problem of database bias was mitigated through the inclusion of literature from a variety of respected sources across the realms of computer science, healthcare informatics and security. Additionally, the databases selected, which included IEEE Xplore, Springer Link, Science Direct, PubMed, ACM Digital Library and arXiv, offered a good indexing system.

Even though a quality assessment matrix was not used, care was taken to ensure that the process of selecting the articles for the study was reliable. Articles for the study were selected after they were screened for their relevance to the topic of quantum cryptography and its relation to the healthcare system. Articles that received positive reviews from the reviewers of the study were selected. The process of using multiple reviewers and the collection of articles from different sources acts as a mechanism that prevents bias.

11.4 Results and discussions

11.4.1 Quantum cryptographic mechanisms in healthcare

This section evaluates the quantum-resilient security solutions designed for intelligent healthcare networks in relation to quantum cryptography methods, post-quantum cryptography and the use of quantum technology. The discussion reveals various methods using quantum key distribution and quantum random number generation that have improved key management for health-related functionalities. In addition, lattice cryptography solutions, hash-based cryptography solutions and quantum-assisted solutions to various other cryptography schemes have been discussed to assess the impact of quantum computer attacks on health-related data.

11.4.1.1 Quantum key distribution

QKD enables two remote end points to establish symmetric cryptographic keys with information-theoretic security by encoding classical information into quantum states whose measurement properties reveal any eavesdropping attempt. The integrity of QKD is ascertained through estimation of QBERs, sensing any interfering effects of eavesdropping, thereby ensuring guaranteed security beyond classical computationally secure cryptographic schemes [23,24]. In health cyber-infrastructure networks, QKD-based cryptography is best suited to securing long-term sensitive medical information and critical medical services. Specific application areas amenable to current deployments of QKD are in-hospital networking of electronic health records from cloud to local, telemedicine in tele-ICUs, digital imaging from PACS to central inference computing in AI-enabled DICOM streaming to inference clusters and healthcare networking of remote clinics to tertiary-care facilities over long-distance national health networks. Recent long-distance demonstrations of QKD using over 200 km of commercially available optical fibres have suggested that such architectures are technically feasible at a national-level scale [26,27].

In terms of QKD protocols, BB84 remains the most established and proven protocol in terms of experimental implementation to date. In BB84, random bit sequences are encoded in two conjugate polarisation states, and measurements are done randomly in different basis choices at the receiving end, finally distilled to secrete shared keys through sifting, QBER measurements, error correction and privacy amplification [28,29]. In terms of telemedicine settings, BB84-generated shared keys would allow for periodic resetting of TLS or IPsec encryption tunnelling the secure transfer of EHRs, medical imagery and live video consultations on metropolitan fibre networks. Evaluations using BB84 within medical imagery applications have proven the protocol to be incorporated within established crypto architectures employing hardware security modules and TLS offload engines to facilitate quantum-safe key management without significantly changing the settings within established healthcare practice streams at present [28,29].

Entanglement-based QKD, as in the E91 and BBM92 protocols, is indeed an extension of the previous security model, as it now involves the distributed correlations of the entangled-photon pairs, as certified by the violation of Bell inequalities and the monogamy of entanglement [30]. From an architectural point of view, when the carrier-grade node is equipped with an entangled-photon source, the correlations can naturally be distributed from the carrier-grade node to several healthcare providers as well as the EHR data centre, in accordance with the healthcare hub-and-spoke architecture [31]. In order to overcome the risks inflicted on the QKD systems by the presence of multi-photon emissions, the decoy-state QKD approach introduces randomness in the intensity level, enabling the analysis of the photon-number-splitting attacks via the statistics, thereby improving the key rates and transmission distances remarkably [32].

As evident in [32], QKD interoperates with telemedicine networks through a key management stack in which keys generated by QKD are distilled, safely archived within HSM key infrastructure and then used by application-level services via standard cryptographic protocols. As such, a hybrid architecture enables QKD to improve the cryptographic underpinnings of clinical data communications while maintaining interoperability with networking and security infrastructures in place today. However, the infrastructure and economic limitations imposed by quantum channels and communications nevertheless highlight the need and relevance of complementary post-quantum strategies in this setting. Within this new landscape, lattice cryptography-based encryption stands out as a natural progression to a topic section and provides quantum resistance through software implementations to offer a complete and scalable post-quantum cybersecurity solution to the healthcare realm in combination with QKD systems.

11.4.1.2 Quantum random number generation

Firstly, the security of any cryptographic system depends upon its randomness in a fundamental way. In addition, low entropy could break even the most secure cryptographic approaches. Several challenges in this area were addressed with the advent of QRNGs, which utilise and leverage basic quantum events like path choice in single-photon transmission, arrival times of photons or other particles and quantum vacuums. Then, following randomness extraction procedures, highly entropic keys could be generated efficiently [28,29]. In healthcare environments, QRNGs would function as hardware security modules in hospital and national data centres, head ends of QKD systems, VPN gateways and critically important medical equipment like defibrillators, insulin pumps and pacemaker monitors. They would provide highly entropic keys and seeds for DRBG. Then, with their output seeded at DRBG, healthcare environments would acquire superior resistance to partial breakage or malicious control of classical random number generators. In reality, QRNGs provide their users with a safety 'fabric' of enhanced entropy in all hospital-based cryptographic activities and operations.

This multidisciplinary integration of quantum entropy sources helps to fill the gaps in post-quantum and quantum-sustainable security solutions adopted by the healthcare sector by offering a comprehensive comparison of QKD solutions, lattice cryptography solutions for public-key encryption, post-quantum cryptography signatures and quantum computing-assisted analytics, as provided in Tables 11.3 and 11.4. QRNGs, along with QKD networking infrastructure and the use of PQC-TLS for Internet-facing services, help to provide a quantum-sustainable networking and security platform for scaling the healthcare sector, which can range anywhere from hospital micro-controller solutions to national healthcare centres. The action plan to implement QRNGs in a healthcare setting prioritises the use of quantum entropy sources, along with the use of FHE for encrypted analytics solutions and QML-assisted SOCs, along with developing quantum-sustainable signature solutions on the firmware, healthcare applications and high-speed analytics of the healthcare sector.

11.4.1.3 Quantum-safe and quantum homomorphic encryption

Quantum-safe or homomorphic encryption (QHE) can therefore further extend the confidentiality, anonymity or privacy assurances provided by classical cryptography into both post-quantum and quantum realms, where computations are done directly on encrypted files. Already, lattice-based fully homomorphic encryption schemes CKKS and BGV, with post-quantum settings, are able to perform encrypted risk

Table 11.3 Quantum-era mechanisms in healthcare networks [17,28,29]

Protocol/ scheme	Type	Security basis	Main use in healthcare network	Key advantages	Key limitations/challenges
BB84	Prepare-and-measure QKD	No-cloning, measurement disturbance	Point-to-point telemedicine links over fibre	Conceptually simple, widely analysed	Requires quantum channel; distance and loss constraints
E91	Entanglement-based QKD	Bell inequality violations, entanglement	Backbone links between major hospitals/data centres	Strong security model; DI-QKD trajectory	Entangled photon sources; complex implementation
BBM92	Entanglement-based QKD	Monogamy of entanglement	Multi-site health clusters with centralised sources	Flexible topology; improved trust model	Similar engineering overhead to E91
Decoy-state QKD	Practical QKD variant	Statistical detection of PNS attacks	High-rate metropolitan links for imaging/PACS	Works with weak lasers; higher key rates	Requires precise intensity control and monitoring
NTRU	Lattice-based PKC	Hardness of lattice problems	VPNs, device-to-gateway encryption, EHR at rest	Efficient; suitable for constrained devices	Larger keys vs classical ECC; careful parameter tuning needed
Kyber like Module-LWE KEMs	Lattice based KEM	Module-LWE hardness	PQC-TLS, cloud APIs, inter-hospital backbones	Standardised; good performance/size trade-offs	Requires migration of existing PKI and software stacks
Hash-based signatures	Signature	Hash preimage/ collision resistance	e-signatures, firmware signing, audit logs	Security based on well-understood hashes	Large signatures/keys; state management issues
Multivariate signatures	Signature	MQ problem hardness	High-volume internal signing (logs, sensor data)	Very fast signing/ verification	Very large public keys; evolving security landscape
Quantum-safe (lattice FHE)	Homomorphic encryption	LWE/RLWE-type lattice problems	Encrypted analytics on EHR/ genomic data in the cloud	Computation on encrypted data	High computational overhead; parameter complexity
QRNG	Randomness source	Intrinsic quantum randomness	Key generation in HSMs, devices, QKD, VPNs	Strong entropy guarantees	Requires specialised hardware; certification processes

Table 11.4 Suitability of quantum mechanisms for healthcare applications [16,28,29,33]

Mechanism	Confidentiality	Integrity	Scalability (networks)	Latency sensitivity	Typical healthcare use cases
BB84/ Decoy-state QKD	High	Via classical MACs	Medium (metro/ regional)	Medium	Securing telemedicine trunks, EHR replication links
E91/BBM92	High	Via classical MACs	Medium (backbones)	Medium	National health backbones, inter-provider key exchange
NTRU/ Kyber KEMs	Medium	With PQ signatures	High (Internet scale)	Low	PQC-TLS for portals, EHR APIs, cloud integration
Hash-based signatures	N/A	High	High	Medium	e-prescriptions, clinical documents, firmware signing, audit trails
Multivariate signatures	N/A	Medium	Medium	Very low	High-throughput internal logging and device attestation
FHE	High	Indirect via protocols	Medium	High	Privacy-preserving analytics, national registries and cohort studies
QML	Context-dependent	Supports detection	Medium	Medium–High	Encrypted traffic anomaly detection, secure federated AI
QRNG	Foundation	Indirect	High	Very low	Entropy for all cryptosystems across the health infrastructure

analyses on EHRs, medical-image analyses and joint institution-wide AI training on encrypted features, both with computation precision approaching that of plaintext at scales now viable on contemporary hardware [34]. Contemporary QHE experiments, using both chip-based quantum optics and EPR, have thus similarly provided early verification that quantum gates are able to perform computations on encrypted qubits, supporting a plausible road map for fully encrypted, quantum-based diagnostics, such as quantum-improved MR Imaging or spectroscopy [35].

From a medical utility perspective, therefore, there now clearly remains a rapidly emerging technical road map, where so forth, classical-quantum-safe encryption schemes will enable secure cloud-based medical record registries, AI analytics and official regulatory analyses, while longer-term implementation of

QHE will provide fully encrypted computations on quantum-based files, so that clinical information stored on EHR databases, on PACS imagery or on IoMT devices will thereby immediately permit encryption before ingestion on public clouds, so that computations on those encrypted files are only able to release decrypted results, risk analyses or statistical analyses, while thus seamlessly integrating with quantum-based RRGs for files accessing cryptographically secure key files.

11.4.2 Post-quantum cryptography for healthcare security

In this section, we discuss post-quantum cryptography in the security of healthcare, including lattice-based cryptographic protocols such as NTRU and Kyber, hash-based signatures and multivariate signatures. The cryptographic systems highlight security reinforced by protection against quantum computer threats. A hybrid quantum-classical cryptographic system is also described, integrating classical and quantum cryptographic systems to enhance the security of data privacy, security integrity and overall system security. Recent developments in post-quantum cryptography have moved towards developing an approachable, scalable system supporting global high-volume healthcare systems, as outlined in Table 11.5.

11.4.2.1 Lattice-based encryption: NTRU and Kyber

Although QKD offers information-theoretic security over quantum channels, the majority of the healthcare IT infrastructure is expected to be in classical settings, making it imperative for PQC to be implementable over conventional computer software and hardware without needing quantum communication channels. Lattice-based cryptography, which is based on the difficulty of problems like Ring/Module-LWE, SVP and CVP, is the foundation of the NIST-proposed key encapsulation mechanisms, specifically the CRYSTAL-Kyber scheme, as well as the well-studied NTRU algorithm [38]. NTRU relies on a polynomial ring architecture, which allows efficient convolution computations even for larger key sizes, making it appropriate for protecting VPN concentrators handling clinician access, EHR servers as well as API servers over PQC-TLS communication, as well as in resource-limited medical applications in the medical IoT domain, specific to patient monitors, ambulance routers etc. [35]. Recent research has shown the feasibility of NTRU-based schemes in satisfying the latency requirements of typical medical IoT applications, along with satisfying the required energy efficiency.

Kyber-based module-LWE KEMs offer a standardised, scalable solution for securing Internet-facing healthcare services, including cloud-served EHR APIs, telemedicine services and healthcare data streams. More specifically, a hybrid PQC-TLS handshake, consisting of Kyber paired with classical ECDHE (elliptic curve Diffie–Hellman exchange), is used to obtain session keys that are subsequently employed with symmetric AEAD encryption (AES-GCM or ChaCha20/Poly1305) in protecting JSON/FHIR data, DICOM imaging and telemetry. For storage security, symmetric encryption keys are wrapped with Kyber or NTRU,

Table 11.5 Classical vs post-quantum cryptographic algorithms in intelligent healthcare [9,36,37]

Algorithm	Type	Quantum-safe	Key size	Signature size	Performance	Standardisation status	Simulation environment	Evaluation criteria	Use cases
RSA-2048	Public-key (classical)	No	256 bytes	256 bytes	Medium	Yes	Classical cryptographic libraries (OpenSSL); simulations in network security testbeds	Encryption/ decryption latency, key generation time, vulnerability under Shor's algorithm	Securing legacy EHR systems, authentication in hospital networks
ECC (P-256)	Public-key (classical)	No	32 bytes	64 bytes	High	Yes	IoT simulation platforms (NS-3), lightweight crypto libraries	Throughput on IoMT devices, energy use, key exchange latency	Wearable device communication, secure telemedicine
DH/ECDH	Key exchange (classical)	No	256 bytes	–	High	Yes	Secure channel simulation; TLS performance benchmarking	Key exchange speed, forward secrecy strength, resistance to MITM	Device-to-device secure communication, IoMT pairing
AES-128	Symmetric (Classical)	Partially	16 bytes	–	Very High	Yes	Hardware encryption modules (AES-NI), embedded simulations	Throughput, resistance to Grover-based quantum attacks	Fast encryption for imaging data, wearable sensor streams
SHA-256	Hash Function	Partially	–	32-byte output	Very High	Yes	Hash collision testing environments; blockchain simulators	Collision resistance, hashing speed	Integrity of medical logs, tamper-proof audit trails

(Continues)

Table 11.5 (Continued)

Algorithm	Type	Quantum-safe	Key size	Signature size	Performance	Standardisation status	Simulation environment	Evaluation criteria	Use cases
Kyber-1024	KEM (Post-Quantum)	Yes	~1.5 KB	(ciphertext)	High	NIST Finalist	PQC simulation toolkits (open quantum safe, PQC lean)	Key encapsulation latency, ciphertext expansion, quantum-attack resilience	PQC-secured EHR sharing, future-proof telehealth encryption
Dilithium-3	Digital Signature	Yes	~1.3 KB	~2.7 KB	High	NIST Finalist	NIST PQC test suites, signature verification profiling	Signing/ verification time, signature size impact on bandwidth	Secure consent forms, audit logs, blockchain-based clinical records
Falcon-512	Digital Signature	Yes	897 bytes	666 bytes	High	NIST Finalist	Lattice-crypto simulation frameworks; hardware optimisation tests	Floating-point stability, verification speed	Lightweight secure authentication for IoMT devices
SPHINCS+	Hash-based Signature	Yes	32 bytes	8–30 KB	Low	NIST Finalist	Hash-based crypto simulators; secure boot chain testing	Signature size overhead, verification cost	Secure firmware updates for medical devices
AES-256	Symmetric (Quantum-Safe)	Yes	32 bytes	–	Very High	Yes	High-security encryption testing on HPC clusters	Quantum-resistant symmetric strength; throughput	Encrypting genomic datasets, long-term research archives

with these managed within healthcare organisation KMS-HSM systems. Existing industrial implementations of comparable hybrid PQC architectures have demonstrated that these can readily be deployed with low overhead, thus providing a secure, quantum-resistant foundation for healthcare information technology. This lattice cryptography-instilled security layer is seamlessly integrated with hash-based digital signatures or multivariate signatures that offer post-quantum message, API or clinical transaction integrity and authentication services, thus enabling a full-spectrum quantum-secure e-health system.

11.4.2.2 Hash-based and multivariate signatures

In today's healthcare environments, the digital signature represents a cornerstone technique for securing e-prescriptions, diagnostic reports, firmware updates and data exchange agreements between organisations. Hash-based signature schemes like XMSS, LMS and SPHINCS+ benefit from collision and preimage resistances provided by cryptographic hash functions and thus resist quantum attacks in both unbroken and Grover-accelerated models with suitably sized hash digest output [26]. Stateful schemes such as XMSS and LMS are particularly suited to devices that sign infrequent artefacts, such as firmware and calibration records, while stateless schemes like SPHINCS+ accommodate microservice-rich hospital information systems (HIS) where multiple concurrent signers operate. Experimental studies in remote attestation for medical IoT devices demonstrate that XMSS-class signatures can be feasibly deployed on constrained microcontrollers such as the ESP32, enabling quantum-resilient attestation for sensors, gateways and edge devices.

Multivariate quadratic (MQ) signature schemes complement hash-based approaches by offering extremely fast signing and verification at the expense of large public keys and the need for ongoing cryptanalytic evaluation [35]. These properties make MQ signatures suitable for high-throughput environments, including bandwidth-rich data centres and internal hospital networks, where fine-grained sensor attestations, audit logging and other volume-intensive signing tasks are required. By overlaying hash-based and multivariate post-quantum signature schemes, healthcare systems can tailor security to workflow demands: high-assurance artefacts such as firmware, EHR documents and critical audit logs can leverage hash-based signatures for maximal integrity, while multivariate signatures support scalable, high-frequency signing of IoMT telemetry and internal system logs. This complementary integration establishes a robust, quantum-resilient authentication framework, naturally bridging towards emerging applications in QML for secure healthcare data exchange.

11.4.2.3 Hybrid quantum-classical cryptographic framework design

Architectural overview

The rapid advancement of quantum computing has escalated concerns about the long-term security of classical public-key cryptography systems, which form the backbone of today's secure communication infrastructures [27]. As alternative advances in quantum algorithms, such as Shor's algorithm, have shown, a sufficiently powerful quantum computer will be able to solve factoring large integers and discrete logarithm problems in polynomial time, a prelude that effectively compromises RSA, Diffie–Hellman and elliptic curve cryptography, which are thus far the foundational mechanisms deployed in healthcare, finance and critical infrastructure [35]. In preparation for this impending shock, two main research directions have taken shape: post-quantum cryptography, which aims at computational security against quantum adversaries, and quantum key distribution, which offers information-theoretic security from non-classical physical principles rather than complexity-based assumptions [39]. Table 11.6 in this context shows the urgency of the situation regarding quantum-capable adversaries challenging the integrity of encrypted data flows and authenticated channels in next-generation healthcare networks.

However, neither PQC nor QKD provides, in isolation, a comprehensive and scalable security solution for practical environments. QKD suffers from fundamental distance limitations, photon loss, atmospheric turbulence and requirements for trusted nodes or emerging quantum repeaters, all of which severely restrict its wider use in heterogeneous healthcare infrastructure [61]. By contrast, the schemes devised under PQC boast strong performance and interoperability while still relying on hardness assumptions susceptible to erosion through future algorithmic advances, especially given the maturation of quantum hardware [39]. Static security configurations also continue to fall short for environments marked by dynamic resource fluctuation, changing adversarial tactics and distributed clinical processes, especially in IoMT-enabled and cloud-based healthcare systems [27].

These challenges have, in turn, ensured that state-of-the-art research increasingly supports the adoption of multi-tiered hybrid architectures comprising QKD for ultra-secure key establishment, classical and post-quantum encryption for high-throughput data protection and AI-driven decision systems to dynamically orchestrate protocol selection, authentication strategies and resource optimisation across complex medical networks [62]. The synergistic models developed in this area have shown improved resilience in IoT infrastructures, smart hospital ecosystems, federated learning pipelines and large-scale healthcare data-sharing frameworks.

Classical encryption layer

In the quantum-resilient security framework, the role of classical encryption algorithms remains as imperative as ever before, as the speed requirements of large-scale data encryption cannot be met through the use of QKD exclusively [3,17]. Symmetric algorithms such as AES-128/256 and ChaCha20 remain the best options due to their ability to support high-throughput encryption at speeds unachievable through QKD today, thus guaranteeing the security of large medical data volumes, constant IoMT data streams and real-time medical communications without being limited by unacceptably high latency and power consumption costs [15,23]. Their

Table 11.6 Summary of relevant studies on hybrid quantum-classical cryptographic frameworks

Ref	Hybrid cryptographic approach	Core techniques (quantum + classical)	Main contributions	Evaluation/results	Relevance to hybrid framework design
[40]	Hybrid quantum-classical image-encryption framework	QKD-derived shared keys, OTP, 6D Hyperchaotic Chen, Ikeda map, DCT/DWT diffusion	Robust medical-image encryption using QKD-protected symmetric keys; multi-stage confusion–diffusion pipeline	Strong entropy, correlation reduction; resilient to brute-force, clipping, noise attacks	Demonstrates QKD-anchored classical encryption; highlights need for quantum-secured key life cycle in multimedia systems
[41]	Multilayer hybrid quantum–AI–blockchain architecture	Quantum cryptography, classical orchestration, transformer-based contextual AI, blockchain	Unified architecture combining quantum processing, AI-guided co-ordination and blockchain trust	Conceptual evaluation in healthcare, supply chain, finance	Provides modular blueprint integrating quantum, classical crypto, AI orchestration relevant for hybrid systems with autonomous control
[42]	Hybrid quantum-classical protocol resisting photon number-splitting	Quantum channel asymmetry + classical authentication	Enhanced security vs BB84-like QKD; resilient under partial key access	Working prototype; improved resistance to number-splitting	Enhances hybrid QKD robustness useful for key management components in hybrid frameworks
[43]	Hybrid post-quantum + classical image encryption	PQ key establishment + classical image encryption	AKQ system tailored for post-quantum image security	Theoretical evaluation	Provides PQ-secure key establishment applicable to hybrid multimedia encryption
[44]	Quantum-resilient hybrid cloud framework	Kyber + McEliece + classical crypto	Dynamic threat-adaptive PQ cryptographic layer for cloud	420 MB/s throughput; <4 ms latency; resistant to MITM, side-channel, Grover attacks	High-performance PQ crypto integration; valuable for hybrid orchestration layers

(Continues)

Table 11.6 (Continued)

Ref	Hybrid cryptographic approach	Core techniques (quantum + classical)	Main contributions	Evaluation/results	Relevance to hybrid framework design
[45]	Hybrid TLS with classical + code-based PQ	McEliece + X25519 + QRNG	Dual-handshake TLS; adaptive parameter negotiation	30% lower latency than pure PQ TLS	Demonstrates practical hybrid TLS; essential for secure transport layer in hybrid architectures
[46]	AI-driven hybrid QKD	Mixed BB84/B92/E91/GHZ with RL adaptation	Dynamic protocol selection; AI-driven QBER optimisation	QBER=0.02; 12 bits/round; 100 rounds simulation	Introduces adaptive intelligence layer—key for scalable hybrid quantum networks
[47]	Hybrid QKD + Quantum ML	QKD-secured QML pipeline	Secure qSVM workflow QKD-based protection	Maintains model accuracy; low overhead	Shows integration of quantum cryptography with computational quantum tasks
[48]	Hybrid quantum-classical HPC	QEC-ready hybrid nodes; C++ pragma-based orchestration	Q-Pragma hardware-agnostic hybrid HPC framework	Scalable and extensible	Useful for hybrid computing infrastructure supporting cryptographic workloads
[49]	Server-based hybrid quantum-classical authentication	QKD + classical network protocol	Detects compromised servers; avoids timestamps	Theoretical + functional demonstration	Early hybrid design showing server-mediated QKD coexistence with classical authentication
[50]	Quantum-classical teleportation-based hybrid	Quantum teleportation + classical channels	Ensures data integrity and confidentiality via hybrid channel	IBM Q experiments	Demonstrates hybrid secure transmission using classical-quantum interplay
[51]	Hybrid PQ KEM + QKD without sifting	CRYSTALS-Kyber + BB84 modifications	Eliminates BB84 sifting; 100% raw key rate	+42% secure distance; BO optimisation	Strong hybrid KEM-QKD model enhancing QKD productivity

(Continues)

[52]	Hybrid quantum-classical performance-optimised encryption	Quantum-enhanced encryption via hybrid compute	470% latency improvement vs classical	Field experiments	Reinforces benefits of blending quantum-accelerated operations with classical algorithms
[53]	Hybrid AKE combining PQ + classical + QKD	Symmetric pre-shared QKD keys + PQ Diffie–Hellman	One-round AKE; strong forward/post-compromise security	Low-latency LAN/ WAN performance	Mature hybrid AKE framework; cornerstone for protocol-level design
[54]	Generic hybrid KEX for minimal code change	Classical + PQ KEX in parallel	Protocol accelerator via parallel hybrid execution	Minimal overhead; faster than sequential	Practical migration model for hybrid schemes
[55]	Quantum-classical hybrid workflow execution	Graph-based hybrid workflow architecture	Supports distributed hybrid cryptographic workloads	Deployable on cloud	Useful for orchestration of hybrid crypto workflows
[56]	Hybrid quantum-classical block cipher	Quantum key + classical block encryption	Secure against quantum + classical attacks	Reusable hybrid keys	Early proof of secure hybrid keying for qubit encryption
[57]	Triple-layer hybrid crypto	QRNG + QKD + PQC + classical	First real-world hybrid quantum-safe data-centre solution	Stable, low-cost deployment	Demonstrates mature industry-scale hybrid deployment
[58]	RL-driven adaptive hybrid authentication	Classical + PQ + QKD	MDP-based protocol-switching engine	99.9% uptime; threat-driven protocol selection	Intelligent hybrid security management – highly relevant for dynamic systems
[59]	Hybrid QKD + classical crypto management	Central KMS + QKD	Scalable hybrid crypto service architecture	Efficient large-scale deployment	Supports large-scale hybrid KMS integration
[60]	Multi-node hybrid IoT crypto	Multi-node QKD + ECC	Low-latency IoT authentication with QKD backbone	99.5% resistance; +30% gen rate	Illustrates hybrid model for resource-constrained networks

backward compatibility remains equally vital to ensure a phase-out transition from the classical to the quantum environment without compromising the defence-in-depth approach employed in the clinical setting [21,14].

A crucial part of the hybrid quantum-classical models above is the classical encryption layer itself, which also provides the classical interface handy in the implementation of the QKD-derived keys through structured key hierarchies and authenticated encryption [31,35]. QKD-based master keys are used through pre-defined derivation algorithms, HKDF in order to ensure the secure usage of AES-GCM, ChaCha20-Poly1305 and the various authenticated encryption algorithms in the secure protection of the clinical data itself through constant cycles of quantum master key refreshes [32]. This fusion of the QKD and classical encryption layers together not only eliminates vulnerabilities in the usage of classical encryption algorithms susceptible to quantum attacks but also provides the scalable operational framework for the secure line-speed protection of clinical communications [35]. However, due to the ever-dynamic state of the threat landscape itself and especially the requirements of resource-deficient environments demanding intelligent adaptation according to the landscape itself, the classical encryption layer becomes ineffective in deriving the best possible quantum security in the context of next-generation clinical communication security [39].

AI-driven decision layer

However, the rising complexity of the distributed healthcare framework involving IoMT devices, edge processing units, federated learning infrastructures and cloud analytics has also highlighted the limitations of static security policy models that fail to dynamically adapt to the increasing levels of clinical volatility [12]. As evident from the study of Robert *et al.* illustrating medical communication pattern volatility and the corresponding windows of vulnerability, static encryption models fail to react dynamically against quantum threats and adapt to system conditions [63]. This has necessitated the advent of the AI decision-layer platform as an enabling complement to the quantum-secure framework that can shift encryption strategies dynamically at the speed of computation and detect deviant network activities through AI-driven telemetry pattern analytics.

Machine learning (ML) plays the role of the computation substrate in this decision layer because of the reasoning about threats, system loading, communication priorities and the availability of cryptographic resources that require the ability of ML to extract patterns in high-dimensional spaces and select the best direction according to the deviation detection described through abnormality patterns [28,29]. In addition to detection tasks, multi-objective optimisation algorithms are used to optimise latency, energy efficiency, performance and security levels of the cryptography due to the crucial role of the latter in the healthcare sector, where urgency of care and safety of patients require medical equipment performance and security at the same time [24]. The AI level captures the adaptation of the cryptography strategies through the multidimensional environment mentioned above.

Reinforcement learning (RL) takes these abilities a step further by enabling the learning-driven decision-making of protocol orchestration based on the iterative interaction of the learning agents with the network environment to refine the security policy [36]. The concept of RL models the system conditions of QKD key availability, communication priority levels, threat levels, network bandwidth variations and device constraints through learning state models that allow the decision-making of the learning agents to consider actions of the QKD system protocol selection, the adjustment of the protocol's authentication technique and variations of the QKD protocol's keys exchange decision [36,37]. The above-mentioned empirical studies mentioned in this section state the contribution of the reinforcement learning-based protocol orchestration solution in improving the system availability and minimising the unnecessary usage of the QKD System by a value of up to 30–50 per cent compared to the existing protocol orchestration baselines, establishing its importance in QKD system management [37].

11.4.3 Quantum-assisted cryptographic mechanisms

This section focuses on quantum-assisted cryptographic systems for healthcare security and the impact of such quantum technology on encryption and data security. Also discussed will be the application and implementation of blockchain technology to ensure data integrity and traceability. Finally, the section will cover challenges in healthcare IoT and the impact that such quantum technology can have on data security.

11.4.3.1 Quantum machine learning for secure data exchange

QML utilises quantum algorithms such as quantum support vector machines, quantum kernels and variational circuits to improve classification and anomaly detection in complex healthcare networks and IoT data [36,37]. The metadata of telemedicine, EHRs and IoMT communications in hybrid security operations centres in healthcare networks are converted into quantum states using amplitude encoding to leverage QML in detecting anomalies such as exfiltration data transfer, lateral movement and ransomware attacks [64].

In addition to detection, QML enables secure federated and encrypted analytics for multi-centre clinical and genomic research, with raw patient data never shared and only encrypted/differentially private summaries and model gradients exchanged. Paired with lattice-based homomorphic encryption and/or secure aggregation, QML brokers can then co-ordinate updates, detect backdoor poisoning and carry out policy-based mitigation, including key revocation and QKD path re-routing, thereby synergistically integrating the intelligence offered by quantum computing with end-to-end secure data sharing and facilitating the transition towards more extensive quantum-safe and homomorphic encryption schemes, including confidentiality-preserving computing within dispersed healthcare networks [64].

11.4.3.2 Blockchain integration for data integrity and traceability

The integration of blockchain in the healthcare IoT has received increased interest as a strategy of guaranteeing secure sensing, logging and exchange of clinical and device data while countering the challenges posed by the conventional architecture [6]. In the existing prototypes and architecture models of blockchain in the healthcare field, the applications of permissioned ledger, off-chain storage augmentation and fog edge comprise the regimen of countering the security limitations of blockchain in the computing environment of IoT [21,61].

In the context of guaranteeing the integrity of health data, the applications of anchoring of hash functions in conjunction with tamper-evident ledger and the concept of on-chain witnessing can guarantee the security of health data against various tampering possibilities, and the results of existing research studies can support low errors in the verification of the independence of wireless measurements concerning the attestations of wearable sensors [6,15]. Authentic encryption alongside shared-key encryption can guarantee protection against the vulnerability of medical data prior to its rebroadcast using the concept of re-encryption in the context of the healthcare field, while the applications of the shared-key encryption regimen can guarantee the security of the payloads of the off-chain storage at the device and gateway levels [19,24].

Verification and validation practices can be extended from these assurances through the addition of algorithmic witness selection, smart contract-based access control management and automated enforcement of provenance and consent constraints in decentralised systems [14,21]. These practices can be applied across a wide range of applications, from patient telemetry, EHR anchoring and device behaviour verification to prescribing provenance verification, though there is only limited research coverage of clinical trials and pharmaceutical distribution [63].

However, threat models have shown that resource-constrained sensor nodes, smart contract vulnerabilities and the addition of storage dependencies can create novel vulnerabilities that must be explicitly modelled in practical settings [39]. Privacy-preserving innovations in the form of anomaly detection driven by ML models on resource-constrained nodes can be combined with the security afforded by the blockchain but do not offer provably consistent coverage of regulatory requirements in HIPAA and GDPR across the reported prototypes [21].

The performance metrics from research studies ensure the effectiveness of fog-based blockchain architecture in improving the latency of remote EHR access, and the hybrid blockchain architecture has improved the performance of cloud storage and mitigated the storage requirements of large-size medical files [6,21]. However, scalability issues, energy efficiency problems, challenges of integration of heterogeneous EHR systems and unexplored economic factors remain as impediments [62,65]. Some of the emerging trends in the field of blockchain-based EHR access include the role of AI-enabled verification concepts, the concept of multi-blockchain separation of concerns, edge computing-based trust models and reduced-consensus algorithms suitable for the healthcare environment of the IoT set-up, though the road map of its integration at the level 2 sidechain remains unexplored [14,21,36].

11.4.3.3 Interoperability and workflow in healthcare IoT

Interoperability within healthcare IoT has increasingly become a fundamental enabler for continuous monitoring, cross-system data exchange and intelligent clinical decision support. The literature defines interoperability as seamless integration among wearables, bedside monitors, gateways and hospital information systems guided by priorities of patient safety, real-time responsiveness and operational efficiency [12,62,63]. Throughout the literature review, HL7 FHIR is emphasised as the key to such integration; it acts as a RESTful translation layer, enabling the conversion of heterogeneous device streams into structured clinical resources suitable for both EHR and HIS environments [1,31].

Complementary standards-including HL7 v2/v3, open EHR and DICOM/PACS-are portrayed as synergistic elements supporting messaging, semantic consistency and imaging workflows, respectively, thus reinforcing the broader enterprise-level engineering framework required for interoperable healthcare ecosystems [1,6,18]. Examples of protocol stacks, such as BLE for device pairing, MQTT for telemetry transport and HTTPS-based FHIR APIs for clinical interfaces, further provide real-world implementations as showcased by smart bed prototypes that express both practical feasibility and system-level complexity [7,12].

Despite this maturing consensus, the literature also recognises that persistent barriers still exist, such as the variability and legacy constraints inherent in most clinical devices, necessitating the use of translation middleware and adaptive integration layers in order to achieve consistent interoperability at scale [5,63]. Key contributions, which include the MeDIC framework's collaboration edge translation model and extensive platform-wide evaluations of open EHR-based IoMT architectures, reinforce the shift towards distributed, vendor-agnostic solutions capable of reducing cloud dependence and enabling multi-standard coexistence [18,21]. Experimental prototypes of smart beds and wearable systems provide concrete mappings from BLE and MQTT streams to FHIR resources to reveal the practical viability of standardised data exchange while exposing the remaining gaps in device compliance. Analyses of policy further emphasise the need for urgent advances in interoperability as a clinical safety imperative, framing its strategic relevance in national digitisation and standardisation initiatives [2,4,62].

11.4.4 Unified quantum security architecture for intelligent healthcare

This section introduces a comprehensive quantum security model architecture for smart health care, focusing on overall workflows and system architecture to enable smooth operation. Risk evaluation strategies and 'system resilience' are reviewed, including an explanation of how the model offers better security, data integrity and reliable operation in health care scenarios.

11.4.4.1 Architecture and workflow integration

Contemporary breakthroughs in the integration of healthcare IoTs have resulted in the development of various models of architectural designs that target the direct integration of device-derived data streams within clinical settings, as exemplified in the architecture designs and flow models described in this section [14,18]. Cloud-neutral edge/fog computing architecture models also target the provision of micro-services involving latency-sensitive, scalable micro-services that target the direct translation of device-derived data within clinical settings [21]. Triple-layer architecture models incorporating device-level BLE/HTTP communication components, fog-layer gateways incorporating MQTT brokers along with translation components and cloud-level FHIR-enabled EHR components represent the predominant architecture model widely used today in the development of bidirectional inter-operation in heterogeneous clinical settings [7,12,63].

Micro-services architecture layers incorporating translation layers can provide additional value through the normalisation of diverse device HIC vendor API outputs from various vendors directly to the unified FHIR data entities, facilitating the interoperation of consumer wearables and conventional clinical software directly in clinical settings [1,5]. Precedents of edge-first architecture models of the MeDIC example imply the importance of edge computing architecture designs that distribute device-level sensing and translating functionality directly at the edge of the network architecture model to minimise the latency of clinical device interaction [5].

The number of pilot studies and real-world applications provides the quantitative data needed to understand the impact of such architecture designs. A smart infusion pump integration involving multiple hospitals resulted in the prevention of 238 patient–drug mismatches and the saving of around 3,749 nursing hours because of automated documentation support and EHR validation, and a smart dental chair system that utilised the support of the blockchain achieved a relative intensity of a 38 per cent relative daily clinic volume enhancement [2]. Studies representative of the above applications BLE-to-MQTT-to-FHIR smart bed pipelines, micro-services supporting wearables through various integration frameworks and MeDIC's usage of multi-case simulations covering traffic mitigation and response time clearly convey its research expressiveness and readiness of today's clinical workflows involving IoT architecture designs [12,18,21].

11.4.4.2 Risk assessment and system resilience metrics

IoT-enabled healthcare environments increasingly rely on risk assessment frameworks that integrate static medical-device safety principles with dynamic, metrics-driven analyses capable of capturing the heterogeneity and real-time variability of modern IoT deployments [21,39]. Across the literature, resilience metrics such as availability, integrity, throughput, device-level integrity, energy consumption and dynamic risk scores form the quantitative basis for evaluating system trustworthiness and operational continuity [7,21,63]. Similarly, static conformance frameworks based on medical-device risk management coexist with adaptive, context-sensitive methodologies that compute device-specific and environment-specific risk scores, enabling a more precise understanding of system behaviour under evolving workloads and threat conditions [62]. This methodological shift is reinforced by works illustrating dynamic device profiling in simulated emergency-room settings, continuous event-driven risk updating in modern IoT networks and matrix-based optimisation of maintainability, safety and security for remote serviceable clinical monitors [36,37]. Collectively, these approaches underscore a growing need for resilience metrics that support both baseline dependability and real-time adaptive control in clinical IoT infrastructures.

Classic dependability indicators combined with application-specific quantitative scores are considered while measuring resilience in these settings. Availability remains the dominant clinical resilience metric, while device integrity, throughput and energy consumption are some of the usual operational benchmarks in edge and IoMT deployments [7]. The interpretable numerical outputs from risk scoring models, which span fuzzy and FAHP-based weighting to dynamic scoring, are used for mitigation prioritisation [63]. A sample example would be that of the FAHP-derived smartwatch injection-attack risk score of 8.57 [66]. In addition, context-aware Markov game frameworks enable quantifying the influence of adaptive security controls in runtime environments, while event-driven analyses iteratively and continuously refine a system's risk posture from live cybersecurity telemetry [36,37]. Such techniques together indicate how multi-criteria metrics and probabilistic models can enable strategic planning and responsive risk reduction across complex clinical IoT ecosystems.

Threat models and demonstrated vulnerability analyses in the corpus give insight into the practical challenges related to wireless protocol attacks, device heterogeneity and supply-chain exposures. Experimental BLE injection and jamming attacks on oximeters, smartwatches and peak-flow metres provide empirical grounding for device-specific risk rankings, while studies on edge-driven IoMT architectures reveal how untrusted or resource-constrained edge components can increase attack surfaces and degrade performance unless enhanced by trust-aware models that improve throughput and resilience [21,67]. Additional analyses show that remote maintenance pathways introduce information-leakage risks unless rigorously controlled and that heterogeneous device fleets with uneven update histories create weakest-link vulnerabilities capable of compromising availability and clinical data integrity [6]. These findings reinforce the need for adaptive, device-profiled assessments able to accommodate the scale, diversity and operational constraints characteristic of healthcare IoT.

Despite these methodological advances, the literature identifies a number of persistent challenges: dynamic threshold tuning, resource constraints management, heterogeneous metric scales harmonisation and large-scale operational validation, among others [62,65]. Dynamic scoring frameworks rely on expert-configured thresholds in order to avoid false alarms or undetected risks, whereas constrained edge devices limit the possibility of computing-intensive runtime metrics [21]. The coexistence of FAHP scales, custom dynamic scores and probabilistic measurements creates issues in cross-study comparability most evaluations remain confined to simulations or single-site case studies, with limited longitudinal validation [36,37]. As a result, emerging best-practice recommendations point towards alignment with established medical-device risk standards, the adoption of dynamic and context-aware scoring, security objectives decomposition into measurable adaptive sub-metrics and controls quantitative prioritisation via FAHP and/or matrix-based optimisation [37]. These recommendations set a structured, metrics-driven baseline that naturally flows into the next section, Implementation Scenarios and Simulation Results, where the operational feasibility and empirical performance of these risk assessment and resilience-measurement approaches are demonstrated by means of concrete deployment scenarios and quantitative simulations.

11.4.5 Implementation scenarios and simulation results

The complexity of intelligent healthcare systems stemming from large data transactions, the IoMT, cloud-based analytics and adversarial cyber environments has propelled the quest for the development of quantum-secure encryption algorithms. Traditional encryption algorithms are already being threatened because of the advent of quantum algorithms such as Shor's and Grover's algorithms that target the security of both RSA and ECC and symmetric-key algorithms [9]. This section aims to compile studies that use empirical and simulation models to assess the performance of quantum and post-quantum algorithms in the healthcare sector.

11.4.5.1 Performance metrics and evaluation criteria

Quantum performance metrics of interest for quantum-enabled healthcare systems include throughput, speed of encryption and decryption, latency, resistance to quantum attacks, key exchange reliability and the system's scalability, as detailed in Table 11.7. Real-world experiments prove that a dramatic performance improvement is achieved over traditional cryptosystems: according to [24], the throughput and speed of decryption achieved through quantum encryption are a remarkable 66.67 per cent and 71.88 per cent, respectively, over classical systems, whereas the time required to encrypt the medical data of 14 KB size using quantum encryption is merely 7 ms, a tremendous performance over AES of 12 ms. The performance indices of the quantum models of secure images GNEQR, NEQR and GNEQR-5D chaotic systems include very high entropy of ~7.9, a change rate of pixels over 99.82 per cent and MSE exceeding 10,000, expressing robustness beyond classical measures [68]. Parallel to that, the performance of the quantum model of the healthcare blockchain shows that the computational and communication overhead could be reduced by a remarkable 30 per cent and 25 per cent, respectively, whereas the hybrid architecture of PQC and blockchain preserves the data integrity of 99.8 per cent [70].

The robustness of quantum cryptography systems continues to serve as the foundation of comparison, especially adversarially performed Shor/Grover attacks. The work of Bala *et al.* shows the privacy and integrity confidence level to be over

Table 11.7 Performance metrics and evaluation criteria for quantum-enabled healthcare cryptography

Metric category	Specific metric	Classical performance	Quantum performance	Relative improvement	Healthcare relevance	Ref
Encryption and decryption efficiency	Encryption time	~12ms	7ms	+41.6% faster	Enables real-time encryption in EHR/IoMT flows	[23]
	Decryption speed	1.0 s	0.58 s	+71.88% faster	Reduces clinical system latency	[24]
	Throughput	100 Mbps	166 Mbps	+66.67% increase	Supports high-volume diagnostic imaging transfers	[24]
Image security metrics	Entropy	7.2	7.9	Higher statistical randomness	Prevents medical-image leakage	[68]
	Pixel change rate	95%	99.82%	+4.82%	Strong differential attack resistance	[68]
	MSE (image encryption stability)	—	>10,000	Substantially higher	Ensures robust image obfuscation against reconstruction	[68]
Key management and exchange	Key exchange success	89.7%	99.8%	+11%	Reliable key distribution for telemedicine	[69]
	QKD key success rate	90–95%	99.8%	+5–10%	Detects eavesdropping in long-range medical networks	[19]
Security and resilience	Attack resistance	45.3%	98.7%	+118%	Protection against Shor/Grover-based adversaries	[19]
	Data integrity	87.5%	99.8%	+14%	Maintains HIPAA/GDPR-compliant data fidelity	[70]
	Teleportation-based key security	Moderate	100% (AVISPA-verified)	Fully resilient	Quantum-secure channel establishment	[71]
ML security integration	AUC (ML model security)	0.80	>0.90	+10%	Secures diagnostic ML against poisoning/ eavesdropping	[72]
	F1-score	~0.80	>0.90	+12–15%	Ensures robust AI-driven clinical decision-making	[72]
Resource efficiency and scalability	Execution time/energy efficiency	Variable	PQC faster & more energy-efficient (Kyber > NewHope)	Measured gains across metrics	Suitable for IoMT microcontrollers	[73]
	Memory footprint	Moderate	XMSS more memory-efficient	—	Facilitates low-power medical device deployment	[73]
	Integration scalability	Legacy system constraints	High interoperability in hybrid networks	Improved compatibility	Supports secure large-scale hospital systems	[72]

97 per cent [19], while the QSMAH protocol of Chawla and Mehra shows 100 per cent resistance to assaults using the AVISPA AVA-Vulner and AVA-LAN tests for the teleportation key exchange scheme [71]. The reliability of the MLD models fortified using quantum and PQC frameworks further amplifies practical usability, showing adequate resistance to poisoning and eavesdropping attacks with high precision and recall (AUC > 0.90 and over 90 per cent, respectively), according to the work of [72]. Further overall performance tests of quantum efficiency, operability and compliance suggest that Kyber is more efficient than New Hope in terms of the speeds and power consumption ratio measured [73].

11.4.5.2 Use cases in quantum healthcare cryptography

The application of quantum cryptography is now being developed into targeted and useful applications of healthcare information systems ranging from secure data at rest, gadget verification, imaging privacy and ledger security to telemedicine over long range [23]. Quantum-resistant cryptography in EHR and archiving; the focus is placed on lightweight quantum primitives: CRYSTALS-Kyber and Dilithium, both facilitating protective covers under regulations, strengthening the resistance against classical and quantum attacks, respectively [19].

Enterprise architectures in the ledger system make use of quantum key distribution and quantum-consensus approaches; these approaches are known to be optimised, less burdensome and efficient by as much as 30 per cent; they assure superior, low-latency, non-tamper-resistant features, optimised for multi-hospital architectures [74]. On IoT gadget authenticity for mhealth, quantum mutual-key agreements have been proposed, for example, QSMAH and QKA, seeming to improve mutual two-way fairness and trust without requiring the involvement of centralised trusted parties, hence securing against key reuse and other insider threats [75].

Among all, the protection of medical images and medical-image pipelines is an especially appealing use case, whereby NEQR/GNEQR quantum image representations mapped onto quantum XOR/chaotic maps enable efficient encryption. These have been shown to maintain integrity at efficiency rates of 99.8 per cent or greater while increasing both entropy and resistance against chosen-plaintext and reconstruction attacks [76]. The blockchain modifications optimised for the healthcare sector, with quantum-chain and entangled BloQs in EQMR, contribute even more to the immutability and management control over sensitive medical data [74].

Finally, the application of QKD systems based on satellites, SQHMT, for remote diagnostics and telemedicine applications, provides secure quantum communication channels for distant diagnostics and telemedicine applications [77]. This enables private and efficient long-distance key exchange and exchange in telemedical settings. Each of the stated applications shows how quantum-classical hybrid applications can confer real-world benefits for an improvement in the health sector, embedding into existing infrastructure to significantly enhance security and robustness, as detailed Table 11.8.

Table 11.8 Quantum computing healthcare use cases [19,23,24,74,76,77]

Use case	Short description	Expected benefit	Required quantum algorithms	Security implications	Data sensitivity	Deployment scale	Evaluation criteria
Secure EHR transmission	End-to-end protection of patient records between providers	Long-term confidentiality prevents 'harvest now, decrypt later'	PQC and QKD + OTP	High: protects the lifetime value of records; key management is critical	Very high	Inter-hospital links, regional health networks	Latency, end-to-end throughput, key renewal rate, compliance with data protection regulations
Telehealth and remote monitoring	Secure streaming and telemetry from wearables/telemedicine	Preserves privacy in real-time consults; integrity of vitals	PQC for device links; QKD for core links where available	Protects live data; authentication needed to prevent spoofing	High	Device → edge → cloud (large scale)	Encryption/decryption time, jitter, packet loss, power per op
Secure genomic data sharing	Confidential exchange/ analysis of genomic datasets	Protects long-lived genomic information; enables collaborative research	QKD + OTP (for highest privacy) or PQC	Extremely sensitive; requires forward secrecy	Extremely high	Research consortia, multi-data centre	Throughput (TB/day), error rate, key rate, successful MPC ops
Medical-image protection and QC	Secure storage and transmission of large imaging (MRI/CT)	Integrity and confidentiality in imaging pipelines; supports secure AI	PQC for storage keys; QKD for inter-data centre links; FHE for encrypted processing	Large files → bandwidth and storage implications	High	Hospital PACS systems, cloud	Encryption/decryption time, storage overhead, image fidelity, processing latency
Authentication for IoMT devices	Strong device identity and firmware update authenticity	Prevents device takeover; secures OTA updates	PQ signatures (Dilithium/Falcon) + QRNG for seeds; hybrid schemes	Mitigates supply-chain and update attacks	Medium–High	Millions of constrained devices	Signature verification time, memory footprint, energy per verification
Secure clinical trials/data integrity	Protect trial data and consent logs across organisations	Ensures data provenance and prevents tampering	PQC for signatures; QKD for high-value links; blockchain for audit	Non-repudiation, tamper evidence	High	Multi-site trials	TPS, commit latency, auditability measures
Drug discovery and molecular simulation	Accelerated simulation for molecular interactions	Much faster candidate screening, lower cost/time	Quantum algorithms (VQE, QPE, quantum chemistry) on QPUs + classical HPC	IP protection required; secure data exchange for partners	Medium–High	Pharma R&D clusters	Time-to-solution, fidelity of simulation, resource (qubits) utilisation
Quantum-enhanced diagnostics (QML)	QML models for image classification /predictive analytics	Improved accuracy, faster training/inference for complex data	Hybrid quantum-classical ML (variational circuits)	Model confidentiality and patient privacy during training	High	Research hospitals, cloud +QPU access	Accuracy, AUC/F1, training time, convergence rate
Radiotherapy planning (Monte Carlo)	Faster, more accurate dose calculations	Better targeting, fewer side effects	Quantum-accelerated Monte Carlo or hybrid acceleration	Need integrity and explainability for clinical use	High	Cancer centres	Compute time per plan, dose accuracy and latency for adaptive therapy
Secure IoMT at scale (edge/ cloud)	Scalable security stack for massive device fleets	Maintain performance while providing quantum resistance	Lightweight PQC at edge, PQ-aware TLS, edge key offload, QKD at backbone	Key distribution and life-cycle management at scale	Varies by device	National/ regional deployments	Scalability (devices managed), handshake size, CPU/memory overhead

11.4.6 Lessons learned

Lesson One: Quantum readiness in healthcare is a Strategic, not merely technical, imperative

One of the key takeaways from the study is the necessity of considering quantum resilience from a strategic, rather than upgrade-focused, cryptography perspective. The confidentiality lifetime of healthcare data, especially genomic data, longitudinal electronic health records and clinical trial databases, is unusually large, making the healthcare sector especially susceptible to 'harvest now, decrypt later' attacks. The analysis in the study makes clear that the healthcare infrastructure will be vulnerable to retroactive attacks if the sector waits to develop the ability to handle mature quantum computers.

Lesson Two: Hybrid quantum-classical security architectures are essential for practical deployment

The study shows how the requirements of real-world healthcare systems, namely scalability, cost and interoperability, can never be met by either post-quantum cryptography or quantum key distribution alone. Rather, a hybrid approach, in which quantum key distribution protects high-value backbone connections, while PQC safeguards either internet-scale or legacy infrastructures, appears to represent the most viable pathway forward. A hybrid approach will help healthcare providers achieve a level of quantum security commensurate with these real-world requirements.

Lesson Three: Performance and resource constraints shape cryptographic viability in healthcare

Another important takeaway is the need to analyse cryptographic performance alongside latency, energy consumption and bandwidth costs. The study points out the heterogeneity of healthcare settings, including high-performance data processing centres, latency-tolerant telemedicine and resource-limited medical devices. Examples of lattice-based cryptographic primitives, such as the NTRU and Kyber schemes, emphasise how post-quantum cryptography can be optimised for these demands, underscoring the importance of empirically informed algorithm selection based on concrete performance considerations rather than theoretical safety margins.

Lesson Four: Cryptographic transition requires robust key management and operational governance

These results highlight that cryptography will only ever be as strong as its key management infrastructure in a quantum-resilient environment, which must include secure key generation, management, rotation, auditing and revocation, encompassed by robust, healthcare-grade HSMs or centralised KMS systems. The takeaway message here clearly relates to post-quantum migration, which is more than simply replacing existing cryptography with organisational change on many fronts.

Lesson Five: Cryptographic agility is critical in an evolving quantum threat landscape

Finally, the manuscript demonstrates the futility of trusting the security of any individual post-quantum solution for more than a short period of time, owing to ongoing cryptanalytic progress, rapidly advancing standards and the rapid development of post-quantum hardware, which demand cryptographic agility. Lastly, the importance of post-quantum agility is emphasised in potential post-quantum security architectures that prepare for known and unknown post-quantum attacks.

11.5 Conclusion and future scope

The transition towards a quantum-resilient healthcare ecosystem represents more than an incremental technological upgrade; it signifies a fundamental realignment of how security, intelligence, trust and interoperability are conceived within digital health infrastructures. As healthcare systems increasingly depend on interconnected IoMT devices, AI-assisted diagnostics, distributed learning models and expansive biomedical repositories, the attack surface expands in parallel with the sophistication of quantum-enabled adversaries. Emerging quantum computers pose direct threats to classical public-key cryptography, quantum-accelerated malware complicates anomaly detection and quantum-level eavesdropping challenges foundational assumptions of secure communication.

These developments require a paradigm shift in security design, one that integrates PQC, QKD, blockchain-anchored data integrity and AI-driven threat intelligence into a cohesive defensive framework. The findings presented in this chapter demonstrate that PQC, when harmonised with intelligent analytics and verifiable distributed trust mechanisms, provides a viable and forward-compatible pathway for securing sensitive clinical workflows, protecting longitudinal patient data and mitigating quantum-era cyber risks across heterogeneous medical environments.

This chapter's evaluation of quantum-safe algorithms, layered communication models and hybridised security architectures confirms that healthcare institutions cannot rely solely on classical security primitives. Instead, sustainable resilience depends on early adoption of PQC standards, systematic transition planning and integration of adaptive AI-based monitoring capable of identifying quantum-influenced anomalies. Likewise, distributed ledger technologies complement these protective layers by ensuring immutability, auditability and decentralised governance, features that are increasingly essential for federated research, patient-centric data sovereignty and cross-border clinical collaborations. These findings emphasise that a unified quantum-resilient architecture is not simply desirable but indispensable for safeguarding next-generation healthcare operations.

11.5.1 Future research directions

The next decade will usher in the convergence of quantum computing, neuromorphic processors and advanced AI ecosystems, each contributing unprecedented capabilities and equally unprecedented security challenges for healthcare, thereby creating an urgent need for unified research directions that collectively strengthen quantum resilience across clinical environments. Central to this agenda is the advancement of quantum homomorphic AI, where the development of FHE schemes compatible with post-quantum cryptographic primitives will enable AI models to analyse patient data entirely in encrypted form, eliminating exposure risks during genomic computation, epidemiological simulation and cross-border collaborative research.

Alongside this, neuromorphic encryption for IoMT emerges as a critical research frontier, leveraging brain-inspired hardware to enable energy-efficient, ultra-fast cryptographic operations ideal for medical implants, biosensors, prosthetic interfaces and continuous remote monitoring devices, particularly in contexts where traditional PQC schemes may impose excessive computational overhead. Complementing these developments is the evolution of quantum biometric authentication, a domain exploring single-photon retinal signatures, quantum fingerprints and quantum random walk-based identity proofs to produce identity mechanisms that are mathematically and physically unforgeable, an essential capability for securing organ-matching registries, precision-medicine infrastructures and high-security clinical control systems where identity compromise could have life-threatening consequences.

As threat complexity escalates, adaptive quantum threat intelligence becomes indispensable, requiring AI-driven defence platforms capable of modelling quantum-accelerated attack vectors, forecasting adversarial behaviour, autonomously adjusting security policies and reinforcing PQC parameters in real time through reinforcement learning-based adaptive agents. Finally, hybrid QKD-PQC network integration represents another priority as future hospital communication infrastructures will combine quantum key distribution for mission-critical communication with PQC-secured traffic for general clinical operations; research must therefore address interoperability challenges, key orchestration strategies, dynamic routing and large-scale deployment under real clinical workloads to ensure seamless coexistence of classical, quantum and post-quantum communication layers.

List of abbreviations

Abbreviation	Full meaning
AES	Advanced Encryption Standard
AES-GCM	Advanced Encryption Standard – Galois/Counter Mode
AES-NI	Advanced Encryption Standard – New Instructions
AI	Artificial Intelligence
AKE	Authenticated Key Exchange
APIs	Application Programming Interfaces
AUC	Area Under the Curve
AVA-LAN	Automotive Audio Video Bridging Local Area Network
BB84	Bennett–Brassard 1984 Protocol
BBM92	Bennett–Brassard–Mermin 1992 Protocol
BGV	Brakerski–Gentry–Vaikuntanathan Scheme
BLE	Bluetooth Low Energy
BloQ	Blockchain-enabled Quantum-secure Framework

(Continues)

(*Continued*)

Abbreviation	Full meaning
ChaCha20	Stream Cipher Algorithm ChaCha20
CKKS	Cheon–Kim–Kim–Song Homomorphic Encryption Scheme
CrypTool	Cryptographic Learning and Simulation Platform
CT	Computed Tomography
CVP	Closest Vector Problem
DICOM	Digital Imaging and Communications in Medicine
ECC	Elliptic Curve Cryptography
ECDHE	Elliptic Curve Diffie–Hellman Ephemeral
EHR	Electronic Health Record
EQMR	Entanglement-based Quantum Memory Register
FHE	Fully Homomorphic Encryption
FHIR	Fast Healthcare Interoperability Resources
GDPR	General Data Protection Regulation
GNEQR	Generalised Novel Enhanced Quantum Representation
GNEQR-5D	Generalised Novel Enhanced Quantum Representation in Five Dimensions
HIC	Healthcare Information Communication
HIPAA	Health Insurance Portability and Accountability Act
HKDF	HMAC-based Key Derivation Function
HIS	Hospital Information System
HTTP	HyperText Transfer Protocol
IoMT	Internet of Medical Things
IoT	Internet of Things
IPsec	Internet Protocol Security
JSON	JavaScript Object Notation
KEMs	Key Encapsulation Mechanisms
KEX	Key Exchange
KMS-HSM	Key Management System – Hardware Security Module
LAN	Local Area Network
LWE	Learning with Errors
MAC	Message Authentication Code
MeDIC	Medical Data Integration and Communication
MITM	Man-In-The-Middle (Attack)
ML	Machine Learning
MLD	Machine Learning-based Detection
MQ	Multivariate Quadratic
MQTT	Message Queuing Telemetry Transport
MRI	Magnetic Resonance Imaging
MSE	Mean Squared Error
NEQR	Novel Enhanced Quantum Representation
NIST	National Institute of Standards and Technology
NTRU	Nth Degree Truncated Polynomial Ring Unit
OIPT	Optimised Intelligent Privacy Technique
OTP	One-Time Pad
PACS	Picture Archiving and Communication System
Poly1305	Cryptographic Message Authentication Code Algorithm
PQC	Post-Quantum Cryptography
PRISMA	Preferred Reporting Items for Systematic Reviews and Meta-Analyses
QBER	Quantum Bit Error Rate

(Continues)

(*Continued*)

Abbreviation	Full meaning
QC	Quantum Computing
QHE	Quantum Homomorphic Encryption
Qiskit	IBM Quantum Development Framework
QKA	Quantum Key Agreement
QKD	Quantum Key Distribution
QML	Quantum Machine Learning
QNNs	Quantum Neural Networks
QPE	Quantum Phase Estimation
QRNGs	Quantum Random Number Generators
QSMAH	Quantum Secure Multi-Agent Healthcare Framework
QSVM	Quantum Support Vector Machine
RL	Reinforcement Learning
RLWE	Ring Learning With Errors
RQ1–RQ4	Research Question 1 to Research Question 4
RSA	Rivest–Shamir–Adleman Encryption
SPHINCS	Stateless Hash-based Digital Signature Scheme
SQHMT	Secure Quantum Healthcare Management Technique
SVP	Shortest Vector Problem
TLS	Transport Layer Security
VQE	Variational Quantum Eigensolver
VPN	Virtual Private Network
WAN	Wide Area Network
XMSS	eXtended Merkle Signature Scheme
XOR	Exclusive OR Logic Operation

References

[1] C. Barbieri, L. Neri, S. Stuard, F. Mari and J.D. Martín-Guerrero, From electronic health records to clinical management systems: How the digital transformation can support healthcare services. *Clinical Kidney Journal*, 2023. **16**(11): p. 1878–1884. https://doi.org/10.1093/ckj/sfad168.

[2] S. Al-Otaibi, S. Ayouni, N. Sarwar, A. Irshad and F. Ullah, AI-driven security framework for medical sensor networks: enhancing privacy and trust in smart healthcare systems. *Cluster Computing*, 2025. **28**(6): p. 408. https://doi.org/10.1007/s10586-024-05049-3.

[3] M. Mosca, Cybersecurity in an era with quantum computers: Will we be ready? *IEEE Security and Privacy*, 2018. **16**(5): p. 38–41. https://doi.org/10.1109/MSP.2018.3761723.

[4] M.V. Devarajan, Improving security control in cloud computing for healthcare environments. *Journal of Science and Technology*, 2020. **5**(6): pp. 178–189. https://doi.org/10.46243/jst.2020.v5.i06.

[5] K. Chava, *Revolutionizing Healthcare Systems with Next-Generation Technologies: The Role of Artificial Intelligence, Cloud Infrastructure, and Big*

Data in Driving Patient-Centric Innovation. 2025: Deep Science Publishing. https://doi.org/10.70593/978-81-988918-5-3.

[6] S. Khezr, M. Moniruzzaman, A. Yassine and R. Benlamri, Blockchain technology in healthcare: A comprehensive review and directions for future research. *Applied Sciences*, 2019. **9**(9): p. 1736. https://doi.org/10.3390/app9091736.

[7] S.R.Sitaraman, Optimizing healthcare data streams using real-time big data analytics and AI techniques. *International Journal of Engineering Research and Science and Technology*, 2020. **16**(3): p. 9–22. https://ijerst.org/index.php/ijerst/article/view/58.

[8] M. SaberiKamarposhti, K.-W. Ng, F.-F. Chua *et al.*, Post-quantum healthcare: A roadmap for cybersecurity resilience in medical data. *Heliyon*, 2024. **10**(10). https://doi.org/10.1016/j.heliyon.2024.e31406.

[9] C.K. Gitonga, The impact of quantum computing on cryptographic systems: Urgency of quantum-resistant algorithms and practical applications in cryptography. *European Journal of Information Technologies and Computer Science*, 2025. **5**(1): p. 1–10.

[10] M. Sharma, V. Choudhary, R. Bhatia, S. Malik, A. Raina and H. Khandelwal, Leveraging the power of quantum computing for breaking RSA encryption. *Cyber-Physical Systems*, 2021. **7**(2): p. 73–92. https://doi.org/10.1080/23335777.2020.1811384.

[11] Q.A. Memon, M. Al Ahmad and M. Pecht, Quantum computing: navigating the future of computation, challenges, and technological breakthroughs. *Quantum Reports*, 2024. **6**(4): p. 627–663. https://doi.org/10.3390/quantum6040039.

[12] Y.A. Qadri, A. Nauman, Y.B. Zikria, A.V. Vasilakos and S.W. Kim, The future of healthcare internet of things: A survey of emerging technologies. *IEEE Communications Surveys & Tutorials*, 2020. **22**(2): p. 1121–1167. https://doi.org/10.1109/COMST.2020.2973314.

[13] G.Alagic, M. Bros, P. Ciadoux *et al.*, Status Report on the Fourth Round of the Nist Post-quantum Cryptography Standardization Process. 2025: *US Department of Commerce, National Institute of Standards and Technology*. https://doi.org/10.6028/NIST.IR.8545.

[14] E. Clark, Integrating quantum networks with AI architectures: A framework for next-gen secure healthcare systems. *Frontier Robotics and Artificial Intelligence Journal*, 2025. **4**(1): p. 24–37.

[15] A. Aljaedi, S. Shaukat Jamal, A. Aljuhni, A.R. Alharbi and T. Shah, Fusion cryptography for secure medical data transmission using mathematical quantum computing operations. *Scientific Reports*, 2025. **15**(1): p. 1–30. https://doi.org/10.1038/s41598-025-12982-9.

[16] A. Marengo and V. Santamato, Quantum algorithms and complexity in healthcare applications: A systematic review with machine learning-optimized analysis. *Frontiers in Computer Science*, 2025. **7**: p. 1584114. https://doi.org/10.3389/fcomp.2025.1584114.

[17] D. Dhinakaran, L. Srinivasan, S.U. Sankar and D. Selvaraj, Quantum-based privacy-preserving techniques for secure and trustworthy internet of medical things an extensive analysis. *Quantum Information and Computation*, 2024. **24**(3&4): p. 227–266.

[18] L. Muralidhar, H. Swapna, N. Sathyanarayana *et al.*, Quantum cryptography for biomedical image security in next-generation telemedicine networks. In *Advanced Secure Transmission of Telemedicine-Based Bio-Medical Images*. 2025, Hershey, PA, USA: IGI Global Scientific Publishing. p. 273–306. https://doi.org/10.4018/979-8-3693-9821-0.ch013.

[19] P.M. Bala, S. Usharani and A. Balachandar, Securing patient data in healthcare with quantum cryptography in the quantum era. In *Quantum Computing*. 2025, Boca Raton, FL, USA: Auerbach Publications. p. 268–282. https://doi.org/10.1201/9781003499459-16.

[20] X. Li, B. Yi, X. Xue, Z. Wang and J. Yang, Quantum-resistant privacy preservation for mobile healthcare services in connected transportation systems via deep neural architectures. *IEEE Journal of Biomedical and Health Informatics*, 2025. https://doi.org/10.1109/jbhi.2025.3612210.

[21] M.A. Hazber, A. Albarrak, M. Altamimi *et al.*, Securing IoMT-based healthcare systems with meta-heuristic quantum cryptography and honeybee optimization: MAG Hazber *et al. The Journal of Supercomputing*, 2025. **81**(17): p. 1586.

[22] A.R. Nangraj, A revolutionary approach using artificial intelligence and quantum cryptography–A review. *International Journal of Innovations in Science & Technology*, 2025. **7**(3), p. 1422–1436. https://doi.org/10.33411/ijist/20257314221436.

[23] C.B. Basha, K. Murugan, T. Suresh, V. SrirengaNachiyar, S. Athimoolam and C. Pappa, Enhancing healthcare data security using quantum cryptography for efficient and robust encryption. *Journal of Electrical Systems*, 2024. **20**(5s): p. 2070–2077. https://doi.org/10.52783/jes.2535.

[24] E. Jhessim and V. Anku, Quantum computing for cybersecurity in healthcare systems: A multi-modal approach. *International Journal of Science and Research Archive*, 2025. **14**(1): p. 612–622. https://doi.org/10.30574/ijsra.2025.14.1.0127.

[25] M.J. Page, J.E. McKenzie, P.M. Bossuyt *et al.*, The PRISMA 2020 statement: An updated guideline for reporting systematic reviews. *British Medical Journal*, 2021. 372. https://doi.org/10.1136/bmj.n71.

[26] S. Kumari, M. Singh, R. Singh and H. Tewari, Signature based Merkle Hash Multiplication algorithm to secure the communication in IoT devices. *Knowledge-Based Systems*, 2022. **253**: p. 109543 https://doi.org/10.1016/j.knosys.2022.109543.

[27] G. Thiyagarajan, V. Bist and P. Nayak, AI-driven configuration drift detection in cloud environments. *International Journal of Communication Networks and Information Security (IJCNIS)*, 2024. **16**(5): p. 721–743.

[28] T. Attema, J.W. Bosman and N.M. Neumann, Optimizing the decoy-state BB84 QKD protocol parameters. *Quantum Information Processing*, 2021. **20** (4): p. 154. https://doi.org/10.48550/arXiv.2006.15962.

[29] Y. Begimbayeva and T. Zhaxalykov, Research of quantum key distribution protocols: BB84, B92, E91. *Scientific Journal of Astana IT University*, 2022. **10**(10): p. 4–14. https://doi.org/10.37943/QRKJ7456.

[30] D. Roosan, R. Khan, S. Nirzhor and F. Hai, Post-quantum cryptography resilience in telehealth using quantum key distribution. *Blockchain in Healthcare Today*, 2025. **8**(1) https://doi.org/10.30953/bhty.v8.379.

[31] K. Hu, D. Ma and S. Qiu, SecureTeleMed: Privacy-preserving volumetric video streaming for telemedicine. *Electronics*, 2025. **14**(17): p. 3371. https://doi.org/10.3390/electronics14173371.

[32] I. Almosallam, Overcoming intensity limits for long-distance quantum key distribution. *Entropy*, 2025. **27**(6): p. 568. https://doi.org/10.3390/e27060568.

[33] O. Amer, S. Chakrabarti, K. Chakraborty *et al.*, Applications of certified randomness. *Nature Reviews Physics*, 2025. **7**(9): p. 514–524. https://doi.org/10.48550/arXiv.2503.19759.

[34] J. Qiu, L. Li, J. Sun *et al.*, Large AI models in health informatics: Applications, challenges, and the future. *IEEE Journal of Biomedical and Health Informatics*, 2023. **27**(12): p. 6074–6087. https://doi.org/10.1109/jbhi.2023.3316750.

[35] Q. Wu, J. Zhang and Z. Li, Generalized NTRU algorithms on algebraic rings. *Electronics*, 2024. **13**(21): p. 4293. https://doi.org/10.3390/electronics13214293.

[36] S. Temara, L. Bhagyalakshmi, S.K. Suman, M. Shakunthala, N.T. Chitra and K. Golla, Cryptography innovations for securing data in the quantum computing era: Integrating machine learning for enhanced security. In *2025 International Conference on Computer, Electrical & Communication Engineering (ICCECE)*. 2025. IEEE. https://doi.org/10.1109/ICCECE61355.2025.10940245.

[37] Y. Seok, J.-B. Kim, Y.-H. Han, H.-K. Lim, C. Lee and W. Lee, Deep reinforcement learning-driven optimization of end-to-end key provision in QKD systems. *Journal of Network and Systems Management*, 2025. **33**(2): p. 1–32. https://doi.org/10.1007/s10922-025-09902-7.

[38] A. Joshi, P. Bhalgat, P. Chavan, T. Chaudhari and S. Patil, Guarding against quantum threats: A survey of post-quantum cryptography standardization, techniques, and current implementations. In *International Conference on Applications and Techniques in Information Security*. 2024. Springer. https://doi.org/10.1007/978-981-97-9743-1_3.

[39] N. Sood, *Cryptography in post quantum computing era*. https://ssrn.com/abstract=4705470.

[40] A. Shafique, S.A.A. Naqvi, A. Raza *et al.*, A hybrid encryption framework leveraging quantum and classical cryptography for secure transmission of medical images in IoT-based telemedicine networks. *Scientific Reports*, 2024. **14**(1): p. 31054. https://doi.org/10.1038/s41598-024-82256-3.

[41] A. Kumar, S. Kiran and R.S. Kumar, Hybrid quantum-classical frameworks for IoT security: bridging AI, federated learning, and cybersecurity. *Journal Publication of International Research for Engineering and Management*, 2025. **10**(3).

[42] S. Vamshi, Introduction to S25-A scalable asymmetry-amplified hybrid quantum-classical protocol. *Discover Quantum Science*, 2026. **2**(1). https://doi.org/10.1007/s44464-025-00004-6.

[43] N.T. Khanh, Truc, N.T. Thanh and V.T. Hai, AKQ: A hybrid quantum-classical image encryption system. In *2023 RIVF International Conference on Computing and Communication Technologies (RIVF)*. 2023. IEEE. https://doi.org/10.1109/RIVF60135.2023.10471803.

[44] K. Samunnisa, S. Gaddam and K. Madhavi, Design and evaluation of a quantum-resilient cryptographic framework for enhancing security and efficiency in distributed cloud environments. *International Journal of Electrical and Electronics Engineering*, 2024. **11**(7): p. 1–22. https://doi.org/10.14445/23488379/IJEEE-V11I7P101.

[45] A. Zafar and S.S. Iqbal, Integrating code-based post-quantum cryptography into SSL TLS protocols through an interoperable hybrid framework. *Discover Computing*, 2025. **28**(1): p. 202. https://doi.org/10.1007/s10791-025-09735-7.

[46] K. Dehingia and N. Dutta, Hybrid quantum key distribution framework: Integrating BB84, B92, E91, and GHZ protocols for enhanced cryptographic security. *Concurrency and Computation: Practice and Experience*, 2025. **37** (21–22): p. e70221. https://doi.org/10.1002/cpe.70221.

[47] B.S. Lokesh and N. Kaulgud, Secure quantum machine learning via quantum cryptography: Theoretical framework and implementation insights. *Journal of Information Systems Engineering & Management*, 2025. 10(49s): pp. 1255–1265. https://doi.org/10.52783/jisem.v10i49s 10122.

[48] A. Gazda and O. Koska, A pragma based C++ framework for hybrid quantum/classical computation. *Science of Computer Programming*, 2024. **236**: p. 103119. DOI: https://doi.org/10.48550/arXiv.2309.02605.

[49] D.R. Kuhn, A quantum cryptographic protocol with detection of compromised server. arXiv preprint quant-ph/0311085, 2003. https://doi.org/10.48550/arXiv.quant-ph/0311085.

[50] M.-I. Pleşa, Hybrid scheme for secure communications using quantum and classical mechanisms. In *2017 9th International Conference on Electronics, Computers and Artificial Intelligence (ECAI)*. 2017. IEEE. https://doi.org/10.1109/ECAI.2017.8166458.

[51] D.J. Mala, A.K. Rastogi, A. Raj and V. Manori, Hybrid KEM-QKD integration with parameter optimization for secure underwater networks via netsquid simulations. *IEEE Access*, 2025. DOI: DOI:10.1109/ACCESS.2025.3607474.

[52] B. Dan, B. Daniel and I.R. GEORGE, Quantum-enhanced algorithms for real-time processing in cryptographic systems: A path towards post-quantum

security. *WORLD*, 2024. **13**(2): p. 193–204. https://doi.org/10.30574/wjaets.2024.13.2.0561.

[53] B. Dowling, T.B. Hansen, and K.G. Paterson, Many a mickle makes a muckle: A framework for provably quantum-secure hybrid key exchange. In *International Conference on Post-Quantum Cryptography*. 2020. Springer. https://doi.org/10.1007/978-3-030-44223-1_26.

[54] R.K Zhao, N.H. Sultan, P. Yialeloglou, D. Liu, D. Liebowitz and J. Pieprzyk, MIKA: A minimalist approach to hybrid key exchange. In *2024 21st Annual International Conference on Privacy, Security and Trust (PST)*. 2024. IEEE. https://doi.org/10.1109/PST62714.2024.10788054.

[55] S. Sivarajah, L. Heidemann, A. Lawrence and R. Duncan, Tierkreis: A dataflow framework for hybrid quantum-classical computing. In *2022 IEEE/ACM Third International Workshop on Quantum Computing Software (QCS)*. 2022. IEEE. https://doi.org/10.48550/arXiv.2211.02350.

[56] N. Zhou, Y. Liu, G. Zeng, J. Xiong and F. Zhu, Novel qubit block encryption algorithm with hybrid keys. *Physica A: Statistical Mechanics and its Applications*, 2007. **375**(2): p. 693–698. https://doi.org/10.1016/j.physa.2006.09.022.

[57] L. Huang, K. Feng and C. Xie, A practical hybrid quantum-safe cryptographic scheme between data centers. In *Emerging Imaging and Sensing Technologies for Security and Defence V; and Advanced Manufacturing Technologies for Micro-and Nanosystems in Security and Defence III*. 2020. SPIE. https://doi.org/10.1117/12.2573558.

[58] A.G. Zaman and M. Qasymeh, Adaptive quantum-classical hybrid authentication: dynamic protocol switching for real-time threat mitigation. 2025. https://doi.org/10.20944/preprints202506.1815.v1.

[59] W. Xue, C. Wang and J. Wang, Design of Cryptosystem architecture based on quantum key distribution. In *2023 IEEE 6th International Conference on Information Systems and Computer Aided Education (ICISCAE)*. 2023. IEEE. https://doi.org/10.1109/ICISCAE59047.2023.10393681.

[60] R. Chaturvedi, D. Sahu, B.P. Singh *et al.*, A hybrid multi-node QKD-ECC architecture for securing IoT networks. *Scientific Reports*, 2025. **15**(1): p. 32831. https://doi.org/10.1038/s41598-025-17184-x.

[61] D. Marchsreiter and J. Sepúlveda, A PQC and QKD hybridization for quantum-secure communications. In *2023 26th Euromicro Conference on Digital System Design (DSD)*. 2023. IEEE. https://doi.org/10.1109/DSD60849.2023.00081.

[62] S. Saeed, S.A. Altamimi, N.A. Alkayyal, E. Alshehri and D.A. Alabbad, Digital transformation and cybersecurity challenges for businesses resilience: Issues and recommendations. *Sensors*, 2023. **23**(15): p. 6666. https://doi.org/10.3390/s23156666.

[63] W. Robert, A. Denis, A. Thomas *et al.*, A comprehensive review on cryptographic techniques for securing internet of medical things: A state-of-the-art, applications, security attacks, mitigation measures, and future research direction. *Mesopotamian Journal of Artificial Intelligence in Healthcare*, 2024. **2024**: p. 135–169. https://doi.org/10.58496/MJAIH/2024/016.

[64] S. Bhowmik and H. Thapliyal, Quantum machine learning for anomaly detection in consumer electronics. In *2024 IEEE Computer Society Annual Symposium on VLSI (ISVLSI)*. 2024. IEEE. https://doi.org/10.48550/arXiv.2409.00294.

[65] A.O. Ezeogu, Post-quantum cryptography for healthcare: Future-proofing population health databases against quantum computing threats. *Research Corridor Journal of Engineering Science*, 2025. **2**(1): p. 29–56. https://doi.org/10.13140/RG.2.2.31914.89281.

[66] P.S. Ng, J. Ignatius, M. Goh, A. Rahman and F. Zhang, The state of the art in FAHP in risk assessment. *Fuzzy Analytic Hierarchy Process*, 2017: p. 11–44. https://doi.org/10.1201/9781315369884-2.

[67] W.S. Admass, Y.Y. Munaye and A.A. Diro, Cyber security: State of the art, challenges and future directions. *Cyber Security and Applications*, 2024. **2**: p. 100031. https://doi.org/10.1016/j.csa.2023.100031.

[68] A.J. Kadhim and T.S. Atia, Quantum encryption of healthcare images: Enhancing security and confidentiality in e-health systems. *Security and Privacy*, 2024. **7**(5): p. e391. https://doi.org/10.1002/spy2.391.

[69] A.R.G. Yallamelli, V. Mamidala, R.K.M.K. Yalla, T. Ganesan, M.V. Devarajan and B. Bhasker, Hybrid cloud computing with post-quantum cryptography and advanced hashing for robust user authentication in healthcare systems. In *2025 5th International Conference on Pervasive Computing and Social Networking (ICPCSN)*. 2025. IEEE. https://doi.org/10.1109/ICPCSN65854.2025.11035631.

[70] S. Poorani and R. Anitha, Quantum-resilient cryptographic framework for cloud-based IoMT data in healthcare enterprises (H-ERPs). *Enterprise Information Systems*, 2025. **19**(10): p. 2534404. https://doi.org/10.1080/17517575.2025.2534404.

[71] D. Chawla and P.S. Mehra, QSMAH: A novel quantum-based secure cryptosystem using mutual authentication for healthcare in the internet of things. *Internet of Things*, 2023. **24**: p. 100949. https://doi.org/10.1016/j.iot.2023.100949.

[72] S. Adeoye and R. Adams, IoT-enabled secure and scalable cloud architecture for multi-user systems: A hybrid post-quantum cryptographic and block chain-based approach towards a trustworthy cloud computing. 2024. 10.47760/cognizance.2024.v04i10.013.

[73] T.B. Maqsood, K. Hameed, M. Junaid and N. Tahir, Enhancing IoT healthcare security with post-quantum cryptography: A performance and case study analysis. In *2025 6th International Conference on Advancements in Computational Sciences (ICACS)*. 2025. IEEE. https://doi.org/10.1109/ICACS64902.2025.10937815.

[74] A. Balasubramaniam and B. Surendiran, QUMA: Quantum unified medical architecture using blockchain. In *Informatics*. 2024. MDPI. https://doi.org/10.3390/informatics11020033.

[75] T.-H. Chen, A Grover search-based quantum key agreement protocol for secure internet of medical things communication. *Future Internet*, 2025. **17**(6): p. 263. https://doi.org/10.3390/fi17060263.
[76] A.J. Kadhim, and T.S. Atia, Strengthening security and confidentiality in e-health systems through quantum encryption of healthcare. *AL-Iraqia Journal for Scientific Engineering Research*, 2023. **2**(3): p. 9–21. https://doi.org/10.58564/IJSER.2.3.2023.83.
[77] K. Sharma and A. Kumar, SQHMT: Satellite-based quantum healthcare monitoring toolkit. In *2025 2nd International Conference On Multi-disciplinary Research and Innovations in Engineering (MRIE)*. 2025. IEEE. https://doi.org/10.1109/MRIE66930.2025.11156306.

Chapter 12

Cryptography in 6G optical wireless-based Internet of Medical Things: potential and state of the art

Jupeng Ding[1,2], *Chih-Lin I*[3], *Jintao Wang*[4] *and Hui Yang*[4]

Abstract

With the exciting development and efficient evolution of 6G-oriented optical wireless Internet of Medical Things (IoMT), relevant investigations in each branch domain of medical optical wireless networks have illustrated tremendous application potential. To fill the gap of the essential security level of these optical wireless IoMT technologies in the future 6G era, more and more investigators are aggressively introducing diverse cryptography into the domain of medical and healthcare optical wireless communications and have derived a series of impressive research advancements. To present the current state of the art of cryptography and security in medical optical wireless networks, this article comprehensively reviews and discusses the cryptography-enabled optical wireless IoMT in distinct branch domains, such as secret key techniques-enabled secure optical wireless Internet of Medical Things, quantum key distribution-enabled secure optical wireless Internet of Medical Things, chaos-enabled secure optical wireless Internet of Medical Things and physical-layer security-enabled secure optical wireless Internet of Medical Things. Finally, this article summarises the potential research opportunities and future prospects of the cryptography-enabled secure optical wireless IoMT technology in the coming 6G era.

Keywords: Optical wireless communications; cryptography; visible light communications; secret key techniques; healthcare wireless communications; optical chaos; physical-layer security; 6G Internet of Medical Things

[1]College of Technology and Data, Yantai Nanshan University, China
[2]Key Laboratory of Signal Detection and Processing in Xinjiang Uygur Autonomous Region, School of Computer Science and Technology (School of Cyberspace Security), Xinjiang University, China
[3]China Mobile Research Institute, China
[4]Department of Electronic Engineering, Beijing National Research Center for Information Science and Technology, Tsinghua University, China

12.1 Introduction

The emerging optical wireless Internet of Medical Things (IoMT) is becoming one promising candidate technology in the diverse applications of the envisioned 6G era [1–4]. Unlike the traditional frequency (RF) technology paradigm, optical wireless IoMT could offer several inherent advantages, including natural immunity to conventional electromagnetic interference, abundant and unregulated optical spectrum resources and higher security performance [5–8]. Nevertheless, optical wireless IoMT techniques have to face a series of urgent challenges, for instance, the shortage of confidentiality and security. Simultaneously, cryptography, especially secret key, has obtained tremendous attention in the mentioned conventional RF-based wireless community, thanks to its amazing capability of reducing secure link outages, increasing secure capacity, enhancing system confidentiality and avoiding potential security vulnerability [9–11]. For addressing the mentioned issues in the medical optical wireless domain, more and more researchers with different academic backgrounds actively leverage various customised cryptography schemes to significantly enhance the security performance indicators in the diverse branch domains of optical wireless IoMT [12–15].

Actually, up to now, cryptography enabling and the related variant designs have been predominantly explored in almost all branch directions of the medical optical wireless community [16–20], including but not limited to secret key generation-enabled medical optical wireless communications [5–8], secret key agreement-enabled medical optical wireless communications [10–13], physical-layer encryption-enabled medical optical wireless communications [15–18], quantum key distribution-enabled medical optical wireless communications [19–22], chaos-enabled optical wireless communications [23–30] and physical-layer security-enabled medical optical wireless transmissions [31–35]. Nevertheless, it must be admitted that these numerous independent publications or articles are far from providing a full overview of the research status of cryptography-enabled optical wireless IoMT techniques. Inevitably, this situation limits the efficient propagation and application of cryptography-related techniques in the optical wireless IoMT domain and, in turn, is negative to pushing the sufficient commercialisation of optical wireless IoMT in the 6G era. To address this issue, this article reviews the advent of cryptography in optical wireless IoMT and then discusses the cryptography enabling distinct branch domains of cryptography in optical wireless IoMT, such as secret key generation-enabled secure medical optical wireless communications, quantum key distribution-enabled secure medical optical wireless communications, chaos-enabled secure medical optical wireless communications and physical-layer security-enabled secure medical optical wireless power transmissions. Finally, this work summarises the potential research directions to further develop cryptography-enabled medical optical wireless communications (OWC) technology in the coming 6G era.

12.1.1 Key contributions of the chapter

Particularly, up to now, to the best of our knowledge, no overview of cryptography-enabled optical wireless Internet of Medical Things has been reported. Therefore, it is definitely challenging for scientists and engineers from the relevant 6G industries and the community to time-efficiently leverage the latest advancements and future directions of cryptography-enabled medical OWC technologies. To fill this key gap, in this article, a brief review of representative designs and methods for cryptography-enabled medical OWC technologies is accomplished in terms of (1) secret key generation-enabled secure medical OWC techniques, (2) quantum key distribution-enabled secure medical OWC techniques, (3) chaos-enabled secure medical OWC techniques, (4) physical-layer security-enabled secure medical OWC techniques and so on, which paves the essential way for finding new potential opportunities and an evolution paradigm in this emerging secure optical wireless IoMT.

12.1.2 Chapter organisation

In this article, the overview of secret key generation-enabled optical wireless communication techniques of IoMT is presented in Section 12.2. A comparative investigation of quantum key distribution-enabled optical wireless techniques of IoMT is included in Section 12.3. The key development advances in chaos-enabled optical wireless sensing techniques and physical-layer security-enabled optical wireless techniques of IoMT are presented in Sections 12.4 and 12.5, respectively. As for Section 12.6, the potential evolution directions and challenges for all the mentioned branches of the IoMT community are briefly summarised. Finally, Section 12.7 concludes this paper.

12.2 Secret key techniques-enabled secure optical wireless Internet of Medical Things

Objectively, the emerging optical wireless IoMT and its branch domains are highly reliant on new-generation solid-state light sources and relevant illumination infrastructures, with special emphasis on the application scenarios of green medical communication, high-precision medical positioning and ubiquitous medical sensing. In view of this, all the above novel applications are expected to face an ever-increasing number of medical users and medical wireless terminals in the envisioned future 6G IoMT ecosystem, and a tremendous enhancement of the security level could be derived by introducing customised cryptography to enable these diverse optical wireless IoMT services.

12.2.1 Secret key generation-enabled secure optical wireless Internet of Medical Things

Specifically, in this subsection, we mainly research the representative advances of secret key generation-enabled optical wireless IoMT techniques. Up to now, secret key generation empowering has been introduced to the optical wireless IoMT research from several main directions, including but not limited to multiple-input multiple-output (MIMO) medical optical wireless systems [5], deep learning-based secret key generation of optical wireless IoMT [6], secret key generation protocol of optical wireless IoMT [7] and the characterisation of secret key generation for optical wireless IoMT [8].

Figure 12.1 shows the schematic of secret key generation-enabled secure optical wireless Internet of Medical Things: (a) the envisioned 4 × 4 secure MIMO optical wireless Internet of Medical Things system and (b) the envisioned block diagram of 4 × 4 secure MIMO optical wireless Internet of Medical Things systems.

With the main aim of addressing the common security concern while ensuring the high data rates of optical wireless IoMT networks, the work in [5] provided one secure multiple-input multiple-output (MIMO) optical wireless system for Internet of Medical Things that relies on the position of the user, as shown in Figure 12.1. Specifically, for encrypting the transmitted data, this system incorporated one novel modified version of the Rivest–Shamir–Adleman (RSA) technique in the media access control MAC layer. The reported simulation results verify that the generation of a number of public and private keys with different key lengths is enough to distribute them on encrypted medical optical wireless cells that have different sizes.

At the same time, it must be noted that multi-user mobility is a key obstacle for future optical wireless IoMT in the realistic indoor environment. To address this critical issue, the deep learning framework that combines channel probing and quantisation has been studied for the secure optical wireless IoMT [6]. Particularly, by tracing mobility to generate the channel impulse response (CIR) and accounting for misalignments and blockages due to movement, this ensemble-based deep long short-term memory (LSTM) recurrent neural network (RNN) model could provide optimal quantised keys at the access point (AP) of the optical wireless IoMT. Numerical results demonstrate that between the uplink and downlink, the proposed deep learning model is capable of generating the initial secret keys with a minimal key disagreement rate (KDR) of 16 per cent, which leads to a key generation rate (KGR) of 79 bits/s after information reconciliation.

As for the vulnerability of optical wireless IoMT to eavesdropping by malicious users, increasing research interest has been paid to the key extraction protocol of medical OWC by leveraging optical orthogonal frequency-division multiplexing (OFDM) schemes in the indoor environment. Particularly, the work in [7] explored the bipolar real signals produced at the output of an optical OFDM system to obtain encryption keys to secure medical OWC links against passive eavesdroppers. In particular, for key extraction throughout the session duration, it is proposed to allocate a portion of cyclic prefix samples placed in the area of low channel impact, and the introduced protocol is capable of extracting keys for the physical level and the medium access control (MAC) level

For clarity, the critical literature review of secret key generation techniques-enabled secure optical wireless Internet of Medical Things is summarised in Table 12.1.

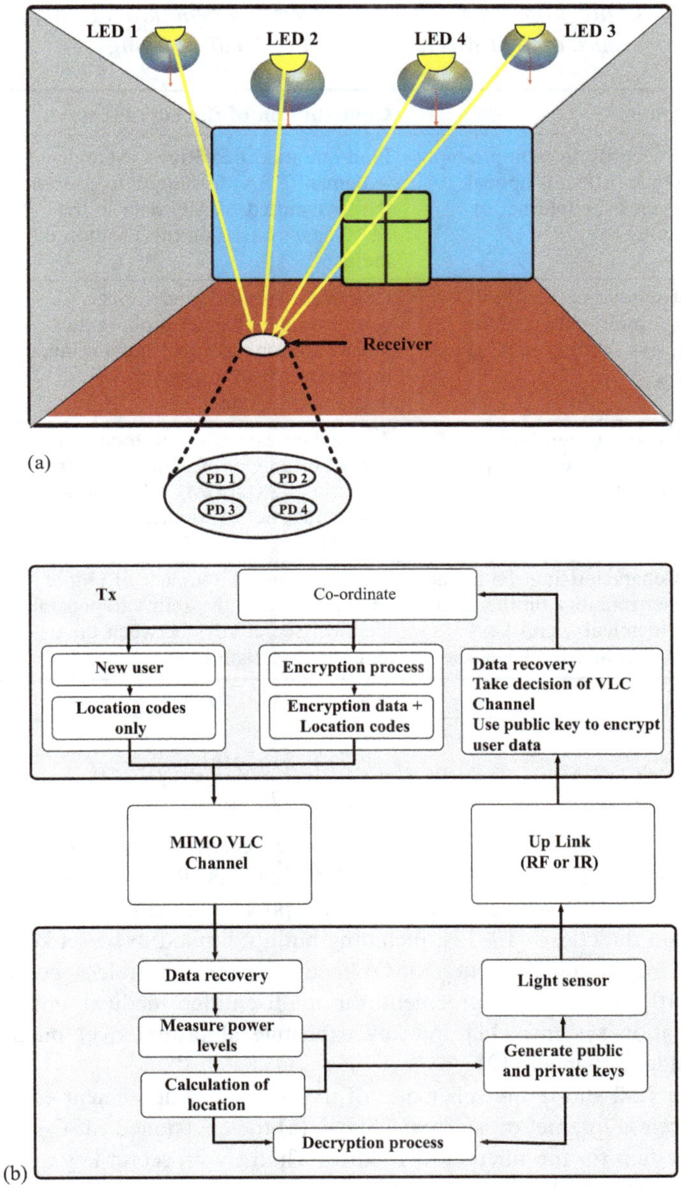

Figure 12.1 Schematic of secret key generation-enabled secure optical wireless Internet of Medical Things: (a) the envisioned 4 × 4 secure MIMO optical wireless system for Internet of Medical Things and (b) the envisioned block diagram of 4 × 4 secure MIMO optical wireless system for Internet of Medical Things

Table 12.1 Critical literature review of secret key generation techniques-enabled secure optical wireless Internet of Medical Things

Identified gaps	Contribution of the current work	References
Absence of security in multiple-input multiple-output (MIMO) optical wireless systems for Internet of Medical Things	Find one modified Rivest–Shamir–Adleman (RSA) technique to encrypt the transmitted MIMO data in the MAC layer based on the location of the user.	[5]
Absence of wireless secret key generation in multi-user mobile optical wireless Internet of Medical Things	One deep learning model that combines the channel probing and quantisation phases to generate initial secret keys with a minimal key disagreement rate.	[6]
The common key generation algorithms could be very complex and costly in many Internet of Medical Things set-ups	A new key extraction protocol for optical orthogonal frequency-division multiplexing (OFDM) schemes in an indoor Internet of Medical Things environment.	[7]
Absence of characterising the impact of indoor user mobility on the ability to generate identical secret keys between the user and the base station	Characterising the impact of indoor user mobility on the ability to generate identical secret keys between the user and the base station.	[8]

12.2.2 Secret key agreement-enabled secure optical wireless Internet of Medical Things

In this subsection, we mainly investigate the representative advances of secret key agreement-enabled optical wireless IoMT techniques. Recently, secret key agreement empowering has been introduced to optical wireless IoMT research from several main directions [10–13], including but not limited to secret key agreement for single-input single-output (SISO) medical optical wireless communication systems [10], secret key agreement for multi-emitter medical optical wireless communication systems [11] and key exchange algorithm over medical optical wireless indoor channel [12].

Figure 12.2 shows the schematic of the secret key agreement-enabled secure optical wireless Internet of Medical Things: (a) the envisioned SISO secure optical wireless system for the Internet of Medical Things with secret key agreement and (b) the envisioned MISO secure optical wireless CW model with an LED array at Alice.

The secret key agreement is becoming a promising technology to deal with the critical security challenge of optical wireless IoMT. In the work of [10], for the conceived SISO medical optical wireless communication systems, one scheme for secret key agreement is proposed to establish the analytical bounds on secret key capacity, as shown in Figure 12.2 (a). Accordingly, the numerical results illustrate

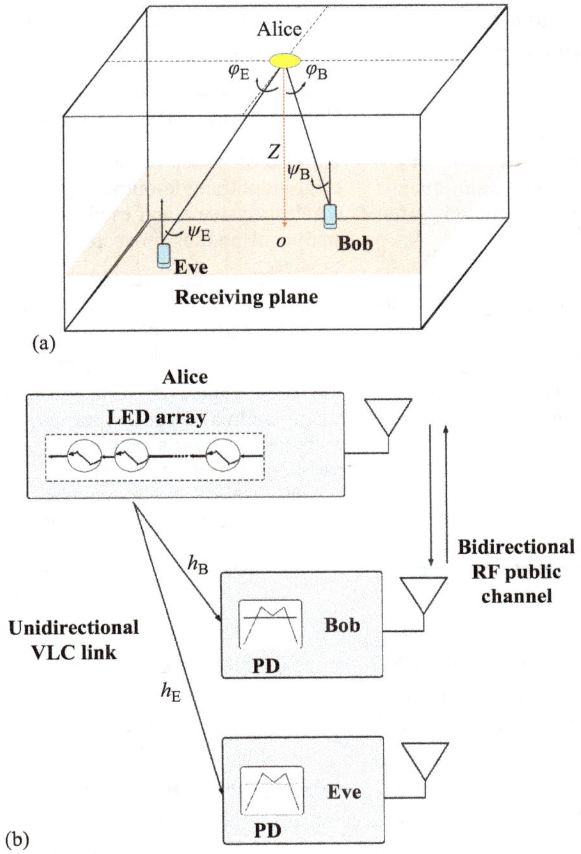

Figure 12.2 Schematic of the secret key agreement-enabled secure optical wireless Internet of Medical Things: (a) the envisioned SISO secure optical wireless system for Internet of Medical Things with secret key agreement and (b) the envisioned MISO secure optical wireless channel-type with wiretapper (CW) model with LED array at Alice

the impact of employing a protected zone on the average secret key capacity and verify that one protected zone can obviously improve the secret performance.

Currently, multiple-input single-output (MISO) is becoming a promising area of medical optical wireless research, whose security techniques include exploiting the optical wireless communication medium to develop secret keys over public channels. In the work of [11], the feasibility of secret key agreement is examined over an optical wireless channel with one light-emitting diode (LED) array at the legitimate transmitter. Besides deriving analytical upper and evaluating lower bounds for secret key rates, the work presents a fractional programming-based two-stage numerical approach for the construction of a secret key MISO beamforming

Table 12.2 Critical literature review of secret key agreement-enabled secure optical wireless Internet of Medical Things

Identified gaps	Contribution of the current work	References
Absence of analysing a secret key agreement scheme in single-input single-output optical wireless systems	The use of secret key agreement for single-input single-output optical wireless systems and establishing the analytical bounds on secret key capacity.	[10]
Absence of secret key agreement for multiple-input single-output optical wireless communications systems	Examining the feasibility of secret key agreement in multiple-input single-output optical wireless channels.	[11]
Absence of a novel key exchange algorithm over the optical wireless indoor channel	One new cryptosystem that combines Diffie-Hellman hyperelliptic curves cryptography and a public-key exchange based on neural synchronisation in a cascade model over an optical wireless indoor channel.	[12]

vector. And the envisioned MISO secure optical wireless CW model with an LED array at Alice is shown in Figure 12.2 (b).

As for the aspect of overcoming a weakness of neuronal synchronisation, in the work of [12], one novel key exchange algorithm is proposed over an optical wireless indoor channel in optical wireless IoMT networks. Moreover, the reported results of this new cryptosystem demonstrate that the involved combination of Diffie-Hellman hyperelliptic curves cryptography could upgrade the neural synchronisation model and could give the certainty that an attack is not possible since the attacker could not implement one viable form of synchronisation with any of the parties in the communication. For clarity, the critical literature review of secret key agreement-enabled secure optical wireless Internet of Medical Things could be summarised in Table 12.2.

12.2.3 *Physical-layer encryption-enabled secure optical wireless communication Internet of Medical Things*

In this subsection, we focus on the representative advances of physical-layer encryption-enabled optical wireless IoMT-relevant works [15–18]. Actually, physical-layer encryption empowerment has been introduced to optical wireless IoMT research from several main directions, including but not limited to temporal ghost imaging-based encryption for optical wireless IoMT [15] and image encryption and watermarking in medical optical wireless system communications [16].

Since temporal ghost imaging (TGI) has been proposed and illustrated to enable fast signal retrieval using slow detectors, the study in [15] explores the effectiveness of a TGI-based optical encryption scheme for optical wireless IoMT with a micro-light-emitting diode (micro-LED), as shown in Figure 12.3. The reported experiment demonstrates that the decryption process is realised by correlated calculation between the secret key according to computational TGI and the encrypted signal received by the single-pixel detector (SPD), and the system could resist attacks of up to 40 per cent occlusion or 80 per cent Gaussian noise and obtain an error-free signal.

In terms of exploring the physical-layer image encryption/watermarking scheme, the study in [16] proposes a new hybrid technique based on discrete memristive chaos for physical-layer encryption and watermarking (PL-EW) and asymmetrically clipped optical orthogonal frequency-division multiplexing (ACO-OFDM). Specifically, in the encryption process, a novel two-dimensional hyperchaotic modular cascaded memristive (2D HC-MCM) map is introduced for a large secret key space. Furthermore, it is identified that compared with the original memristive map, the proposed 2D HC-MCM map could extend the secret space by a factor of 100.

Figure 12.3 shows the schematic of the envisioned secure optical wireless communication encryption and decryption process based on computational temporal ghost imaging.

For clarity, the critical literature review of physical-layer encryption-enabled secure optical wireless Internet of Medical Things is summarised in Table 12.3.

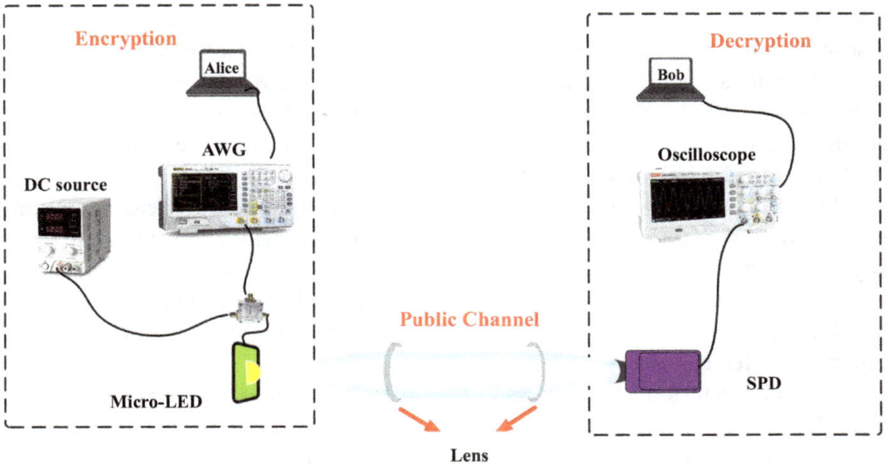

Figure 12.3 Schematic of the envisioned encryption and decryption process based on computational temporal ghost imaging for physical-layer encryption-enabled secure optical wireless communication

Table 12.3 Critical literature review of physical-layer encryption-enabled secure optical wireless Internet of Medical Things

Identified gaps	Contribution of the current work	References
Failure to explore further development of optical wireless encryption universal schemes	One optical encryption scheme for optical wireless Internet of Medical Things is proposed based on temporal ghost imaging (TGI) with a micro-light-emitting diode (micro-LED).	[15]
Failure to explore the physical-layer image encryption/watermarking scheme in an optical wireless system employing asymmetrically clipped optical orthogonal frequency-division multiplexing	Image encryption and watermarking in an asymmetrically clipped optical orthogonal frequency-division multiplexing optical wireless system employing a novel memristive hyperchaotic map.	[16]
Failure to explore the optical wireless communication system based on the data encryption method using asymmetric encryption	An optical wireless communication technique using asymmetric data encryption for Internet of Medical Things.	[17]

12.3 Quantum key distribution-enabled secure optical wireless Internet of Medical Things

In this section, we mainly discuss the representative advances of quantum key distribution-enabled optical wireless IoMT techniques. Actually, quantum key distribution has been introduced to the optical wireless IoMT research from several main directions, including but not limited to high-dimensional quantum key distribution in medical optical wireless access networks [19], optical wireless quantum communication coding [20] and continuous-variable quantum key distribution in medical optical wireless systems [21].

Figure 12.4 shows the schematic of quantum key distribution-enabled secure optical wireless Internet of Medical Things: (a) the envisioned exchanging secret keys between an indoor wireless user, Alice and the central office (CO) within one secure optical wireless Internet of Medical Things and (b) the envisioned relevant decoding scheme for time-phase encoding technique in secure optical wireless Internet of Medical Things.

Recently, thanks to the increased demand for overcoming some practical issues faced by conventional quantum key distribution (QKD) systems, high-dimensional quantum key distribution (HD-QKD) has gained increasing attention in terms of providing a promising and efficient platform. Tremendous attention has been paid to high-dimensional QKD IoMT systems, which are interesting options for boosting error tolerance and for improving the secret key generation rate. In the work of [19], d-dimensional time-phase encoded states are used to exchange a secret key between an indoor wireless user and the central office on the other end of the access network.

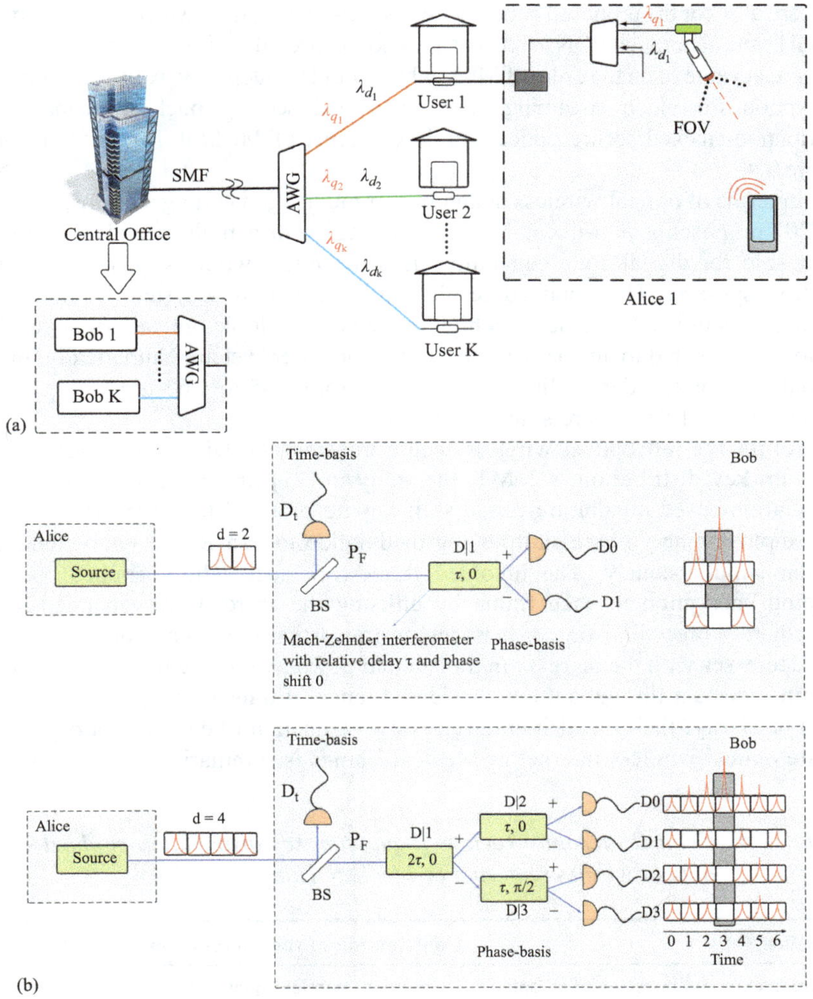

Figure 12.4 Schematic of quantum key distribution-enabled secure optical wireless Internet of Medical Things: (a) the envisioned exchanging secret keys between an indoor wireless user, Alice and the central office (CO) within one quantum key distribution-enabled secure optical wireless Internet of Medical Things and (b) the envisioned relevant decoding scheme for time-phase encoding technique in quantum key distribution-enabled secure optical wireless Internet of Medical Things

Numerically, for the proposed optical wireless HD-QKD system with a repetition rate of 2.5 GHz and at nominal PON lengths of a few kilometres, d = 4 is sufficient for obtaining a secure key rate of on the order of Mb/s, which could be adopted to realise one-time pad encryption for video streaming. The envisioned scene model of quantum key distribution-enabled secure optical wireless Internet of Medical Things is shown in Figure 9.4.

In terms of optical wireless quantum communication coding systems, the study in [20] proposes a non-linear MRR system to form a multi-wavelength system applicable for digital code generation used in optical wireless quantum communication systems. The obtained results illustrate that this system of micro-ring resonators is utilised to generate a high capacity of multiple optical bright and dark solitons, connected to an analogue-to-digital converter system. Numerically, ranges of multi-bright and dark soliton wavelengths from 1.45 to 1.65 μm with a central wavelength of 1.55 μm are simulated as well.

As for the left optical wireless secure transmission with continuous-variable quantum key distribution in IoMT, the study in [21] implements a quantum key distribution-based quadrature phase shift keying protocol to share the secret key and employs binary phase shift keying modulation to transfer the encrypted information simultaneously. The involved framework successfully applies one-time padding encryption to each frame by utilising the secret keys generated in the immediately preceding frame. It is demonstrated that the overall bit error rate of Bob decreases with the increase in transmitted power, while the bit error rate of Eve remains constant throughout the considered range of transmitted power.

For clarity, the critical literature review of quantum key distribution-enabled secure optical wireless Internet of Medical Things is summarised in Table 12.4.

Table 12.4 Critical literature review of quantum key distribution-enabled secure optical wireless Internet of Medical Things

Identified gaps	Contribution of the current work	References
Failure to explore the design of an optical wireless quantum communication coding system using a decimal converter	One optical wireless quantum communication coding system using a decimal converter	[20]
Failure to provide a paradigm to simultaneously facilitate key distribution and secure information transmission in an intensity-modulated/direct-detection optical wireless system	One secure transmission with one-time pad encryption using frequency-division multiplexing-based continuous-variable quantum key distribution in an optical wireless system	[21]
Failure to use the Kerr non-linear type of light in the micro-ring resonators to generate multiple wavelengths for the desired application in quantum network cryptography	The proposed micro-ring resonators quantum dense coding for an optical wireless communication system using a decimal converter	[22]

12.4 Chaos-enabled secure optical wireless Internet of Medical Things

Figure 12.5 shows the schematic of chaos-enabled secure optical wireless Internet of Medical Things: (a) the envisioned application scenario of the proposed architecture for chaos-enabled secure optical wireless Internet of Medical Things and (b) the envisioned modified key generation mechanism in chaos-enabled secure optical wireless Internet of Medical Things.

In this section, we summarise the main advances in chaos-enabled optical wireless IoMT techniques from the views of optical chaos-based remote health monitoring [23], optical chaos-based secure optical body area network [24] and medical optical wireless communication based on chaos encryption [25], separately.

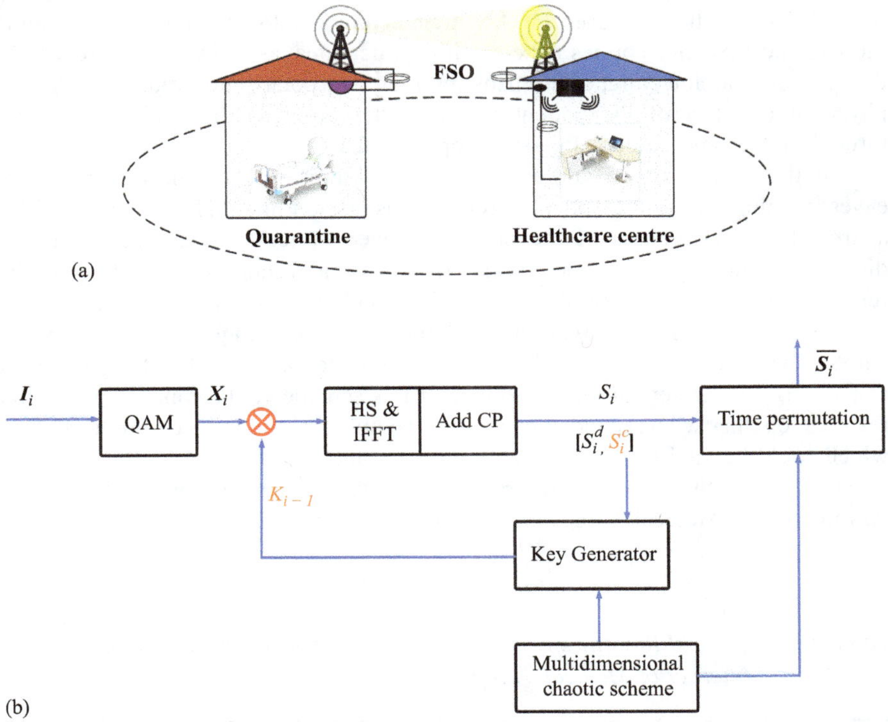

Figure 12.5 Schematic of chaos-enabled secure optical wireless Internet of Medical Things: (a) the envisioned application scenario of the proposed architecture for chaos-enabled secure optical wireless Internet of Medical Things and (b) the envisioned modified key generation mechanism in chaos-enabled secure optical wireless Internet of Medical Things

Note that the transmission of electroencephalogram (EEG) signals to locations where a trained neurophysiologist is present has become critical for ensuring accurate and timely treatment of patients with serious brain disorders. Nevertheless, this patient data, including highly sensitive information, must be transmitted but not eavesdropped on or changed.

To address this issue, the study in [23] proposes a secure EEG signal transmission system for remote health monitoring using optical chaos. In the proposed scheme, one semiconductor laser source is used to generate optical chaos, which hides the EEG signal before its transmission over the optical fibre medium. Specifically, the scheme is implemented for long-haul communication in which the linear impairments of optical fibre are controlled for the successful transmission of the secure signal. The results demonstrate that after 120 km of fibre length, the Q-factor of the received signal is drastically reduced to 5.72, which is about 66.67 at a 5 km fibre length.

In terms of optical chaos-secured optical body area network (OBAN), the study in [24] detailed the proposed OBAN architecture employing polarisation multiplexing and free-space optics for e-Health applications, as shown in Figure 9.5. In this proposed architecture, after chaos masking and polarisation multiplexing, the physiological data of the patient is transmitted towards the healthcare centre through indoor and outdoor free-space optics links.

At the same time, to address the potential threat of unauthorised receivers eavesdropping during the optical wireless transmission of IoMT, the work in [25] affirms the effectiveness of the chaos-improved optical wireless gadget under diverse chaotic signal parameters and transmission distances. In particular, the related design paradigm exploits one novel method of combining chaotic indicators with optical wireless to defend data transmission. Accordingly, the two involved codes (Lyapunov exponent code and ANFIS code) are adopted to generate a chaotic signal. Numerical results show that this scheme could improve protection features together with expanded encryption complexity and reduced predictability, which could be vital in stopping eavesdropping and fact theft.

For clarity, the critical literature review of chaos-enabled secure optical wireless Internet of Medical Things is summarised in Table 12.5.

Table 12.5 Critical literature review of chaos-enabled secure optical wireless Internet of Medical Things

Identified gaps	Contribution of the current work	References
Failure to design a chaos-based security model for IEEE 802.15.7 optical wireless Internet of Medical Things	The design of a chaos-based security model for IEEE 802.15.7 optical wireless Internet of Medical Things	[28]

(Continues)

Table 12.5 (*Continued*)

Identified gaps	Contribution of the current work	References
Absence of a physical-layer encryption method to effectively enhance the security of a discrete cosine transform precoded orthogonal frequency-division multiplexing-based medical optical wireless system	The chaos-based encryption scheme for discrete cosine transform precoded orthogonal frequency-division multiplexing-based medical optical wireless communication systems	[29]
Failure to support secure image transmission in orthogonal frequency-division multiplexing medical optical wireless communication systems	One chaos theory-based physical-layer two-level encryption for orthogonal frequency-division multiplexing medical optical wireless systems	[30]

12.5 Physical-layer security-enabled optical wireless Internet of Medical Things

In this section, the main advances of physical-layer security-enabled optical wireless IoMT techniques include but are not limited to physical-layer security in multi-user medical optical wireless systems [31], physical-layer security of reconfigurable intelligent surfaces (RIS)-aided medical optical wireless communications [32] and secure optical body area network based on relay position optimisation [33], separately.

Figure 12.6 shows the schematic of a physical-layer security-enabled secure optical wireless Internet of Medical Things: (a) the envisioned reflected indoor relay-aided secure optical wireless system in which one transmitter communicates with two legal receivers and an eavesdropper and (b) the envisioned non-orthogonal multiple access (NOMA) secure optical wireless system with a single eavesdropper.

To develop a sustainable IoMT, the concept of physical-layer security has arisen to address the security issue in a downlink multi-user optical wireless system, where the performance of the optical wireless legitimate users in the presence of a single eavesdropper and multiple eavesdroppers has been taken into account separately. As for facilitating the secure application of optical wireless IoMT, the work in [31] investigates the security performance of multi-user optical wireless systems with non-orthogonal multiple access (NOMA), as shown in Figure12.6 (a). And it is verified that for given a NOMA legitimate user, the secrecy outage probability eventually settles at around 0.2 while the semi-angle at half illuminance of the light-emitting diode (LED) transmitter varies between 15° and 60°.

In terms of physical-layer security of reconfigurable intelligent surfaces (RIS)-aided optical wireless IoMT, the study in [32] proposes one optimisation problem to maximise the achievable secret key rate of a secure optical wireless system with liquid crystal (LC) RIS-aided receiver, where the refractive index of LC acts as the decision variable. Simulation results illustrate that the system's achievable secret

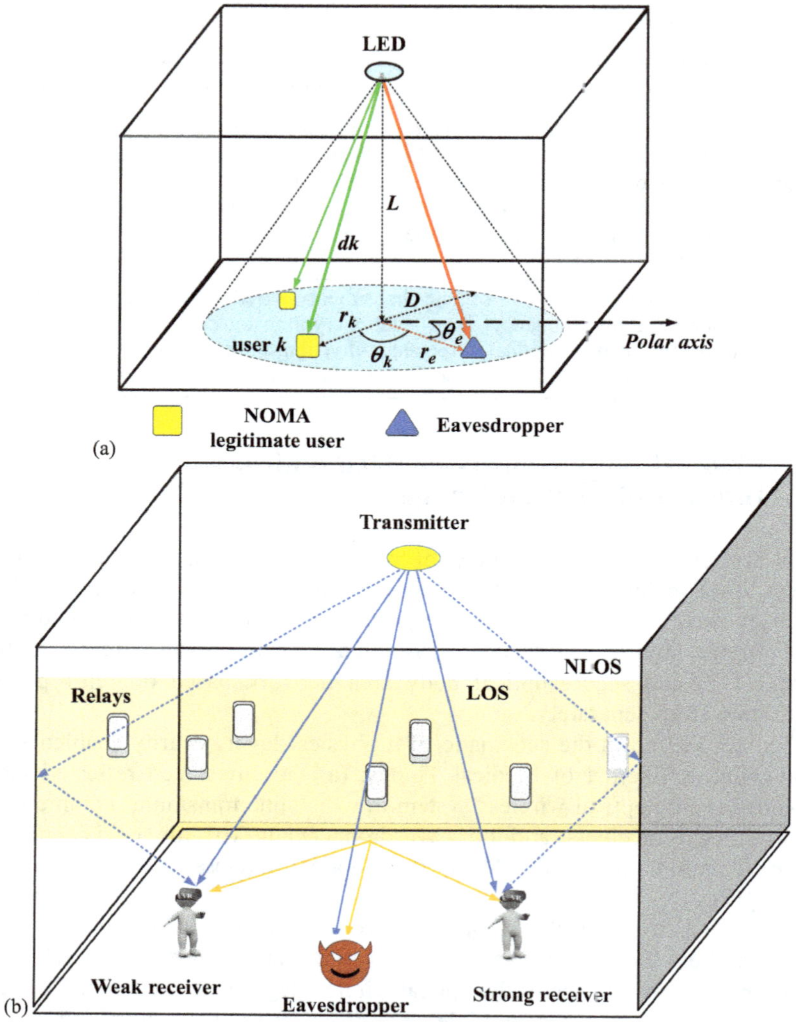

Figure 12.6 Schematic of physical-layer security-enabled secure optical wireless Internet of Medical Things: (a) the envisioned non-orthogonal multiple access (NOMA) secure optical wireless system with a single eavesdropper and (b) the envisioned reflected indoor relay-aided secure optical wireless system in which one transmitter communicates with two legal receivers and an eavesdropper

key rate could be significantly enhanced by the integration of an LC RIS-aided receiver. In particular, this optimisation framework is capable of maximising the achievable secret key rate and employs the Sine-Cosine Algorithm (SCA) to solve it.

Table 12.6 Critical literature review of physical-layer security-enabled secure optical wireless Internet of Medical Things

Identified gaps	Contribution of the current work	References
Absence of securing the relay transmission of patient health data in optical body area networks	The proposed relay-aided physical-layer security scheme for optical wireless Internet of Medical Things	[33]
Failure to identify the potential data sniffing over an open optical wireless channel	Accomplishing the data sniffing over an open optical wireless channel	[34]
Absence of demonstrating secret message transmission using a light-emitting diode	The demonstrated secret message transmission using a light-emitting diode	[35]

As for the left-relay-based secure optical body area network, the work in [33] proposes a novel relay-aided physical-layer security scheme in a reflected indoor optical wireless IoMT system, as shown in Figure 12.6 (b). Based on the formulated problems, the proposed iterative scheme is capable of jointly optimising the spatial positions of relays and beamforming vectors to maximise the sum data rate of the involved indoor optical wireless IoMT while defending against eavesdropping. Numerically, for the amplify-and-forward and decode-and-forward case, the achievable maximum data rates of the legal receivers could be increased by 1.1 and 1.2 times, respectively.

For clarity, the critical literature review of physical-layer security-enabled secure optical wireless Internet of Medical Things is summarised in Table 12.6.

12.6 Future research opportunities and challenges

Although the above representative progresses have been reported in the emerging domain of cryptography-enabled optical wireless IoMT, it must be noted that many challenges and limitations of cryptography-enabled optical wireless IoMT must be carefully investigated and addressed, which mainly include limited mobility support, constrained energy efficiency, a shortage of link robustness and random device orientation. To address these main challenges, a series of distinct branch directions of the optical wireless IoMT family are waiting for the exploration of customised cryptography techniques [36–39]. As potential and fundamental near-future studies, the suggested branch directions include but are not limited to optical beam search and recovery for secure optical wireless IoMT, hybrid RF and optical wireless IoMT, optical beam predictions for secure optical wireless IoMT, uniform coverage for secure optical wireless IoMT, ultraviolet secure optical wireless IoMT and so on. It must be emphasised that the novel non-Lambertian optical beam could provide commercially available and distinct spatial emission characteristics, typically as shown in Figure 9.7, which potentially presents research opportunities in the hybrid light beam and cryptography dimension for the cutting-edge secure optical wireless IoMT. Following the work of I. Moreno, for convenience, the respective spatial radiation intensity could be profiled by a sum of Gaussian functions [40,41].

$$R_{\mathrm{NSPW}}(\phi,\alpha)=\sum_{m=1}^{2} g_{1m}^{\mathrm{NSPW}}\exp\left[-(\mathrm{In}2)(|\phi|-g_{2m}^{\mathrm{NSPW}})^2\left(\frac{\cos^2\alpha}{(g_{3m}^{\mathrm{NSPW}})^2}+\frac{\sin^2\alpha}{(g_{4m}^{\mathrm{NSPW}})^2}\right)\right] \tag{12.1}$$

where $\alpha_{i,j}$ denotes the azimuth angle from the i-th LED optical source and the j-th PD receiver, the coefficient values of the Gaussian functions are given as $g_{11}^{\mathrm{NSPW}}=0.13$, $g_{21}^{\mathrm{NSPW}}=45°$, $g_{31}^{\mathrm{NSPW}}=g_{4_1}^{\mathrm{NSPW}}=18°$, $g_{1_2}^{\mathrm{NSPW}}=1$, $g_{2_2}^{\mathrm{NSPW}}=0$, $g_{3_2}^{\mathrm{NSPW}}=38°$ and $g_{4_2}^{\mathrm{NSPW}}=22°$. Correspondingly, for the future OWN, the 3D representation of the non-Lambertian NSPW optical beam is shown in the right part of Figure 9.7 (a).

Figure 12.7 shows the schematic of novel non-Lambertian optical beams for the developing artificial intelligence-enabled optical wireless network technique:

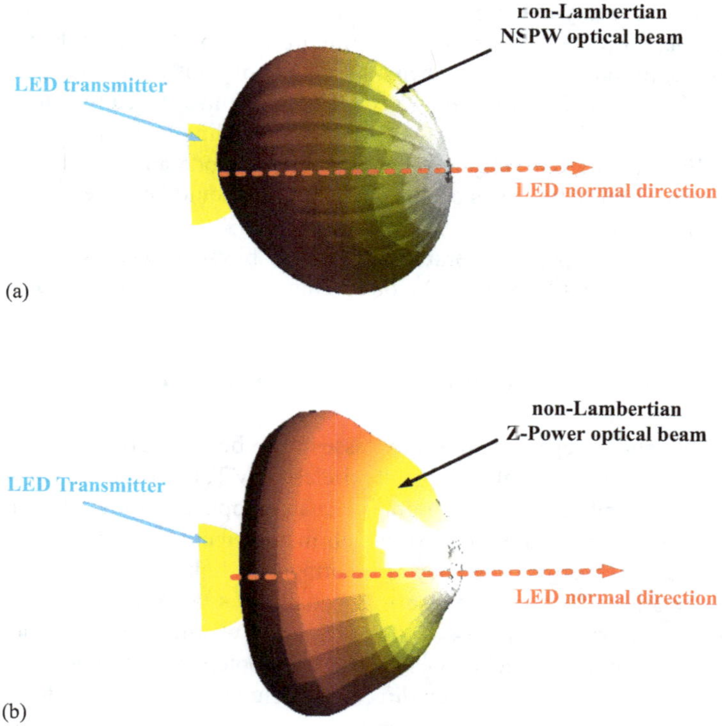

Figure 12.7 Schematic of distinct non-Lambertian light beam patterns for developing secret key techniques-enabled secure optical wireless Internet of Medical Things: (a) 3D representation of non-Lambertian NSPW optical beam and (b) 3D representation of non-Lambertian Z-Power optical beam

(a) 3D representation of non-Lambertian NSPW optical beam and (b) 3D representation of non-Lambertian Z-Power optical beam.

Following the work of Yao Sen *et al.*, for convenience, for the ultraviolet beam pattern 1 of the optical source for optical wireless IoMT, the spatial radiation intensity could be profiled by a sum of two Gaussian functions [40,41].

$$R_{\text{Z-Power}}(\phi) = \sum_{k=1}^{K} g_{1k}^{\text{Z-Power}} \exp\left[-\ln 2\left(\frac{|\phi| - g_{2k}^{\text{Z-Power}}}{g_{3k}^{\text{Z-Power}}}\right)^2\right] \tag{12.2}$$

where $K = 3$ is the number of Gaussian functions. Specifically, the values of the coefficients in this expression are: $g_{11}^{\text{Z-Power}} = 0.542$, $g_{21}^{\text{Z-Power}} = 22.75°$, $g_{31}^{\text{Z-Power}} = 49.96°$, $g_{12}^{\text{Z-Power}} = 0.573$, $g_{22}^{\text{Z-Power}} = 77.84°$, $g_{32}^{\text{Z-Power}} = 23.7°$, $g_{13}^{\text{Z-Power}} = 0.279$, $g_{23}^{\text{Z-Power}} = 86.67°$ and $g_{33}^{\text{Z-Power}} = 8.43°$. where θ denotes the irradiance angle from the optical source to the receiver. Correspondingly, for the future optical wireless IoMT, the 3D representation of non-Lambertian Z-Power optical beams is shown in the right part of Figure 12.7 (b).

Up to now, actually, the published articles of cryptography-assisted optical wireless IoMT have mainly been limited to performance evaluation and numerical simulation, while relevant hardware testing and prototype implementation are still unavailable, which means that more diverse research activities are needed to push the solid development of cryptography-assisted optical wireless IoMT for the upcoming 6G era.

12.7 Conclusions

This work aims to provide a thorough overview of cryptography-enabled optical wireless IoMT techniques for the upcoming 6G era. It is started with of the state of the art in secret key-enabled medical and healthcare optical wireless communications. This work discusses the most representative advances in secret key generation, secret key agreement and physical-layer encryption in the medical optical wireless community. Then, the representative approaches and algorithms are comparatively investigated for quantum information-enabled medical optical wireless techniques, particularly quantum key distribution in the IoMT. Moreover, the rapidly growing methods and implementations of chaos-enabled optical wireless communication techniques and physical-layer security-enabled medical optical wireless transmission techniques are also presented in different aspects, followed by a key discussion of potential research opportunities and challenges for cryptography-enabled IoMT in the more challenging 6G era.

Acknowledgement

This work was supported in part by the Shandong Provincial Natural Science Foundation (Grants No. ZR2026MS1120), National Natural Science Foundation of

China (Grants No. 62061043), Yantai Nanshan University Research and Development Fund Project (Grants No. 2025KJ01001), Tianshan Cedar Project of Xinjiang Uygur Autonomous Region (Grants No. 2020XS27), and High-level Talents Introduction Project in Autonomous Region (Grants No. 042419004).

References

[1] H. Herfandi, O. S. Sitanggang, M. R. A. Nasution, H. Nguyen and Y. M. Jang, "Real-time patient indoor health monitoring and location tracking with optical camera communications on the internet of medical things," in *Applied Sciences*, vol. 14, no. 3, p. 1153, 2024. https://doi.org/10.3390/app14031153.

[2] S. Caputo, G. Borghini, S. Jayousi, A. Rashid and L. Mucchi, "Visible light communications for healthcare applications: Opportunities and challenges," in *2023 IEEE 17th International Symposium on Medical Information and Communication Technology (ISMICT)*, Lincoln, NE, USA, 2023, pp. 1–6, https://doi.org/10.1109/ISMICT58261.2023.10152290.

[3] P. D. P. Adi, R. Sariningrum, R. Rahmatullah, *et al.*, "Light fidelity and internet of things for medical applications," in *2024 4th International Conference of Science and Information Technology in Smart Administration (ICSINTESA)*, Balikpapan, Indonesia, 2024, pp. 347–352, https://doi.org/10.1109/ICSINTESA62455.2024.10748065.

[4] J. Song, X. Wang, J. Wang, *et al.*, "The Converged Internet of Lights Network for Telecommunication, Positioning, Illumination, and Medical Therapy," in *IEEE Communications Standards Magazine*, vol. 4, no. 3, pp. 70–75, 2020, https://doi.org/10.1109/MCOMSTD.001.2000011.

[5] F. I. K. Mousa, N. Al Maadeed, K. Busawon, A. Bouridane and R. Binns, "Secure MIMO visible light communication system based on user's location and encryption," in *Journal of Lightwave Technology*, vol. 35, no. 24, pp. 5324–5334, 2017, https://doi.org/10.1109/JLT.2017.2761301.

[6] E. Mahalal, M. Ismail, Z. -Y. Wu, M. M. Fouda, Z. M. Fadlullah and N. Kato, "Robust deep learning-based secret key generation in dynamic LiFi networks against concept drift," *2024 IEEE 21st Consumer Communications & Networking Conference (CCNC)*, Las Vegas, NV, USA, 2024, pp. 899–904, https://doi.org/10.1109/CCNC51664.2024.10454770.

[7] Y. M. Al-Moliki, M. T. Alresheedi and Y. Al-Harthi, "Secret key generation protocol for optical OFDM systems in indoor VLC networks," in *IEEE Photonics Journal*, vol. 9, no. 2, pp. 1–15, Art no. 7901915, 2017, https://doi.org/10.1109/JPHOT.2017.2667400.

[8] E. Mahalal, M. Ismail, Z. Wu and M. M. Fouda, "Characterization of secret key generation in 5G+ indoor mobile LiFi networks," in *GLOBECOM 2022 - 2022 IEEE Global Communications Conference*, Rio de Janeiro, Brazil, 2022, pp. 01–06, https://doi.org/10.1109/GLOBECOM48099.2022.10001158.

[9] E. Mahalal, E. Hasan, M. Ismail, *et al.*, "Concept drift aware wireless key generation in dynamic lifi networks," in *IEEE Open Journal of the*

Communications Society, vol. 6, pp. 742–758, 2025, https://doi.org/10.1109/OJCOMS.2024.3524497.

[10] D. Guo, J. Zhang, G. Xin, K. Tian and X. Xiao, "Secret key agreement for SISO visible light communication systems." in *Photonics*, vol. 12, no. 5, p. 498, 2025, https://doi.org/10.3390/photonics12050498.

[11] A. Mukherjee, "Secret-key agreement for security in multi-emitter visible light communication systems," in *IEEE Communications Letters*, vol. 20, no. 7, pp. 1361–1364, 2016, https://doi.org/10.1109/LCOMM.2016.2558562.

[12] A. Araya, I. Jirón and I. Soto, "A new key exchange algorithm over a VLC indoor channel," in *2017 First South American Colloquium on Visible Light Communications (SACVLC)*, Santiago, Chile, 2017, pp. 1–5, https://doi.org/10.1109/SACVLC.2017.8267603.

[13] T. Araki and T. Suzuki, "Fuzzy timing passwords for providing easy user authentication to disable persons and their application to visible light communication," *World Automation Congress 2012*, Puerto Vallarta, Mexico, 2012, pp. 1–5.

[14] Y. M. Al-Moliki, M. T. Alresheedi, Y. Al-Harthi and A. H. Alqahtani, "Robust lightweight-channel-independent OFDM-based encryption method for VLC-IoT networks," in *IEEE Internet of Things Journal*, vol. 9, no. 6, pp. 4661–4676, 2022, https://doi.org/10.1109/JIOT.2021.3107395.

[15] Y. Wang, H. Chen, W. Jiang, *et al.*, "Optical encryption for visible light communication based on temporal ghost imaging with a micro-LED," in *Optics and Lasers in Engineering*, vol. 134, no. 1, p. 106290, 2020.

[16] A. A. Elsadany, A. Elsonbaty and E. A. A. Hagras, "Image encryption and watermarking in ACO-OFDM-VLC system employing novel memristive hyperchaotic map," in *Soft Computing*, vol. 27, pp. 4521–4542, 2023. https://doi.org/10.1007/s00500-023-07818-5

[17] Y. U. Lee, "Secure visible light communication technique based on asymmetric data encryption for 6G communication service," in *Electronics*, vol. 9, no. 11, p. 1847, 2020, https://doi.org/10.3390/electronics9111847

[18] S. Abedin, T. Tasbin and A. Hira, "Optical wireless data transmission with enhanced substitution Caesar Cipher WHEEL encryption," in *2017 International Conference on Electrical, Computer and Communication Engineering (ECCE)*, Cox's Bazar, Bangladesh, 2017, pp. 552–556, https://doi.org/10.1109/ECACE.2017.7912967.

[19] O. Elmabrok, M. Razavi, T. Eltaif and K. A. Alaghbari, "High-dimensional quantum key distribution in quantum access networks," in *IEEE Photonics Journal*, vol. 16, no. 3, pp. 1–7, Art no. 7301407, 2024, https://doi.org/10.1109/JPHOT.2024.3383780.

[20] A. Shahidinejad, A. Nikoukar, T. Anwar *et al.*, "Optical wireless quantum communication coding system using decimal convertor," in *Optical and Quantum Electronics*, vol. 45, pp. 449–457, 2013, https://doi.org/10.1007/s11082-013-9656-6

[21] R. Verma, A. Jaiswal, V. Saini, M. Rawat and M. Goyal, "Secure Transmission with one-time pad encryption using FDM-based CV-QKD in VLC

system," *2025 National Conference on Communications (NCC)*, New Delhi, India, 2025, pp. 1–6, https://doi.org/10.1109/NCC63735.2025.10983674.
[22] A. Nikoukar, I. S. Amiri, A. Shahidinejad, A. A. Shojaei, J. Ali and P. P. Yupapin, "MRR quantum dense coding for optical wireless communication system using decimal convertor," in *2012 International Conference on Computer and Communication Engineering (ICCCE)*, Kuala Lumpur, Malaysia, 2012, pp. 770–774, https://doi.org/10.1109/ICCCE.2012.6271321.
[23] R. Shahzadi, S. M. Anwar, F. Qamar, M. Ali, J. J. P. C. Rodrigues and M. Alnowami, "Secure EEG signal transmission for remote health monitoring using optical chaos," in *IEEE Access*, vol. 7, pp. 57769–57778, 2019, https://doi.org/10.1109/ACCESS.2019.2912548.
[24] B. Kanwal, F. Kanwal, T. Kausar, N. Habib, A. Atieh and J. Mirza, "Optical chaos based secure optical body area network employing polarization multiplexing and free space optics for e-health applications," in *2022 19th International Bhurban Conference on Applied Sciences and Technology (IBCAST)*, Islamabad, Pakistan, 2022, pp. 1055–1060, https://doi.org/10.1109/IBCAST54850.2022.9990449.
[25] H. K. Rumuh and I. A. Murdas, "Co-simulation of visible light communication based on chaos encryption using MATLAB® and OptiSystem," in Swaroop, A., Virdee, B., Correia, S.D., Polkowski, Z. (eds.), *Proceedings of Data Analytics and Management. ICDAM 2024. Lecture Notes in Networks and Systems*, vol 1298. Springer, Singapore, 2025. https://doi.org/10.1007/978-981-96-3355-5_27.
[26] Y. M. Al-Moliki, M. T. Alresheedi and Y. Al-Harthi, "Improving availability and confidentiality via hyperchaotic baseband frequency hopping based on optical OFDM in VLC networks," in *IEEE Access*, vol. 8, pp. 125013–125028, 2020, https://doi.org/10.1109/ACCESS.2020.3007429.
[27] P. Canyelles-Pericas, P. A. Haigh, Z. Ghassemlooy *et al.*, "Chaos synchronization in visible light communications with variable delays induced by multipath fading," *Applied System Innovation*, vol. 1, no. 4, p. 45, 2018, https://doi.org/10.3390/asi1040045.
[28] T. T. Son, H. B. Hung, H. Le-Minh, N. Aslam and P. Canyelles-Pericas, "Chaos-based physical layer security model for IEEE 802.15.7 visible light communications," in *2017 Seventh International Conference on Information Science and Technology (ICIST)*, Da Nang, Vietnam, 2017, pp. 85–90, https://doi.org/10.1109/ICIST.2017.7926497.
[29] Z. Wang and S. Chen, "A chaos-based encryption scheme for DCT precoded OFDM-based visible light communication systems," in *Journal of Electrical and Computer Engineering*, vol. 2016, 2326563, 7 pages, 2016. https://doi.org/10.1155/2016/2326563.
[30] Z. Wang and J. Lv, "Secure Image Transmission in Orthogonal Frequency Division Multiplexing Visible Light Communication Systems," in *IEEE Access*, vol. 7, pp. 107927–107936, 2019, https://doi.org/10.1109/ACCESS.2019.2932908.

[31] X. Zhao, H. Chen and J. Sun, "On physical-layer security in multiuser visible light communication systems with non-orthogonal multiple access," in *IEEE Access*, vol. 6, pp. 34004–34017, 2018.

[32] D. Guo, G. Xin, Q. Wu and J. Zhang, "Physical layer secret key security of liquid crystal RIS-aided visible light communication systems," in *2025 IEEE 7th International Conference on Communications, Information System and Computer Engineering (CISCE)*, Guangzhou, China, 2025, pp. 1072–1076, https://doi.org/10.1109/CISCE65916.2025.11065715.

[33] K. Wu, Y. Dong, X. Yu, J. Song and X. -P. Zhang, "Physical layer security based on 2-D spatial position optimization of relays in reflected indoor VLC system," in *IEEE Wireless Communications Letters*, vol. 13, no. 8, pp. 2235–2239, 2024, https://doi.org/10.1109/LWC.2024.3409582.

[34] I. Marin-Garcia, A. M. Ramirez-Aguilera, V. Guerra, J. Rabadan and R. Perez-Jimenez, "Data sniffing over an open VLC channel," in *2016 10th International Symposium on Communication Systems, Networks and Digital Signal Processing (CSNDSP)*, Prague, Czech Republic, 2016, pp. 1–6, https://doi.org/10.1109/CSNDSP.2016.7573963.

[35] A. Mammuthu, L. John, P. T. Habeebullah, C. Mathilakath, K. Jithisha and M. R. Manu, "Cryptolight: Secret message transmission using LED," in *2016 Online International Conference on Green Engineering and Technologies (IC-GET)*, Coimbatore, India, 2016, pp. 1–3, https://doi.org/10.1109/GET.2016.7916846.

[36] J. Ding, C.-L. I, J. Wang, H. Yang and L. Wang, "Multiple optical beam switching for physical layer security of visible light communications," in *IEEE Photonics Journal*, vol. 14, no. 1, pp. 1–9, Art no. 7306209, 2022.

[37] J. Ding, C.-L. I, J. Wang and H. Yang, "Optical beams switching-based coverage enhancement scheme for compact visible light communications," in *Journal of Lightwave Technology*, vol. 40, no. 18, pp. 6139–6150, 2022.

[38] J. Ding, C.-L. I, J. Wang and H. Yang. "Link characteristics comparison of lambertian & non-lambertian MIMO-based 6G vehicular visible light communications," in *Inventions*, vol. 10, no. 1, p. 1, 2025.

[39] J. Ding, C.-L. I, J. Wang, and H. Yang, "NOMA visible light communications with distinct optical beam configurations," in *Photonics*, vol. 11, p. 944, 2024, https://doi.org/10.3390/photonics11100944.

[40] I. Moreno and C. Sun, "Modeling the radiation pattern of LEDs," in *Optics Express*, vol. 16, no. 3, pp. 1808–1819, 2008.

[41] J. Ding, C.-L. I, and Z. Xu, "Indoor optical wireless channel characteristics with distinct source radiation patterns," in *IEEE Photonics Journal*, vol. 8, no. 1, pp. 1–15, 2015.

Chapter 13

Cryptography-enabled multi-network emergency communication framework for intelligent healthcare systems

David Panebianco[1], Roberta Avanzato[1], Francesco Beritelli[1], Agbotiname Lucky Imoize[2], Gaetano Costa[3] and Cristian Randieri[3]

Abstract

The progressive digitalisation of healthcare systems, the spread of interconnected medical devices and the expansion of Internet of Medical Things (IoMT) environments are transforming the way clinical information is generated, transmitted and used in emergency situations. In this context, communication reliability, security and continuity play a central role, since delays, infrastructure failures, network congestion or cyberattacks may directly compromise the timeliness of intervention and the clinical outcome.

This chapter proposes EMES-H (Emergency Mobile Europe System - Health), a secure multi-network communication framework designed to support, rather than replace, existing emergency infrastructures. The objective of the framework is to enable the protected and continuous transmission of urgent medical information across heterogeneous telecommunication environments by combining cryptographic data protection mechanisms with adaptive network selection and resilient routing strategies.

The contribution of the chapter develops along three main lines. First, it examines the role of intelligent healthcare systems, the evolution of IoMT and the communication requirements specific to medical emergencies. Second, it discusses the main security and reliability issues affecting healthcare networks, with particular attention to clinical data protection, operational continuity and the vulnerabilities typical of critical scenarios. Finally, it presents the EMES-H architecture

[1]Department of Electrical, Electronic, and Computer Engineering (DIEEI), University of Catania, Italy
[2]Department of Electrical and Electronics Engineering, Faculty of Engineering, University of Lagos, Nigeria.
[3]Department of Theoretical and Applied Sciences, eCampus University, Italy.

together with its operational workflow, functional modules and principal application scenarios.

Overall, EMES-H can be regarded as a modular architecture aimed at strengthening the resilience of emergency healthcare communications. By integrating cryptographic protection, multi-network co-ordination and adaptive transmission logic, the framework provides a coherent basis for the development of safer, more interoperable and more operationally robust intelligent healthcare systems.

Keywords: Emergency healthcare communications; Internet of Medical Things; cryptography; multi-network communication; network resilience; healthcare cybersecurity; interoperability

13.1 Introduction

The digital transformation of healthcare has profoundly reshaped the organisation of clinical services, patient monitoring processes and the co-ordination mechanisms among the actors involved in care delivery. Contemporary healthcare infrastructures increasingly rely on communication networks, digital platforms, data analytics systems and connected medical devices capable of making clinical information available in real time across different operational contexts. In this scenario, communication infrastructures no longer represent a mere technological support but rather an enabling element for the routine delivery of healthcare services and, even more critically, for the management of medical emergencies.

The notion of an intelligent healthcare communication system encompasses the set of digital platforms, connected medical devices, sensorised environments and telecommunication infrastructures that enable continuous, timely and efficient information exchange among patients, healthcare professionals, emergency operators and medical institutions. These systems support remote diagnosis, continuous monitoring, co-ordinated intervention and rapid access to clinically relevant information, thereby extending care beyond the traditional boundaries of the hospital. Their relevance is especially evident in emergency settings, where the quality and speed of data transmission may directly affect both response effectiveness and clinical outcomes [1,2].

13.1.1 Evolution of digital healthcare infrastructures

Healthcare infrastructures have evolved from largely isolated clinical environments into complex digital ecosystems in which multiple communication channels and different types of medical data converge. This transformation has been driven by advances in computing, networking, sensing and data analysis techniques, which have made a more integrated management of healthcare information possible. As a result, modern healthcare increasingly depends on the ability to collect, share and process clinical data within distributed and interconnected environments [1,2].

Digital healthcare ecosystems can be understood as environments in which hospitals, clinics, emergency services, patient devices and remote platforms co-operate through secure communication infrastructures. Their value lies not only in data availability but also, above all, in the co-ordination they enable across different levels of care, fostering continuity, timeliness in decision-making and greater operational coherence among the actors involved [1,3].

Telemedicine and remote monitoring play a particularly important role in this evolution. Telemedicine has expanded the possibility of conducting consultations, follow-up activities, diagnostic support and therapeutic supervision without requiring the patient to be physically present at the healthcare facility. Remote monitoring systems have further reinforced this model by enabling the continuous observation of physiological parameters through connected sensors and devices. These solutions have proved especially relevant in chronic disease management, elderly care, post-treatment surveillance and emergency contexts, in which early intervention can be decisive [1,2].

The COVID-19 pandemic made the strategic value of these infrastructures particularly evident, showing that the resilience of healthcare systems depends not only on the clinical capacity of institutions but also on the robustness and adaptability of the communication ecosystem that supports them.

13.1.2 Internet of Medical Things (IoMT)

The development of intelligent healthcare systems is closely linked to the spread of the Internet of Medical Things (IoMT), namely the set of connected medical devices, wearable sensors, monitoring systems and digital platforms capable of collecting, transmitting and processing health data through communication networks. IoMT is significant because it enables a more continuous, distributed and data-oriented management of health, expanding the possibilities for clinical observation and timely intervention [1,2].

Within IoMT environments, medical sensors play a central role in collecting physiological parameters such as heart rate, blood pressure, oxygen saturation, body temperature and other clinically relevant indicators. These sensors may be integrated into portable devices, hospital equipment, ambulances, patients' homes or assisted-care facilities. Once acquired, the data can be transmitted to healthcare platforms for real-time or near-real-time analysis in support of both clinical assessment and operational decision-making [1,2,4].

Wearable devices also represent a significant component of IoMT. Smartwatches, fitness trackers and wearable medical devices support the continuous observation of patient conditions in everyday settings, contributing to the early identification of anomalies and to a more proactive, patient-centred model of care. Although levels of accuracy and reliability may vary depending on the device type, their widespread adoption has reinforced the idea of healthcare that extends well beyond traditional clinical spaces [1,2].

This dimension is complemented by the role of internal hospital networks, which integrate medical devices, monitoring systems, electronic health records, diagnostic platforms and administrative services into a unified digital environment. The convergence of IoMT and advanced communication infrastructures has thus made healthcare systems more interconnected, more responsive and more strongly oriented towards the operational use of data. At the same time, it has increased the importance of issues such as network reliability, communication security and interoperability among heterogeneous systems [1,2,5].

13.1.3 Communication requirements in medical emergencies

Medical emergencies impose particularly stringent requirements on communication infrastructures. In such contexts, the rapid and reliable transmission of clinical information is not merely a technical objective but an essential condition for ensuring effective and co-ordinated intervention. Responders, physicians, hospitals and control centres must be able to receive and process critical information even in unstable or heavily congested communication environments [3,4].

A first requirement is low latency. In emergency situations, even limited delays may slow clinical decision-making and reduce the effectiveness of the response. Information such as vital signs, patient location, preliminary diagnostic data and alerts must therefore be transmitted as quickly as possible to the actors responsible for co-ordination and intervention [3,5].

Alongside latency, reliability is equally important. Emergency communication systems must remain operational even in the presence of network congestion, partial failures, coverage limitations or degraded operating conditions. This entails the adoption of redundancy mechanisms, alternative paths and resilient transmission strategies capable of preserving service continuity even when one or more channels become unstable or unavailable [5–7].

A further requirement is the ability to support the real-time or near-real-time transmission of continuous clinical data streams. In many emergency scenarios, communication does not end with the sending of a single alert but also includes electrocardiograms, ambulance telemetry, physiological parameters and dynamic updates on patient status. The value of such information depends largely on its timeliness and integrity, since delays or partial loss may compromise its usefulness for pre-hospital assessment and for preparing the receiving facility [2–4].

13.2 Security challenges in healthcare communication networks

The rapid digitalisation of healthcare infrastructures and the spread of interconnected medical devices have significantly expanded the exposure surface of cyber risks. The more healthcare services depend on digital platforms, distributed networks and continuous data flows, the more crucial it becomes to protect sensitive clinical information and ensure the continuity of critical operations. From this perspective, cybersecurity can no longer be regarded as a secondary issue but must be understood as a structural condition for patient protection, institutional trust and the resilience of healthcare services [4,8,9].

Healthcare communication networks represent a particularly attractive target for attackers because they combine two highly critical elements: valuable personal data and essential public services. Medical information has long-term relevance, can be exploited in multiple ways and is often stored and transmitted within complex digital ecosystems that include hospitals, remote platforms, monitoring devices and emergency systems. As a result, a vulnerability in healthcare communication may compromise not only confidentiality and privacy but also the operational reliability of the system as a whole [4,8,9].

13.2.1 Cyber threats in healthcare systems

Among the most significant threats is unauthorised access to health information, including medical records, diagnostic reports, clinical histories and other confidential patient data. Such breaches may stem from weaknesses in information systems, insecure links, inadequate credential management or vulnerabilities affecting connected devices. The particular sensitivity and persistent value of clinical data make this form of threat especially critical [4,8,9].

Another major risk is represented by ransomware attacks, which may encrypt archives, hospital systems or network resources, rendering them unavailable until a ransom is paid or the systems are restored through alternative procedures. In healthcare, service interruption may directly affect continuity of care and the response capacity of institutions, thereby significantly increasing the severity of such attacks [4,9].

A further source of concern is network spoofing, namely the falsification of the identity of devices or communication nodes. In such cases, unauthorised actors may intercept data, alter transmitted information, send false alerts or gain access to sensitive systems by impersonating legitimate entities. In emergency contexts, manipulated messages or messages originating from non-authentic sources may compromise co-ordination, delay response and generate incorrect clinical or operational decisions [8,9].

13.2.2 Privacy and protection of medical data

The protection of patient information is a fundamental requirement of contemporary healthcare systems. Medical data includes highly sensitive elements such as clinical history, diagnostic reports, treatments, biometric parameters and other information directly related to an individual's health condition. Their unauthorised disclosure or misuse may lead to privacy violations, identity theft, discrimination and loss of trust in healthcare institutions and digital health services [3,9].

Protecting such information requires the adoption of robust measures capable of preventing unauthorised access, interception, tampering and misuse. In this context, tools such as encryption, digital signatures, integrity checks and secure

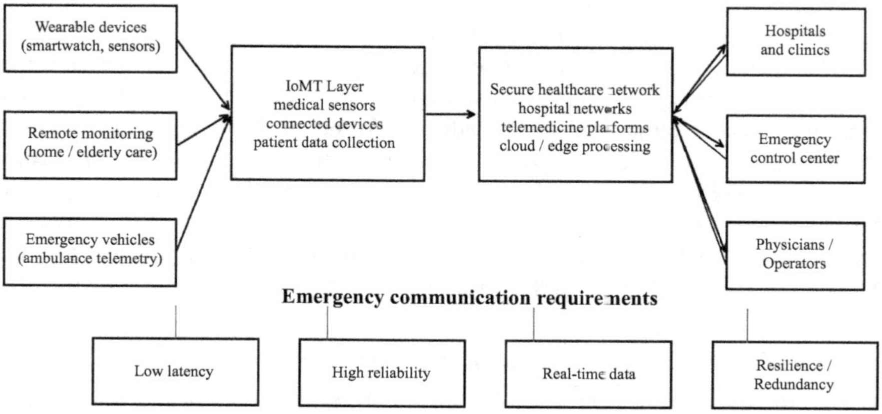

Figure 13.1 Digital healthcare ecosystem and communication requirements

authentication protocols play an essential role in ensuring that medical data remains accessible only to authorised subjects and that transmitted information remains trustworthy throughout the communication process.

At the same time, healthcare systems must operate in compliance with stringent legal and regulatory frameworks such as HIPAA in the United States and GDPR in Europe, which govern the collection, processing, storage and transfer of personal data, including health data. Compliance with these principles not only responds to legal and organisational requirements but also constitutes a condition of credibility for digital healthcare systems, since controlled access, accountability, traceability and transparency are central elements in the management of sensitive information.

Figure 13.1 illustrates the main elements of the digital healthcare ecosystem, showing how IoMT devices, remote monitoring systems, ambulance telemetry and healthcare network infrastructures interact with hospitals, emergency control centres and medical personnel. It also highlights the main communication requirements associated with emergency healthcare services.

13.2.3 Vulnerabilities in emergency communications

Emergency scenarios introduce additional vulnerabilities that may severely affect the performance and reliability of healthcare communication systems. A typical problem is network congestion, which arises when numerous devices, platforms and operators attempt to exchange information simultaneously during a critical event. Under such conditions, increased latency, packet loss, service degradation or temporary interruptions may occur, with immediate consequences for the speed and effectiveness of the response [3,5,6].

Another critical issue concerns infrastructure failures, which may result from natural disasters, blackouts, physical damage, cyberattacks or malfunctions affecting access networks and communication facilities. When such events occur during an emergency, the loss of connectivity may limit co-ordination among ambulances, hospitals, control centres and public authorities, thereby reducing the overall effectiveness of intervention [5–7].

Large-scale emergencies such as pandemics, major incidents or mass-casualty scenarios amplify these vulnerabilities. In such contexts, healthcare communication infrastructures are required to sustain simultaneous information exchange among multiple institutions, field teams and authorities, with increased organisational complexity and a greater risk of interruptions or security compromise. For this reason, emergency healthcare communication architectures must integrate data protection, service continuity, dynamic network management and rapid response capabilities in the face of changing operational conditions [3,6,7].

13.3 Multi-network communication architectures for emergency systems

In emergency healthcare, reliable communication infrastructures are essential to support rapid co-ordination among patients, responders, hospitals and public safety authorities. Traditional systems have long been based on a single network managed by a specific operator; although such architectures may prove adequate under ordinary conditions, they do not always provide the flexibility and resilience required to face large-scale emergencies or degraded network environments [5–7].

Advances in communication technologies have fostered the development of multi-network architectures designed to co-ordinate multiple heterogeneous infrastructures dynamically. Their underlying rationale is to improve reliability, coverage and continuity of information exchange while reducing dependence on a single path or provider. Through adaptive network selection and the use of alternative channels, such systems can remain operational even in the presence of congestion, partial failures or service degradation.

13.3.1 Limitations of single-operator communication systems

Reliance on a single telecommunication operator constitutes a first structural limitation. Any problem affecting that network – congestion, local outage, hardware failure or service degradation in a given area – may immediately compromise the ability to transmit critical medical information. The issue becomes especially significant in emergencies, when communication demand rises rapidly and may exceed the actual capacity of the available network [5,6].

A further limitation concerns geographic coverage. Operators do not guarantee uniform service levels across the territory, and performance may vary substantially across urban areas, rural zones, mountainous regions or isolated contexts. Under these conditions, dependence on a single provider may reduce communication continuity precisely where it is most needed.

Finally, single-operator models often lack genuine adaptive routing. When network conditions deteriorate, communication flows cannot easily be redirected towards alternative infrastructures, with the result that delays, losses or interruptions may persist at the moment of greatest need. These limitations justify the adoption of more flexible architectures capable of operating in environments marked by high variability and uncertainty [6,7].

13.3.2 Heterogeneous telecommunication networks

Contemporary communication environments are characterised by the coexistence of multiple telecommunication technologies, each with its own performance characteristics, coverage properties and operational advantages. This heterogeneity offers important opportunities to increase the resilience of emergency communication systems, provided that the different infrastructures are properly integrated and co-ordinated [5–7].

Cellular networks, including 4G, 5G and future developments towards 6G, constitute the main communication infrastructure for many mobile healthcare applications. They provide broad coverage, high speeds and mobility, making them particularly suitable for ambulances, wearable devices, portable medical equipment and other tools used in the field. Next-generation networks also offer features such as network slicing and ultra-reliable low-latency communication, which may be highly beneficial for critical healthcare services.

Alongside terrestrial cellular networks, satellite systems represent a complementary resource of major importance, especially in remote or underserved areas and in situations in which ground infrastructures are damaged or unavailable. A similar role may be played by specialised emergency networks, dedicated radio systems and fibre backbones intended for critical services. The possibility of combining these technologies makes it possible to build integrated architectures that are more robust and less dependent on a single communication path [5,6].

13.3.3 Resilient multi-network communication approaches

Resilient multi-network architectures aim to improve system robustness through the dynamic use of multiple available infrastructures. Instead of relying on a single network path, they evaluate the status of different communication options and select the most suitable transmission strategy on the basis of parameters such as latency, reliability, congestion level and service availability.

A particularly widespread strategy is adaptive network selection, through which the system continuously observes the conditions of the available channels and can migrate from one interface to another when performance degradation is detected. A healthcare monitoring device may, for example, use a cellular connection under ordinary conditions and switch to a satellite link or another alternative channel when the primary connection becomes unstable or unavailable [5–7].

A further approach is multipath communication, in which the same information is transmitted simultaneously over different networks in order to increase the probability of successful delivery even in the presence of local failures, congestion or temporary interruptions. Such architectures may also integrate orchestration mechanisms capable of co-ordinating communication flows, assigning priority to critical traffic and dynamically allocating available resources. Taken together, these capabilities provide the basis for secure and interoperable frameworks suited to supporting critical healthcare services in complex and variable environments [5–7].

Table 13.1 presents the main cybersecurity threats, privacy-related risks and communication vulnerabilities affecting emergency healthcare communication systems, with representative impacts and mitigation strategies. Table 13.2 summarises the main cryptographic techniques and secure communication mechanisms adopted in healthcare systems, together with their advantages, limitations and typical application contexts.

13.4 EMES-H framework: a secure multi-network architecture for emergency communications

The growing complexity of healthcare communication environments, together with the spread of connected medical devices and distributed monitoring systems, calls for architectures capable of combining security, resilience and operational continuity in emergency situations. EMES-H (Emergency Mobile Europe System - Health) is positioned in this context as a multi-network communication framework designed to support the protected transmission of emergency medical information across heterogeneous telecommunication environments [3,6,9].

EMES-H is conceived as a healthcare-oriented co-ordination structure that operates alongside existing emergency infrastructures without replacing them. Its aim is to strengthen the communication dimension of emergency healthcare management by enabling the secure and reliable transmission of medical information across multiple available networks, even under adverse operating conditions. In this respect, the framework is consistent with the broader European context of emergency response.

From a technical standpoint, EMES-H is based on the principle that critical healthcare communication should not depend exclusively on a single network, a single transmission path or a single operator. The framework, therefore, introduces a co-ordination architecture capable of dynamically orchestrating communication flows across different infrastructures while integrating cryptographic mechanisms for data protection. This approach strengthens service continuity, reduces dependence on individual providers and preserves the confidentiality, integrity and authenticity of transmitted medical information.

Table 13.1 Main cybersecurity threats

Threat/vulnerability	Description	Potential impact on healthcare systems	Representative mitigation strategies
Data breaches	Unauthorised access to patient records, diagnostic information and other sensitive medical data through system vulnerabilities, insecure connections or compromised credentials.	Privacy violations, identity theft, loss of trust and legal and regulatory consequences.	Encryption, secure access control, strong authentication, network segmentation and continuous monitoring.
Ransomware attacks	Malicious software that encrypts hospital systems, medical records or operational platforms, making them unavailable until recovery actions are taken.	Service disruption, delayed treatments, suspension of clinical activities and diversion of emergency patients.	Backup strategies, incident response plans, end point protection, segmentation and timely patch management.
Network spoofing	Impersonation of legitimate devices or network entities in order to intercept, manipulate or inject medical information.	False alerts, corrupted medical data, unauthorised access and incorrect emergency co-ordination.	Mutual authentication, digital certificates, message integrity verification and trusted device identity management.
Unauthorised data manipulation	Alteration of medical information during storage or transmission, whether accidental or malicious.	Incorrect diagnoses, inappropriate treatment decisions and reduced reliability of emergency information.	Hash-based integrity checks, digital signatures, secure logging and controlled write access.
Privacy exposure	Inadequate protection of highly sensitive medical and biometric information exchanged across digital healthcare systems.	Confidentiality loss, discrimination risks, reputational damage and reduced patient trust.	End-to-end encryption, privacy-by-design policies, role-based access control and regulatory compliance mechanisms.
Network congestion	Overload of communication infrastructures caused by simultaneous data exchange during emergencies or crisis situations.	Delays, packet loss, reduced service quality and impaired emergency co-ordination.	Traffic prioritisation, adaptive routing, redundancy and multi-network communication support.
Infrastructure failure	Loss or degradation of communication services caused by outages, disasters, cyberattacks or physical damage to network facilities.	Communication interruption, reduced co-ordination among responders and loss of service continuity.	Redundant infrastructures, failover mechanisms, multipath transmission and satellite or backup communication support.
Large-scale emergency stress	Exceptional pressure on healthcare communication systems during pandemics, natural disasters or mass-casualty incidents.	Simultaneous overload, co-ordination complexity, increased risk of communication breakdowns and security weaknesses.	Resilient communication architectures, dynamic orchestration, scalable co-ordination layers and interoperability across networks.

Table 13.2 Main cryptographic techniques and secure communication mechanisms

Mechanism/technique	Main function	Advantages	Limitations	Typical use in healthcare
AES	Symmetric encryption	High speed, low overhead	Requires secure key distribution	Real-time medical data protection
RSA	Asymmetric encryption	Secure key exchange, digital signatures	Large key size, higher overhead	Secure authentication and key management
ECC	Asymmetric encryption	Strong security with smaller keys	More complex implementation than symmetric methods	IoMT devices and wearable sensors
TLS	Secure communication protocol	End-to-end confidentiality and integrity	May introduce latency/handshake overhead	Device-to-platform and hospital communications
Digital signatures	Integrity and authentication	Prevent tampering, verify sender identity	Computational cost	Medical reports and authenticated transmissions
Certificates/PKI	Device authentication	Trusted identities across networks	Certificate management complexity	Secure IoMT onboarding

13.4.1 Overview of EMES-H

The guiding principle of EMES-H is that critical healthcare communications must remain available even when network conditions become unstable, degraded or heterogeneous. In emergency situations, medical information may need to traverse environments characterised by partial failures, overload, variable coverage or uneven service quality. In such scenarios, a rigid model based on a single operator or a fixed path is often inadequate.

EMES-H is therefore built around three main objectives: communication resilience, secure information exchange and interoperability across different communication environments. Resilience enables the system to monitor available networks and adapt transmission decisions to operating conditions; security protects medical data against unauthorised access, alteration and misuse; interoperability allows the framework to operate across multiple infrastructures and operators, avoiding confinement within a single communication domain.

A distinguishing feature of the framework is that it is not conceived as a closed system but rather as a co-ordination layer that supports emergency healthcare communications within a broader ecosystem including public emergency services, healthcare institutions, mobile networks and specialised infrastructures. Its value thus lies in its ability to integrate, strengthen and co-ordinate existing resources rather than proposing the replacement of systems already in use [5–7].

13.4.2 System architecture

The framework adopts a layered architecture in which components are organised according to their role in the emergency communication process. This structure promotes modularity, interoperability and extensibility, thereby facilitating integration with different network infrastructures and healthcare operating environments [6,7].

The first layer is the terminal layer, which includes devices and end points capable of generating, acquiring or transmitting emergency medical information. It comprises wearable medical sensors, portable monitoring devices, ambulance communication systems, mobile applications and other technologies used by patients, carers and healthcare personnel. The function of this layer is to detect significant events and initiate requests containing data such as patient status, physiological indicators, event type and location.

The second layer is the network co-ordination layer, which represents the functional core of EMES-H. It analyses communication conditions, manages transmission priorities, evaluates available network resources and decides how emergency information should be routed so as to maximise continuity, protection and delivery reliability. This layer concentrates the main adaptive and cryptographic functions of the framework.

The third layer is the emergency authority layer, which includes the operational entities responsible for receiving and using emergency medical information: hospitals, control centres, healthcare institutions, ambulance co-ordination systems and public safety organisations. At this layer, the received information is transformed into concrete decisions and interventions in support of healthcare and institutional response.

Figure 13.2 presents the cryptographic protection framework for intelligent healthcare and IoMT communications, showing the interaction among authentication mechanisms, encryption techniques, secure communication protocols and emergency operational constraints.

13.4.3 Functional modules of the EMES-H framework

Within the co-ordination architecture, EMES-H manages emergency communications through a set of specialised and closely integrated modules. The emergency detection module identifies the occurrence of a clinically significant event, which may result from direct user activation, automatic analysis of physiological data or alerts generated by external monitoring systems.

Once the event has been recognised, the emergency request generation module translates it into a structured message containing elements such as event type, geographic location, timestamp, device identity and selected medical data relevant to emergency assessment. The structured format improves consistency, readability and co-ordination at the receiving entities.

The orchestration and control module constitutes the main decision-making centre of the framework: it assigns priorities, applies communication rules and adapts system behaviour to network conditions. Supporting this function is the network evaluation module, which monitors parameters such as latency, packet loss, availability, congestion and quality of service. On the basis of this analysis, the path selection module identifies the channel or set of channels most suitable for information delivery, with the possibility of dynamically updating the decision when network conditions change significantly.

Taken together, these modules enable EMES-H to operate as a secure and adaptive communication infrastructure in which event detection, data structuring, network-aware orchestration and resilient transmission are integrated into a single process.

Figure 13.3 illustrates the conceptual architecture of the proposed EMES-H framework, highlighting the network co-ordination layer, secure multi-network routing mechanisms and the interaction among healthcare institutions, emergency operators and heterogeneous communication infrastructures. In particular, Figure 13.3 provides a conceptual overview of the proposed EMES-H framework and its main functional components.

13.5 Secure and adaptive multi-network communication workflow in emergencies and application scenarios

EMES-H integrates cryptographic protection, adaptive network selection and resilient multi-network transmission into a unified communication model designed for

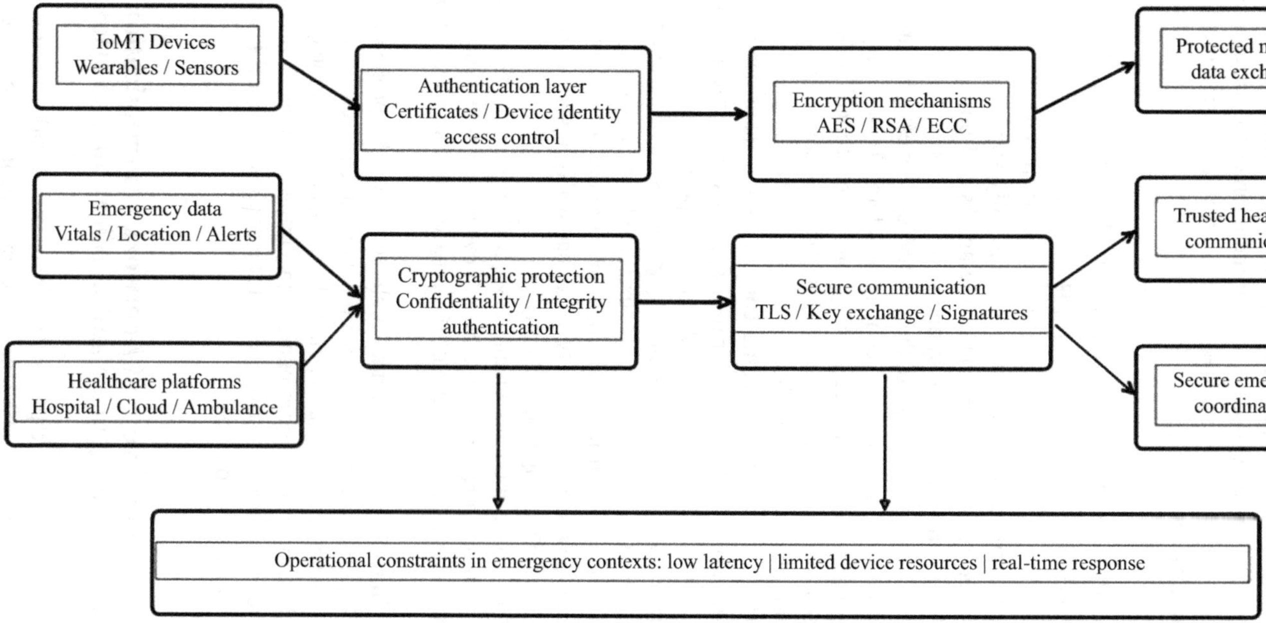

Figure 13.2 Cryptographic protection framework for intelligent healthcare and IoMT communications

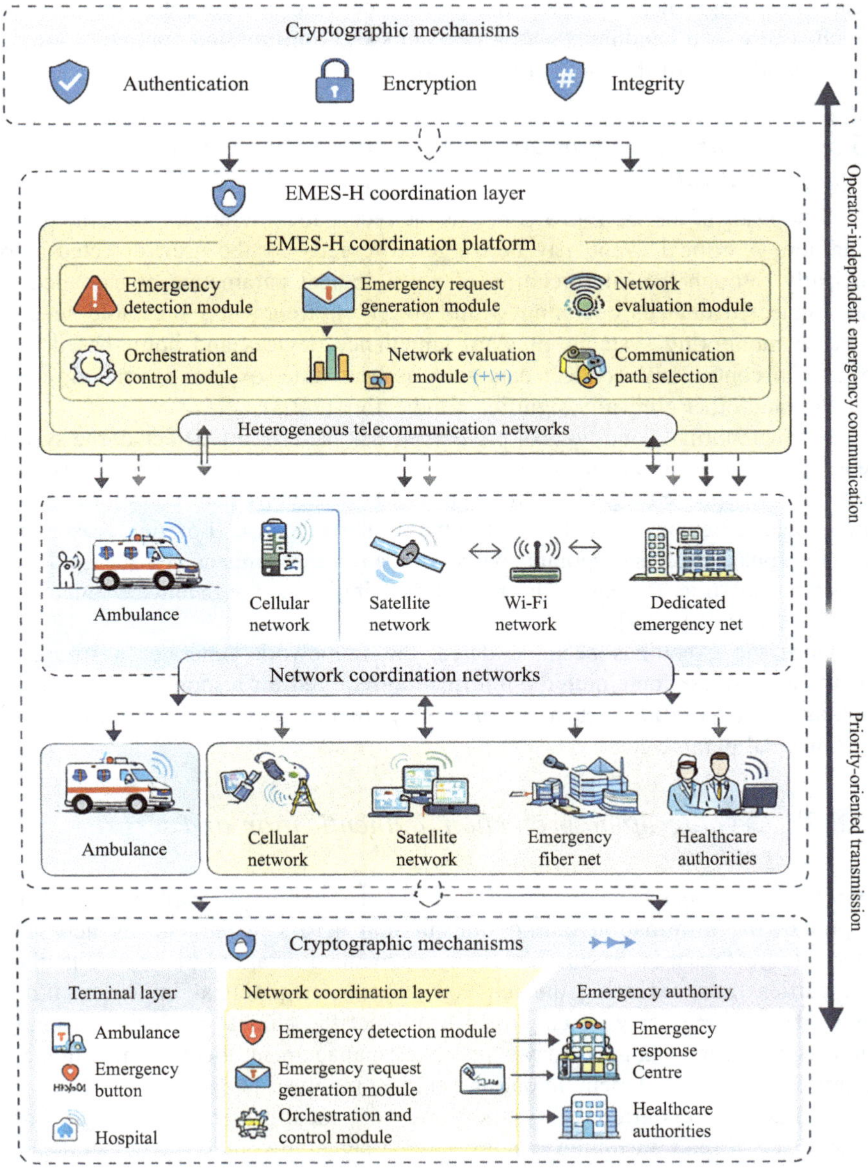

Figure 13.3 Overview of the proposed EMES-H framework

emergency healthcare contexts. In such scenarios, communication infrastructures must simultaneously satisfy requirements of confidentiality, integrity, authentication, reliability and timely delivery of critical information. These requirements cannot be adequately addressed through isolated measures or conventional models based on a single network path; rather, they demand an integrated workflow in which secure data handling, routing decisions and transmission continuity are co-ordinated as parts of the same process [3,6,7].

13.5.1 Secure detection and generation of emergency requests

The first phase of the workflow concerns the secure identification of an emergency condition. A critical event may be triggered directly by the user, detected automatically through the interpretation of physiological parameters or signalled by interconnected devices operating within IoMT environments. Wearable sensors, portable monitoring systems, onboard ambulance devices and home-based platforms can continuously collect data such as heart rate, oxygen saturation, blood pressure and other clinically significant indicators [1,2,4].

When abnormal conditions or predefined risk patterns are detected, the system generates an emergency signal. At this stage, source trustworthiness is essential: false alarms, spoofed signals or manipulated notifications may compromise both operational efficiency and the credibility of the response. For this reason, the detection phase must incorporate source-validation mechanisms and secure device identification so as to reduce the risk of reacting to non-authentic, erroneous or malicious signals [4,8,9].

Once the event has been validated, the framework generates a structured emergency request containing the information required for response co-ordination, such as patient location, detection time, preliminary event classification and relevant medical parameters.

13.5.2 Cryptographic protection, authentication and secure routing

After the request has been generated, EMES-H applies cryptographic mechanisms to protect the information during transmission across heterogeneous networks. Encryption safeguards the confidentiality of medical data and prevents unauthorised parties from accessing the message content. In practical implementations, symmetric schemes may be employed to protect the clinical payload efficiently, while asymmetric techniques may support key management, identity validation and secure session establishment among the entities involved [3,8,9].

In addition to confidentiality, the framework must preserve the integrity and authenticity of the transmitted request. Digital signatures, message authentication mechanisms and hash-based controls make it possible to verify that the content has not been altered and that the message actually originates from a legitimate source. In emergency contexts, even minimal modifications to clinical data, event type or patient location may lead to assessment errors, delayed intervention or incorrect allocation of resources.

Secure routing is entrusted to the co-ordination layer, which selects paths not only on the basis of performance but also according to their reliability and trustworthiness. The cryptographic protection provided by EMES-H is therefore not limited to message content but extends to the entire communication flow, including the choice of the path through which critical medical data travels [3,6,7].

13.5.3 Adaptive network evaluation and path selection

A central feature of EMES-H is its ability to evaluate multiple communication networks in real time and to adapt the transmission strategy accordingly. Emergency healthcare scenarios often unfold in environments marked by congestion, partial failures, variable signal quality or progressive infrastructure degradation. Under such conditions, fixed strategies are inadequate; what is needed instead is a system capable of continuously observing the status of available resources and selecting the path best suited to the current operational context [5–7].

Network evaluation is based on a set of performance indicators. Among these, latency is fundamental because it directly affects the speed with which alerts, patient data and diagnostic information reach their destination. Packet loss is equally relevant, since it may compromise the completeness of alerts and the reliability of continuous clinical data streams. A third parameter is the overall availability of the resource, understood not merely as instantaneous performance but as the probability that a given network will remain stable throughout the interval required for transmission.

13.5.4 Redundancy, multipath transmission and dynamic adaptation

To further increase resilience, EMES-H supports redundancy and multipath transmission strategies. Rather than relying exclusively on a single channel, the framework may send critical medical information simultaneously across multiple infrastructures, such as cellular networks, satellite links or dedicated emergency systems. This increases the probability that at least one copy of the message will reach its destination even in the presence of congestion, degradation or temporary unavailability affecting some channels [5–7].

Multipath transmission is especially useful in large-scale emergencies and geographically complex environments, in which the performance of individual infrastructures is difficult to predict. At the same time, redundancy should not be applied indiscriminately: excessive duplication of flows may generate overload and contribute to communication stress. For this reason, the co-ordination layer defines the degree of redundancy according to the severity of the event, the nature of the data, the state of the network and the availability of resources.

A further defining element of the workflow is dynamic adaptation. If the initially selected path becomes unreliable owing to increased latency, packet loss or partial failures, EMES-H can redirect the flow towards alternative infrastructures without interrupting the ongoing information exchange. This capability is essential for preserving continuity and reliability in uncertain and changing operational scenarios [5–7].

13.5.5 Priority-based transmission in emergencies

Because emergency medical communications are strongly time-sensitive, EMES-H adopts priority-based transmission policies. Emergency requests, medical alerts and other clinically relevant data must be handled before non-critical traffic. The co-ordination layer, therefore, classifies flows on the basis of urgency and assigns communication resources according to the importance of the transmitted information.

This approach is particularly useful when multiple sessions compete for the same infrastructure or when network capacity is reduced because of overload or other critical conditions. By prioritising vital data – such as patient location, physiological parameters, ECG traces or pre-hospital alerts directed to receiving facilities – the framework increases the probability that essential messages will arrive without significant delay. In this way, transmission priority works together with adaptive routing and redundancy mechanisms to support both timeliness and reliability [5–7].

13.5.6 Application scenarios

The operating principles of EMES-H apply to several scenarios in which secure and resilient communication is essential. A first case concerns ambulance-to-hospital communication, in which data on patient status, clinical updates and pre-alert information must be transmitted to the receiving facility during transport so as to support the preparation of personnel and resources before arrival.

A second scenario concerns remote patient monitoring and the reporting of critical events. In intelligent healthcare environments, wearable devices and home-based systems can continuously observe chronic patients older adults or subjects in post-treatment follow-up. When abnormal values or potentially dangerous physiological patterns are detected, alerts must be transmitted promptly to healthcare operators or emergency facilities. In such contexts, EMES-H contributes to ensuring data protection, delivery reliability and the availability of alternative paths in the event that the primary network becomes unavailable [1,2,4].

A third scenario is disaster response and large-scale emergencies. Events such as natural disasters, pandemics or mass-casualty incidents may place exceptional pressure on healthcare communication infrastructures, requiring simultaneous exchanges among institutions, field units and public authorities. In such contexts, EMES-H can support operational continuity through the co-ordinated use of heterogeneous networks, secure routing logic and multipath transmission.

Finally, the framework can be integrated with broader public emergency management systems in which healthcare communication must interact with civil protection agencies, emergency medical services, public safety authorities and co-ordination centres. In this role, EMES-H acts as a healthcare-focused support layer aimed at improving the secure exchange of medical alerts, facility status updates, patient data and contextual information among different organisations.

13.6 Final discussion and research perspectives

The progressive digitalisation of healthcare systems, together with the spread of interconnected medical devices and distributed data-exchange platforms, has made communication security and reliability an essential requirement of contemporary healthcare. In this context, EMES-H proposes an architecture intended to support emergency healthcare communications through the integration of cryptographic protection and adaptive multi-network communication strategies [3,5,6].

One of the most significant features of the framework lies in the fact that security is conceived not as an accessory component but as a structural part of the entire communication process. Encryption, authentication and integrity-verification mechanisms make it possible to protect clinical data, emergency notifications and co-ordination information during transmission, thereby reducing the risk of unauthorised access, content alteration or message falsification. At the same time, the use of multiple communication infrastructures and adaptive routing logic strengthens service continuity even in scenarios characterised by congestion, degradation or partial network unavailability.

Nevertheless, several relevant challenges remain with regard to practical implementation. A first critical issue concerns interoperability among heterogeneous infrastructures, which is necessary for effectively co-ordinating cellular networks, satellite links and dedicated emergency systems within a coherent ecosystem. A second challenge relates to the constraints of IoMT devices, which often have limited computational capacity, memory and energy resources; in such contexts, protection mechanisms must be calibrated so as not to compromise the timeliness and functionality of the system. Further central issues include the scalability of the co-ordination layer in large-scale emergencies and full consistency with regulatory and governance requirements concerning data protection, access control, traceability of operations and accountability in the management of sensitive information.

From an evolutionary standpoint, EMES-H should be regarded not as a closed solution but as a modular architecture open to further development. In this perspective, advanced mechanisms for auditing and traceability of communication events are particularly relevant, as is the evolution of network infrastructures towards scenarios characterised by lower latency, greater flexibility in resource allocation and tighter integration between terrestrial and non-terrestrial segments. In particular, satellite-assisted communication appears especially significant in remote or underserved contexts and in critical situations in which terrestrial infrastructures may be limited or compromised [5,6].

13.7 Conclusion

In conclusion, EMES-H addresses a central issue in contemporary digital healthcare: how to guarantee a secure, continuous and reliable exchange of emergency information across heterogeneous and potentially degraded network environments. Through the combination of cryptographic protection, dynamic path adaptation and co-ordinated use of multiple network infrastructures, the framework provides a coherent foundation for supporting healthcare communications in complex operational conditions. More broadly, the chapter shows that communication resilience must now be recognised as a structural condition for patient safety, institutional reliability and the overall effectiveness of emergency response.

References

[1] S. M. R. Islam, D. Kwak, M. H. Kabir, M. Hossain and K. S. Kwak, "The internet of things for health care: A comprehensive survey," *IEEE Access*, vol. 3, pp. 678–708, 2015.

[2] M. Chen, Y. Ma, Y. Li, D. Wu, Y. Zhang and C. H. Youn, "Wearable 2.0: enabling human-cloud integration in next generation healthcare systems," *IEEE Communications Magazine*, vol. 55, no. 1, pp. 54–61, 2017.

[3] R. Lu, X. Lin and X. Shen, "Secure and efficient data transmission for cloud-based eHealth systems," *IEEE Transactions on Parallel and Distributed Systems*, vol. 25, no. 3, pp. 750–761, 2014.

[4] P. Kumar and H. J. Lee, "Security issues in healthcare applications using wireless medical sensor networks: A survey," *Sensors*, vol. 12, no. 1, pp. 55–91, 2012.

[5] N. H. S. Suhaimi, N. H. Kamarudin, M. N. A. Khalid, I. Tahir and M. A. A. Mohamed, "State-of-the-art authentication measures in satellite communication networks: A comprehensive analysis," *IEEE Access*, vol. 12, pp. 142241–142264, 2024.

[6] T. Taleb, K. Samdanis, B. Mada, H. Flinck, S. Dutta and D. Sabella, "On multi-access edge computing: A survey of the emerging 5G network edge architecture and orchestration," *IEEE Communications Surveys and Tutorials*, vol. 19, no. 3, pp. 1657–1681, 2017.

[7] R. Roman, J. Lopez and M. Mambo, "Mobile edge computing, Fog *et al.*: A survey and analysis of security threats and challenges," *Future Generation Computer Systems*, vol. 78, pp. 680–698, 2018.

[8] J. Granjal, E. Monteiro and J. S. Silva, "Security for the internet of things: A survey of existing protocols and open research issues," *IEEE Communications Surveys and Tutorials*, vol. 17, no. 3, pp. 1294–1312, 2015.

[9] S. Sicari, A. Rizzardi, L. A. Grieco and A. Coen-Porisini, "Security, privacy and trust in internet of things: The road ahead," *Computer Networks*, vol. 76, pp. 146–164, 2015.

Index